THE CHEMISTRY OF
CLAY–ORGANIC REACTIONS

The Chemistry of
Clay–Organic
Reactions

B. K. G. THENG, Ph.D.

Scientist, Soil Bureau,
Department of Scientific and Industrial Research,
Lower Hutt, New Zealand.

738·11
Classified sequence.

ADAM HILGER
LONDON

First Published August 1974

© B. K. G. Theng, 1974

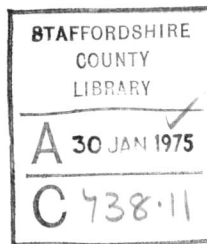

ISBN 0 85274 211 8

Published by Adam Hilger Ltd
Rank Precision Industries
29 King Street, London, WC2E 8JH
Text set in 10/11 pt Monotype Baskerville, printed by photolithography,
and bound in Great Britain at The Pitman Press, Bath

Table of Contents

Preface   ix

Introduction   xi

1 Clay Mineral Structures

1.1 Introductory and General   1
1.2 The 1:1 Type Minerals   3
    1.2.1 Kaolinite   3
    1.2.2 Halloysite   5
1.3 The 2:1 Type Minerals   8
    1.3.1 Montmorillonite   9
    1.3.2 Vermiculite   13
    1.3.3 Mica   15
    1.3.4 Brittle Mica   16

2 Interactions with Uncharged Polar Organic Compounds. General Considerations

2.1 Introduction   17
2.2 The Role of Interlayer Water   18
2.3 Adsorption from Aqueous Solutions   20
2.4 Interlayer Organization of Adsorbed Species   23
    2.4.1 Molecular Orientation   23
    2.4.2 Shortening of Contact Distances   28

3 Interactions with Uncharged Polar Organic Compounds. Complex Formation with some Defined Classes of Compounds

3.1 Complexes with Primary n-Alcohols   32
3.2 Complexes with Polyhydric Alcohols   47
3.3 Complexes with Ketones, Aldehydes, Ethers, Nitriles and Other
    Compounds   63
    3.3.1 General   63
    3.3.2 Ketones   64
    3.3.3 Aldehydes   73
    3.3.4 Ethers and Other Compounds   74
3.4 Complexes with Amines   84
    3.4.1 Aliphatic Monoamines   84
    3.4.2 Aliphatic Polyamines and Cyclic Amines   92
    3.4.3 Aromatic Amines   102
3.5 Complexes with Amides   111
3.6 Complexes with Aliphatic and Aromatic Hydrocarbons   128

4 Interactions with Organic Compounds of Biological Importance

4.1 Complexes with Organic Pesticides	136
4.1.1 Introduction	136
4.1.2 Non-ionic Pesticides	137
4.1.3 Anionic Pesticides	148
4.1.4 Cationic Pesticides	148
4.2 Complexes with Amino Acids and Peptides	158
4.3 Complexes with Antibiotics and Alkaloids	186
4.4 Complexes with Pyrimidines, Purines and Nucleosides	189
4.5 Complexes with Fatty Acids and Fats	198
4.6 Complexes with Saccharides	206

5 Interactions with Positively Charged Organic Species

5.1 Mechanisms of Formation	211
5.2 Interlayer Organization of Adsorbed Species	219
5.3 Some Properties of Cationic Complexes	230

6 Complexes with the Kaolinite Group of Minerals

6.1 Halloysite	239
6.2 Kaolinite	243

7 Organic Reactions Catalysed by Clay Minerals

7.1 Colour Reactions	261
7.2 Polymerization Reactions	269
7.3 Transformation and Decomposition Reactions	281

References	293
Index	327

To the memory of
George F. Walker

Preface

Some years ago, G. F. Walker suggested that I should collaborate with him in writing a monograph on clay-organic systems which was to be a critical assessment and not just a neutral review of the literature. It was not until the spring of 1968, however, when I joined Walker's group at the Division of Applied Mineralogy, C.S.I.R.O., Melbourne, that I seriously began to sift through the published material dealing with clay-organic interactions. The untimely death of G. F. Walker in early 1970 not only deprived me of a co-author who had an intimate and incisive knowledge of the subject but also brought the project to a halt, at least temporarily. Its resumption was prompted by the need for a comprehensive treatment of clay-organic chemistry which would serve as a work of reference. Also, by writing about it I hoped to sort out my own ideas and, in doing so, to see the subject with greater clarity. In style and content the finished product would, of course, have been different had Walker shared in its authorship. Further, by going it alone I am solely responsible for whatever merits and deficiencies the book may have.

The subject of clay-organic reactions has made appreciable advances over the past few years. In common with many other disciplines, both the volume and complexity of information in this area are growing at such a rate that we are facing 'the prospect of instant antiquity'* Some selection of the literature was therefore necessary to avoid excessive length while retaining essential depth and detail. In collating and screening the published material and in preparing the text I have tried to be as objective as possible.

This book is directed to the interests of my fellow workers as well as teachers and students of soil science. Owing to the interdisciplinary nature of the subject, however, it may also be read to advantage by surface and colloid chemists, organic geochemists, and agronomists.

I am indebted to the many colleagues and authors who have sent me reprints of their papers, and whose diagrams have served as a basis for the illustrations. In this connection, I acknowledge the various publishers and councils of scientific societies who gave permission to reproduce figures and diagrams. I wish especially to record my indebtedness to the late Dr. M. Fieldes, formerly Director of the Soil Bureau, for allowing me to work full time on the final stages of the book. Also I thank those who typed the manuscript and Mrs. M. Oliver, who drew the numerous diagrams. Finally I would like to thank my wife for her patient support and encouragement.

Lower Hutt   B. K. G. Theng
November 1972

* Lukasiewicz, J., 1970. The ignorance explosion. *Impact of science on society*, **20**: 251–253.

Introduction

The study of clay-organic interactions is a relatively young scientific discipline, dating back to the early 1940s. This is perhaps hardly surprising, as progress in this field is dependent on the advances in clay mineralogy which is itself only a recent development.[1,2] As Grim[1] has pointed out, the concept that clays are essentially composed of crystalline particles of one or more members of a small group of minerals did not gain general acceptance until the mid 1930s, following the elucidation of the structures of the main groups of clay minerals.

But despite ignorance of the scientific basis of the interaction process, prior to this date people knew that clay-organic complexes existed. For example, the ability of clays to take up certain organic substances preferentially has been known for a very long time, being the basis of the wide use of clays for decolourizing edible oils and clarifying alcoholic beverages.[3,4] Indeed, the fulling process, that is, the removal of grease from raw wool by means of an aqueous slurry of crude clays, is traceable to Biblical times. Such well-known practical applications of the clay-organic interaction have given rise to the familiar terms of 'bleaching earth' and 'fuller's earth'.

Soil scientists were perhaps the first to recognize the importance of the association between organic materials and clays because it had been supposed that the stability and resistance of organic matter in soil were, at least in part, due to complex formation between these two soil constituents.[5-8] As clay-organic systems find applications in agriculture and a variety of industries, an increasingly diverse group of workers has become interested in both the theoretical and practical aspects of the reactions between clays and organic compounds.[9-17] The large increase in published work on clay-organic systems illustrates the growth of this branch of organic geochemistry.

Of the new and improved spectroscopic techniques used in recent years to examine clay-organic complex formation, infra-red spectroscopy is perhaps the most versatile single method. In conjunction with classical X-ray diffractometry, its application to the study of clay-organic reactions has shed much light on the bonding to and arrangement at the clay surface of organic species.[18-20] Considerable progress has been made over the last decade in understanding the behaviour of organic molecules at the surface of layer silicate minerals, and some kind of rational order is emerging from the mass of experimental data. This has meant that, although clay-organic chemistry is largely a descriptive science, the descriptions have become more precise and the terms of reference more quantitative. Nevertheless, the reader will soon become aware of the many gaps that remain in the overall picture and which require further research.

This monograph is therefore an attempt to summarize the data available on the interactions of organic compounds with clay minerals, indicating the classes of compounds which have been studied as complexing agents, and the factors which are known to influence the formation and properties of the resultant complexes.

As indicated before, research on clay-organic systems began with the aim of clarifying the mechanisms underlying the formation of clay-humus complexes. Because of the wide range and complexity of soil organic constituents—many of which are polymeric in nature—together with the scant knowledge about their chemical properties, the focus of attention soon shifted to systems consisting of 'pure' clay mineral species and well-defined 'off-the-shelf' organic compounds.

A wide variety of both low molecular weight organic molecules (micromolecules) and organic polymers are known to react with clay minerals. As might be expected, the interaction of layer silicates with organic macromolecules is considerably more complex than that involving small, relatively simple organic species. A considerable amount of literature dealing with clay-polymer systems has accumulated over recent years. Interest has broadened because of the great potential and practical importance of such systems in agriculture, foundation engineering, and a variety of industrial processes.[12,14] Yet to include these data in the present volume and also to give them sufficient coverage and critical appraisal would not only lead to excessive length but detract from the central theme. This monograph is, in essence, an integrated presentation of the reaction of clays with organic micromolecules, embracing in consequence only those interactions which lead to polymer formation. Clay-polymer systems will form the subject of a separate book.

The subject matter of this book has mostly built itself up from a mass of isolated observations about the behaviour of a variety of organic species at clay mineral surfaces. Its arrangement—with the possible exception of the chapter on clay mineral structures, which logically precedes all else—is somewhat arbitrary. If a principle is discernible, it is one of discussing first those systems about which most information is currently available.

1
Clay Mineral Structures

1.1. Introductory and General

Details of the structures of clay minerals, together with their identification by X-ray diffraction, have been discussed by Grim[1] and Brown[2]. Only the basic structural features of the main groups of clay minerals and the more important species in each group will be mentioned here, emphasizing those peculiarities of architecture and organization which influence the behaviour of clays towards organic compounds.

Hadding[3] and Rinne[4] in the early 1920s were the first to apply X-ray diffraction analyses to the study of clays. They were able to show that clays, in the main, were composed of crystalline materials, a conclusion which ran counter to opinions currently held at that time. By the late 1920s evidence for the essentially crystalline nature of clay minerals was accumulating, largely due to the work of Ross, Hendricks, and their respective co-workers.[5-7] In 1930 Pauling[8,9] published the results of his classical X-ray diffraction studies on the micas, talc, pyrophyllite, chlorite, and kaolinite together with those on cristobalite (SiO_2), gibbsite $(Al_2(OH)_6)$, and brucite $(Mg_3(OH)_6)$. Although these minerals had widely different chemical compositions, the dimension of their unit cell in the plane of cleavage was remarkably similar, approximating to 0·51 nm by 0·88 nm. This led Pauling to conclude that these minerals had a layer structure composed of sheets of cristobalite and either gibbsite or brucite, thus laying the foundation for a scheme of classification and for correlation of structure with properties. By 1940 the crystallinity and structures of the main groups of layer silicates related to clay minerals had been established. Subsequent efforts were directed principally to the elucidation of the partial order and disorder in these structures. The early history and development of clay mineralogy culminating in the 'clay mineral concept', that is, the general acceptance of the crystalline nature of clays make fascinating reading.[1,10,11]

As the name suggests, the layer silicates or phyllosilicates are essentially made up of layers formed by condensation of sheets of linked $Si(O,OH)_4$ tetrahedra with those of linked $M_{2-3}(OH)_6$ octahedra, where M is either a divalent or trivalent cation. Condensation in a 1:1 proportion gives rise to the two-sheet or dimorphic minerals with a general layer formula $M_{2-3}Si_2O_5(OH)_4$ of which kaolinite is, perhaps, the best-known example. Similarly, the three-sheet

1

or trimorphic clays are formed by a $2:1$ condensation, the octahedral sheet being sandwiched between two sheets of inward-pointing tetrahedra (the mica-type layer structure) giving a layer formula of $M_{2-3}Si_4 O_{10} (OH)_2$. Four-sheet or tetramorphic types also occur in which trimorphic units alternate with $M(OH)_{2-3}$ sheets of octahedrally coordinated M^{2+} or M^{3+} ions, exemplified by chlorite.

There is scope in these structures for isomorphous replacement, that is, for substitution of Si^{4+} and/or $M^{2+/3+}$ for cations of similar size but different (usually lower) valency. Thus if $3 Mg^{2+}$ ions replace $2 Al^{3+}$ ions in the octahedral sheet of pyrophyllite (Fig. 7), the mineral is talc. Since in such a situation all (or three out of three) octahedral positions are filled, talc is referred to as trioctahedral whereas pyrophyllite, with only two out of three octahedral positions occupied, is a dioctahedral mineral. As a result of isomorphous substitution the structure of many phyllosilicates is negatively charged. This positive charge deficiency may, to some extent, be compensated by internal substitution but for the most part electrical neutrality is maintained by sorption of extraneous cations, which may or may not be exchangeable. The amount as well as the site of isomorphous replacement influence the surface and colloidal properties (e.g. swelling in water) of the layer silicates, since they determine the surface density of charge and the cation-silicate layer attraction. The charge per formula unit (x) is thus an important parameter which enters into the classification scheme of the phyllosilicates. This scheme, presented in Table 1, follows the recommendations submitted by the Nomenclature Committee of the Clay

TABLE 1

Classification scheme for phyllosilicates related to clay minerals.[12]

Type	Charge per formula unit (x)	Group	Subgroup		Species*
1:1	~0	Kaolinite-serpentine	Dioctahedral	Kaolinites	Kaolinite, Halloysite
			Trioctahedral	Serpentines	Antigorite, Chrysotile
2:1	~0	Pyrophyllite-talc	Dioctahedral	Pyrophyllites	Pyrophyllite
			Trioctahedral	Talcs	Talc
	~0·25—0·6	Smectite or Montmorillonite-saponite	Dioctahedral	Smectites or Montmorillonites	Montmorillonite, Beidellite, Nontronite
			Trioctahedral	Smectites or Saponites	Saponite, Hectorite Sauconite
	~0·6—0·9	Vermiculite	Dioctahedral	Vermiculites	Dioctahedral Vermiculit
			Trioctahedral	Vermiculites	Trioctahedral Vermiculi
	~1	Mica†	Dioctahedral	Micas	Muscovite, Paragonite
			Trioctahedral	Micas	Biotite, Phlogopite
	~2	Brittle Mica	Dioctahedral	Brittle micas	Margarite
			Trioctahedral	Brittle micas	Clintonite
2:1:1	variable	Chlorite	Dioctahedral	Chlorites	Donbassite
			Di, Trioctahedral	Chlorites	Sudoite
			Trioctahedral	Chlorites	Pennine, Clinochlore

* Only the more common species are given.
† Illite is not included in the scheme pending further recommendations.

Minerals Society.[12] Each group of minerals is divided into two subgroups, di- and trioctahedral, containing a number of mineral species.

Because a crystal is composed of discrete layers which can be stacked in a number of different ways, structural varieties or polymorphs are found in many of the groups.[13-15] Additionally, interstratification of two or more types of layers within a single crystal is possible. Interleaving of this kind may be regular or random, or a crystal may show regions of both types.[16]

It is also of interest to note that the basic symmetry, for example, in the arrangement of the basal oxygens, is ditrigonal rather than hexagonal. This departure from true hexagonal symmetry was ascribed by Radoslovich[17] to the opposed rotation of alternate tetrahedra due to a misfit between the larger tetrahedral sheet and the smaller octahedral sheet.

Brief mention must be made of palygorskite and sepiolite, for which there are only few references on reactivity towards organic compounds. These minerals are composed of trimorphic layers arranged in chains (bands) which are joined through oxygen ions.[18,19] The tetrahedral sheets are continuous but the apices of the SiO_4 tetrahedra in adjacent chains point in opposite directions, giving rise to a corrugated surface with open channels running parallel to the chains. Palygorskite (attapulgite) has a smaller trimorphic unit, intermediate between di- and trioctahedral in character, than has sepiolite (meerschaum) which appears to be all trioctahedral. These minerals have been considered to belong to the category of chain-lattice silicates. However, because they bear a closer relationship to the phyllosilicates than to the chain silicates, the term *pseudo-layer silicates* has been recommended by the AIPEA (Association International pour l'Étude des Argiles) Nomenclature Committee.[20]

The major part of the clay fraction (<2 μm equivalent spherical diameter) of soils derived from volcanic ash consists of non-crystalline or highly disordered minerals which collectively are termed *allophane*.[21-23] These materials are probably made up of (Si, Al) $(O,OH)_4$ tetrahedra randomly linked with $Al_2 (OH)_6$ octahedra, although some sort of structural regularity has been proposed.[24,25] The potential of allophane as a cheap and effective sorbent of organic wastes, e.g. meat works and dairy effluents, has been recognized.[26] However, because these minerals lack long-range order and show variations in composition, work on the interaction of allophane with organic compounds has only relatively recently appeared in the literature.[27-29]

1.2. The 1 : 1 Type Minerals

Of these, only the dioctahedral species kaolinite and halloysite belonging to the subgroup kaolinites (Table 1) will be discussed. The reactions of the minerals in the serpentine subgroup with organic substances have not received much attention.

1.2.1. Kaolinite

The structure of the kaolinite layer, together with the manner of stacking of successive layers within a crystal, are shown in Fig. 1. Each layer occupies a

3

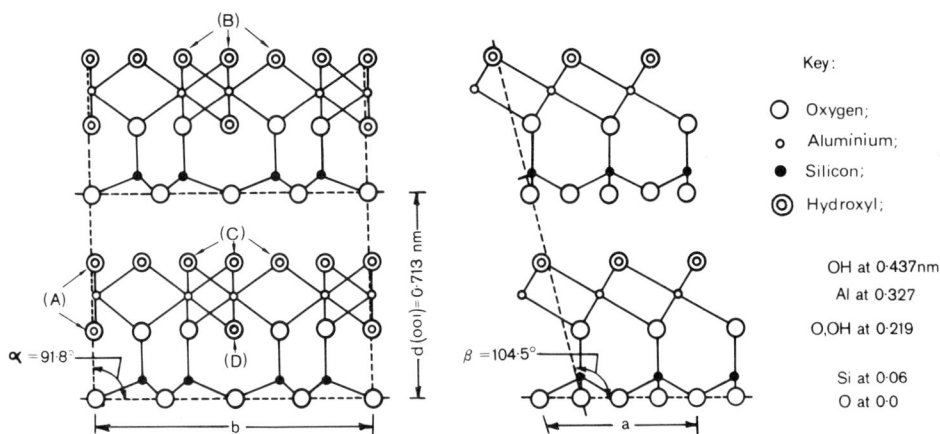

Fig. 1 Kaolinite layer viewed along the *a* axis (left) and along the *b* axis (right). Height of atoms above the basal oxygen plane are given in nanometres (nm), from Brindley[13]

thickness of $\sim$0·72 nm, a value equal to the basal (d(001)) spacing of kaolinite. As depicted in Fig. 1, the kaolinite layer is electrically neutral but in reality it carries a small negative charge due to a small amount of isomorphous replacement.[30-32] This 'permanent'—as opposed to the 'pH-dependent'—negative charge[33] is responsible for the small but measurable ($<$0·1 mol kg^{-1}, monovalent cations) exchange capacity of kaolinite samples under acid conditions. Recently, Hofmann *et al.*[34] have proposed that the exchange sites are located on only the tetrahedral surface of the kaolinite crystal. This concept, with its important implications in the surface and colloid chemistry of kaolinite, has subsequently been supported by the results of electron-optical studies.[35,36]

The superposition of oxygen and hydroxyl planes of successive layers within a single kaolinite crystal gives rise to pairing of O and OH ions and interlayer O . . . HO H-bond formation (Fig. 2). Clearly, the forces arising from H bonding and those due to non-specific van der Waals interactions holding adjacent layers together, must be overcome if interlayer sorption (intercalation) of extraneous species is to occur. For this reason, penetration of the kaolinite interlayers by organic compounds is difficult to achieve and adsorption is generally confined to the external crystal surfaces. In this connection, the edges of kaolinite crystals are of particular importance. This is so because the edges containing unsatisfied valencies ('broken bonds') occupy an appreciable proportion (10–20 per cent) of the total crystal area ((15–40).10^3 m^2 kg^{-1}).[32,37] On the other hand, in montmorillonite-type minerals, less than 10 per cent and more usually only 2–3 per cent, of the total (external crystal and interlayer) area of $\sim$760·10^3 m^2 kg^{-1} is apportioned to the crystal edges.[38] Hence, the influence of the crystal edges on the pH-dependent charge and sorption of anions,[32,33,39] electron-transfer reactions involving organic compounds,[40,41] and the initiation and/or inhibition of polymerization of organic monomers[40,42,43] is much more in evidence with kaolinite than with montmorillonite and mica-type minerals. Under suitable conditions, however, kaolinite crystals

4

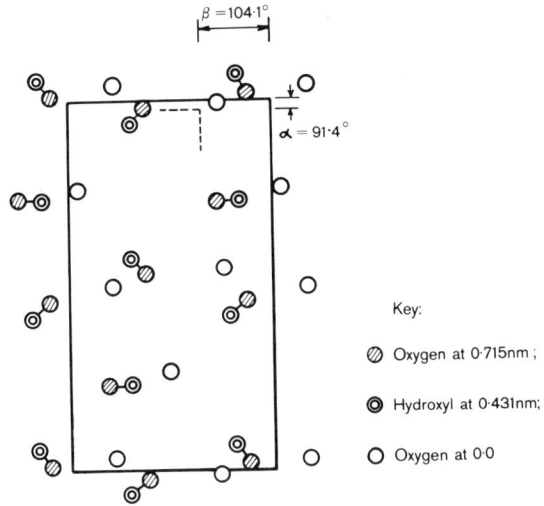

Fig. 2 Relation of oxygens to hydroxyls between two adjacent layers of kaolinite viewed in projection on (001) plane. Distorted layer structure with $\alpha = 91.4°$ and $\beta = 104.1°$, from Brindley[13]

$\beta = 104.1°$

$\alpha = 91.4°$

Key:

⊘ Oxygen at 0·715nm ;

◉ Hydroxyl at 0·431nm;

○ Oxygen at 0·0

are capable of intercalating certain inorganic salts[44] and polar organic molecules.[45,46]

1.2.2. Halloysite

This mineral may be regarded as being composed of kaolinite layers between which a single layer of water molecules is interposed. As compared with kaolinite, stacking of successive layers within a single crystal is disordered[13] and the basal spacing is increased from ~0·72 nm to ~1·01 nm. Hendricks and Jefferson[47] have suggested that the interlayer water molecules are arranged in a hexagonal network, linked to each other and to adjacent halloysite layers by H bonding (Fig. 3).

Under the electron microscope kaolinite particles are seen as flat, hexagonal plates whereas those of halloysite often appear as tubes.[1,48] This tubular morphology is attributed to a difference in the b dimension of the octahedral

Fig. 3 Hendricks and Jefferson's[47] conception of the arrangement of water molecules in the interlayer space of halloysite

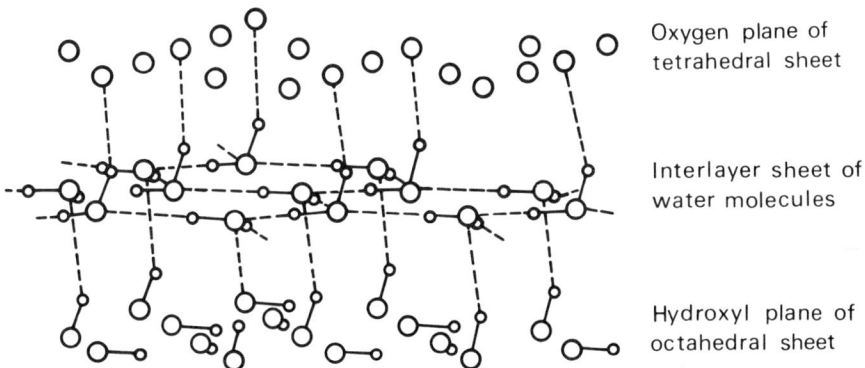

Oxygen plane of tetrahedral sheet

Interlayer sheet of water molecules

Hydroxyl plane of octahedral sheet

and tetrahedral sheets[48] analogous to Radoslovich's[17] concept referred to earlier. The inherent strain causes the layer to curl when the restraint imposed by the close proximity of adjacent layers (as in kaolinite) is removed by the presence of interlayer water. An alternative explanation for the tubular form based on the existence of two opposing forces, those of Al-Al repulsion and OH-OH contraction within the octahedral sheet, has been tentatively suggested by Radoslovich.[49]

A series of partially dehydrated forms of halloysite (metahalloysites) with basal spacings lying between the two end members (1·01 nm and 0·72 nm) is obtained by dehydration of the mineral in atmospheres of different relative humidities or by mild heat treatment.[13] Thus, heating halloysite at 373° K reduces the basal spacing from 1·01 nm to about 0·73–0·75 nm, but random stacking is apparently preserved. The residual water molecules, corresponding to one water layer for every 8–10 halloysite layers, are presumably entrapped as 'islets' in the interlayers[13] (Fig. 4).

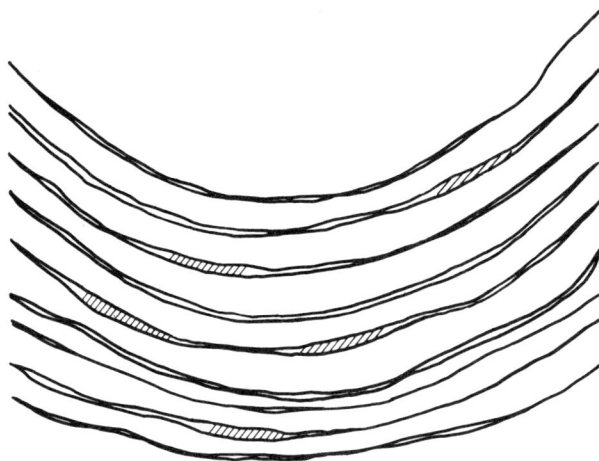

Fig. 4 Tubular structure of metahalloysite with entrapped water 'islets' (hatched areas) in the interlayer space, from Brindley[13]

Halloysite is capable of intercalating a variety of organic liquids by replacement of the interlayer water.[50] As might be expected, intercalation is easier here than in kaolinite because the halloysite layers are already separated. Thus, halloysite readily intercalates a single layer of ethylene glycol (d(001) ~1·08 nm) whereas the meta-form does so with difficulty and then only incompletely, presumably because most of the layers in the partially dehydrated material have been effectively sealed off (Fig. 4).

Another noteworthy feature is that one side of the halloysite layer consists of an open 'hexagonal' network of oxygen ions carrying negative charges whereas the other side may be likened to an α-hydroxide surface and is positively charged. This amphoteric nature of halloysite layers is the source of the difference between halloysite and montmorillonite in their sorption properties. Thus, if an organic compound (such as methanol or ethylene glycol) is capable of being intercalated, montmorillonite tends to form a double-layer complex, but only a single layer of the organic is adsorbed in the halloysite interlayers. Similarly, halloysite tends to form no complex with those organic liquids which are adsorbed as a single layer by montmorillonite (Fig. 5). MacEwan[50] has rationalized these observations by proposing that the negatively charged oxygen surfaces of opposing montmorillonite layers tend to orient polar molecules, and in so doing, individually adsorb a monolayer of the organic compound. Halloysite, being amphoteric, tends to collect only one layer of such molecules in its interlayer space. By the same token, weakly polar and non-polar molecules being held by non-specific van der Waals (dispersion) forces only, give single-layer complexes with montmorillonite but fail to be intercalated by halloysite (Fig. 6).

Fig. 5 Relationship between Δ value (observed basal spacing of complex–assumed thickness of individual silicate layer) and chain length for interlayer complexes of montmorillonite and halloysite with different organic compounds
○   montmorillonite–primary
    n-alcohol complexes
●   montmorillonite–nitrile complexes
×   halloysite–primary n-alcohol complexes
(From Grim[1] based on MacEwan's data)

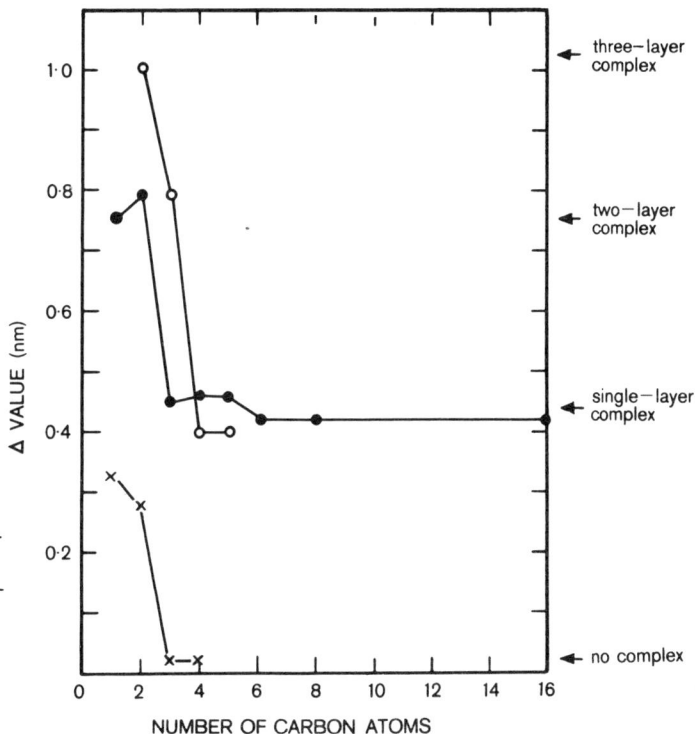

Fig. 6 Relationship between layer charge characteristics of montmorillonite and halloysite and number of intercalated organic layers, from MacEwan[50]
(a)   polar molecules in montmorillonite;
(b)   mainly non-polar groups with some polar groups in montmorillonite;
(c)   polar molecules in halloysite

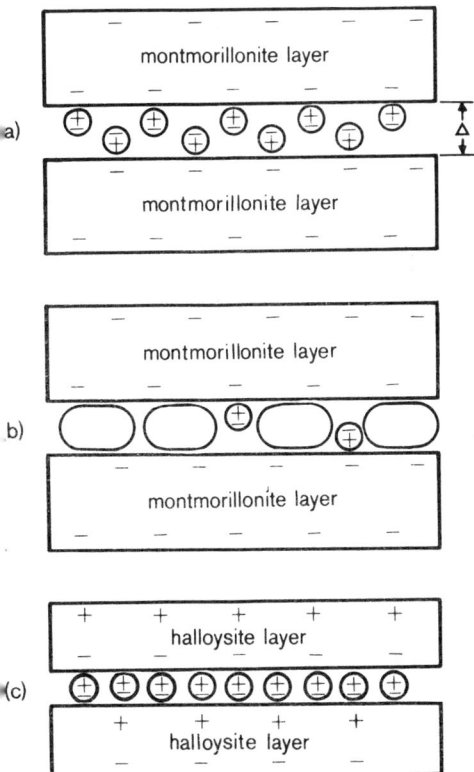

7

1.3. The 2 : 1 Type Minerals

Of the dioctahedral smectites (Table 1) montmorillonite has held the limelight for many years. The choice of montmorillonite as a sorbent of organic compounds is dictated by its large surface area ($760 \cdot 10^3$ m^2 kg^{-1}), its high cation exchange capacity ($\sim$1 mol kg^{-1} monovalent cations) which is largely independent of salt concentration and pH, and the relative ease by which it forms an interlayer complex with a wide variety of organic molecules.[1,51-53] An added advantage is that changes in basal spacing accompanying interlayer sorption may readily be followed by X-ray diffraction, which also allows deductions of the probable orientation of the interlayer species. The other minerals in the dioctahedral subgroup (beidellite and nontronite) and the trioctahedral species (saponite, hectorite, and sauconite) have received comparatively little attention.

Vermiculite, like montmorillonite, is capable of intercalating many organic compounds.[52,54] Because of its higher surface charge density (x value of Table 1), the extent to which the interlayer space of vermiculite may be expanded by intercalated organic liquids is generally less than that observed for the lower-charged montmorillonite. However, because of the larger crystal size and the more ordered layer stacking, vermiculite-organic complexes offer greater scope for conformational analysis by X-ray diffraction of the intercalated species[55,56] than similar complexes with montmorillonite.

Similarly, successive layers within a crystal of mica (with its high value for

TABLE 2

Influence of surface charge density of 2 : 1 type minerals on their ability to intercalate water and organic compounds.[51]

Group	Charge per formula unit	Main interlayer cation	Organic Complexes	Interlayer swelling in water
Pyrophyllite-talc	0	—	no complex	no swelling
Smectite	0·25—0·6	Na, Ca, etc.	wide variety of simple and polymeric organics, charged and uncharged	extensive with some monovalent cations
Vermiculite	0·6—0·9	Mg	wide variety of charged and uncharged organics	may be extensive with special treatment
Mica	1	K(Na, Li)	long-chain mono- and di-alkylammonium ions	may swell slightly with special treatment
Brittle mica	2	Ca	long-chain mono- and di-alkylammonium ions	no swelling

8

x) are held together by strong electrostatic attractive forces. Crystals of mica therefore show very little tendency to swell in water (Table 4) and exchange of the interlayer K^+ ions for other inorganic ions is somewhat difficult.[57-59] On the other hand, some long-chain alkylammonium cations, particularly the dodecylammonium ions, can rapidly and effectively displace K^+ ions from the mica interlayers.[57,60] Once penetration and expansion of the interlayer space have been effected by this means, further interchange of material between the bulk and the interlayer phase is possible.[61]

As MacEwan[51] has pointed out, the distinction between the mica, vermiculite, smectite, and the pyrophyllite–talc groups has become rather tenuous since the minerals of the 2:1 type share a common basic layer structure. Nonetheless, differences in charge characteristics as well as in the kind of interlayer cation normally present between these groups, and hence in their behaviour towards water and defined organic compounds, provide a workable basis for contrast and comparison. Table 2 modified from MacEwan[51] illustrates these points.

1.3.1. Montmorillonite

The structure of montmorillonite was first given by Hofmann, Endell, and Wilm[62] on the basis of its similarity with that of pyrophyllite (Fig. 7). This basic structure into which modifications by Marshall,[63] Maegdefrau and Hofmann,[64] and Hendricks[65] were subsequently incorporated, is now generally accepted.[11,51] The montmorillonite layer differs from that of pyrophyllite in that substitution of Al^{3+} for other cations (e.g. Mg^{2+}, Fe^{2+}) in octahedral positions, and less

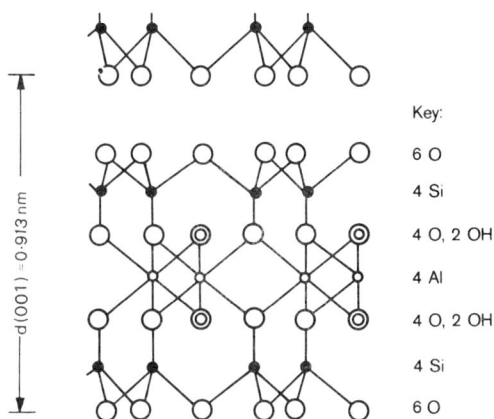

Fig. 7   The pyrophyllite layer structure viewed along the *a* axis. The basal spacing (d(001)) is given in nanometres

Key:

6 O
4 Si
4 O, 2 OH
4 Al
4 O, 2 OH
4 Si
6 O

$d(001) = 0.913$ nm

frequently of Si^{4+} for Al^{3+} in the tetrahedral sheet, always occurs. Although some internal compensating substitution may occur, the final result of isomorphous replacement in the pyrophyllite structure is a layer which carries a permanent negative charge. This positive charge deficiency is balanced by

9

sorption of exchangeable cations which, apart from those associated with external crystal surfaces, are situated between the randomly superposed layers within a crystal.[64] Water is also readily adsorbed in the interlayer space. These concepts are illustrated in Fig. 8. Water appears to enter the interlayer region as an integral number of complete layers of molecules, this number being dependent on the nature of the exchangeable cation.[1,51,66,67] The d(001) spacing of

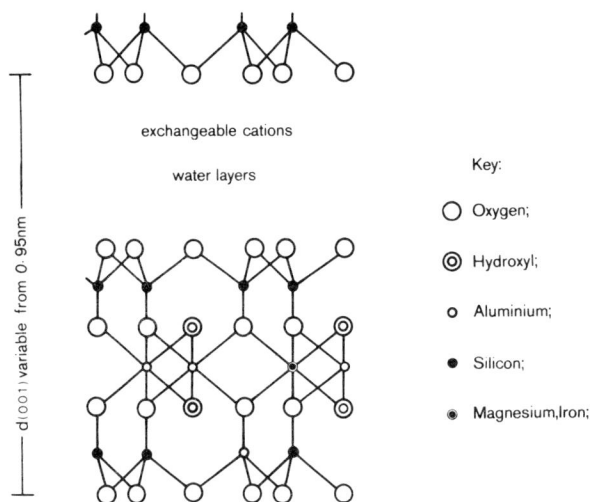

exchangeable cations

water layers

Key:

○ Oxygen;

◎ Hydroxyl;

o Aluminium;

● Silicon;

⊛ Magnesium,Iron;

d(001) variable from 0.95nm

Fig. 8 The Hofmann-Endell-Wilm-Marshall - Maegdefrau - Hendricks[62-65] structure of a montmorillonite layer viewed along the a axis. Basal spacing is given in nm units

montmorillonite can thus vary over a wide range, the minimum corresponding to the fully collapsed state being 0·95 nm. With large monovalent and divalent cations occupying interlayer exchange sites, interlayer (or intracrystalline) swelling is limited to a d(001) spacing of about 1·9 nm. These points are summarized in Table 3. On the other hand, montmorillonite samples saturated

TABLE 3

Basal spacings (d(001)) in nm, of montmorillonite samples saturated with various inorganic cations during water uptake.[67]

	Interlayer cation								
H_3O^+**	Li^+*	Na^+*	K^+**	NH_4^+**	Cs^+	Mg^{2+}	Ca^{2+}	Sr^{2+}	Ba^{2+}
1·00	0·95	0·95	1·00	1·00	1·20	0·95	0·95	0·95	0·98
1·24	1·24	1·24	1·24	?	—	—	—	1·20	1·20
1·54	1·54	1·54	1·50	1·50	—	1·54	1·54	1·55	1·55
1·90	1·90	1·90	—	—	—	1·92	1·89	?	1·89
2·24	2·25	—	—	—	—	—	—	—	—

increasing humidity

* may swell extensively (d(001) > 4 nm) in dilute aqueous solutions of their respective cation salt and in distilled water.
** may swell extensively with special pretreatment.
 ? the spacing is probably close to that of the samples on the same line.

10

with small, monovalent cations (Li^+, Na^+) may show extensive interlayer expansion in dilute aqueous solutions (<0.3 N) of their respective cation salt and in water, and under optimum conditions the layers can dissociate completely.[67] The influence of surface density of charge and the nature of the exchangeable cation on the hydration properties of 2:1 type phyllosilicates is shown in Table 4. The important conclusion that has emerged from studies on

TABLE 4

Influence of charge per formula unit, x, and nature of interlayer cation on the maximum value of basal spacing (nm) of different 2:1 type clay minerals on hydration.[67]

Interlayer cation	Clay mineral group				
	Pyrophyllite-talc	Smectite	Vermiculite	Mica	Brittle mica
	$x \sim 0$	$x \sim 0.25$—0.6	$x \sim 0.6$—0.9	$x \sim 1$	$x \sim 2$
Li^+		2·20*	1·50*		
Na^+		1·90*	1·48	1·43	
Al^{3+}	no interlayer cation	1·90	1·40		
Mg^{2+}	no swelling reported	1·90	1·46	1·43	
Ca^{2+}		1·90	1·54		1·00
Ba^{2+}		1·90	1·54		
K^+		1·50	1·10	1·00	
NH_4^+		1·50	1·10		
Cs^+		1·38	1·20		

* May give extensive interlayer swelling ($d(001) > 4nm$) in water and in dilute aqueous solutions of the respective cation salt.

clay-water systems is that the water properties are affected more by the interlayer cation than by the silicate surface.[67-69] Since uncharged polar organic molecules are adsorbed essentially by replacement of the interlayer water, the behaviour of such molecules is likewise strongly influenced by the exchangeable cation. Evidence is accumulating to show that, at least at low water contents, cation-dipole interactions are of paramount importance in their effect on the adsorption of polar organic species by clay minerals.[70]

An alternative and radically different structure for montmorillonite, in which every alternate SiO_4 tetrahedron in the tetrahedral sheet is inverted, has been proposed by Edelman and Favejee.[71] The apical oxygens of such inverted tetrahedra, now pointing away from the surface, are replaced by hydroxyl groups which also fill the gaps left in the octahedral sheet (Fig. 9). No isomorphous replacement within the structure is envisaged, the observed cation exchange capacity being solely ascribed to the dissociation of apical hydroxyl groups. A somewhat similar structure in which excess hydroxyls are present owing to the replacement of some Si^{4+} ions by $4H^+$ or $Al^{3+}H^+$ ions in the

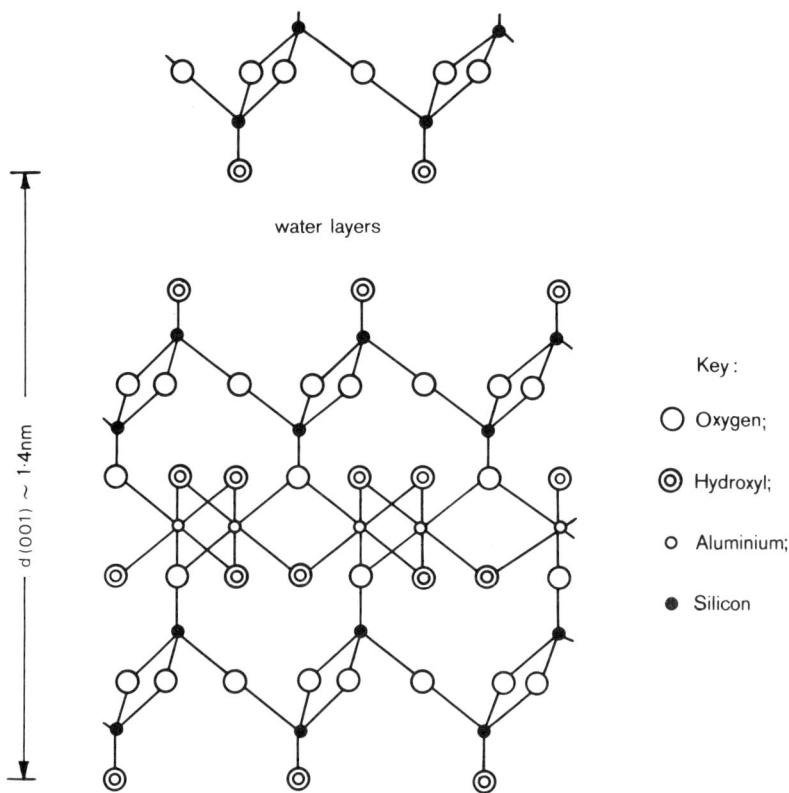

water layers

Key :

O Oxygen;

◎ Hydroxyl;

o Aluminium;

● Silicon

d(001) ~ 1·4nm

Fig. 9 The montmorillonite layer structure viewed along the *a* axis according to Edelman and Favejee.[71] Basal spacing in nm units

tetrahedral sheet, has been suggested by McConnell.[72] That (at least) two types of montmorillonite may exist, one conforming to the Hofmann-Endell-Wilm-Marshall-Maegdefrau-Hendricks structure (the *Wyoming* type, Fig. 8) and the other to the Edelman-Favejee structure, (the *Cheto* type) has, in fact, been postulated by Grim and Kulbicki.[73] It needs pointing out that the X-ray diffraction patterns of montmorillonite samples are generally broad and diffuse because of the poorly crystalline nature of the materials. This, together with problems of obtaining homogeneous and uncontaminated samples, makes it difficult to discriminate between the various possibilities using X-ray analyses alone. By a combination of X-ray, differential-thermal (DTA), thermal-gravimetric (TGA), and chemical analyses of some eighty montmorillonite samples, Schultz[74] has been able to distinguish between four types of mont-morillonite (Wyoming, Tatatilla, Otay, and Chambers) and 'non-ideal' types, and between 'ideal' and 'non-ideal' beidellites. The 'ideal' type has 4 OH groups per layer unit cell (a unit cell being twice the formula unit, Table 1). Table 5 gives these results.

   The results of studies by Berger[75] and more especially by Deuel and co-workers[76,77] on surface esterification of montmorillonite and related silicate

12

TABLE 5
Types of smectites with their distinguishing properties, according to Schultz.[74]

Type	Net negative layer charge		MgO (%)	Fe_2O_3* (%)	DTA Dehydroxylation temperature (°K)	TGA Structural OH per unit layer
	Amount	Location				
Ideal montmorillonites						
Wyoming	low $x < 0.425$	oct. > tet.	2—3	3—4	973—998	=4
Tatatilla	high $x > 0.425$	oct. > tet.	2—4	1	973—1008	=4
Otay	high	oct. ≥ tet.	3·5—5	1—2	923—963	=4
Chambers	high	oct. > tet.	3—4·5	1—4	933—973	=4
Non-ideal beidellite	mostly high	oct. < tet.	0—2	0—8	823—873	>4
Ideal beidellite	variable	tet.	0	0	993—1033	=4
Non-ideal montmorillonite	mostly high	oct. > tet.	2—4	5—10	823—863	<4?

* net layer charge depends on oxidation state of iron; can be changed by oxidation or reduction of structural iron.

minerals have been explained in terms of the existence of surface silanol groups, thus favouring the Edelman-Favejee concept. Although this interpretation has subsequently been questioned,[78,79] the formation of covalently bonded (homopolar) organic complexes of montmorillonite and clay minerals in general, referred to as *clay-organic derivatives*, is by no means unlikely if proper precaution is taken to eliminate side reactions (due to traces of water) and the physical sorption of the reagent or its decomposition products.[80,81] As shown in Table 5 some smectite samples may have more than 4 OH groups per unit cell. It seems probable that—using special techniques and under suitable conditions—part of this excess hydroxyl is capable of being esterified or of participating in covalent-bond-type interactions. There is less uncertainty about minerals in the kaolinite group since one surface of their crystal is composed of hydroxyl groups (Fig. 1). Apart from restrictions imposed by accessibility and steric factors, these hydroxyls are capable of reacting with certain organic compounds (e.g. CH_2N_2, $SOCl_2$, alkylchlorosilanes) to yield covalently bonded organic complexes.[82] However, because of the difficulties associated with their preparation and characterization, organic derivatives of montmorillonite have received comparatively little attention, although such complexes have great potential as fillers and selective adsorbents.[81]

1.3.2. Vermiculite

The complete structure of vermiculite was elucidated by Mathieson and Walker[68] from single-crystal X-ray data which show conclusively that the vermiculite layer is of the trimorphic type in which an octahedrally coordinated sheet of (Mg, Fe) ions is sandwiched between two inward-pointing sheets of linked $(Si, Al)O_4$ tetrahedra. Since vermiculite behaves in many respects like montmorillonite, the work of Mathieson and Walker argues for the Hofmann structure of montmorillonite. The amount of isomorphous substitution in vermiculite, however, is generally greater than that in montmorillonite and much of this occurs in the tetrahedral sheet (Al^{3+} for Si^{4+}); but as Walker[54] has

13

pointed out, it seems probable that the highly charged members of the smectite group grade into the low-charged vermiculites (cf. Table 2).

Because of the light it throws on the relationship between the exchangeable cations, water molecules, and the surface oxygens of the silicate layers in the interlayer region, the results of studies by Walker and co-workers[54,68,83] on vermiculite will be briefly discussed. In Mg vermiculite the interlayer water molecules occupy two sheets of sites arranged in a distorted hexagonal pattern, each site being equivalently related to a single oxygen ion of the silicate surface. The water molecules in the same sheet are linked to each other and to the surface oxygens by H bonding, and the pair of water sheets is held together by Mg^{2+} ions which lie midway between opposing silicate layers (Fig. 10). The water molecules in the hydration shell around the Mg^{2+} ion are referred to as

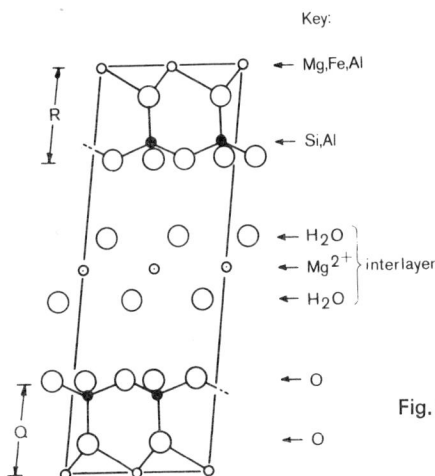

Fig. 10 The structure of (hydrated) magnesium-vermiculite according to Mathieson and Walker.[68] Q and R refer to silicate half-layers

bound water; those not enclosing the interlayer exchangeable cation are *unbound*. The location of water sites is clearly determined by the surface configuration of the silicate layers, while the mutual relationship of the water sheets in a single water-cation layer is determined by the requirement for octahedral coordination of the water molecules around the exchangeable cations. It would appear that direct electrostatic interaction between cation and surface oxygens is not important in determining the distribution of the interlayer cations.

Only in the fully hydrated phase (d(001) = 1·48 nm) are all the available sites occupied by water molecules. As dehydration progresses, some of the unbound water is removed (creating vacant sites) before the bound water, and several phases, each displaying a characteristic basal spacing, can be distinguished. Thus, the 1·16 nm phase corresponds to a structure in which there are 8 (instead of the original 16) water molecules for each Mg^{2+} ion present as a single sheet. The mechanisms of dehydration of Mg vermiculite, together with the accompanying changes in basal spacing, are illustrated in Fig. 11.

14

Basal d(001) spacing, nm

Fig. 11 Mechanism of dehydration of magnesium-vermiculite according to Walker and Cole.[83] Basal spacings (nm) at each stage of interlayer contraction are indicated

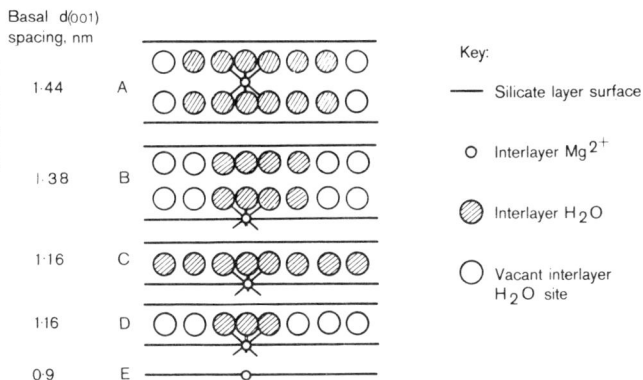

Key:

— Silicate layer surface

O Interlayer Mg^{2+}

◐ Interlayer H$_2$O

O Vacant interlayer H$_2$O site

1·44	A
1·38	B
1·16	C
1·16	D
0·9	E

It will be appreciated from the above remarks that interlayer expansion and collapse of vermiculite are influenced by the nature of the exchangeable cation (Table 3) as well as by that of the interlayer liquid.[84] In this respect and in its ability to form interlayer complexes with organic compounds, vermiculite bears a striking resemblance to montmorillonite.[51,54,55] Any subtle differences between vermiculite and montmorillonite, such as their respective response to contact with organic liquids, are usually ascribable to differences in the amount and location of isomorphous replacement. Thus, Mg-saturated vermiculites fail to expand beyond a basal spacing of ~1·45 nm with glycerol (single-layer complex) whereas all Mg montmorillonites appear to give a double-layer complex with glycerol (d(001) ~ 1·78 nm).[51,54] By the same token, the replacement of interlayer inorganic cations by alkylammonium ions in montmorillonite is complete after bringing the clay into contact with an aqueous solution of the appropriate alkylammonium chloride for two hours at room temperature[85] whereas much longer periods of contact and elevated temperatures are usually required to effect a similar interchange in vermiculite crystals.[56,86]

1.3.3. Mica

The macrocrystalline micas cannot strictly be regarded as clay minerals although clay-sized mica (illite or hydromica) is of widespread occurrence in soils and, of course, both montmorillonite and vermiculite have a mica-type layer structure. Much substitution of Si^{4+} for Al^{3+} in the tetrahedral sheet occurs and many polymorphs have been described.[87] The positive charge deficiency due to isomorphous replacement is balanced by sorption of interlayer K$^+$ ions giving a basal spacing of ~1·0 nm (Fig. 12). The K$^+$ ions are in twelve-fold coordination, being situated centrally on the lines joining the centres of the ditrigonal network of oxygen ions of adjacent silicate layers.[1,87] This arrangement of the interlayer cation together with the high charge per formula unit (Table 1) gives rise to strong electrostatic attraction (~0·7 J m^{-2})[67,88] between successive layers of a crystal. Accordingly, cations of large size and low hydration energy (NH$_4^+$, Rb$^+$, Cs$^+$) cannot expand the interlayer space sufficiently to release the interlayer K$^+$ ions.[59] As we have noted, dodecylam-

15

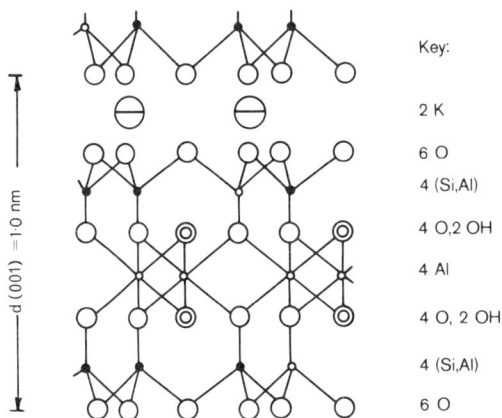

Key:

2 K

6 O

4 (Si,Al)

4 O,2 OH

4 Al

4 O, 2 OH

4 (Si,Al)

6 O

d (001) = 1·0 nm

Fig. 12 The muscovite layer structure viewed along the *a* axis. Basal spacing is given in nm units

monium ions form a notable exception in that they can replace the K^+ ions from mica with relative ease.[60]

Muscovite is dioctahedral; biotite and phlogopite with Mg^{2+} and Fe^{2+} replacing Al^{3+} in the octahedral sheet are trioctahedral. In the biotites the Mg/Fe ratio is less than 2 and in the phlogopites this ratio exceeds 2. Sometimes the term phlogopite is used to denote the end member of the series in which all the octahedral positions are occupied by Mg^{2+} ions.

The micaceous minerals occurring in clays tend to have less isomorphous replacement and a proportion of the interlayer K^+ ions is replaced by Ca^{2+}, Mg^{2+}, and H_3O^+.[1,87] In addition, the stacking of the layers in a crystal is apparently more disordered than in the micas, as indicated by a broadening and a tailing off of the 1·0 nm line to higher spacings in the diffractogram. Illite does not appear to intercalate organic compounds and failure of ethylene glycol to expand illite samples beyond a basal spacing of 1·0 nm is diagnostic. Because soil micas are variable in composition and in degree of crystallinity and frequently occur as interstratified complexes with montmorillonite-type minerals, they are less attractive as sorbents for organic compounds than are montmorillonite and vermiculite.

1.3.4. Brittle mica

These minerals are not known to occur in clays. The large negative charge (Table 1) is balanced by Ca^{2+} rather than by K^+ ions as it is in the micas. Complex formation between margarite and long-chain alkylammonium ions has been reported.[57]

16

2

Interactions with Uncharged Polar Organic Compounds General Considerations

2.1. Introduction

Until the 1960s, X-ray diffractometry was used almost exclusively to study the structure of the complexes formed between expanding 2:1 type layer silicates and organic compounds. While this method has yielded much useful information on the arrangement and conformation of the organic molecule in the interlayer space of such minerals,[1-3] its usefulness in investigating those changes that may take place in the structure of the adsorbed compound is limited.

The recent development of infra-red spectroscopy, together with improved methods of sample preparation, has provided the means to observe structural changes on a molecular scale and so give insight into the nature of the clay-organic bond. The general theory, experimental techniques, and instrumentation used in examining the infra-red spectra of adsorbed species have been fully discussed in a recent monograph by Little.[4]

The principle behind the application of infra-red spectroscopy to the study of clay-organic systems is that the vibrational spectrum of the adsorbed molecule is directly comparable with that of the 'free' species, that is, of the compound in the solid (crystalline), liquid, or dissolved state. Thus, interactions and perturbations occurring at the clay surface of certain structural groups of atoms in the organic molecule often give rise to changes in the position and/or in the intensity of the corresponding vibrational bands in the spectrum of the complex, as compared with similar bands in the spectrum of the unadsorbed compound. These observations enable conclusions to be drawn regarding the mode of bonding of the intercalated organic molecule to the mineral surface. Because interatomic interactions are directly observable and do not depend on the existence of the long-range regularity required for a detailed analysis by X-ray diffraction, such conclusions are often unambiguous.[5] Infra-red spectroscopy is thus singularly suited for the study of clay-organic interactions. Its use in conjunction with and complementary to classical X-ray diffractometry has been instrumental in clarifying many of the mechanisms underlying the formation of complexes between layer silicates and organic substances.[5,6]

As pointed out in Chapter 1, montmorillonite and vermiculite minerals readily form interlayer complexes with a variety of uncharged polar organic molecules. In addition, these minerals and their complexes with organic materials can yield thin, self-supporting films with good infra-red transmission properties. Not surprisingly, much of our understanding of the behaviour of polar organic compounds at the clay surface has come from examination of their complexes with montmorillonite, and to a lesser extent, of those with vermiculite. This chapter discusses the general modes of formation and properties of complexes with these minerals. The interactions of polar organic compounds with 1:1 type minerals (kaolinite and halloysite) are treated in a later chapter.

2.2. The Role of Interlayer Water

At air temperature and humidity, water is perhaps the most common polar compound present in the interlayer space of montmorillonites and vermiculites. Water is also intimately involved in the binding and transformation of polar organic compounds at the surface of these minerals.[5-8] A description of the properties of interlayer water, at this point, would thus seem relevant and illuminating.

In the previous chapter we noted that water uptake and retention by montmorillonite and vermiculite minerals at various levels of humidity (as indicated by the concomitant changes in basal spacing) are profoundly influenced by the nature of the interlayer cation (cf. Table 3). Using X-ray diffraction, Walker and co-workers[2] demonstrated the presence of two distinct types of interlayer water in vermiculite crystals (Fig. 11).

The first type (I) constitutes the inner (primary) hydration shell around the exchangeable cation, that is, the water which is directly coordinated to the cation. The second type (II) forms the outer (secondary) coordination sphere of the cation; being indirectly linked to the cation and possessing greater mobility, this water is more labile than that of type I. The existence of both types of water (at ambient humidity) in smectites and vermiculites has since been substantiated by infra-red spectroscopy. Clearly, the size and valency and, hence, the polarizing power of the exchangeable cation plays a decisive role in water adsorption and this is also evident in the infra-red spectra of water adsorbed by layer silicates.[9-14] The more labile water (type II) gives a spectrum similar to that shown by liquid water with a broad, strong OH stretching band near 3,400 cm^{-1} and a much weaker H—O—H bending maximum near 1,630 cm^{-1}. The OH groups of water directly coordinated to the cation (type I) are apparently involved in weaker hydrogen bonding giving a maximum near 3,600 cm^{-1} and an unusually strong absorption near 1,630 cm^{-1}.

The infra-red pattern of interlayer water in the OH stretching region, however, is often complicated by overlapping of the bands due to both types of

18

water, but methods for resolving such spectra have been suggested.[10,12] In montmorillonite, structural OH groups also absorb near 3,600 cm^{-1} but interference from this source is less serious in the trioctahedral smectites, such as saponite and hectorite, whose structural OH groups give only weak bands lying at higher frequencies than those shown by adsorbed water. For example, the spectrum of hydrated hectorite saturated with the non-polar tetramethylammonium ions shows two distinct bands in the hydroxyl stretching region; the high frequency band at 3,630 cm^{-1} can be ascribed to OH groups of type I water weakly hydrogen bonded to surface oxygens of the silicate layer, and that at 3,425 cm^{-1} to OH groups of type II water intermolecularly hydrogen bonded.[11] The infra-red spectra of adsorbed water in smectites and vermiculites saturated with various cations show similar features typified by the presence of a high-frequency (HF) band or shoulder between 3,650 and 3,570 cm^{-1}, a principal maximum in the region of 3,430 to 3,350 cm^{-1}, and a low-frequency (LF) shoulder near 3,230 cm^{-1}. A band comparable to the HF maximum also occurs in the infra-red spectra of concentrated aqueous solutions and crystalline hydrates of perchlorate salts[15,16] for which weak hydrogen bonding (of the order of 8,368 J mol^{-1}) has been shown to exist between water and the perchlorate anions. However, the HF band is notably absent from the spectra of pure water and of aqueous solutions of anions with stronger electron-donating properties. These observations, and the apparently high angle at which the OH groups responsible for the HF band are directed to the silicate layer, indicate that hydrogen bonding to surface oxygens is, at best, only weak. This interpretation also accords with the observation that the position of the HF maximum is influenced by the site of isomorphous replacement in the structure. Thus, where this replacement occurs principally in the octahedral sheet, such as in hectorite and montmorillonite, the HF band appears at a higher frequency (3,630 cm^{-1} for H_2O in hectorite[11,12] and near 2,690 cm^{-1} for D_2O in montmorillonite[10,14]) than where the negative layer charge chiefly arises from Al- for Si-substitution in the tetrahedral sheet such as in vermiculite (below 3,600 cm^{-1} for H_2O and at about 2,655 cm^{-1} for D_2O[14,17,18]). The main broad absorption below 3,500 cm^{-1} is assigned to OH groups, lying nearly parallel to the silicate layer, of water molecules forming intermolecular hydrogen bonds. This is true at least in montmorillonite and hectorite. In vermiculite as well as in saponite, OH groups of water, weakly hydrogen bonded to surface oxygens, may also contribute to this broad band.

That the surface oxygens of the anionic framework of layer silicates are only weak electron donors is further supported by the high frequencies of the N—H stretching modes of ammonium and substituted ammonium compounds adsorbed in montmorillonite and vermiculite.[11,19-22] These frequencies are comparable with those shown by the corresponding perchlorates but are much higher than in the halide salts.[23,24] More recently, Laby and Theng[25] have come to similar conclusions from a comparison of the infra-red spectra of ammonium in montmorillonite, vermiculite and in the dichromate salt recorded at different temperatures.

Infra-red spectroscopy also provides compelling evidence that adsorbed polar organics, at least short-chain compounds, interact with the exchangeable cations rather than with the silicate surface. Uncharged organic molecules, like those of water, are evidently either directly coordinated to the cation, or indirectly linked to the cation by means of (bridging) water molecules.[5,6] Many examples of each type of ion-dipole interactions are known and they are described in the relevant sections of Chapter 3. What is important to appreciate here is that polar organic compounds must successfully compete with water for essentially the same ligand positions or sites around the exchangeable cation for adsorption to occur. For this reason, the adsorption process is largely dependent on the polarizing power of the cation. Ion-dipole interactions are also influenced by the basicity of the organic molecule and its mode of packing in the inter-layer space.[26] Thus exchangeable transition metal cations having unfilled d orbitals can form stable coordination complexes with bases capable of donating electrons.[21,26] All things being equal, the water in the outer coordination spheres (type II) of the cation is less strongly held than that in the primary hydration shell (type I). Many polar molecules such as pyridine, nitrobenzene, and benzoic acid can therefore displace type II water in montmorillonite at ambient humidity.[26,27] If the saturating cation is weakly polarizing (e.g. Na^+, Ba^{2+}) type I water may similarly be displaced.

As might be expected, displacement of the water directly linked to strongly polarizing cations (e.g. Mg^{2+}, Al^{3+}) is more difficult, although it can be achieved by special treatments which dehydrate the cation.[28] Removal of much of the water of hydration, however, may mean that the cation then polarizes the residual water molecules in its primary hydration shell, causing them to dissociate.[7,8] The protons derived from this source are then available for protonation of at least a proportion of the intercalated organic base. Thus in the magnesium-montmorillonite-pyridine system to which we have already referred, dehydration gives rise to the formation of magnesium hydroxide and pyridinium ions.[26] Protonation of organic compounds adsorbed at clay mineral surfaces is involved or implicated in the initiation of colour,[29] polymerization,[30] and decomposition[7,31] reactions catalysed by layer silicates. Each of these topics is dealt with elsewhere.

2.3. Adsorption from Aqueous Solution

In keeping with the concept that uncharged organic molecules compete with water for ligand positions around the exchangeable cation, little or no adsorption of short-chain compounds by montmorillonite occurs from dilute (<0.5 mol 1^{-1}) aqueous solutions. Using a range of non-ionic aliphatic compounds and calcium-montmorillonite, Hoffmann and Brindley[32] observed that a chain length of at least five units was required for appreciable adsorption to occur from dilute aqueous solutions (Fig. 13). This effect of chain length extends to about ten units, beyond which it becomes less pronounced. Recently, German and Harding[33] have measured the adsorption of a series of primary n-alcohols

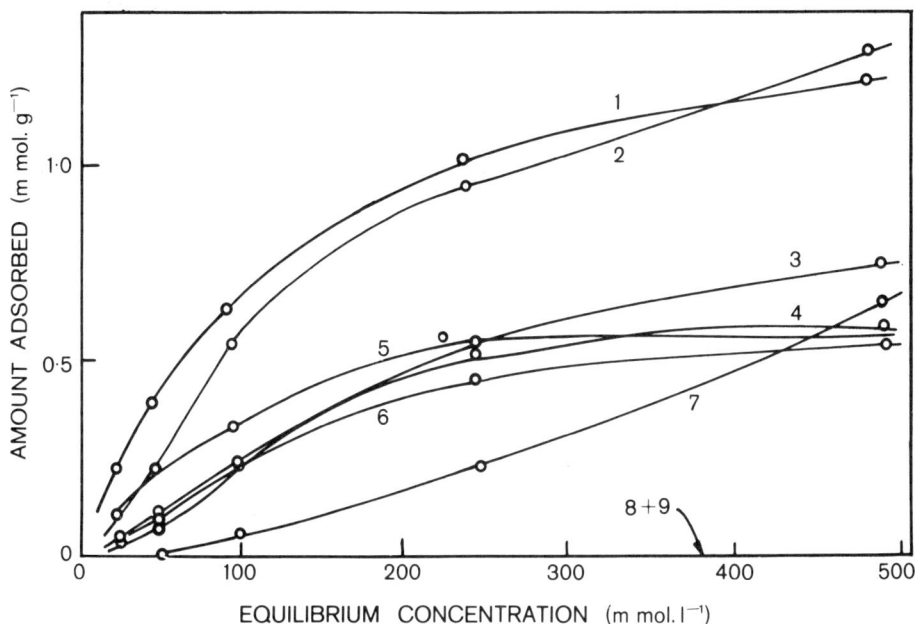

Fig. 13 The influence of chain length on the adsorption of some polar organic molecules from aqueous solution. (1) nonanetrione-2:5:8; (2) bis-(2-ethoxyethyl)-ether; (3) hexanedione-2:5; (4) bis-(2-methoxyethyl)-ether; (5) ethyleneglycoldiglycidether; (6) triethyleneglycol; (7) hexanediol-1:6; (8) diethyleneglycol; (9) pentanediol-1:5. After Hoffmann and Brindley[32]

showing that ethanol, *n*-propanol, and *n*-butanol were taken up by calcium- and sodium-montmorillonite. A similar departure from the simple chain length 'rule' was reported earlier by Bradley[34] for some aliphatic amines.

As might be expected, adsorption is usually increased by raising the solute (organic) concentration[35] or, in the extreme case, by elimination of water from the system.[35-36] Thus ethylene glycol, glycerol, and pentanediol-1:5, none of which shows preferential uptake over water, form stable, well-ordered interlayer complexes with montmorillonite by evaporating the aqueous clay suspension containing these substances.[32,37] Such a treatment, which removes the more volatile component (water) leaving the clay and the organic compound behind, may be likened to immersion of the clay in the corresponding organic liquid.[35] If, on the other hand, the organic compound has a low boiling point relative to water, solute and solvent will evaporate together, leaving the pure clay behind. For example, acetylacetone, which gives high adsorption from aqueous solution, fails to form a complex with calcium-montmorillonite when the suspension is allowed to dry in a desiccator.[32]

Larger molecules with more than five units—both aliphatic and aromatic—may be adsorbed to an appreciable extent by montmorillonite in the presence of excess water.[32-34,38,39] That is, they can displace the water molecules asso-

21

ciated with the exchangeable cations. The increase in affinity with molecular size or chain length can be generally applied to the adsorption of organic compounds by clays[3] and is attributed to the increased contribution of van der Waals forces to the adsorption energy. As the size of the molecule increases, van der Waals interactions become important because these forces are essentially additive and tend to orient the molecule so that the maximum number of contact points is established.[40,41]

If, in addition, the adsorption of one organic molecule is accompanied by the desorption of a number of water molecules initially coordinated to the cation, an appreciable amount of entropy is gained by the system, favouring adsorption. Thus, entropy effects arising from multiple bond formation between the organic compound and the water molecules in the primary hydration shell contribute to the strong adsorption of some uncharged linear polymers by montmorillonite.[42,43] Besides chain length (molecular size) the chemical character of the organic molecule influences adsorption behaviour. If the molecule is sufficiently large to be adsorbed from dilute aqueous solution the adsorption process is

Fig. 14   The influence of chemical character on the adsorption from aqueous solution of some polar organic molecules with a constant chain length of 6 to 7 units. (1) α-methoxy-acetylacetone; (2) acetoacetic-ethylester; (3) β-ethoxypropionitrile; (4) hexanedione-2:5; (5) hexanediol-1,6; (6) 2,4-hexadiynediol-1,6. Data from Hoffmann and Brindley[32]

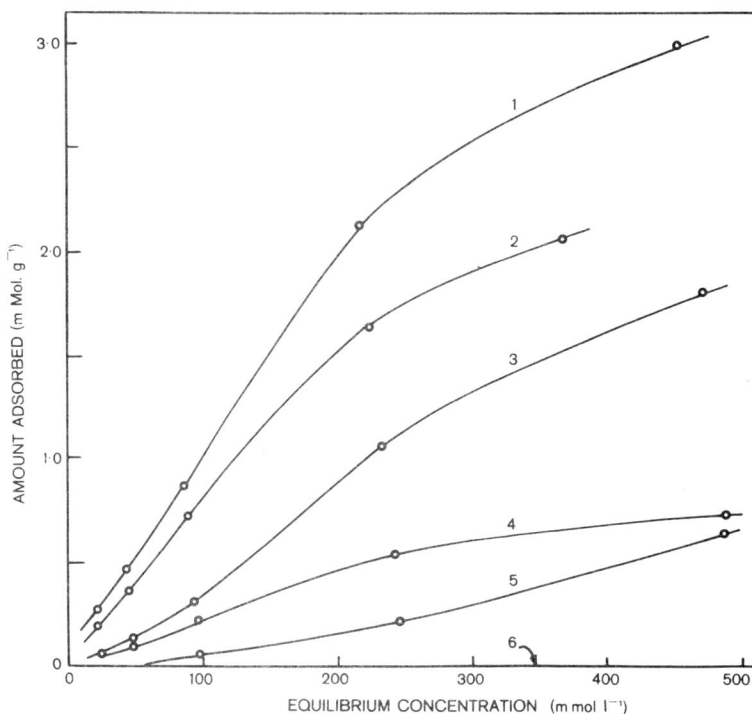

22

determined by its 'character', that is, the presence or absence of certain structural groupings.

For many aliphatic chain molecules, a useful index of character is their so-called CH activity arising from the activation of methylene groups by neighbouring electron-withdrawing structures, such as $C=O$ and $C\equiv N$.[32] Molecules possessing many carbonyl or nitrile groupings adjacent to methylene groups clearly have greater CH activity and accordingly would be more strongly adsorbed than those compounds in which such groupings are few or absent (Fig. 14). Although CH activity is related to the polarity by which it is induced, the polarity of the molecule *per se*, as measured by its respective dipole moment, does not appear to be closely and consistently correlated with adsorption behaviour.[32]

Chain length and CH activity appear to exert their influence independently.[32] Thus pentanediol-1:5 ($OHCH_2$—CH_2—CH_2—CH_2—CH_2OH) is not measurably adsorbed by calcium-montmorillonite from dilute aqueous solution (Fig. 13), whereas acetylacetone (CH_3—CO—CH_2—CO—CH_3) with high CH activity is taken up by the clay under similar conditions. By the same token, low CH activity may be overcome by an increase in chain length. Thus diethyleneglycol ($HOCH_2$—CH_2—O—CH_2—CH_2OH), like pentanediol-1:5, shows no preferential uptake over water, but triethyleneglycol with six units to the molecule is adsorbed.

The importance of CH activity in relation to adsorption behaviour was recognized by Bradley[34] and MacEwan[35] in their classical studies on interlayer complexes of montmorillonite (and halloysite) with a variety of organic liquids. They suggested that the activated methylene groups formed C—H . . . O hydrogen bonds with the surface oxygens of the silicate layer, which would account for the apparent contraction in contact distances of the adsorbed molecules. However, the results of subsequent X-ray diffraction studies,[3,44,45] complemented by more recent infra-red data,[3,5,6] have cast doubt on the importance of C—H . . . O type interactions in clay-organic systems. These studies and their conclusions are discussed below.

2.4. Interlayer Organization of Adsorbed Species

2.4.1. Molecular Orientation

The orientation and spatial arrangement of polar organic molecules in the interlayer space of montmorillonite and vermiculite can ideally be deduced from X-ray diffraction data using one-dimensional electron density projections (Fourier synthesis) based on the *00l* reflections of the respective complexes. Thus the arrangement of ethylene glycol in the interlayers of these minerals, together with the manner in which the molecules are related to one another and to the silicate surface, has been determined reasonably accurately.[45−49] Similar Fourier sketches have been given for complexes of montmorillonite with some aromatic compounds such as benzidine.[34,44] For this method to be of much value, however, the minimum recorded d$(00l)$ spacing should be less than 0·1 nm. More recently, Steinfink and co-workers[50-52] have used two- and three-dimensional Fourier syntheses in an attempt to locate the position of hexamethylene diamine molecules and pyridinium ions in the interlayers of vermiculite. Their results emphasize the limitations of X-ray diffraction methods in studies of this kind. Vermiculite complexes are, of course, inherently more amenable to study by such methods because single crystals can be obtained for the purpose. On the other hand, as Brindley and Hoffmann[45] have pointed out, the small particle size and poorly crystalline nature of montmorillonite seldom permit sufficient resolution for structural analysis to be carried out. Hence, the orientation and arrangement of interlayer organic molecules in montmorillonite has mostly been deduced from basal spacing measurements.[1,3,44,45] Infra-red spectroscopy has also proved of much value in this area of investigation. Thus the arrangement of some aromatic molecules such as pyridine and benzonitrile in the interlayer space of montmorillonite, deduced earlier from basal spacing examination, has been unequivocally confirmed by infra-red measurements of the corresponding complexes.[53,54]

From an analysis of published basal spacing data of montmorillonite complexes containing a single layer of uncharged organic molecules in the interlayer space (single-layer complexes), Brindley and Hoffmann[45] have concluded that aliphatic chain molecules, adsorbed with their shortest axis perpendicular to the silicate surface, may adopt two kinds of orientation. The first type, designated α_I, refers to an interlayer arrangement in which the plane of carbon zig-zag is perpendicular to the silicate layer. Molecules which adopt an α_I orientation give rise to complexes with a basal spacing in the range of 1·325 to 1·365 nm. In the second type, α_{II} arrangement, this plane is parallel to the silicate layer and the corresponding complexes give basal spacings of 1·30 to 1·31 nm. Organic compounds possessing strongly polar groups, such as ketones and esters, usually prefer an α_{II} orientation so as to present the dipole parallel to the surface. But molecules without such groups, for example many alcohols, tend to favour the α_I arrangement.

An analogous situation obtains with aromatic ring compounds. Thus an orientation in which the plane of the ring is perpendicular to the silicate layer may be regarded as α_I while the arrangement in which this plane is parallel to the clay surface may be likened to an α_{II} orientation. Saturated ring molecules having a puckered or zig-zag (non-planar) structure, such as piperidine, are more difficult to classify in this way and a 'long-hand' description of their orientation in the interlayer space may be required. The α_I arrangement is usually observed with saturated ring compounds.[55] Unsaturated ring compounds, on the other hand, tend to adopt the α_{II} configuration[45] giving basal spacings which are 0·08–0·10 nm less than those obtained for complexes in which the adsorbed molecules assume an α_I arrangement.

The type of orientation adopted by many aromatic compounds is also dependent on their concentration. When the concentration is low the α_{II} arrangement is favoured. Rearrangement of the molecules to give an α_I

24

orientation, as indicated by an increase in basal spacing, may occur at high concentrations.[44] Thus montmorillonite saturated with nitrobenzene and pyridine gives a basal spacing of 1·52 and 1·48 nm, respectively, which on heating to 373°K (in order to evaporate a part of the intercalated compound) reduces to 1·25 nm. The low-spacing complex with pyridine is also produced when the molecule is introduced as the cation. The existence of both types of orientation with pyridine,[44] that is, α_I with a d(001) spacing of 1·48 nm and α_{II} with a basal spacing of 1·25 nm, has been elegantly confirmed by Serratosa using infra-red spectroscopy.[53] His evidence is based on the change in intensity of certain absorption bands in the infra-red spectrum of the complex as the angle that the incident beam makes with the normal to the sample is varied. The sample of the montmorillonite complex with pyridine (or pyridinium ion) is presented as a thin, self-supporting film (usually prepared by evaporating the suspension of the complex) so that the clay platelets are in parallel orientation in the plane of the film.

Both pyridine and the pyridinium ion belong to the point group C_{2v}, that is, there is a principal axis (z axis) about which (C_2) rotation occurs and which contains two reflection or symmetry planes, the xz and yz planes (Fig. 15). The infra-red active vibrations fall into three symmetry classes A_1, B_1, and B_2 arising from dipole moment changes parallel to the z, x, and y axis, respectively.[56]

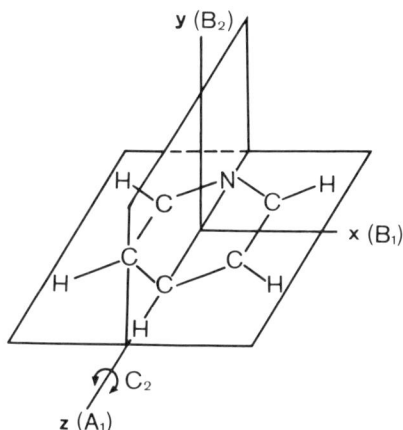

Fig. 15   Diagram showing symmetry elements of the pyridine molecule belonging to the point group C_{2v}, after Serratosa[53]

Clearly, vibrational modes in the A_1 and B_1 classes are in-plane and those in B_2 are out-of-plane. The infra-red pattern of the high-spacing (d(001) = 1·48 nm) pyridine complex is similar to that shown by liquid pyridine.[57] Absorption at 703 cm^{-1} and 748 cm^{-1} in the spectrum of the complex may thus be assigned to CH out-of-plane deformation vibrations (B_2), and those at 1,443 cm^{-1} (B_1) and 1,492 cm^{-1} (A_1) to in-plane ring-stretching vibrations (Fig. 16). Of the latter, only the 1,443 cm^{-1} band shows an appreciable increase in intensity as the angle of incidence is changed from 0 to 0·698 rad (40°) by rotating the

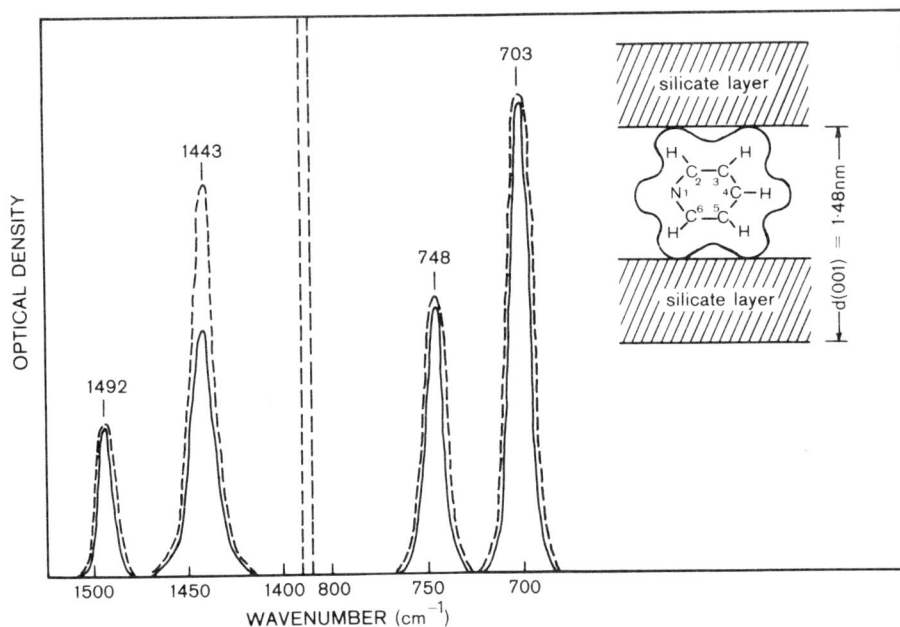

Fig. 16 Selected bands in the infra-red spectra of pyridine intercalated by sodium-montmorillonite. The spectra were recorded at two angles of incidence: 0 rad (0°) ——— and 0.698 rad (40°) - - -. The deduced orientation of pyridine in the interlayer space is also shown, together with the observed basal spacing (after Serratosa[53])

film. Evidently, the intercalated pyridine molecule is oriented with the plane of its ring more or less perpendicular (possibly slightly tilted) and the NC_4 axis parallel to the silicate layer, as depicted in Fig. 16. On the other hand, the montmorillonite complex with pyridinium ion (d(001) = 1·25 nm) gives a spectrum closely resembling that of the unadsorbed cation[56,57] (Fig. 17). By analogy with the spectrum of the pyridine complex, the bands appearing at 677 cm^{-1} and at 748 cm^{-1} may be assigned to out-of-plane vibrations (B_2). Since these bands intensify when the film is similarly rotated, Serratosa[53] concludes that the pyridinium ion adopts an α_{II} orientation with the plane of the ring parallel to the silicate layer (Fig. 17).

One other possible arrangement is also of the α_I type but in which the NC_4 axis is perpendicular to the silicate layer. This configuration is given by inter-layer pyridinium ions in vermiculite on the evidence that the 1,491 cm^{-1} band (A_1) in the spectrum of the complex intensifies as the incidence angle is increased[53] (Fig. 18). It is also of interest to note that the basal spacing of the vermiculite-pyridinium complex at 1·38 nm is 0·10 nm less than that of the (high-spacing) montmorillonite complex with uncharged pyridine. This observation reflects the importance of electrostatic attractive interactions (between —NH$^+$ groups of the pyridinium ion and the negatively charged

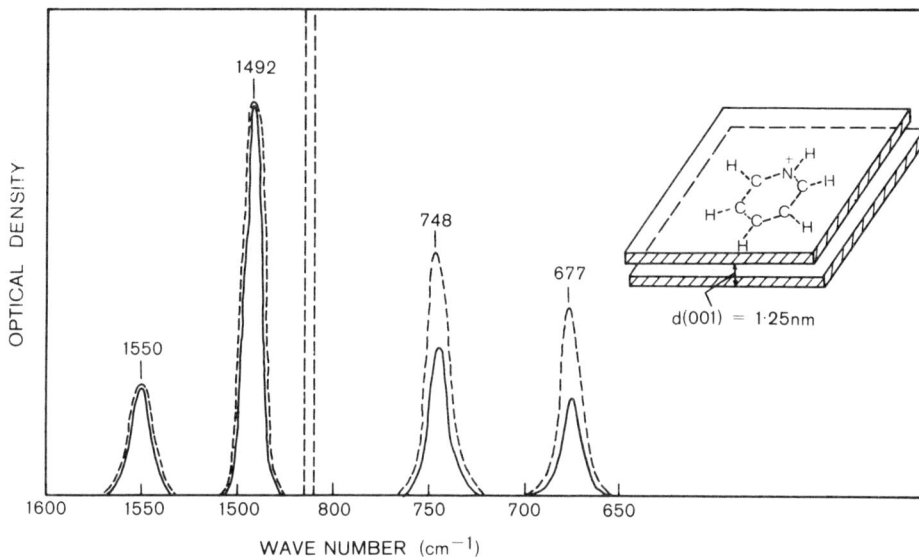

Fig. 17 Infra-red spectra of pyridinium ions intercalated by montmorillonite recorded at two angles of incidence: 0 rad (0°) ——— and 0.698 rad (40°) - - -. The orientation of the pyridinium ion in the 1.25 nm complex is indicated (after Serratosa[53])

Fig. 18 As for Fig. 17 but with vermiculite. The pyridinium ion is shown to be keyed into the silicate surface (d(001) = 1.38 nm), after Serratosa[53]

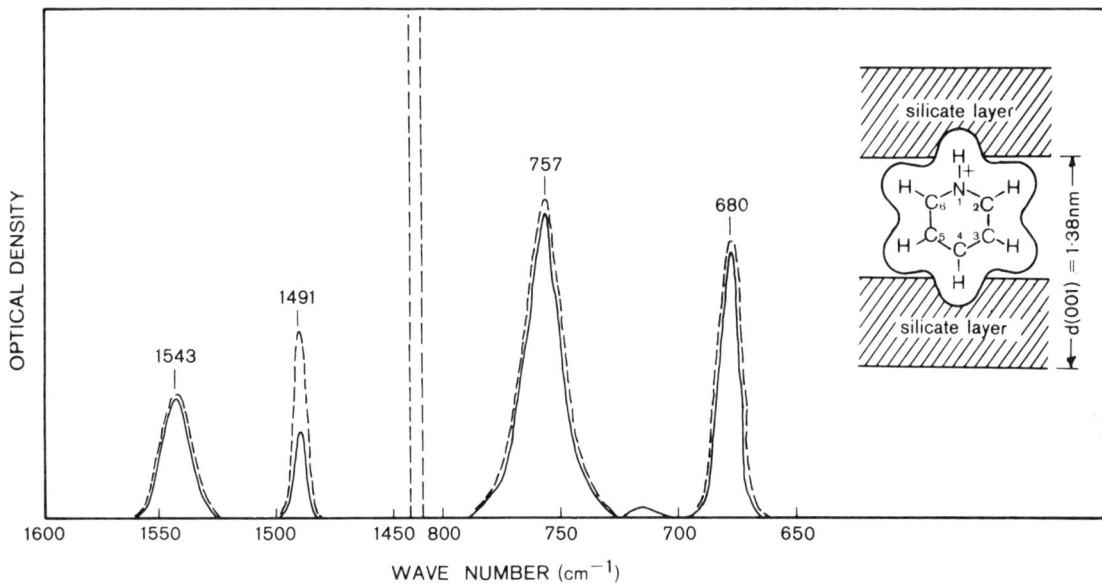

surface oxygens) in the vermiculite system. The reason why the pyridinium ion in montmorillonite prefers the 'flat' configuration while it adopts an 'upright' posture in vermiculite may be sought in the difference in the charge per formula unit between the two mineral species (Table 1) and, hence, in the space requirement of the intercalated organic cation. It can be shown[53,58] that in the flat (α_{II}) configuration, all the pyridinium ions can be accommodated in the interlayer space of montmorillonite at full exchange. However, if the organic ions were to adopt the α_{II} orientation in vermiculite, there would only be room for 60 to 70 per cent of the total amount which may enter the interlayer region by exchanging for the inorganic cations (e.g. Na^+) initially present.[53] On the other hand, with the plane of the ring perpendicular to the silicate surface (α_I orientation) the projected area of the pyridinium ion (~ 0.24 nm^2) is substantially less than that of the ion in the flat configuration (~ 0.4 nm^2)[53,58] which thus allows the full complement of the ions to be intercalated. This suggests that in vermiculite, a reorientation—and apparently a more intimate 'keying' into the ditrigonal depressions of the silicate surface as well—occurs at high surface coverage by pyridinium ions. As we have already remarked, such an effect is observed during the intercalation of many uncharged polar aromatic compounds.[44] Other examples of the influence of surface charge density of the mineral, and that of molecular size (chain length) and concentration on the orientation of defined organic compounds adsorbed at the clay surface, are described in the sections dealing with these compounds.

2.4.2. Shortening of Contact Distances

It is customary to discuss the basal spacing of clay-organic complexes in terms of Δ value,[35] that is, the difference between the observed (expanded) spacing of the complex and the (assumed) thickness of the individual silicate layer. MacEwan[35], who first coined the term Δ value, took 0.94 nm for this thickness, a figure which may be obtained by adding twice the van der Waals radius of oxygen (0.14 nm)[59] to the distance between centres of oxygen ions of the tetrahedral sheets in muscovite (0.662 nm).[60] Values of 0.95 and 0.96 nm have also been used in the literature. Montmorillonite, saturated with different inorganic cations of less than 0.12 nm radius, gives a minimal basal spacing of 0.95 nm on heating or evacuation.[44] Since we are mostly dealing with montmorillonite complexes, we shall use the value of 0.95 nm—which also represents an average of published figures for the van der Waals thickness of a trimorphic silicate layer. The error in this value would not exceed 0.01 nm.

The dimension of organic molecules may be estimated from the generally accepted values of bond distance and bond angle, and of the van der Waals radius of the constituent atoms which are probably reliable to between 0.005 and 0.01 nm.[59]

Neglecting any specific packing at or chemical bonding to the clay surface of the intercalated organic compound, the basal spacing of the complex should be equal to the thickness of the silicate layer plus the cross section of the organic molecule. Hence, the uncertainty in basal spacing values is probably not greater than about 0·03 nm.[45]

Bradley[34] and MacEwan[35] observed that the C—H . . . O distances in montmorillonite complexes with different aliphatic compounds are significantly shorter than would be expected from the van der Waals dimension of the molecules. This led them to suggest that under the influence of activating atoms or groups, the hydrogen atom of methyl or methylene groups formed a C—H . . .O bond with the surface oxygen ion of the silicate layer. Subsequent studies using diverse organic compounds have fully confirmed Bradley's and Mac-Ewan's earlier observations that the Δ values are less than the normal van der Waals diameter of the intercalated organic molecules.[3,44,45] The picture that has emerged from these studies, however, is that C—H . . . O hydrogen bonding is not an important mechanism causing the contraction in apparent contact distances.

Greene-Kelly[44] has examined the basal spacing of 61 complexes of montmorillonite with aromatic compounds. Assuming a certain interlayer orientation of the adsorbed molecules, that is, either flat or upright, C . . . O distances ranging from 0·29 to 0·33 nm were calculated. These values may be compared with the sum of the normal van der Waals radii of C and O (0·31 nm) and with the normal van der Waals distance between C and O atoms with hydrogen interposed in the orientation of Fig. 16 (0·36 nm). Greene-Kelly concluded that the shortening did not vary greatly between complexes, that is, the amount of shortening (~0·05 nm per contact) for small, strongly polar molecules was similar to that for large, weakly polar compounds. This behaviour would not be expected if C—H . . . O bonding were operative.

Shortening of this order has been observed with some organic compounds and was earlier ascribed to C—H . . . O hydrogen bonding.[61] However, as Donohue[62] has convincingly pointed out, this shortening need not represent hydrogen bond formation. More recently, Goel and Rao[63] have carried out theoretical calculations on hydrogen bonding in acetylene, hydrogen cyanide and fluoroform. The equilibrium C . . . donor distances in these systems lie between 0·29 and 0·33 nm (comparable with Greene-Kelly's[44] data and much larger than those found for hydrogen bonds of medium strength) corresponding to dissociation energies of 1,674 J mol^{-1} (fluoroform dimer) to 11,715 J mol^{-1} (hydrogen cyanide dimer). These results indicate that even in situations for which both hydrogen-bond donor and acceptor atoms would favour the formation of C—H . . . O bonding, the resultant interaction is weak. The surface oxygens in layer silicates, however, are weak electron donors[5,6] so that C—H . . . O type interactions between adsorbed organic molecules and clay surfaces would not be expected to occur; if they do, such bonding would at best be very weak.

Perhaps the most compelling evidence against the C—H . . . O bonding concept has come from infra-red spectroscopic studies. If C—H . . . O hydrogen bonding is an important mechanism in causing the shortening of contact distances, the C—H stretching bands in the infra-red spectra of the organic compound would be expected to shift to lower frequencies on adsorption.

Using a high-resolution spectrophotometer, Laby[64] has examined the spectra of methanol, dioxane, and ethylene glycol adsorbed in montmorillonite between 3,800 cm^{-1} and 2,700 cm^{-1} but failed to observe a lowering in C—H stretching frequencies as compared with the spectra of the corresponding organic liquids. Earlier, Tensmeyer *et al.*[65] compared the infra-red spectra of 2,5,8-non-anetrione

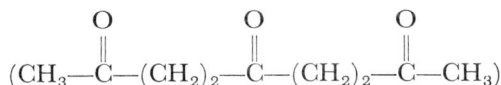

$$(CH_3—\overset{\overset{\displaystyle O}{\|}}{C}—(CH_2)_2—\overset{\overset{\displaystyle O}{\|}}{C}—(CH_2)_2—\overset{\overset{\displaystyle O}{\|}}{C}—CH_3)$$

and 2, 5-hexanedione

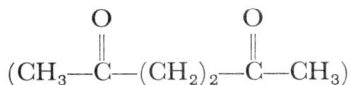

$$(CH_3—\overset{\overset{\displaystyle O}{\|}}{C}—(CH_2)_2—\overset{\overset{\displaystyle O}{\|}}{C}—CH_3)$$

dissolved in carbon disulphide and in carbon tetrachloride with those of the compounds in the solid state and when adsorbed by calcium-montmorillonite. Each methyl or methylene group in both ketones is activated by an adjacent carbonyl group and both compounds are strongly adsorbed by the clay (Fig. 13). No decrease in C—H stretching frequencies was observed; rather, the spectrum of 2,5,8-nonanetrione in the single-layer complex closely resembled that of the ketone in the solid state. Intermolecular interactions are evidently more important than those operating between the organic molecule and the clay surface, and hence CH activity is related more to organic-organic than to clay-organic interactions. Similar conclusions were reached by Mortland[66] and by Farmer and Ahlrichs[67] for urea-montmorillonite systems.

Since C—H . . . O type interactions are not important in causing the shortening of contact distances and N—H . . . O or O—H . . . O hydrogen bonding is weak,[5,6,11,22] 'keying' of the organic molecule into the silicate surface is likely to be the source of the discrepancy between Δ values of montmorillonite complexes and the van der Waals cross-sectional dimension of the intercalated organic compounds.

The geometrical packing of the adsorbed molecules at the clay surface has been examined by Brindley and Hoffmann.[45] They pointed out that packing of aliphatic chains parallel to lines of surface oxygens is favoured since the repeat distance of a carbon chain (about 0·25 nm) is close to the interval between such lines (about 0·26 nm). Using scale models of the organic molecule and of the silicate layer (assuming hexagonal symmetry) it can be shown that in the α_{II} orientation the aliphatic chain can be close packed to the extent of about 0·04 nm at each organic-silicate interface. In the α_I arrangement, the corresponding figure ranges from 0·03 to 0·05 nm depending on whether a methylene group, in Brindley and Hoffmann's[45] phrase, 'sits' astride an oxygen ion or 'rides' in the hollow between two adjacent oxygens.

Aromatic rings lying on the silicate surface may likewise be keyed to an appreciable extent (0·02–0·03 nm) at each clay-organic interface. Cycloaliphatic compounds with each hydrogen atom positioned between two oxygen ions would give a contraction of about 0·05 nm; with each hydrogen placed adjacent to one oxygen ion, the contraction is about 0·025 nm.

It can therefore be said that geometrical packing or keying to give contractions of about 0·025 to 0·05 nm per contact is more the rule rather than the exception. This observation is in accord with Greene-Kelly's[44] earlier suggestion that about 0·1 nm be added to the observed basal spacings of aromatic complexes to give values within a few hundredths of a nanometre of the expected spacing if normal van der Waals contact were to occur.

As Radoslovich[68] has pointed out, the symmetry of the surface oxygen ions is actually ditrigonal rather than hexagonal. But the above conclusions are not invalidated, and contractions of similar magnitude are observed when a model of the montmorillonite surface incorporating the rotation of the silica tetrahedra is used.[69,70]

3

Interactions with Uncharged Polar Organic Compounds Complex Formation with some Defined Classes of Compounds

3.1. Complexes with Primary *n*–Alcohols

The adsorption of primary *n*-alcohols (C_2–C_9) from dilute aqueous 'solutions' (<0.6 mol l^{-1}) by calcium- and sodium-montmorillonite has been investigated by German and Harding.[1] They observed that the lower members of the series (ethanol, *n*-propanol, and *n*-butanol) were adsorbed to an appreciable extent ($\sim$3.5–5 mol kg^{-1}). This behaviour seems to be at variance with the earlier findings of Hoffmann and Brindley[2] who reported that polar compounds with chain lengths less than six units were not measurably adsorbed by calcium-montmorillonite.

German and Harding[1] did not offer an explanation for this departure from the simple chain-length requirement 'rule'. It is perhaps significant that in the concentration range used by them, the limit of complete miscibility with water of the alcohols is reached with *n*-butanol. Infra-red spectroscopic studies by Dowdy and Mortland[3] have shown that ethanol, at least, is capable of displacing the water in the primary hydration shell around such cations as calcium, copper (II), and aluminium. Dehydration to this extent of such highly polarizing cations in montmorillonite can only be otherwise achieved by heating to high temperatures ($\sim$600°K).[4] This ability of ethanol—which presumably applies also for *n*-propanol and *n*-butanol—to compete successfully for ligand sites around the cation may partly account for its strong adsorption in an aqueous environment.

German and Harding[1], rather unexpectedly, observed that whereas the lower *n*-alcohols (C_2–C_4) were adsorbed in comparable amounts by calcium- and sodium-montmorillonite, much larger quantities of the partially water-miscible higher alcohols (C_5–C_{10}) were taken up by the calcium clay. They suggested that this might be due to the fact that the structure of the calcium clay-water system encloses a greater intercrystalline—or more likely inter-domain[5]—pore space in which the organic molecules can be accommodated by phase separation. Further support for this suggestion was provided by the observation that the isotherm for the adsorption of n-heptanol by sodium-montmorillonite, after conversion from the calcium form, resembled the corresponding individual isotherms for the separate calcium and sodium systems, although in two different regions and with reduced adsorption.

If partition or phase separation of alcohol molecules between bulk and interstitial water were constant, the resulting isotherms would be expected to be of the 'C', or linear, type[6] as has been observed in calcium-montmorillonite-amino acid[7] and in allophane-alkylammonium[8] systems. Since the isotherms obtained by German and Harding[1] were of the 'L', or Langmuir, type[6] the partition process was either concentration-dependent, or factors other than those related to the structural status of the relevant clay-water systems were involved. Because alcohols are apparently adsorbed by displacement of water molecules in the primary hydration shell of the cations,[3,9,10] the entropy gain in this process would be greater for the calcium than for the sodium system since the 'structure-breaking entropy' for calcium is larger than for sodium ions.[11] This entropy effect may contribute to the greater adsorption by calcium montmorillonite as compared with the sodium clay. The extensive swelling in water of montmorillonite, containing about equal amounts of calcium and small n-alkylammonium ions on the exchange complex, has been ascribed to a similar effect.[12]

Because of the competition between water and alcohol molecules for adsorption sites at the clay surface, complexes of montmorillonite with n-alcohols are usually prepared in the absence of water, by presenting the appropriate alcohol as a vapour,[3,13,14] a liquid,[13-16] or dissolved in an (inert) organic solvent.[9]

MacEwan[15] was the first to study the formation and properties of complexes between ammonium-montmorillonite and an homologous series of primary n-alcohols. He noted that except for methanol and ethanol, both of which gave a double-layer complex ($d(001) \sim 1.7$ nm), primary aliphatic alcohols formed single-layer complexes ($d(001) = 1.35 - 1.40$ nm) with the alkyl chain lying parallel to the silicate layer. Subsequently, Barshad[16] and Glaeser[17] reported similarly for calcium-montmorillonite, but Barshad also observed that n-nonanol (C_9) and n-decanol (C_{10}) gave 'long-spacing' ($d(001) > 2.5$ nm) complexes. He suggested that the dielectric constant of the alcohol might be important in producing the observed variations in basal spacing.

Following up these studies, Brindley and Ray[13] examined complexes of calcium-montmorillonite with even-numbered n-alcohols (C_2 to C_{18}) and observed four series of basal spacing. The two series of short spacing complexes correspond to the presence in the interlayer space of a single ($d(001) \sim 1.4$ nm) and a double layer ($d(001) \sim 1.7$ nm) of alcohol molecules oriented with their

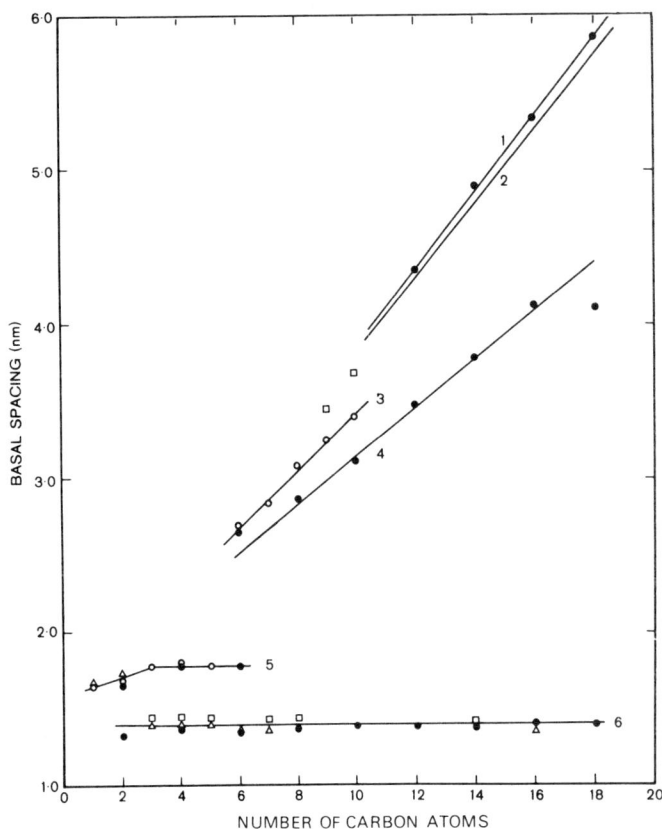

Fig. 19 Relationship between basal spacing of calcium-mont-morillonite complexes with primary *n*-alcohols and number of carbon atoms in the alcohol molecule. Data from Brindley and Ray[13] (●), German and Harding[14] (○), MacEwan[15] (△), and Barshad[16] (□)

alkyl chain parallel to the silicate surface in general agreement with Mac-Ewan's[15] earlier finding. In addition, two series of long-spacing complexes were obtained for C_6 to C_{18}, one above and one below the melting point of the respective pure alcohol (Fig. 19). The temperature at which the transition from one to the other type of complex occurs coincides within 275–277°K with the melting point of the alcohol used. The basal spacings suggest that the interlayer alcohol molecules are adsorbed end-on, that is, with their alkyl chain oriented away from and making a high angle with the silicate surface. More recently, Weiss[18] reported similar long-spacing complexes using an identical set of *n*-alcohols (C_6–C_{18}) and four smectite samples, but failed to observe the formation of the (short-spacing) single- and double-layer complexes. Weiss' data for *n*-tetradecanol are given in Table 6, showing that layer charge is of less importance than the nature of the exchangeable cation in influencing interlayer expansion. Table 6 also illustrates the effect of temperature on the basal spacing of the complex. Thus, for any given cation the d(001) spacing decreases slightly as the temperature of the system is increased until a point (transition temperature) is reached where there is an abrupt fall in basal spacing. Unlike

34

TABLE 6

The effect of charge per formula unit (x), nature of the exchangeable cation, and sample temperature on the basal spacing of complexes between montmorillonite and n-tetradecanol.[18]

Basal spacing (nm) of complexes

°K	Montmorillonite from Geisenheim ($x = 0.33$)		Montmorillonite from Cyprus ($x = 0.42$)		Nontronite from Kropfmühl ($x = 0.66$)						Beidellite from Unterrupsroth ($x = 0.44$)			
	Mg^{2+}	Ca^{2+}	Mg^{2+}	Ca^{2+}	Li^+	K^+	Mg^{2+}	Ca^{2+}	Sr^{2+}	Ba^{2+}	Mg^{2+}	Ca^{2+}	Sr^{2+}	Ba^{2+}
283	—	—	—	—	5·13	—	—	—	—	—	—	—	—	—
293	5·17	4·88	4·82	4·66	5·13	4·88	5·14	5·15	4·96	4·95	5·17	4·83	4·81	4·82
303	5·16	4·87	4·81	4·65	5·09	4·80	5·14	5·07	4·83	4·96	5·13	4·71	4·80	4·81
313	5·10	4·84	4·78	4·57	5·08	4·73	5·03	4·96	4·82	5·01	5·10	4·69	4·76	4·76
323	4·91	4·06	4·68	4·44	5·13	4·67	4·97	4·86	4·51	4·97	5·03	4·63	4·66	4·63
333	4·42	3·93	4·16	4·00	5·13	4·61	4·61	4·38	3·96	4·85	4·95	4·47	4·36	4·58
343	4·29	3·84	4·07	3·81	4·85	3·74	4·17	4·05	3·83	3·97	4·19	4·05	4·09	4·43
353	4·24	3·76	3·98	3·80	4·40	3·58	4·08	3·90	3·78	3·83	4·00	3·94	4·06	4·10
363	4·24	3·95	—	—	4·37	3·38	4·02	3·85	3·75	3·72	3·83	3·90	4·05	4·06
373	—	—	3·94	—	—	3·41	3·93	—	3·71	3·62	3·67	3·88	4·04	—
383	—	—	—	—	—	3·41	3·91	—	3·69	—	3·50	3·86	—	—

35

Brindley and Ray's[13] observations, however, the transition temperature recorded by Weiss lies considerably above the melting point of *n*-tetradecanol (312–313°K). Another point of note is that the variations in basal spacing observed by Brindley and Ray[13] as the temperature was taken above and below the melting point of the respective pure alcohol were essentially reversible, so that these changes were probably related to molecular rearrangements of the intercalated species. Weiss[18], on the other hand, stated that there was an actual loss of material above the transition temperature; the amount adsorbed per formula unit decreased from about 2 for the 'low-temperature' series to about 1·25 for the 'high-temperature' complexes. However, in the absence of adsorption data it is difficult to demonstrate unequivocally whether or not an apparent loss of adsorbed material has occurred. It seems probable that such a loss occurs at the high end of the temperature scale, or is sustained as a result of repeated cooling and heating[13] and this would give rise to irreversible changes in basal spacing.

By adopting suitable models for the arrangement of the intercalated organic molecules and making certain assumptions, an equation to fit the observed basal spacing data can be derived. As an example, the d(001) spacing for the 'below-melting-point' series of long spacings increases by increments of 0·248 nm per carbon atom in the complex, whereas the corresponding value in an alkyl

Fig. 20   The various models proposed for the arrangement of primary *n*-alcohol molecules in the montmorillonite interlayers. (a) Emerson's[19] model for a single-layer complex; (b) Brindley and Ray's model for maximum spacing, below-melting-point complexes; (c) as in (b) but for above-melting-point complexes allowing rotation around OC_I or around $C_{II}C_{III}$; (d) German and Harding's model for their complexes also incorporating OC_I bond rotation; (e) German and Harding's model with a non-linear hydrogen bond

(a)

(b)

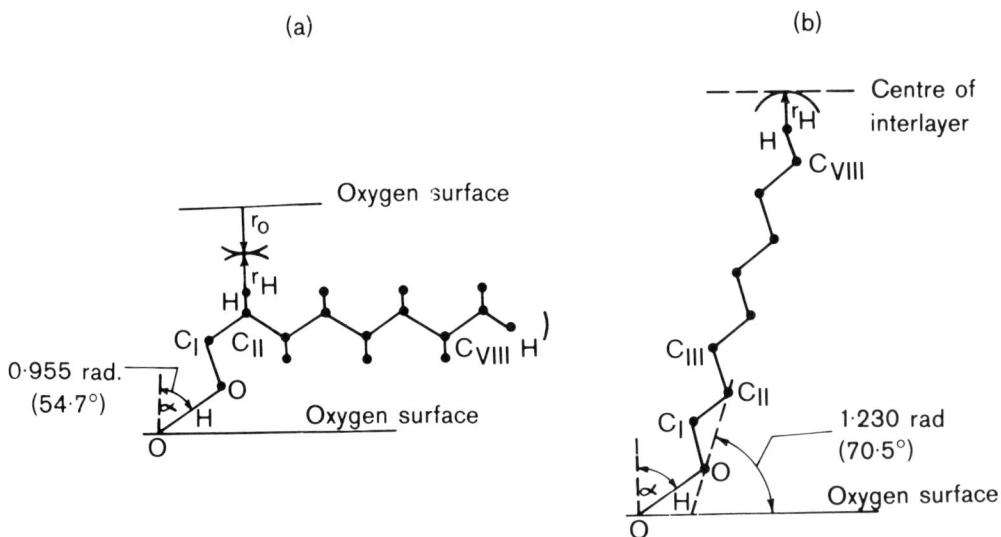

chain is 0·127 nm. Possibly this is because the alcohols are intercalated as a double layer of extended molecules inclined at an angle of $\sin^{-1}(0·248/0·254) = 1·344$ rad (77°) to the silicate surface. Assuming further that the terminal methyl groups of one alcohol layer make van der Waals contact with those of the opposite organic layer, and taking the generally accepted values of bond length and bond angle, the basal spacing may be calculated from the equation[13]

$$d(001) = 1·916 + (n - 2)\,0·248 \text{ nm} \qquad (1)$$

where n is the number of carbon atoms in the molecule. The line of Eq. (1) corresponds to line 1 of Fig. 19, exactly agreeing with the observed d(001) values.

Some time ago, Emerson[19] proposed a model in which the plane of carbon zig-zag is perpendicular to the silicate surface and the terminal hydroxyl groups of the alcohol are hydrogen-bonded to surface oxygens of the silicate layer; the O—H . . . O bond is co-linear, making an angle of 0·955 rad (54·7°) to the normal of the silicate layer (Fig. 20a). If Eq. (1) is calculated on the basis of this model, the inclination angle of the alkyl chain is 1·230 rad (70·5°). This is depicted in Fig. 20b. The corresponding basal spacing equation is represented by line 2 in Fig. 19. A similar method of analysis may be applied to the 'above-melting-point' series of long-spacing complexes. Here the incremental increase per carbon atom is 0·157 nm so that, if the presence of a double layer of alcohol molecules in the interlayer space is assumed, the angle of inclination is $\sin^{-1}(0·157/0·254) = 0·679$ rad (38·9°). To fit this angle to Emerson's model, however, necessitates bond rotation, for example around OC_I (Fig. 20c). The basal spacing equation then becomes[13] (Fig. 19, line 4)

$$d(001) = 2·176 + (n - 4)\,0·157 \text{ nm} \qquad (2)$$

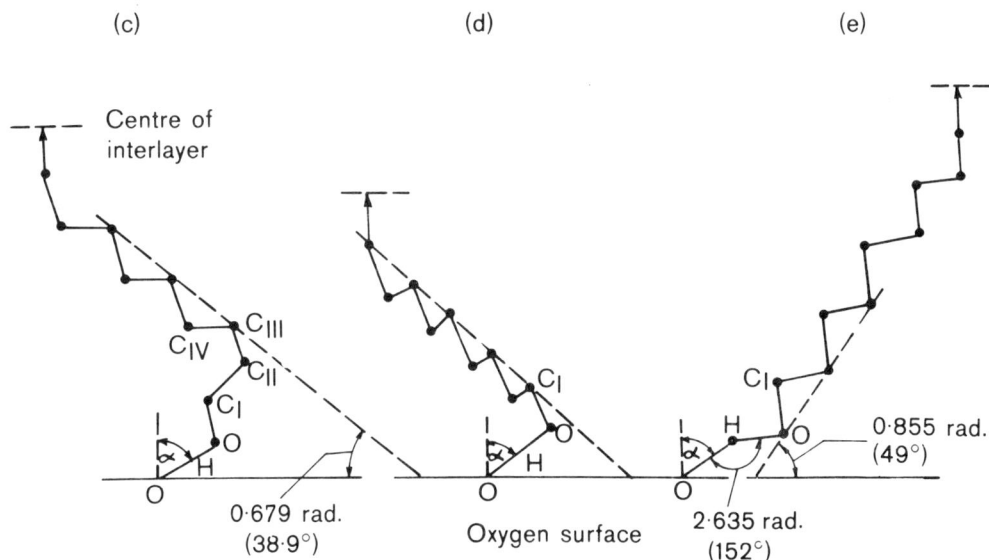

(c) (d) (e)

Centre of interlayer

C_{IV} C_{III} C_{II} C_I O H O 0·679 rad. (38·9°)

C_I H O Oxygen surface

C_I H O 0·855 rad. (49°) 2·635 rad. (152°)

Recently, German and Harding[14] reported long-spacing complexes of calcium-montmorillonite with C_6 to C_{10} primary n-alcohols (Fig. 19, line 3), analogous to the above-melting-point series discussed above. Their incremental spacing of 0·192 nm per carbon atom, however, is notably greater than that observed by Brindley and Ray,[13] and corresponds to an inclination angle of 0·855 rad (49°). Following a similar line of argument, they concluded that an angle of this magnitude is incompatible with the presence of 'fully' extended double layers of alkyl chain. However, bond rotation around OC_I, as previously suggested, would satisfy all requirements (Fig. 20d). As German and Harding have pointed out, it is possible to fit almost any angle between the aliphatic chain and the silicate layer by assuming that bond rotation of this type occurs. An inclination angle of 0·855 rad may also be fitted if co-linearity of the O—H . . . O hydrogen bond and formation of this bond at the tetrahedral angle at the surface oxygen ion[19] are not maintained, as depicted in Fig. 20e. Although a non-linear or 'bent' hydrogen bond is not uncommon and, indeed, appears to be more the rule rather than the exception,[20] the amount of deviation from linearity of 0·489 rad (28°) is close to the generally considered maximum allowable extent of bending ($\sim$0·524 rad = 30°) observed in many organic systems.[20] In the absence of further data from independent measurements, the proposed orientations for the above-melting-point complexes must remain somewhat speculative.

To account for the decrease in basal spacing of the complexes as the temperature is raised from below to above the melting point of the corresponding pure alcohols, Brindley and Ray[13] have offered the following explanation. While the alcohol molecules may still be attached to the silicate surface by O—H . . . O bonds, a rise in temperature would impart a greater mobility to the adsorbed compounds owing to an increase in thermal energy and the weakening of intermolecular van der Waals attraction. This, coupled with the tendency of the silicate layers to contract owing to cation-surface attraction, would result in bending of the molecules and, hence, a decrease in inclination angle.

The following comments may be made at this point. Firstly, the interlayer cation exerts a profound influence on the adsorption process. Indeed, for the lower n-alcohols, such as ethanol, the nature of the cation determines the amount adsorbed, the type of complex formed (that is, whether a single or a double layer of flatly oriented molecules is taken up in the interlayer space), and hence the packing of the organic compound.[3,9] The saturating cation is less important in the adsorption of long-chain n-alcohols (above C_6). These compounds are present as double layers with steeply inclined hydrocarbon chains in the interlayer space, an orientation which is determined by van der Waals interactions between adjacent alkyl chains rather than by cation-dipole or hydrogen bonding to the surface.[21] This explanation would account for the observation that the adsorption of n-hexanol and higher n-alcohols by calcium-montmorillonite in a non-aqueous environment tends to yield long-spacing complexes.[13,14,18] On the other hand, the same compounds adsorbed from an aqueous medium usually give rise to short-spacing complexes (d(001)

~ 1·9 nm), that is, they form double layers with the alkyl chains lying parallel to the silicate surface.[1] Presumably the interposition of water molecules in this instance prevents the alignment of the alkyl chains in the extended conformation. However, on removal of interlayer water from the n-octanol complex, for example, the basal spacing increases from about 1·9 nm to about 3·0 nm. This observation suggests that, in the absence of water, intermolecular van der Waals attraction becomes predominant and takes over from molecule-water-cation type interactions. This change in molecular conformation is doubtless also promoted by the increase in alcohol concentration and by the creation of additional interlayer space as water is removed from the interlayers. Reorientations of the type described appear to be of general occurrence in clay-organic systems. We have already referred to the work by Greene-Kelly[22], who observed that many aromatic compounds reorient from a flat to an upright conformation when the solute concentration is increased, enabling more solute molecules to be packed in the interlayer space. Greenland and Quirk[23] and Theng[24] found a similar situation for cetylpyridinium and cetyltrimethylammonium cations in montmorillonite and vermiculite, respectively.

On the other hand, short-chain alcohols (C_6 and below) tend to form a double layer of flatly intercalated molecules,[14] presumably because in this orientation maximum molecule-to-molecule and molecule-to-surface van der Waals contact is achieved. By the same token, single-layer complexes with short-chain alcohols are more difficult to prepare than the corresponding double-layer complexes.[14]

Also in this context, the rate of entry into the interlayer region, particularly of long-chain alcohol molecules (C_7–C_{10}), is influenced by whether the alcohol is presented as a vapour or as a liquid.[14] This is another manifestation of the concentration effect in that the frequency of solute-mineral collisions is less in the clay-alcohol vapour system than in the corresponding liquid system. Furthermore, the vapour pressure of the alcohol at ambient temperature decreases with an increase in molecular weight.

The rate of interlayer penetration is also influenced by the nature of the saturating cation when complex formation is attempted in the absence of water. Thus, penetration of liquid C_7–C_{10} alcohols into the interlayer space of sodium-montmorillonite previously dried at 523°K proceeds very slowly, compared with that observed for the calcium clay. German and Harding[14] interpreted this behaviour in terms of the low solvation energy of sodium relative to that of calcium. It seems probable also that some interlayer water remains in the calcium saturated material after heating to this temperature[4], facilitating re-expansion of the clay crystals. On the other hand, all of the interlayer water in sodium-montmorillonite would have been lost at 523°K[4] and only the small, relatively highly polar methanol and ethanol molecules appear to have sufficient solvation energy to effect ready entry to give double-layer complexes.[14]

The second comment concerns the formation of O—H . . . O—Si hydrogen bonds between adsorbed alcohol molecules and surface oxygen ions, as suggested by Emerson.[19] Although such bonding is clearly possible and indeed probable, the evidence from infra-red[3,10] and X-ray[9] studies indicates that, at least for the small n-alcohols, cation-dipole interactions are far more important than surface-molecule O—H . . . O type bonding in determining the adsorption process. By analogy with water[25] and with n-alkylammonium cations[21] in montmorillonite and vermiculite, O—H . . . O hydrogen bonding would be weak (of the order of 8,368 J mol^{-1}) particularly if, as in Emerson's[19] model, the hydroxyl group of the alcohol is directed at a high angle to the plane of the silicate layer (Fig. 20a).

Thirdly, differences in basal spacing for the above-melting-point series of complexes between the values reported by Brindley and Ray[13] and those obtained by German and Harding[14] (cf. Fig. 19, lines 3 and 4) are too large to ascribe to general uncertainties in the van der Waals radii of the constituent atoms. The reason for such discrepancies in these and in similar systems (for example, in complexes with long-chain amines discussed in a later section) remains as yet obscure. Subtle variations in the amount and site of isomorphous replacement between montmorillonite samples and in methods of complex preparation may conceivably be the source of such small, but real, differences in the data obtained by various workers.

Figure 19, line 5 shows that the basal spacing of the two-layer (flat orientation) complexes with methanol (d(001) $\sim$ 1·65 nm) and ethanol (d(001) $\sim$ 1·68 nm) is notably smaller than that observed with n-propanol to n-hexanol ($\sim$1·78 nm). Using the model depicted in Fig. 20a, Emerson[19] calculated that the basal spacing of a two-layer methanol complex would be 1·60 nm (the corresponding value of a single-layer complex is 1·27 nm). On this basis, the observed d(001) spacing is greater than the theoretical value and this is contrary to the general rule that contact distance is shorter than the normal van der Waals thickness of the adsorbed molecule (Chapter 2). This analysis provides further evidence against O—H . . . O bonding.

Dowdy and Mortland[3] have examined the adsorption and retention of ethanol vapour by montmorillonite saturated with copper (II), aluminium, calcium, and sodium ions using infra-red spectroscopic and X-ray diffraction techniques. The infra-red spectra of the copper-clay-ethanol system under different treatments recorded by Dowdy and Mortland are reproduced in Fig. 21, illustrating the behaviour of small polar organic molecules at clay surfaces.

Exposure of the parent copper-montmorillonite containing about 5 per cent water (curve A) to ethanol vapour at increasing relative pressure (P/P$_0$) causes a reduction in the intensity of the 1,632 cm^{-1} deformation band (due to vibrations involving the H—O—H bond angle) of water (curve B). At P/P$_0$ = 1 the water content of the complex fell to 0·6 per cent and the 1,632 cm^{-1} band was hardly detectable. Besides reducing the absorbance at 1,632 cm^{-1}, displacement of adsorbed water by ethanol causes a splitting of this band into one with a maximum at 1,598 cm^{-1} and one absorbing at 1,640 cm^{-1} (curve C). Following Russell and Farmer,[4] Dowdy and Mortland suggested that two types of adsorbed water might be distinguished. The first is water which is

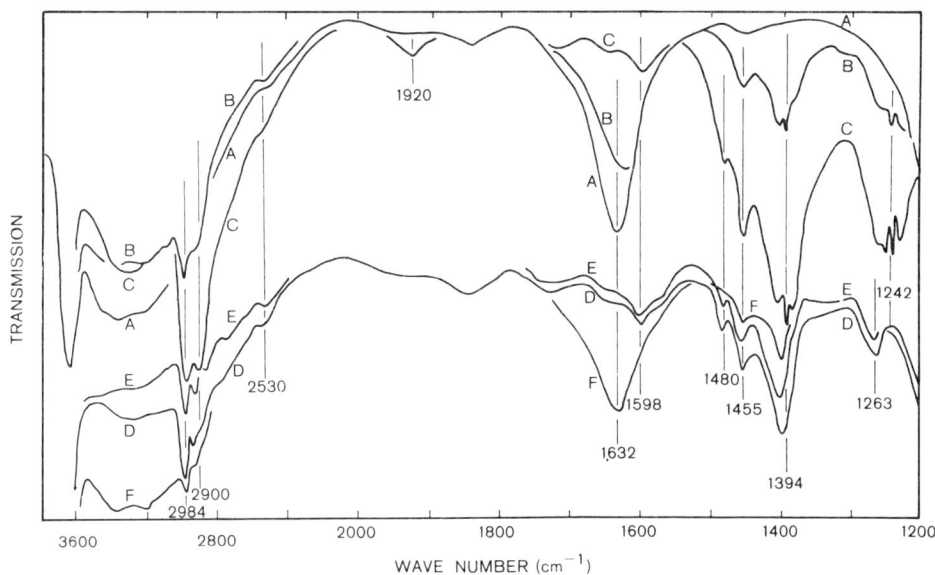

Fig. 21 Infra-red spectra of copper-montmorillonite and its complexes with ethanol taken under the specified conditions. (A) air-dry clay film; (B) film exposed to $P/P_0 = 0.15$ ethanol; (C) film exposed to $P/P_0 = 1.0$ ethanol; (D) film degassed for 1 hour after saturation with ethanol; (E) same as (D) but degassed for 10 hours and then heated to 373°K while degassing; (F) film exposed to atmosphere for 24 hours after ethanol saturation, degassing, and heating. After Dowdy and Mortland[3]

directly coordinated to copper ions and is responsible for the 1,598 cm^{-1} band at very low water contents. The reason for its appearance at such a low frequency is not clear although frequencies as low as 1,550 cm^{-1} have been reported for water in crystalline hydrates, such as $CuCl_2.2H_2O$.[26] The second type of water is less strongly held and is the dominant species at air humidity giving rise to the strong band at 1,632 cm^{-1}.

The hydroxyl bending mode of ethanol in the vapour phase occurs at 1,242 cm^{-1} and shifts to 1,265 cm^{-1} for the adsorbed state. Both the direction and magnitude of this shift suggest that the alcohol molecule is coordinated to the copper ion through the oxygen atom. Persistence of the C—H stretching (between 3,000 cm^{-1} and 2,900 cm^{-1}) and bending (1,480–1,400 cm^{-1}) modes on degassing at 373°K for 10 hours indicates that ethanol is strongly retained (curve E). Parallel desorption experiments showed that copper montmorillonite retained 4·5 ethanol molecules per copper ion under these conditions. This value is consistent with the formation of a square planar copper-ethanol coordination complex in the interlayer space. At low P/P_0 this complex adopts an orientation coplanar with the silicate layer (d(001) = 1·33 nm). Reorientation to a position perpendicular to the clay surface may occur at high relative pressures ($P/P_0 > 0·6$) giving a basal spacing of 1·65 nm.[3]

When the heated and degassed sample was then exposed to the atmosphere of 40 per cent relative humidity for 24 hours some of the copper-coordinated ethanol molecules were displaced by water (curve F). Restoration to the original (parent) clay was essentially complete after 70 hours' exposure to air. Clearly, ligand exchange between water and ethanol is reversible although the kinetics of this process are vastly different, that is, ethanol is adsorbed much more rapidly than it can be desorbed.

A noteworthy feature of the spectra shown in Fig. 21 is that the C—H stretching frequencies of ethanol (at 2,984, 2,936, and 2,898 cm^{-1})[27] are not detectably lowered on adsorption and hence, C—H . . . O hydrogen bonding of the alcohol to surface oxygens seems unlikely.

As opposed to the copper saturated clay, the deformation band of water at 1,635 cm^{-1} in aluminium- and also calcium-montmorillonite is shifted to a higher frequency (near 1,650 cm^{-1}) on exposure to ethanol vapour, while in the sodium- and ammonium-clays the position of this band remains virtually unchanged. The behaviour of the first two forms of montmorillonite may partly be due to the fact that the water molecules exist in octahedral coordination around the Al^{3+} (and Ca^{2+}) ion. Dowdy and Mortland suggested that the residual water molecules in the square planar complex with Cu^{2+} ions are subjected to identical force fields of lower intensity than those in the octahedral complex with Al^{3+} ions. Spontaneous hydrolysis of the Al$(-OH_2)_6^{3+}$ complex to give species of the type Al$(OH)_x(H_2O)_{6-x}^{(3-x)+}$ is also likely and would distort the structure of the complex, exposing the water molecules to different as well as higher intensity levels in the electric field of the clay structure.

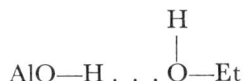

$$\begin{array}{c} H \\ | \\ \text{AlO—H . . . O—Et} \end{array}$$

bonding may also partly be responsible for the observed increase in frequency of the water deformation band.

On the other hand, sodium and ammonium ions interact but weakly with ethanol. On heating and degassing, for example, sodium- and ammonium-montmorillonite retained an average of less than one alcohol molecule for every cation. Further, these samples did not expand beyond a basal spacing of 1·36 nm on exposure to ethanol at $P/P_0 = 1$ so that no more than a single layer of ethanol molecules was present in the interlayer space.[3] The X-ray diffraction data indicate that between certain limits of relative pressure ($P/P_0 = 0.15-0.25$) a single-layer ethanol complex may be formed with calcium-montmorillonite. Beyond $P/P_0 = 0.5$ a double-layer complex is formed and between $P/P_0 = 0.35$ and $P/P_0 = 0.5$ both single- and double-layer complexes may exist. This observation is in line with the unpublished results of Brindley and Bhalla, recently referred to by Brindley.[29] The demarcation between one- and two-layer complex formation is apparently not as sharp or clear-cut as is generally supposed. Earlier, Glaeser and Méring[28] examined the relative stabilities of hydrogen-, calcium-, and sodium-montmorillonite complexes with methanol, ethanol, n-propanol, isopropanol, n-butanol, and cyclohexanol. They found that whereas the hydrogen and calcium forms retained these com-

42

TABLE 7

Isotherm and X-ray diffraction data for the adsorption of ethanol and acetone by montmorillonite saturated with different cations.[9]

Interlayer cation	Ethanol						Acetone					
	Basal spacing (nm)		Number of layers		Amount adsorbed		Basal spacing (nm)		Number of layers		Amount adsorbed	
	(1)	(2)	(1)	(2)	g/g clay	molec/cation	(1)	(2)	(1)	(2)	g/g clay	molec/cation
K$^+$	1·36	1·31	1	1	0·088	2	1·34	1·32	1	1	0·096	2
Na$^+$	1·35	1·34	1	1	0·125	3	1·32	1·33	1	1	0·174	3
Ba^{2+}	1·72	1·67 → 1·36	2	2 → 1	0·160	8	1·73	1·70 → 1·33	2	2 → 1	0·198	8
Ca^{2+}	1·73	1·72 → 1·37	2	2 → 1	0·215	10	1·73	1·72 → 1·35	2	2 → 1	0·203	8

(1) for the clay in equilibrium with 15 per cent ethanol-(or acetone-)dodecane solutions in sealed capillaries.
(2) for flakes of the clay complex with ethanol or acetone under ambient conditions.

43

pounds as a monolayer against prolonged evacuation, the sodium clay did not.

The importance of cation-dipole interactions in montmorillonite complexes with small polar compounds such as ethanol and acetone is further supported by the adsorption and X-ray diffraction data of Bissada et al.[9] These workers measured the uptake of ethanol and acetone from a solution of n-dodecane by oven-dry (498°K) montmorillonite saturated with sodium, potassium, calcium, and barium ions. From such isotherms the numbers of organic molecules associated with each cation were deduced. The coordination numbers at maximum adsorption together with the X-ray data are summarized in Table 7, which clearly shows that the smaller (2–3) and the larger (8–10) of these numbers correspond to single- and double-layer complexes, respectively.

What is striking is that despite the difference in dipole moment, μ_p, and structure (the possibility of O—H . . . O—Si hydrogen bonding, for example, does not exist for acetone) the number of molecules associated with a particular cation is essentially the same for both ethanol and acetone. The conclusion that cation-dipole interactions have a determining influence on adsorption behaviour seems inescapable.

Following Benson and King[30] the cation-dipole interaction energy, E_i, may be estimated from the expression

$$E_i = \frac{z.e.\,\mu_p}{r_i^{\,2}} \tag{3}$$

where z is the valency of the cation and e is the electronic charge; r_i is the interaction distance, that is, from the centre of the cation to the centre of the dipole. We estimate r_i by assuming that the cations make contact with the oxygen atoms of the dipoles. Bissada et al.[9] further assumed that for acetone, the centre of the cation is co-axial with the C = O group, and for ethanol the point of contact coincides with that of the emergence of the sp^3 hybrid orbitals of the oxygen atom. Values for E_i calculated by Bissada et al. from Eq. (3) are shown in Table 8. Using a model proposed by Bernal and Fowler,[31] Mackenzie[32] has

TABLE 8

Interaction Energies for Cation-Organic Dipole (E_i)[9] and Cation-Water (P)[32] Systems

Cation	Ethanol		Acetone		Water
	$(\mu_p = 0.577 \times 10^{-25}$ Cm$)$		$(\mu_p = 0.953 \times 10^{-25}$ Cm$)$		
	r_i (nm)	E_i(J mol^{-1})10^3	r_i (nm)	E_i(J mol^{-1})10^3	P(J g ion^{-1})10^3
K$^+$	0·324	47·7	0·407	49·8	79·9
Na$^+$	0·291	59·0	0·371	60·3	101·3
Ba^{2+}	0·326	94·1	0·409	98·8	146·9
Ca^{2+}	0·294	115·9	0·375	117·5	195·8

44

carried out similar calculations for the electrostatic (Coulomb) attractive energy, P, between water molecules in the primary hydration shell and the interlayer cations in montmorillonite. P values for the potassium-, sodium-, barium-, and calcium-water systems show a similar trend to E_i (Table 8). Ethanol and acetone, like water, can thus be said to 'solvate' the interlayer cation. The solvation energy, the coordination number (molecules/cation, Table 7), and hence the stability of the complex, increase in the order K < Na < Ba < Ca.

The feasibility of packing the ethanol and acetone molecules in the available interlayer space of montmorillonite, consistent with the coordination number of the respective cation, was considered by Bissada et al.[9] Using models for the organic molecules and the silicate surface they were able to fit in the appropriate number of cations over the available surface although (to avoid significant steric overlap) three-fold and five-fold planar coordination complexes for the sodium- and calcium-ethanol systems, respectively, would necessitate very close packing. In these models, the ethanol molecule is oriented with its plane of symmetry perpendicular to the silicate layer (α_I); for acetone, the double bond is placed parallel to the silicate layer and the plane of symmetry containing both methyl groups is perpendicular to the clay surface. In this orientation the van der Waals cross-sectional thickness for ethanol and acetone is about 0·50 or 0·61 nm, whereas the Δ value is 0·42 or 0·41 nm, respectively. Chemical bonding of the type O—H . . . O and/or C—H . . . O between the organic molecule and the surface oxygens is not considered in the above models and, moreover, seems unlikely; this indicates a slight amount of 'keying' of the organic species into the ditrigonal depressions of the silicate surface.

For the barium and calcium clays which form two-layer complexes, the second layer of ethanol or acetone molecules is superposed on the first layer. Both layers are presumably identical in structure, the cation in the double-layer complex still being centrally located with respect to the solvating organic molecules.[9]

We briefly return now to the complexes with C_3–C_6 n-alcohols which, like methanol and ethanol, give two-layer complexes with calcium-montmorillonite (Fig. 19, line 5). The basal spacing of these complexes (1·72–1·78 nm) however, is about 0·1 nm larger than that of the clay containing methanol or ethanol. For the orientation as described above the expected basal spacing, assuming normal van der Waals contact, is about 1·85 nm. The amount of contraction is consistent with what is generally observed (0·03–0·05 nm per clay/organic interface) in complexes with polar compounds where geometrical packing rather than chemical bonding is a likely cause for the shortening of contact distances.[33]

German and Harding[14] suggested that the larger basal spacings of calcium-montmorillonite complexes with C_3 to C_5 might be due to the increased contribution of van der Waals forces between adsorbed alcohol molecules and the clay surface. Steric factors must also influence the extent of expansion. The interlayer surface may possibly just fit two layers of five-fold planar groups of n-propanol around calcium ions. However, this coordination (solvation) number cannot be maintained for n-butanol and n-pentanol without a considerable amount of steric overlap if the model of Bissada *et al.*[9] for ethanol is to be adopted.

Single-layer complexes between calcium-montmorillonite and C_2 to C_{18} *n*-alcohols with the alkyl chain lying parallel to the silicate surface have also been reported by various workers[13-17] (Fig. 19, line 6). However, there is some difficulty in their formation which indicates that two-layer complexes, where the alcohol molecules adopt an orientation with their hydrocarbon chain either parallel or steeply inclined to the silicate layer, are preferentially formed. To obtain one-layer complexes with even-numbered C_2 to C_{18} of the type described above, Brindley and Ray[13] had to remove part of the alcohol initially adsorbed. Similarly, to achieve the same end German and Harding[41] allowed excess liquid to evaporate from the corresponding two-layer complexes with C_4 to C_6 alcohols, but failed to observe an integral series of basal reflections in their X-ray diffractograms. As Brindley[29] has recently mentioned, a sequence of complexes developed when calcium-montmorillonite saturated with C_1 to C_7 *n*-alcohols was progressively evacuated. To speak of either a one-layer or a two-layer complex may therefore be an over-simplification.

Within the limits of uncertainty regarding basal spacing values, the d(001) spacing of one-layer complexes is about 1·40 nm although a figure as low as 1·32 nm has been given for the methanol complex.[13] If the van der Waals dimension of a hydrocarbon chain in α_I and α_{II} orientations is taken as 0·45 and 0·42 nm, respectively, and the thickness of an individual silicate layer as 0·95 nm, the basal spacing corresponding to the respective orientation is 1·40 and 1·35 nm. An α_I arrangement is therefore indicated for these complexes in accord with Emerson's[19] model but without invoking O—H . . . O hydrogen bonding.

We have already remarked that the electrostatic field strength of the exchangeable cation influences the adsorption of ethanol by montmorillonite (cf. Table 8).[9] Thus, all things being equal, the low-field sodium ion not only complexes fewer alcohol molecules but also retains them more weakly as compared with the calcium ion of high field.[3] The difference in solvation energy is also reflected in the slow rate of intercalation of ethanol in sodium-montmorillonite.[14] Another feature of the sodium system is that complexes with C_1 to C_{10} *n*-alcohols do not give an integral series of basal reflections, with the possible exception of methanol and ethanol. German and Harding also failed to form the short-spacing two-layer complexes with C_3 to C_5 or the long-spacing complexes with C_6 through to C_{10} as they achieved with the calcium saturated clay (Fig. 19, line 3).

MacEwan's[15] failure to prepare short- and long-spacing double-layer complexes of C_3 to C_7 *n*-alcohols with ammonium-montmorillonite may likewise be ascribed to weak cation-dipole interactions. As Dowdy and Mortland[3] have shown by infra-red spectroscopy, the behaviour of ammonium-montmorillonite towards ethanol is very similar to that of the sodium clay.

However, using calcium-montmorillonite, Barshad[16] earlier reported basal spacing values for C_3 to C_8 complexes consistent with the presence of a single

46

layer of flatly oriented alcohol molecules in the interlayer space. The reason for this observation is not readily apparent although the high exchange capacity ($1 \cdot 2$ eq. kg^{-1}) of the montmorillonite (Goldfield) sample used by him was probably responsible for the limited interlayer expansion. The basal spacing of the C_3 and C_4 complex with the calcium-saturated Goldfield clay, for example, was $1 \cdot 44$ nm, whereas those of the complex with a similarly treated montmorillonite sample from Wyoming (MacEwan used a sample from the same locality), having an exchange capacity of $\sim 0 \cdot 95$ eq. kg^{-1}, were $1 \cdot 74$ and $1 \cdot 66$ nm, respectively.

It might be argued that basal spacing measurements, by themselves, cannot provide unequivocal evidence of the probable molecular conformation of the interlayer organic compound. For example, a complex between montmorillonite and n-propanol with a d(001) spacing of about $1 \cdot 78$ nm can be interpreted as a single layer of perpendicularly oriented molecules present in the interlayer space, rather than the superposition of two layers of molecules oriented with their alkyl chain parallel to the silicate surface. The former possibility has recently been raised by Radul and Ovcharenko[34] for n-propanol in a sodium-montmorillonite sample on the basis of infra-red and X-ray data. They suggested that the propanol molecules formed O—H . . . O hydrogen bonds to the surface oxygens and simultaneously interacted through their oxygen atoms with the sodium ions embedded in the ditrigonal 'holes' in the silicate surface. The methyl groups at the other end of the molecules were thought to be keyed into similar holes in the opposite silicate layer.

3.2. Complexes with Polyhydric Alcohols

The interaction of montmorillonite with some monohydric (other than the primary n-alcohols) and polyhydric alcohols, and alcohol ethers was reported by Bradley[35] and MacEwan[15] in their classical papers dealing with interlayer complex formation. Like the primary n-alcohols, branched chain and cyclic isomers of n-pentanol and n-hexanol formed single-layer complexes with ammonium-montmorillonite.[15] The basal spacing of these complexes indicated that the molecules were intercalated with their alkyl chain parallel to the silicate surface, probably in an α_I configuration. On the other hand, the glycols (and their ethers) as well as glycerol tended to be intercalated as double layers of molecules with similar orientation, giving basal spacings between $1 \cdot 67$ and $1 \cdot 81$ nm.

Of the polyhydric alcohols, ethylene glycol (1,2-ethanediol, or simply glycol) and glycerol (1,2,3-propanetriol) have received most attention as complexing agents of expanding 2:1 type minerals. In introducing his excellent summary of this topic, Brindley[36] has stated that complexes of montmorillonites and vermiculites with ethylene glycol and glycerol have been studied more than any other type of organo-silicate complexes. Only the salient features of the formation, properties, and applications of glycol and glycerol complexes with expanding layer silicates are discussed here, together with the work which has been published since Brindley's review.

MacEwan[37] and Bradley[38] in the mid 1940s were the first to suggest that ethylene glycol and glycerol might be useful in identifying montmorillonite minerals, which they showed were capable of intercalating a double layer of these liquids giving basal spacings of 1·7 and 1·77 nm, respectively. Since then both compounds have been widely used for routine identification purposes (e.g. Carroll[39]) and for distinguishing 'clay vermiculites' (that is, vermiculites occurring in the fine fraction of soils) from montmorillonites. The basis for this distinction is that vermiculites and montmorillonites tend to take up a single and a double layer of the organic molecules, respectively, at least under certain prescribed and controlled conditions. Walker[40,41] has pointed out that vermiculites, for example, can form either single- or double-layer complexes depending on such factors as the nature of the interlayer cation and the magnitude of layer charge.

Differentiation of montmorillonites and vermiculites is inherently difficult because, as may be recalled, the former grade into the latter with increasing charge and both groups cover a range of layer charge or charge per formula unit (value of x in Table 1). Low-charge vermiculites may thus have interlayer expansion characteristics which bear a closer resemblance to those of high-charge montmorillonites than to those shown by other vermiculites of high charge.

Techniques have been evolved, however, to differentiate these groups of minerals on the basis of their behaviour towards glycerol and ethylene glycol. Thus Walker[40,41] has suggested saturating the clay with magnesium ions prior to addition of glycerol as an effective means of distinguishing between montmorillonites and vermiculites. Under these conditions, the former form double-layer complexes (d(001) $\sim$ 1·78 nm) while the latter intercalate a single layer of organic molecules (d(001) $\sim$ 1·43 nm). Similarly, Mehra and Jackson[42,43] have described a technique of glycerol sorption which seems capable of determining quantities of montmorillonite and vermiculite in clay samples and of estimating the relative amounts of vermiculite and montmorillonite layers in mixed-layer systems. No exact correspondence in the response of vermiculites to ethylene glycol, however, appears to exist and hence diagnostic tests using glycerol seem preferable to those based on ethylene glycol. For example, macroscopic vermiculites in finely divided form and containing different interlayer cations can expand to a basal spacing of about 1·6 nm[41] in the presence of ethylene glycol. This value is close to the spacing obtained ($\sim$1·7 nm) when montmorillonites are treated with this liquid. Glycol complexes with some montmorillonites may not be stable when exposed to ambient atmospheric conditions, yielding basal spacings in the range of 1·61–1·68 nm instead of the 'normal' 1·70–1·71 nm.[44] The effect of exposure to air on the structure and basal spacing of ethylene glycol-montmorillonite complexes is discussed more fully below.

Ethylene glycol and glycerol treatment for characterizing expanding 2:1 type layer silicates have been employed in conjunction with the *potassium contraction*, that is, the ability of the silicate layers to contract when the minerals are saturated with potassium ions.[45] The normal action of potassium is to inhibit interlayer expansion when these organic liquids are added to the minerals. In the case of ethylene glycol, Jonas and Thomas[46] have shown that the proportion of potassium ions occupying interlayer positions is critical. For a given proportion, ethylene glycol seems to be more effective than water in causing interlayer expansion. However, certain high-charge montmorillonites also fail to expand to ∼1·7 nm on addition of ethylene glycol if they have previously been treated with potassium.[44]

Interlayer expansion may also be inhibited by heating the minerals to high temperatures ($>673\,°K$) prior to treatment with ethylene glycol or glycerol[47]. As might be expected, the temperature to which, for example, montmorillonite samples must be taken in order to prevent interlayer swelling in these liquids depends on the nature of the interlayer cation and layer charge,[48,49] as illustrated in Table 9. Of interest here is the behaviour of montmorillonites containing interlayer lithium and magnesium ions which show irreversible layer collapse on heating to moderate temperatures ($473–573\,°K$). This effect, first observed by Hofmann and Klemen,[50] has been attributed to the migration of

TABLE 9

Basal spacing (nm) of montmorillonite samples (Wyoming) containing different interlayer cations heated at various temperatures prior to treatment with glycerol.[48]

Cation	Ionic radius (nm)	Pre-heating temperature (°K)					
		376	473	603	703	813	913
Li^+	0·060	1·78	(1·78)*	0·95	0·95	0·95	0·95
Mg^{2+}	0·065	1·78	1·78	(0·95)*	0·95	0·95	0·95
Na^+	0·095	1·78	1·78	1·78	(1·78)*	0·95	0·95
Ca^{2+}	0·099	1·78	1·78	1·78	1·78	0·95	0·95
Sr^{2+}	0·113	1·78	1·78	1·78	1·78	0·95	0·95
La^{3+}	0·115	1·78	1·78	1·78	1·78	0·95	0·95
K^+	0·133	(1·8)*	(1·8)*	(1·8)*	(1·8)*	0·99	0·99
Ba^{2+}	0·135	1·78	1·78	1·78	1·78	1·78	0·98

* irregular structures.

these (small) cations from interlayer positions to vacant octahedral sites within the silicate structure (*place exchange*). From this fact, Johns and Tettenhorst[51] have shown that some layers in a given sample of montmorillonite may be distinguished from others by their ability or otherwise to expand when treated with polar liquids, such as water, ethylene glycol, and glycerol. Water appears to be least effective in causing interlayer re-expansion, in agreement with the finding of Quirk and Theng.[52] Glycerol is superior to ethylene glycol in revealing differences between layers in a montmorillonite.

Recently, Brindley and Ertem[53] have examined the swelling ability in different organic liquids of mixed-cation montmorillonites, one of the cations being lithium. After heating the clay samples to 493°K for 24 hours most of the lithium ions occupying exchange positions became non-exchangeable, in general agreement with Hofmann and Klemen's[50] postulate of place exchange. Closer inspection of the exchange capacity data (after heat treatment), however, revealed that not all the lithium ions migrated into vacant octahedral sites, that is, up to the limit of the octahedral charge deficiency. The interlayer expansion of the heated samples (Table 10) showed that the silicate surface also played a part in the retention of polar organic molecules. Thus the swelling in water, acetone, and 3-pentanone may be qualitatively explained in terms of the number and polarizing power of the cations (K < Na < Ca). For these compounds, cation-dipole interactions are apparently the dominant mechanism controlling adsorption. With alcohol, ethylene glycol, and morpholine, however, interlayer swelling was insensitive to the number of cations occupying exchange positions within the range of 0.88 to 0.20 meq per g clay. Adsorption of these compounds is evidently influenced by both cation-dipole and surface-dipole interactions.

With acetone and 3-pentanone, mechanisms other than those involving coordination to the exchangeable cation are difficult to envisage. On the other hand, O—H . . . O—Si (for ethanol and ethylene glycol) and N—H . . . O—Si (for morpholine) can and most probably do occur. Hydrogen bonding to surface oxygen ions of this type, although weak, may nevertheless be important when there is an insufficient number of cations to which the organic molecule can coordinate. For morpholine, Brindley et al.[54] have earlier found it to be strongly preferred over water by calcium montmorillonite surfaces. As we note later, acetone may hydrogen bond to the (residual) water molecules in the primary hydration shell of cations with a high electric field, such as Ca^{2+}. Indeed, in excess water-acetone mixtures and at different mole per cent of acetone in the mixture, calcium-montmorillonite shows a complex swelling behaviour.[54] Below a mole per cent of 20, extensive interlayer expansion occurs, similar to that shown by sodium-montmorillonite in water (cf. Table 10). For mole per cent acetone between 20 and 60, basal spacings of about 2.7 nm obtain. Similar observations were reported by Ruiz-Amil and MacEwan[55] in the montmorillonite-water-acetone-NaCl system.

In connection with layer charge, the question arises whether total charge or source of charge (that is, where isomorphous replacement chiefly occurs) exerts the stronger influence on the intercalation of polar liquids by expanding-layer minerals. In an attempt to answer this question, Harward and Brindley[56] examined the expansion characteristics of (synthetic) montmorillonites and beidellites with varying proportions of octahedral-to-tetrahedral substitution in the structure. With similar total charge, both groups of minerals behaved similarly towards ethylene glycol. However, calcium- and magnesium-saturated

50

TABLE 10

Basal spacing of montmorillonite samples containing sodium, potassium, and calcium ions and increasing proportions of lithium ions after heat treatment at 493°K for 24 hours and solvation by the appropriate liquid. Hatched areas refer to changes in solvation state.[53]

Solvating liquid	Cation other than lithium	Basal spacing (nm) of heated clays — Fraction of exchangeable Li^+ prior to heating					
		0	0.2	0.4	0.6	0.8	1.0
Water	Na^+	∞	∞	0.96(vb)	0.95_5(sh)	0.95	0.96_0
	K^+	∞ ▨	1.05(vb)	0.99_0(b)	0.98_5(sh)		0.96_0
	Ca^{2+}	1.91_0(sh)	1.91_0	1.92_0	0.96_0	0.96_0	
Acetone	Na^+	1.85_0	1.85_0	1.30_0(sh) ▨	1.3(ml) ▨	0.93_5	0.94_5*
	K^+	1.31(sh)	1.30(sh)	1.29_5	▨	▨	0.94_5*
	Ca^{2+}	1.74_5(sh)	1.74_5(sh)	1.75_0(sh)	1.31_5(ir)	1.31(ir)	
3-Pentanone	Na^+	1.32_0(v.sh)	1.31_5(v.sh)	1.31_0(sh)	1.32(sh)	0.94_5 (1.28)**	0.96 (1.32)
	K^+	1.34	1.32_5	1.32_5(sh)	▨	▨	0.96 (1.32)
Ethanol	Na^+	1.68_0	1.70_5	1.71_0	1.70_0	1.70	1.7 (1.85) (ml)
Ethylene glycol	Na^+	1.70_0(v.sh)	1.72_0(v.sh)	1.70_5(v.sh)	1.71_0(v.sh)	1.71_5(sh)	1.71(ir)
	Ca^{2+}		1.68_5(v.sh)	1.69_5(v.sh)		1.70_5(v.sh)	
Morpholine	Na^+	1.48_0 (0.96) 0.99	9.48_0 (0.96)	1.48 (0.96)			1.51_5 (0.95)

v.sh = very sharp; sh = sharp; b = broad; vb = very broad; ir = irregular; ml = mixed layered; ( ) = also present; ∞ = no basal spacing observed; ** = trace of 2·2 nm spacing ($0.945 + 1.28 = 2.225$); * = trace of long spacing.

samples treated with glycerol vapour showed differences. The montmorillonites, where the negative charge arises chiefly from isomorphous substitution in octahedral positions, expanded to 1·67–1·77 nm whereas the beidellites, where the negative charge mostly resides in the tetrahedral sheet, expanded to a basal spacing of 1·42–1·46 nm. This observation led Harward and Brindley to propose that beidellite is intermediate between montmorillonite and vermiculite as regards solvation by polyhydric alcohols.

In a more recent study, Harward et al.[57] have extended the above measurements to a range of naturally occurring montmorillonites, beidellite (1 sample), and vermiculites. They observed that the montmorillonites intercalated two layers of ethylene glycol or glycerol from their respective vapour phase whereas the beidellite sample expanded to give only a one-layer complex. On the other hand, the vermiculites did not yield regular double-layer complexes with either ethylene glycol or glycerol, irrespective of the saturating cation (calcium and magnesium) and of pre-exposure to water vapour. Rather, a series of complexes formed with basal spacings of about 1·36, 1·40, 1·43, 1·50, and 1·53 nm; two or more such complexes may be present in the same sample. These observations, together with potassium-contraction data, indicate two discrete populations, making it possible to distinguish montmorillonites from vermiculites.

Another approach to this question has been suggested by Tettenhorst et al.[58] Using a range of diols and triols, they noted that whereas ethylene glycol and glycerol expanded calcium- and sodium-montmorillonite samples to the same extent ($d(001) = 1·70$ and $1·78$ nm, respectively), polyhydric alcohols which have a methylene group not directly linked to an oxygen atom (e.g. 1-3-butanediol and 1,5-pentanediol) gave complexes with different basal spacing depending on the amount and site of isomorphous substitution in the structure.

The picture which has emerged from these and earlier studies[45,59] is that both total charge *and* source of charge are important in determining the interlayer expansion of layer silicates when treated with polar organic liquids or vapours. To emphasize either total charge[41,60] or source of charge[61] in this connection may therefore give an oversimplified picture of the nature of the intercalation process.

The retention of ethylene glycol[62-68] and glycerol[69,70] by clay minerals has been shown to be dependent on the nature of the cation occupying interlayer positions. Table 11, compiled by Brindley,[36] for ethylene glycol shows the influence of the exchangeable cation on retention values. The amount retained is clearly related to the polarizing power of the cation; cations having a high electrostatic field (Ca^{2+}, Al^{3+}, Mg^{2+}) tend to hold more of the organic compound than those of low field such as the alkali ions and ammonium (cf. Table 8). These observations together with the results of thermal stability studies[68,71-73] strongly indicate that, like ethanol[3] (Section 3.1), ethylene glycol, glycerol, and probably other small polyhydric alcohols interact more strongly with the exchangeable cation, at least if it is of the strongly polarizing type, than with the silicate surface. More direct and clear-cut evidence for this hypothesis

TABLE 11

Amounts of ethylene glycol retained by different montmorillonite samples with different interlayer cations, expressed as kg per kg of dry clay, reported by various authors.[36]

Interlayer Cation	Author reference number					
	63	62	64	65	67	68
Ca^{2+}	(a) 0·280 (b) 0·280 (c) 0·250 (d) 0·230 (e) 0·240 (f) 0·280	0·260 ±0·020	(d) 0·244 (f) 0·261	(f) 0·290 (f*) 0·305 (g) 0·255	(h) 0·255	(h) 0·267
H^+	(a) 0·280 (b) 0·270 (c) 0·265 (d) 0·220 (e) 0·295 (f) 0·290	0·270 ±0·018	(c) 0·236—0·259 (d) 0·240 (f) 0·253—0·262	(f) 0·249		
K^+	(a) 0·240 (b) 0·160 (c) 0·140 (d) 0·135 (e) 0·195 (f) 0·185	0·176 ±0·031	(f) 0·117			(h) 0·125
Na^+			(f) 0·222			(h) 0·193
Mg^{2+}			(f) 0·217			
Other			(f) 0·114 (NH$_4^+$)			(h) 0·237 (Al^{3+})

Locality of samples: (a) Chambers; (b) Cheto; (c) Itawamba; (d) Randsburg; (e) Tehachapi; (f) Wyoming; (g) North Africa; (h) Mississippi.
* with oven drying.

was recently provided by Dowdy and Mortland[74] who measured the uptake and retention of ethylene glycol by montmorillonite saturated with different cations (Cu^{2+}, Al^{3+}, Ca^{2+}) using infra-red spectroscopy in conjunction with X-ray diffractometry. Self-supporting films of the clay were exposed to ethylene glycol vapour at a pressure of 4·7 kN m^{-2} (35 torr) and 388°K for 24 hours. The films were then allowed to rehydrate in air of 40 per cent relative humidity and at 293°K for different lengths of time. At each stage of the treatment the weight, basal spacing, and infra-red spectrum of the montmorillonite films were measured.

The spectra of the copper clay under different treatment regimes are shown in Fig. 22. The band frequencies for ethylene glycol in the liquid and the adsorbed forms together with their assignments, taken from the literature,[75–78] are listed in Table 12.

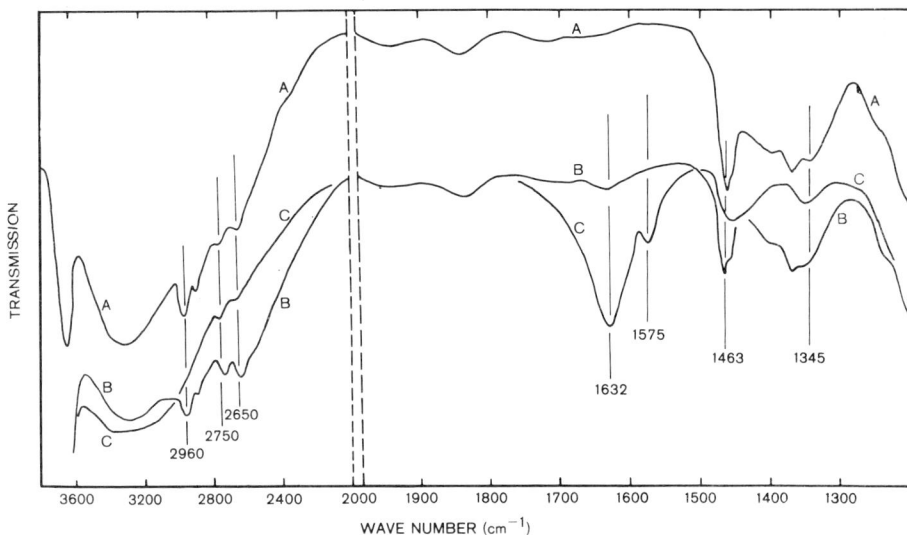

Fig. 22 Infra-red spectra of complexes between copper-montmorillonite and (ethylene) glycol, after Dowdy and Mortland[74]
(A) after adsorbing glycol at 388°K for 24 hours;
(B) sample (A) after exposure to air for 4 hours;
(C) sample (A) after exposure to air for 163 hours

Exposure to ethylene glycol vapour under the stated conditions caused complete dehydration of the clay, as indicated by the absence of the 1,632 cm^{-1} band of vibrations involving the H—O—H angle of water (Fig. 22,A), whereas the control clay film retained about 4 per cent water. Conversely, the

TABLE 12

Infra-red band frequencies and their respective assignments for ethylene glycol in the liquid state and when adsorbed on copper-montmorillonite.[74]

Band frequencies (cm^{-1})		Assignment[75-78]
Adsorbed phase	Liquid phase	
3300	3360	OH stretching mode
2960	2950	asym. CH$_2$ stretching mode
2890	2884	sym. CH$_2$ stretching mode
2750 ⎤	—	OH stretching mode of
2650 ⎦	—	copper-coordinated glycol
1463	1457	CH$_2$ scissoring mode
1370	1370	CH$_2$; 70% wagging, 25% twisting
—	1253	CH$_2$; 40% wagging, 60% twisting
—	1205	CH$_2$; 25% wagging, 70% twisting
1400	1410	Combination mode (CH + OH
1345	1330	bonding)

54

water deformation band at 1,632 cm^{-1} appeared and intensified when the clay-ethylene glycol complex was exposed to the atmosphere for increasing lengths of time; concomitantly, the intensity of the bands due to adsorbed glycol decreased (Fig. 22,B and C). Although these observations indicate that ethylene glycol and water compete for essentially the same ligand sites around the copper ion, it is the appearance of two bands at 2,750 and 2,650 cm^{-1} which points to the involvement of the cations in the adsorption process. Dowdy and Mortland ascribed these bands to the stretching vibration of O—H groups in the glycol molecule which is directly coordinated to the copper ion. It seems reasonable to attribute the 2,750 cm^{-1} band to O—H linked to copper through its oxygen atom, and the band occurring at 2,650 cm^{-1} to the second O—H group forming intramolecular or intermolecular hydrogen bonds. The possibility that these bands arise from perturbed C—H stretching vibrations seems unlikely, because absorption in this region is weak or absent in the spectra of ethylene glycol complexes with montmorillonite containing other cations and of ethanol-montmorillonite systems (cf. Fig. 21). A further point of note is that no lowering of C—H stretching frequencies was observed. If anything, there was a slight (6–10 cm^{-1}) upward shift of these frequencies on adsorption (Table 12). While this observation accords with the suggested bonding mode of glycol, it would exclude the existence of C—H . . . O-clay interactions.[15,35] Nevertheless, the very substantial decrease in the O—H stretching frequencies from the corresponding frequency in liquid ethylene glycol (Table 12) would indicate involvement of the structure of the copper-glycol complex and the way this complex packs in the interlayer space. Thus in cyclic 1,2-glycols examined by Kuhn,[79] the frequency decrease due to intramolecular hydrogen bonding is of the order of 40 cm^{-1} and only in chelate compounds in which resonance stabilization occurs (e.g. acetyl acetone) are O—H stretching frequencies between 3,200 and 2,500 cm^{-1} observed.[80,81]

The infra-red spectra of aluminium- and calcium-montmorillonite complexes with ethylene glycol show features similar to those in Fig. 22 except that absorption between 2,750 and 2,650 cm^{-1} is weak. There is a distinct shoulder at 2,660 cm^{-1} in the spectrum of the aluminium clay (Fig. 23), presumably caused by structural distortion of the Al-glycol complex. Also, in calcium montmorillonite the O—H deformation band of glycol occurs at 1,335 cm^{-1}. This is very close to the value observed for the pure liquid but is some 15 cm^{-1} lower than that recorded for the copper and aluminium complex. Ion-dipole interactions are apparently stronger for the copper and aluminium complexes than for the calcium complex.

Basal spacing measurements at different mole ratios of glycol to water indicate that, beyond a level of 0·25 kg glycol adsorbed per kg of clay, the basal spacing of the montmorillonite-glycol complexes remains constant and a rational series of basal reflections obtain. For the copper and calcium clay systems this spacing is about 1·66 nm; a slightly larger value (1·71 nm) is given by the aluminium clay. Below an amount adsorbed of 25 per cent—which incidentally approximates the amount present in a monolayer spread over the

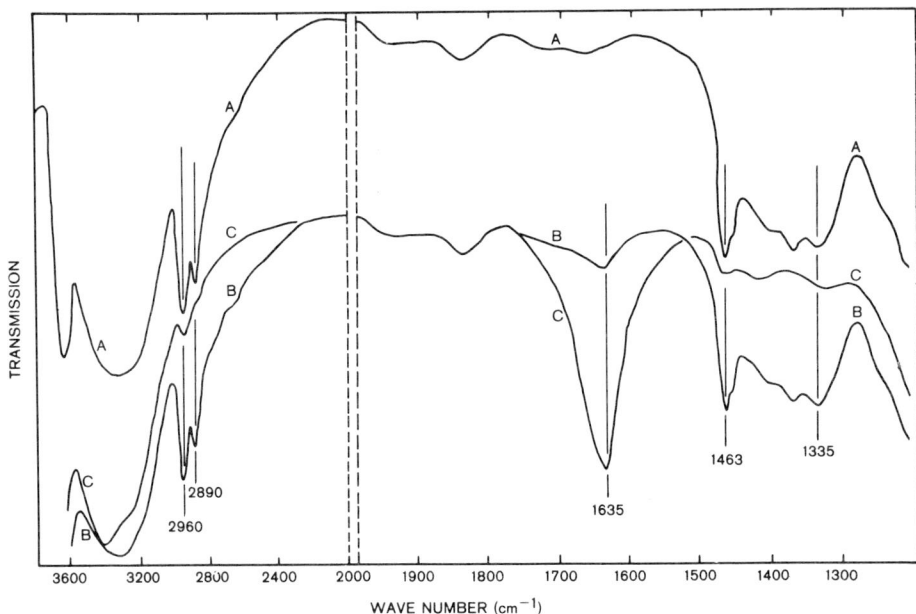

Fig. 23 Infra-red spectra of complexes between aluminium-montmorillonite and (ethylene) glycol
(A) after adsorbing glycol at 388°K for 24 hours;
(B) sample (A) after exposure to air for 4 hours;
(C) sample (A) after exposure to air for 280 hours
After Dowdy and Mortland[74]

total planar surface (a double layer in the interlayer space) of calcium- and aluminium-montmorillonite (see Table 11)—the basal spacing decreases and non-integral series of reflections are observed (Fig. 24). The transition from one phase to another occurs at a glycol:water mole ratio of about 6 for the copper

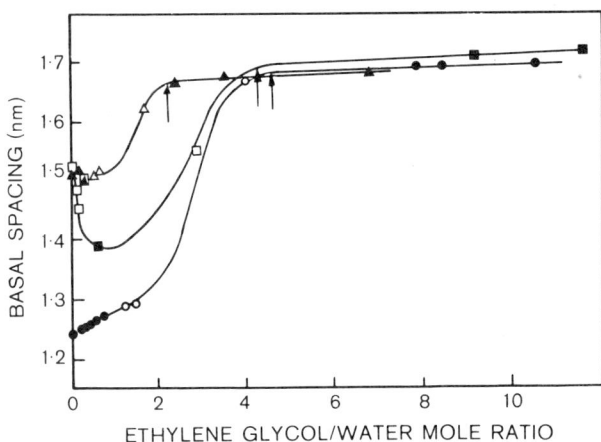

Fig. 24 Variation of basal spacing with (mole) ratio of adsorbed ethylene glycol to water for ($\triangle$, $\blacktriangle$) calcium-, ($\square$, $\blacksquare$) aluminium-, and ($\bigcirc$, $\bullet$) copper-montmorillonites. Solid markers refer to non-interstratified structures giving rational basal reflections; open markers indicate randomly interstratified which 0.25 g glycol per g clay is present. After Dowdy and Mortland[74]

56

and aluminium complexes. The corresponding figure for the calcium system is about 2·3, which accords with MacKenzie's[82] earlier data although he reported basal spacings of 1·71 nm extending to mole ratios of less than 2. Similarly, Tettenhorst *et al.*[58] reported that the basal spacing of montmorillonite complexes with ethylene glycol and glycerol was not influenced by the initial water content.

Dowdy and Mortland[74] explained these observations in terms of structural changes in the respective cation-glycol complexes as glycol molecules were being progressively displaced by water molecules from ligand sites around the cations. For example, the minimum in the basal-spacing-mole ratio curve for the aluminium system (Fig. 24) probably corresponds to Hoffmann and Brindley's[83] one-layer magnesium-ethylene glycol-montmorillonite complex (d(001) = 1·39 nm; mole ratio = 0·57). Non-integral basal reflections above and below a mole ratio of 0·6 may be ascribed to random interstratification of the 1·71 nm (double-layer glycol complex) and 1·39 nm phases and of the 1·39 nm and 1·52 nm (hydrated) phases, respectively. For the copper system, it was suggested that an octahedral Cu-glycol complex was formed which transformed into a square planar complex as water was adsorbed, finally giving (at mole ratios <0·5) an essentially hydrated copper complex. Again, these different phases may be randomly interstratified in a single crystal between a given range of mole ratios of glycol to water and so give rise to non-integral series of reflections.

The ability of atmospheric moisture to compete with ethylene glycol for adsorption sites, and hence to substitute for the organic molecules in the co-ordination complex of the interlayer cation, is probably the source of the apparent discrepancies in the amount of ethylene glycol adsorbed or retained by layer silicates. Martin[65], for example, observed that drying a calcium-montmorillonite over phosphorus pentoxide considerably increased (by 56 per cent) the amount of ethylene glycol retained against evacuation, compared with the moist (17·5 per cent water) sample. On the other hand, Morin and Jacobs[84] reported glycol retention values which were virtually invariant with the water content (0 to 12 per cent) of the clay.

Dowdy and Mortland[74] have shown that the desorption of ethylene glycol from the montmorillonite interlayers is a diffusion-controlled process. As might be expected, the measured diffusion coefficients depend on the nature of the interlayer cation, and actually represent interdiffusion coefficients of glycol and water. Water in the interlayer space reduces glycol adsorption/retention if this water is unable to diffuse out of the system when the organic liquid is added/removed. Possibly, under a given set of conditions more or less glycol is retained because some of the water molecules may be prevented from leaving the interlayers ('trapped')[65] owing to steric factors. Alternatively, the amount of ethylene glycol retained at the point where a monolayer coverage is assumed to occur may be underestimated because the system is only in 'quasi-equilibrium'.[68]

The desorption of glycerol from montmorillonite, vermiculite, kaolinite, and a mixture of montmorillonite and vermiculite systems at 368° and 353°K was examined by Hajek and Dixon.[85] The isotherms for montmorillonite conformed to the Langmuir equation (in the linear form) from which the monolayer glycerol coverage (that is, two layers of the organic molecules were present in the interlayer region) could be derived. Vermiculite behaved similarly, whereas two intersecting straight lines described the data for the montmorillonite-vermiculite admixture and so provided a means for quantitatively determining the two minerals when they are both in a given sample. In a later study Moore and Dixon[86] applied the Brunauer, Emmett, and Teller (BET) equation[87] to their glycerol vapour adsorption data with montmorillonite, vermiculite, and kaolinite. Montmorillonite samples gave BET monolayer values between 0·209 and 0·239 kg glycerol per kg clay, corresponding to surface areas of 738×10^3 to 844×10^3 m² kg⁻¹. Lower monolayer values were obtained for the vermiculites indicating that some of the surfaces might not be accessible to or expandable by glycerol, as pointed out earlier by Mehra and Jackson.[43]

Besides its usefulness in clay mineral identification, complex formation of layer silicates with ethylene glycol and glycerol has been widely employed to estimate the total surface area of these minerals. Dyal and Hendricks[62] were the first to describe a technique for determining both the external crystal and interlayer area of montmorillonite minerals based on the amount of ethylene glycol retained under conditions in which a monolayer of the organic molecules was presumed to be present on all planar surfaces. If, by some means, penetration of the interlayer region by glycol can be prevented, such as by preheating the mineral to 873°K, the external crystal area is measured. The internal (interlayer) area may then be simply obtained by difference. Modifications and refinements of Dyal and Hendricks's method have subsequently been suggested using both ethylene glycol[44,64,65,67,68,84,88] and glycerol.[42,43,69,70,86,89,90] The advantages and limitations of such methods for surface area determination and the assumptions behind them have been discussed by Mortland and Kemper,[91] Brindley,[36] and more recently by Guyot[92] and van Olphen.[93]

The principle behind these techniques is that the rate of desorption of the adsorbed organic molecules levels off when all the 'free' liquid has been removed by evacuation. At this point, the quantity retained by the clay may be assumed to correspond to that required to give a monolayer coverage over the total surface, that is, two layers of, say, ethylene glycol molecules are present in the interlayers and one layer is adsorbed on external crystal surfaces (corrections for the amount adsorbed on crystal edges may be introduced.) This quantity is therefore proportional to the surface area, which may be calculated if the effective area occupied by each organic molecule is known. The molecular area may be estimated by several means, each of which involves different assumptions. Brindley[36] has collated the data for ethylene glycol retention by montmorillonite (Table 13) and has suggested that a figure of 30×10^3 m² kg⁻¹ for each 1 per cent glycol retained is probably the best compromise in calculating surface areas from retention measurements.

Surface areas of montmorillonites may similarly be derived from glycerol retention data, on the assumption that the density of glycerol in the adsorbed

TABLE 13
Ethylene glycol retention by montmorillonite and related parameters.[36]

Parameter	Column				
	I	II	III	IV	V
Glycol retained (kg/kg clay)	0·250	0·250	0·280	0·320	0·338
Surface area of clay (m²/kg)	760 × 10³(1)	810 × 10³(2)	760 × 10³(1)	760 × 10³(1)	760 × 10³(1)
Surface area (m²/kg) for each 1% retained	30·4 × 10³	32·4 × 10³	27·1 × 10³	23·7 × 10³	22·5 × 10³
Surface density of glycol (kg/m²)	0·329 × 10⁻⁶	0·310 × 10⁻⁶	0·368 × 10⁻⁶	0·421 × 10⁻⁶	0·445 × 10⁻⁶
Glycol molecules/(001) unit cell face/layer	1·48(6)	1·40	1·65(3)	1·89	2·0(5)
Area of glycol molecule (nm²)	0·313	0·332	0·281	0·245	0·232
Density of glycol (kg/m³) (7)	870	820	970	1110(4)	1170

Underlined figures form the basis of calculation for each column (I—V).

Basal surface area of calcium-montmorillonite, calculated with $a = 0·517$ nm and $b = 0·895$ nm.
Surface area of hydrogen-montmorillonite, calculated with $a = 0·525$ nm and $b = 0·920$ nm (ref. 62).
Ref. 82.
Normal liquid density of ethylene glycol.
From Fourier synthesis (ref. 104).
A value of 1·5 estimated from Fourier synthesis (ref. 105).
Δ value for each glycol layer is taken as 0·385 nm, corresponding to $d(001) = 1·71$ nm.

and liquid phases is the same and using a given value for the thickness of the silicate layer in calculating Δ values.[36]

Because, as we have seen, ethylene glycol and glycerol, like water and ethanol, tend to solvate the exchangeable cation, another variable is introduced in that for any given sample of montmorillonite (or vermiculite) the amount retained will depend on the nature of the saturating cation (assuming, of course, that the mineral is homogeneous and each interlayer is equally expandable). A similar dependence attaches to the measurement of surface areas of clays using, for example, water as adsorbate.[93,94] As Brindley[36] has concluded, the methods evolved using ethylene glycol (or glycerol) may be suited for comparative rather than for absolute determinations of surface areas. Nevertheless, the glycol retention method is widely and frequently used for such determinations partly because of its ease and simplicity of operation, requiring no elaborate and expensive equipment, and partly because it seems 'to work'[36] reasonably well.

It must be mentioned in this context that other organic compounds have been tried as adsorbates for determining surface areas of layer silicates. Carter *et al.*[95] have proposed ethylene glycol monoethyl ether for this purpose, as this compound has a lower boiling point (332°K) than ethylene glycol (470°K) and hence requires less time to reach the point (in the desorption-time curve) where monolayer coverage is established.

Other compounds examined have the advantage that their respective adsorption isotherms, from which the monolayer value may be derived, can be rapidly and accurately determined using colorimetric or ultra-violet absorption techniques. If the area per molecule is also known or can be determined independently, the surface area of the sample may be computed. An example of such a compound is o-phenanthroline. A method based on the adsorption of this substance has been given by Lawrie[96] who assumed that the organic molecule, lying flat on the clay surface, occupied an area of 0·6 nm². Bower[97] subsequently showed that, in some cases at least, the concentration of o-phenanthroline used by Lawrie might not be sufficient to give a monolayer coverage of all surfaces. Moreover, with some clays (halloysite and vermiculite) not all of the interlayer surface is accessible to the organic molecule. Restricted interlayer penetration of this type is, as we have noted, also encountered in ethylene glycol-vermiculite systems but to a lesser extent because of the relatively small size of the ethylene glycol molecule. However, by applying the Langmuir equation to the adsorption data, Bower derived monolayer coverage values which are useful for comparative determinations of surface areas.

Similarly, Theng[98] has shown that surface areas of clays and soils may be estimated by adsorption of cetylpyridinium bromide from its aqueous solution using the Langmuir equation. Full descriptions of the methods used to determine surface areas of clays and soils with this compound have been given by Greenland and Quirk.[99,100] More recently, Pham and Brindley[101] have proposed that under appropriate conditions, surface areas of clay minerals may be derived from methylene blue adsorption data. The interactions of layer silicates with organic cations, which include cetylpyridinium and methylene blue, are treated in a later chapter.

To conclude this section, we discuss the results of attempts by several workers to determine the orientation and mode of packing of ethylene glycol in the interlayer space of montmorillonite and vermiculite.

Bradley[35] and MacEwan[15] were the first to publish one-dimensional Fourier sketches based on several orders of basal reflections. Their values of minimum basal spacing, d_{min}, for some complexes of montmorillonite with polar organic liquids were about 0·18 and 0·167 nm, respectively. Although the organic molecules were clearly intercalated in layers, their exact organization within such layers was not revealed. MacEwan suggested that the adsorbed molecules existed as a two-dimensional liquid in the interlayer space.

Subsequently, Brown[102] obtained 14 orders of reflection ($d_{min} \sim 0·13$ nm) from an ethylene glycol complex with montmorillonite. Similarly, Bradley et al.[103] and Walker[41] have published a one-dimensional Fourier synthesis of a vermiculite complex containing single and double layers of ethylene glycol, respectively. But these workers reached no definite conclusions about the precise orientation of the organic molecules in the interlayer space. In a later attempt, Bradley et al.[104] established that the ethylene glycol molecules in a single-layer complex of sodium-vermiculite ($d(001) = 1·29$ nm) were oriented with the plane of the carbon chain perpendicular to the silicate layer and parallel

60

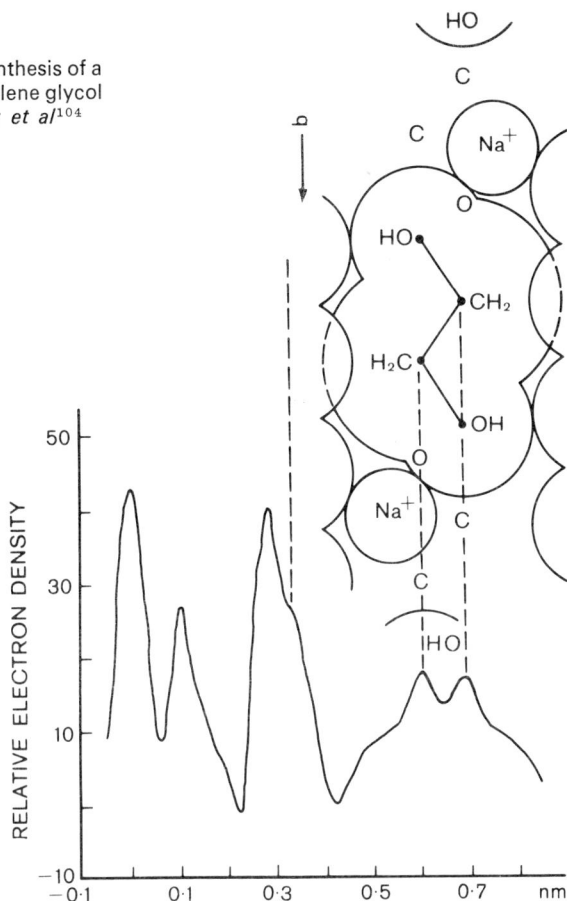

Fig. 25 One-dimensional Fourier synthesis of a sodium-vermiculite complex with ethylene glycol (d(001) = 1.29 nm), after Bradley et al[104]

to the b axis, that is, in an α_I configuration. (Fig. 25). Two glycol molecules are intercalated per unit cell per layer of vermiculite. Using an ethylene glycol complex of allevardite, an expanding mica-like mineral, Brindley[105] arrived at a similar conclusion from 22 orders of reflection with d_{min} of about 0·12 nm (see also Brindley and Hoffmann[33]).

The orientation of ethylene glycol in a two-layer complex with calcium-montmorillonite $(d(001) = 1·69 \text{ nm})$ has been examined by Reynolds,[106] whose one-dimensional Fourier projections obtained from 14 orders of reflections are almost identical with those of Brindley's allevardite complex. The orientation and packing of the organic molecules in the interlayer space of montmorillonite deduced by Reynolds from X-ray data are shown in Fig. 26. The glycol molecules (about 1·7 per unit cell per layer) are oriented with their symmetry plane parallel to the b–c plane, that is, normal to the silicate surface. The second glycol layer is slightly displaced in the b direction to the extent of about half an oxygen diameter between the two opposing oxygen networks of the silicate layers. The calcium and (residual) water molecules are situated between the two ethylene glycol layers, just above and below the a–b plane which

61

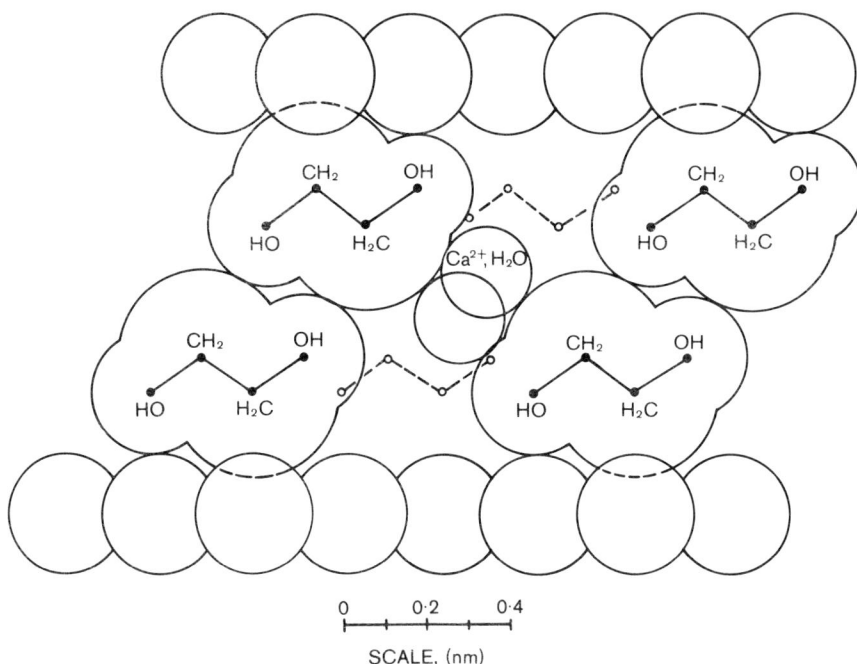

Fig. 26 Orientation and packing of ethylene glycol molecules in the interlayer space of calcium-montmorillonite; the basal spacing of this double-layer complex is 1.69 nm, after Reynolds[106]

bisects the organic layers. The cation-glycol complex depicted is clearly of the octahedral type and weak interactions between one of the hydroxyl groups of each glycol molecule (through its oxygen atom) and the calcium ion is possible in accordance with the infra-red results of Dowdy and Mortland.[74] If ethylene glycol is randomly replaced by water—that is, 6 water molecules substitute for glycol molecules per cation randomly in each interlayer space within a single crystal—a constant basal spacing of about 1·7 nm can clearly be maintained over a wide range of glycol:water mole ratios.[82] On the other hand, cation-glycol-rich and cation-water-rich layers may be randomly interleaved so that below a certain glycol:water mole ratio, the basal spacing decreases and non-rational series of reflections obtain.[74]

Reynolds'[106] structure is consistent with the available interlayer volume and also allows the methylene groups to be keyed into the ditrigonal holes of the silicate surface to account for the observed basal spacing of about 1·69 nm.

Following up this work, Reynolds[107] examined a two-layer complex of sodium-montmorillonite with ethylene glycol monoethyl ether $(d(001) = 1·60$ nm). From a one-dimensional synthesis based on 12 orders of reflections, he deduced that the organic molecules were oriented with the plane of symmetry of the aliphatic chain perpendicular to the silicate surface, essentially similar to that shown by ethylene glycol.

62

3.3. Complexes with Ketones, Aldehydes, Ethers, Nitriles, and Other Compounds

3.3.1. General

Glaeser[108] and MacEwan[15] were among the first to prepare complexes of montmorillonite (and halloysite) with ketones. By treating the clay with a large excess of acetone and nitrobenzene, with or without boiling, MacEwan was able to form two-layer complexes with calcium and ammonium-montmorillonite, respectively. Under the same conditions, acetonitrile and nitromethane gave rise to three-layer complexes with the calcium clay. These observations together with data for other polar organic liquids indicate that the number of organic layers intercalated by montmorillonite is related to the ratio of dipole moment (μ) to molecular size (volume) of the respective compound. However, Hoffmann and Brindley[2] subsequently showed that—at least for adsorption from aqueous solution—it is the CH activity (see Chapter 2) rather than the polarity of the appropriate organic molecule that influences the intercalation process, although the former parameter is, of course, induced by the polar character of the molecule. Thus, if polarity *per se* were important, the nitriles ($\mu = 3.5 - 4.0$ D) should show a higher molar adsorption than, say, the 1,3-diketones ($\mu = 3.0$ D) of comparable chain length and molecular weight. Similarly, one would expect that bis-(2-ethoxyethyl)-ether ($\mu = 2.0$ D) is adsorbed to a lesser extent than, for example, triethylene glycol ($\mu \sim 2.5$ D). In both instances, the opposite behaviour is observed, as shown in Fig. 27.

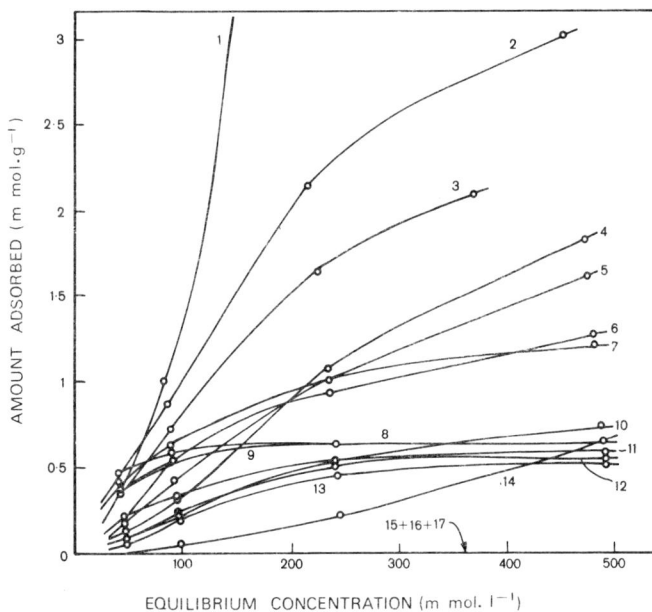

Fig. 27 Isotherms for the apparent adsorption of some non-ionic aliphatic compounds from aqueous solution by calcium-montmorillonite. (1) acetylacetone; (2) α-methoxyacetylacetone; (3) acetoaceticethylester; (4) β-ethoxypropionitrile; (5) β:β′-oxy-dipropionitrile; (6) bis-(2-ethoxyethyl)-ether; (7) nonanetrione-2:5:8; (8) triethyleneglycoldiacetate; (9) diethyleneglycoldiacetate; (10) hexanedione-2:5; (11) bis-(2-methoxyethyl)-ether; (12) ethyleneglycoldiglycidether; (13) triethyleneglycol; (14) hexanediol-1:6; (15) pentanediol-1:5; (16) 2:4-hexadiynediol-1,6; (17) diethyleneglycol. Data from Hoffmann and Brindley[2]

63

3.3.2. Ketones

As explained previously, complex formation with polar organic compounds is profoundly affected by the nature of the exchangeable cation and by the water content (hydration status) of the clay. The influence of adsorbed water on acetone uptake by montmorillonite was demonstrated as early as 1943 by Glaeser[108]. She reported, for example, that sodium- and calcium-montmorillonite previously dried by evacuation at 523°K gave one-layer complexes (d(001) ~ 1·32 nm) when treated with acetone vapour (P/P$_0$ = 0·1 to 1·0) alone. On the other hand, exposure of the clay to saturated acetone vapour and water vapour (50 per cent relative humidity) simultaneously, gave rise to two-layer complexes (d(001) = 1·75 nm). Interstratification of three- and four-layer acetone-water structures seemed to have occurred when the montmorillonite samples were exposed to the saturated vapour of both acetone and water simultaneously. Apparently, hydration of the clay facilitated acetone uptake, presumably by expanding the mineral interlayers. The data also indicated that water and acetone might interchange giving rise to mixed acetone-water (in reality, acetone-water-cation coordination) complexes. A similar situation had earlier been reported by Mackenzie[82] for ethylene glycol and by Dowdy and Mortland[3,74] for ethanol and ethylene glycol. In an attempt to prepare acetone complexes using MacEwan's[15] method, Glaeser found that dehydrated calcium-montmorillonite invariably yielded a two-layer complex whereas the corresponding sodium clay gave either a single- or a double-layer complex. This difference between calcium- and sodium-montmorillonites in their behaviour towards polar organic liquids accords with later findings of German and Harding[14] and Bissada et al.[9] for ethanol- and acetone-montmorillonite systems and is ascribable to the greater solvation energy of the calcium ion compared with that of the sodium ion (cf. Table 8).

The interaction of acetone, 2,4-pentanedione (acetylacetone), and 2,5-hexanedione with montmorillonite containing different interlayer cations has been examined by Parfitt and Mortland[109] using X-ray diffraction and infra-red spectroscopic techniques.

They showed that although the spectrum of adsorbed acetone shared some common features with that of the pure liquid between (that is, containing dissolved) salts, important differences were observed, notably as regards the C=O (ν_2) and C—C (ν_4) stretching bands. In addition, the CH$_3$ deformation bands occurring at 1,440 and 1,424 cm^{-1}, and the band at 1,366 cm^{-1} in the spectrum of liquid acetone, appeared as a single band (1,422 cm^{-1}) and was shifted to 1,372 cm^{-1}, respectively, in the spectrum of the adsorbed species. The infra-red spectra of acetone adsorbed by montmorillonite saturated with different cations are shown in Figs. 28 and 29; the frequencies of the important absorption bands are listed in Table 14.

After degassing and prior to treatment with acetone, the basal spacing of the *sodium* clay stood at 0·97 nm, indicating the presence of only traces of residual water. On exposing this material to acetone vapour followed by degassing, the basal spacing was 1·26 nm, corresponding to the formation of a one-layer com-

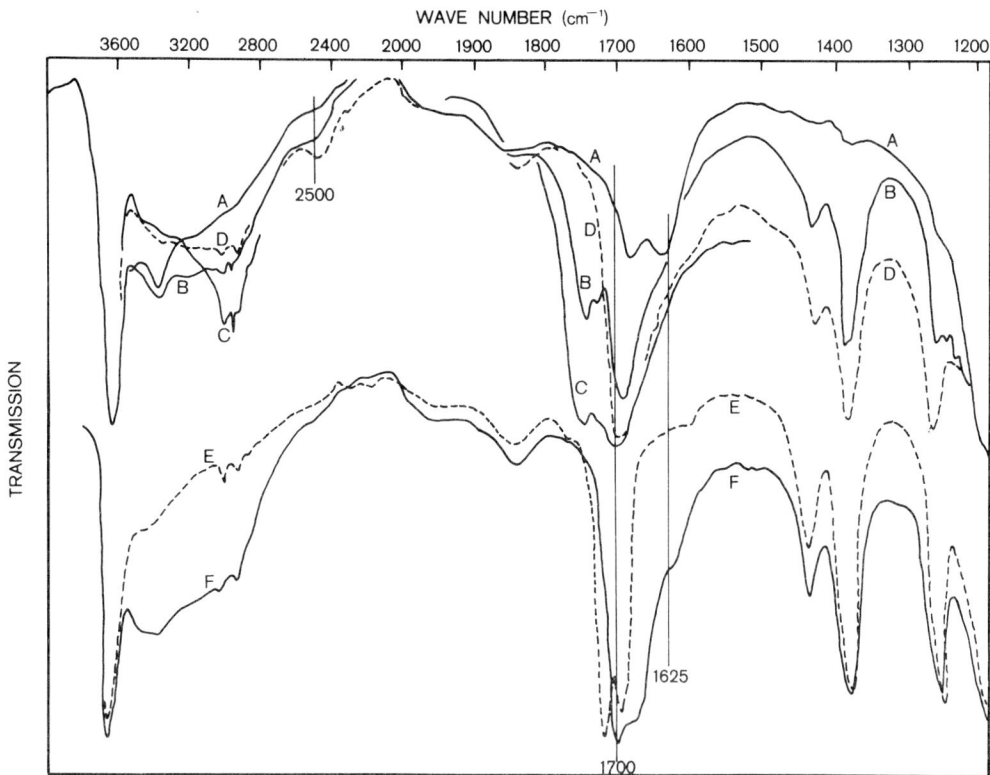

Fig. 28 Infra-red spectra of acetone adsorbed by aluminium-, sodium-, and copper-mont-morillonites under different conditions. (A) Al-montmorillonites degassed (trace of acetone); (B) sample (A) exposed to $P/P_0 = 0.06$ acetone; (C) exposed to $P/P_0 = 0.29$ acetone; (D) sample (C) degassed for 16 hours; (E) Na-montmorillonite exposed to $P/P_0 = 0.84$ acetone and then degassed for 2 hours; (F) Cu-montmorillonite exposed to $P/P_0 = 0.96$ acetone and then degassed for 2 hours, after Parfitt and Mortland[109]

Fig. 29 Infra-red spectra of calcium and magnesium-mortmorillonites and their complexes with acetone. (A) Ca-montmorillonite degassed 2 hours; (B) sample (A) exposed to $P/P_0 = 0.92$ acetone and then degassed for 5 hours; (C) Mg-montmorillonite heated at 353°K and degassed for 2 hours: (D) sample (C) exposed to $P/P_0 = 0.05$ acetone; (E) sample (D) degassed for 2 hours, after Parfitt and Mortland[109]

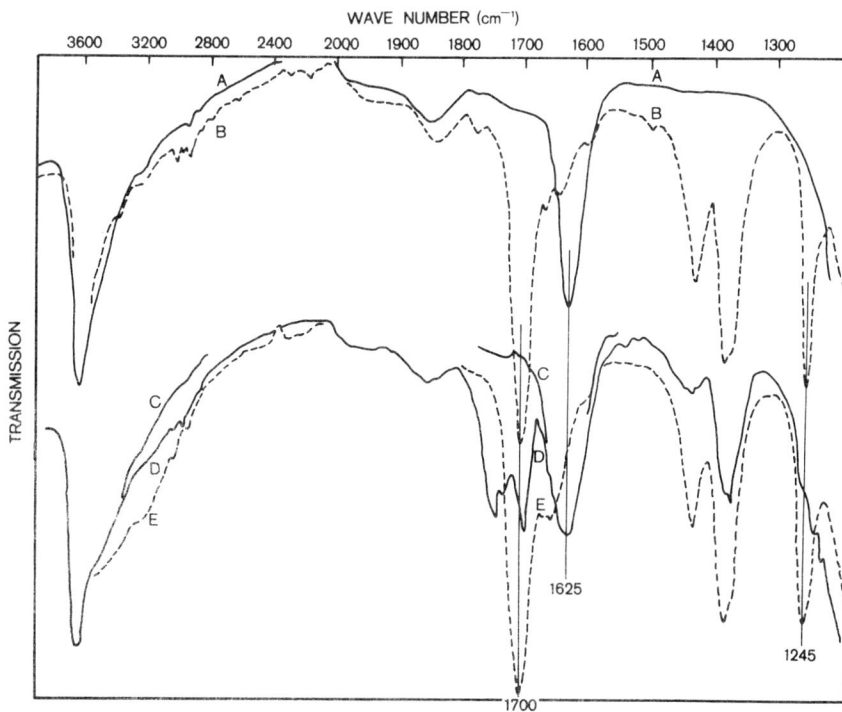

TABLE 14

Infra-red band frequencies (C=O and C—C stretching, H_2O deformation), basal spacings and retention data for complexes between montmorillonite containing different interlayer cations and acetone.[109]

| | Pure Acetone | | Interlayer cation | | | | | | | | | |
	Gas	Liquid	Na⁺ A	Na⁺ B	Mg²⁺ A	Mg²⁺ B	Ca²⁺ A	Ca²⁺ B	Al³⁺ A	Al³⁺ B	Cu²⁺ A	Cu²⁺ B
Frequency, cm⁻¹ C=O(ν_2)	1725	1715	1714(S) 1690(S)	1714(S) 1700(S)	1700(VS)	1701(S)	1700(VS)	1700(S)	1692(S)	1695(M)	1695(VS)	—
H₂O(δ)	—	—	—	1650(M)	1650(M)	1642(M)	1665(M) 1643	1634(S) 1648	Sh	1636(M)	Sh	Sh
C—C(ν_4)	1202	1225	1238(S)	1238(S)	1244(S)	1242(S)	1243(S)	1242(M)	1253(S)	1242(M)	1253(Sh) 1243(S)	1244(S) 1291(W)
Basal spacing (nm)			1·26		1·32		1·32		1·32	1·35	1·24	
Hours for acetone loss to atmosphere			0·3—1·0		0·3—1·0		0·3—1·0		20		1·5—2·0	

A. degassed film exposed to acetone vapour and then evacuated.
B. as for A, then exposed to atmosphere at 40 per cent relative humidity.
(VS) = very strong; (S) = strong; (M) = medium; (W) = Weak; (Sh) = shoulder.

66

plex in which the organic molecules assumed an α_{II} orientation. The C—C stretching band of adsorbed acetone occurred at 1,238 cm^{-1} which was appreciably higher than the corresponding frequency for liquid acetone (1,225 cm^{-1}). The C=O stretching band at 1,715 cm^{-1} in the liquid state was split into two bands at 1,714 and 1,690 cm^{-1} (Fig. 28, E). These observations suggest that the intercalated acetone is physically adsorbed (1,714 cm^{-1}) and also directly linked to the sodium ion by electrostatic interactions probably through the formation of the following resonance structure[110]

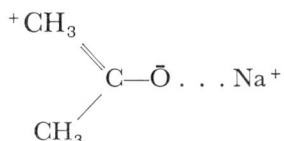

$$\begin{array}{c} {}^+CH_3 \\ \diagdown \\ C{-}\bar{O} \ldots Na^+ \\ \diagup \\ CH_3 \end{array}$$

When the acetone-clay complex was exposed to air (40 per cent relative humidity) the 1,714 cm^{-1} band weakened and the water deformation band shifted from 1,632 to 1,650 cm^{-1}. As more water was adsorbed the 1,690 cm^{-1} band shifted to 1,700 cm^{-1}, indicating a weakening of the electrostatic interactions between acetone and sodium. Parfitt and Mortland postulated that the water molecules formed a hydration shell around the sodium ion so that the carbonyl group was now linked through this water to sodium rather than directly to the cation. Interposition of water molecules (*bridging water*) between the interlayer cations and adsorbed organic species has been shown to occur in montmorillonite complexes with many polar organic molecules such as pyridine[111], nitrobenzene[112], benzoic acid[112], and benzonitrile[113]. Each of these systems is described in the appropriate sections.

The spectrum of acetone intercalated by *calcium-* and *magnesium-*montmorillonite showed similar features to those described above for the sodium clay. However, the parent clay samples contained more (residual) water than the sodium material, as indicated by a larger basal spacing ($\sim$1·25 nm) and by the presence of a strong water deformation band at 1,625 cm^{-1} (Fig. 29, A and C). Direct linkage between the carbonyl group of acetone and calcium or magnesium was therefore not observed. On exposing the calcium clay to acetone vapour and degassing, the 1,625 cm^{-1} water band was split into one absorbing at $\sim$1,640 cm^{-1} and another at $\sim$1,660 cm^{-1} (Fig. 29, B). The latter intensified while the former weakened as more acetone was taken up. The 1,660 cm^{-1} band may therefore be reasonably assigned to the deformation vibration of bridging water molecules; the 1,640 cm^{-1} band is possibly due to intermolecularly hydrogen-bonded water. A shift in the same direction (from 1,632 to 1,652 cm^{-1}) of the water deformation vibration noted earlier by Tensmeyer *et al.*[119] during adsorption of 2,5-hexanedione and 2,5,8-nonanetrione by calcium-montmorillonite may similarly be explained.

If acetone were hydrogen-bonded to water there should be a decrease in the O—H stretching frequency of the residual water molecules on adsorption of acetone. This was indeed observed with magnesium-montmorillonite (Fig. 29, E). Again there was a downward shift of the C=O stretching frequency from

1,715 cm^{-1} for liquid acetone to 1,700 cm^{-1} for the adsorbed molecule, further supporting the formation of

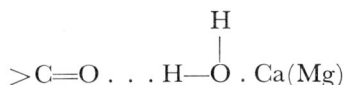

$$\overset{\displaystyle H}{\underset{\displaystyle >C=O\ldots H-O}{\mid}}.\,Ca(Mg)$$

type bonding. A comparable decrease in C=O stretching frequencies has also been reported by Larson and Sherman[114] for acetone (from 1,715 to 1,698 cm^{-1}) and by Kohl and Taylor[115] for diethyl ketone (from 1,709 to 1,683 cm^{-1}) adsorbed by calcium-bentonite. These workers explained this observation in terms of hydrogen bond formation between the carbonyl groups of the respective ketone and the hydroxyls of the silicate structure (*lattice* O—H) but their evidence is not entirely convincing.

The smaller downward displacement of the C=O stretching frequency from the corresponding band in the spectrum of liquid acetone for the calcium and magnesium systems ($\sim$14 cm^{-1}), compared with that for sodium-montmorillonite (25 cm^{-1}), might be because the carbonyl group of the sodium material is directly involved in linkage to the cation, whereas for the divalent cation systems this group is bonded to the cation through a bridging water molecule.

Formation of acetone-water-cation structures in the interlayer space of calcium- and magnesium-montmorillonite is also supported by the upward shift ($\sim$22 cm^{-1}) of the C—C stretching vibration of acetone on adsorption.

In *aluminium*-montmorillonite the acetone molecule appeared capable of linking directly or indirectly (through a water bridge) to the interlayer cation, depending on the partial pressure of the organic vapour. This effect was more clearly seen in the *copper*-montmorillonite system where the C=O stretching vibration of adsorbed acetone at 1,689 cm^{-1} was split into two bands at 1,688 and 1,662 cm^{-1} on standing (Fig. 28, F), absorption at 1,662 cm^{-1} being the result of direct >C=O—Cu bonding. Saturation with acetone followed by degassing did not materially alter the situation. Bands near 1,690 and at 1,243 cm^{-1} may be assigned to acetone linked through water to copper; those occurring at about 1,675 and 1,253 (shoulder) are due to acetone directly coordinated to the copper ion (Table 14).

If the difference ($\Delta\nu$) in C=O stretching frequency between liquid and adsorbed acetone were to be taken as an index of the relative strength of ion-dipole interaction in these systems, sodium, magnesium, and calcium would fall into one class ($\Delta\nu \sim 15$ cm^{-1} for acetone-water-cation type bonding) and aluminium and copper into another class of stronger interactions ($\Delta\nu \sim 22$ cm^{-1}). Although this accords with the difference in the polarizing power of the cations between the two classes, subtle variations between (individual) cations in the extent of interaction as set out in Table 8 are not revealed by infra-red measurements.

Parfitt and Mortland[109] carried out similar measurements with acetylacetone. The infra-red band frequencies and their respective assignments[116-118] are

TABLE 15

Infra-red band frequencies (cm⁻¹) and assignments for acetylacetone in the liquid and solid phases and when adsorbed by montmorillonite containing different interlayer cations together with basal (d(001)) spacing of the respective acetylacetone-montmorillonite complexes.[109]

Assignment	Acetylacetone Liquid Enol	Liquid Keto	Solid	Na⁺ A	Na⁺ B	Na⁺ b	Mg²⁺ A	Mg²⁺ B	Mg²⁺ b	Ca²⁺ A	Ca²⁺ B	Ca²⁺ b	Al³⁺ A	Al³⁺ B	Al³⁺ b	Al³⁺ c	Cu²⁺ A	Cu²⁺ B	Cu²⁺ c
ν(C—O)ₐ	—	1727	—	1726 1753(S)	1750(S)	—	1732(S)	1741(M)	—	1740(VS)	1748(S)	—	1725(M)	1715(Sh) 1740(Sh)	—	—	1734(Sh)	—	—
ν(C—O)ₐ / ν(C—O + C—C)ₐ	—	1707	1616(S)	1703(W) 1627(S)	1700(W) 1635(M)	1618	1700(S) 1615(S)	1694(M) 1637(S)	1618	1703(S) 1622(S)	1694(W) 1628(S)	1629	1695(M) 1615(S)	1695(Sh) 1627(S)	1618	1590	1708(S) 1616(S)	1635(M) 1583	1580
δ(H₂O)	—	—	—	1627(W)	—	1538	1615(S)	—	1538	1622(S)	—	1534	1580(S)	1590(S)	1529	1545	1583(S)	1530(S)	1534
ν(C⋯C)_c	1450(M)	—	1450(S)	1528(W) 1590(Sh)	1538(Sh) 1595(Sh)	1475	1564(S) 1532(M)	1542(Sh) 1590(Sh) 1532(M)	1490	1539(Sh) 1570(S)	1539(Sh) 1575(Sh)	1490	1540(S)	1541(S) 1535(S)	—	1530	1533(S)	—	1464
ν(C⋯O) + ν(C—H)_c	—	—	—	1465(Sh)	1465(Sh)	—	1460(Sh)	1460(Sh)	—	1465(W) 1500(W)	1460(W) 1500(W)	—	1466	1466(Sh)	1471	1466(Sh)	—	—	—
δ(CH₃)_{a,c}	1419(M)	1420	1425(Sh)	1418(W)	1418(W)	1429	1415(Sh)	1419(Sh)	1429	1415(W)	1420(W)	1422	1400(S)	1400(S)	1414	1387	1390(Sh)	1430(W)	1415
δ(CH₃)_{a,c}	1360(M)	1360	—	1367(S)	—	—	1365(S)	1365(M)	—	1366(S) 1390(W)	1370(S) 1390(Sh)	—	1370(Sh)	1387(S)	—	1387	1365(S)	1375(S)	1356
δ(CH₂)	—	1300	—	1310(W)	—	—	1310(M)	1310	—	1310(M)	1310	—	—	—	—	—	—	—	—
ν(C⋯C)_c + (C—CH₃)_{a,c}	1247(M)	1246	1250(S)	1250(W)	1250(W)	1245	1250	1252(W)	—	1245(Sh)	1290	1264	1255(Sh)	1288(W)	1261	1288	1287(M)	1287(W)	1274
ν(C—H)_c or ρ(CH₃)ₐ	1170(M)	1157	1180(W)	—	—	1206	—	—	1206	1215(W)	—	1200	1190(Sh)	1190(Sh)	1208	1191	—	—	1190
d(001), nm				1·21			1·67			1·66				1·70		1·40	1·21	1·245	

A air-dry film exposed to acetylacetone (liquid or vapour), then evacuated with mechanical pump.
B as for A then exposed to the atmosphere at 40 per cent relative humidity.
a = ref. 116;   b = ref. 117;   c = ref. 118.   S = strong;   M = medium;   W = weak;   Sh = shoulder.

summarized in Table 15. As for acetone, the spectra of calcium- and magnesium-montmorillonite containing interlayer acetylacetone were very similar and showed features in common with those displayed by both the *enol* and the *keto* form of the organic liquid. Whether the intercalated acetylacetone was present in a liquid- or solid-like form (the latter was suggested by Tensmeyer *et al.*[119] for 2,5-hexanedione) could not be unequivocally answered, because some bands in the spectra of the adsorbed species also occurred in the spectrum of solid acetylacetone. The *enol* form of the adsorbed molecule, however, appeared to be unstable in air as was the *keto* form ($1,727$ cm^{-1}) which was not complexed to the cation. An upward shift in frequency of the C=O stretching vibrations attending intercalation by *sodium, calcium,* and *magnesium*-montmorillonites was ascribed to the formation of structures, such as (carbonyl groups in *cis*-configuration)

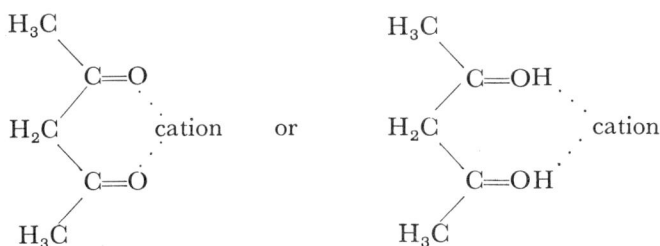

The bands between $1,615$ and $1,622$ cm^{-1} were probably due to C=O and C=C stretching vibrations of the *enol* form ($1,616$ cm^{-1} in the liquid). Other bands in the range of $1,500$ to $1,650$ cm^{-1} may be ascribed to acetylacetone complexed to the cation (as opposed to the physically adsorbed species).

Such complexes formed with *aluminium* and *copper* ions in montmorillonite were characterized by their stability in air (up to several weeks) whereas the corresponding complexes with sodium, calcium, and magnesium ions were relatively unstable, the organic molecule being displaced by water within hours. Acetylacetone which did not complex with aluminium or copper ions gave rise to absorption at $1,725$ and $1,710$ cm^{-1} (in the copper clay) and at $1,695$ cm^{-1} (in aluminium-montmorillonite). This species (*keto* form) was rapidly lost on exposure of the films to air.

The spectrum of the aluminium clay with acetylacetone (after being exposed to air) showed a strong band at $1,615$ cm^{-1} indicating *enol* formation. This and the close resemblance it bore to that of the corresponding aluminium-acetylacetone complex reported by Nakamoto[118], led Parfitt and Mortland to propose that the following reaction took place at the mineral surface:

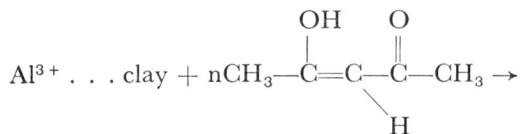

$$(Al(CH_3-\overset{\overset{\displaystyle O^-}{|}}{C}=C-\overset{\overset{\displaystyle O}{\|}}{C}-CH_3)_n)^{(3-n)+} + nH^+ \ldots \text{clay}$$
$$\underset{H}{\diagdown}$$

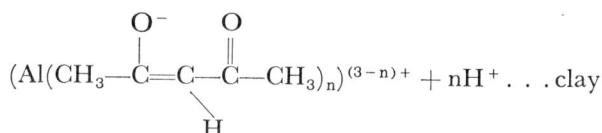

A similar scheme seemed to operate with the copper clay, but proton donation from acetylacetone to the clay is not important in the sodium, calcium, and magnesium systems.

The interaction of 2,5-hexanedione with calcium-montmorillonite was earlier examined by Tensmeyer et al.[119], who reported the formation of single-layer $(d(001) = 1\cdot305$ nm$)$ and double-layer $(d(001) = 1\cdot685$ nm$)$ complexes. They showed that the spectra of the single-layer complexes of this diketone and of 2,5,8-nonanetrione were similar to the spectrum of the solid trione, whereas this type of resemblance was less apparent for the double-layer complexes. Tensmeyer et al. suggested that the spectral changes attending the formation of a two-layer complex were partly due to some replacement of clay-organic contact by organic-organic contact and partly to conformational changes, the one-layer structure being more ordered than the two-layer arrangement.

In extending this study to the sodium-, magnesium-, aluminium-, and copper-montmorillonite systems, Parfitt and Mortland observed that the spectrum of 2,5-hexanedione adsorbed by the *sodium* clay was similar to that of the liquid (Table 16). The 1,415 and 1,404 cm^{-1} bands appeared as a single band at 1,409 cm^{-1} (for all samples); the C=O stretching band shifted downward from 1,723 to 1,710 cm^{-1} and the 1,314 cm^{-1} band intensified.

In the *calcium* and *sodium* systems there was a larger downward shift of the C=O stretching vibration. After introducing the organic vapour, two-layer complexes were formed with basal spacings of $1\cdot67$–$1\cdot68$ nm which decreased to $1\cdot43$ nm (magnesium) and $1\cdot31$ nm (calcium) on exposure to air. This transition from a two-layer to a one-layer complex was accompanied by changes in the bands at 1,313 and 1,274 cm^{-1}, probably reflecting effects due to subtle changes in molecular orientation as water was taken up.

The part played by water in complex formation was more clearly shown by the *aluminium* system where the O—H stretching vibration of water shifted from $\sim$3,400 cm^{-1} to $\sim$2,900 cm^{-1} (Fig. 30). This would be expected if 2,5-hexanedione was bonded to the cation through a water bridge. Changes in the C=O stretching bands on exposure to air of the aluminium- and copper-2,5-hexanedione complexes again indicated the presence of a carbonyl-water-cation type link.

It is clear from the foregoing discussion that ketones may interact with expanding 2:1 type layer silicates in various ways. Coordination of the carbonyl group of the ketones to the exchangeable cation may occur directly, such as obtained with cations of low hydration energy, or indirectly through a bridging water molecule. In addition, electrovalent bonding through proton donation to the clay is indicated for acetylacetone adsorbed by aluminium- and copper-

TABLE 16

Infra-red band frequencies (cm⁻¹) for 2,5-hexanedione in the liquid phase (between salts) and when adsorbed by montmorillonite containing different interlayer cations, together with basal spacing of the respective hexanedione-montmorillonite complexes.[109]

Liquid hexanedione between salts	Na+ A	Na+ B	Mg²⁺ A	Mg²⁺ B	Ca²⁺ A	Ca²⁺ B	Al³⁺ A	Al³⁺ B	Ca²⁺ a	Cu²⁺ A	Cu²⁺ B
1723(S) 1715(S)	1710(S)	1710(S) 1643(M)	1700(VS)	1702(S) 1644(S)	1700(S)	1700(S) 1695(S) 1645(S)	1700(VS)	1694(S) 1645(Sh)	1689	1711(S) 1695(S)	1691(S) 1640(Sh)
			1532(W)				1585(W) 1538(W)	1590(M) 1539(M)		1582(W) 1528(W)	1585(M) 1528(M)
1415(S) 1404(S) 1367(S) 1313(W) 1274(W) 1232(W) 1202	1409(S) 1367(S) 1314(S) 1277(W) 1230(W)	1411(M) 1367(M) 1313(M) 1280(W)	1410(S) 1368(S) 1322(M) 1260(M)	1411(M) 1368(M) 1315(M) 1280(W)	1410(S) 1366(S) 1322(M) 1250(M)	1411(M) 1368(M) 1315(M) 1280(W) 1235(Sh)	1480(W) 1410(S) 1371(S) 1320(M) 1280(W) 1230(br)	1475(W) 1408(S) 1368(S) 1315(M) 1282(W) 1240(Sh)	1424 1404 1365 1311 1277 1242	1412(M) 1370(M) 1320(W) 1266(br) 1236(br)	1410(M) 1363(M) 1315(M) 1286(W)
1163(S)	1161(Sh)		1162(Sh)								
Basal spacing, nm	1·21	1·23	1·67	1·43	1·68	1·31	1·67	1·32		1·64	1·23

A. air-dry film exposed to 2,5-hexanedione vapour, then evacuated with mechanical pump.
B. as for A, then exposed to the atmosphere.
a. ref. 119 ('one-layer complex').
(VS) = very strong; (S) = strong; (M) = medium; (W) = weak; (Sh) = shoulder; (br) = broad.

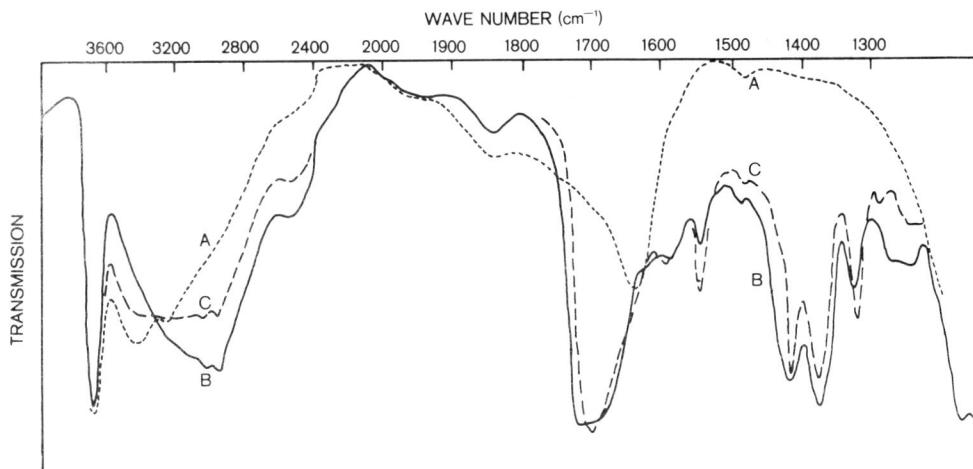

Fig. 30   Infra-red spectra of aluminium-montmorillonite after degassing for 16 hours (A); then exposed to 2,5 hexanedione vapour and degassed (B); and followed by exposure to air for 19 hours (C), after Parfitt and Mortland[109]

montmorillonite, resulting in the formation of the corresponding metal-acetylacetonates.

3.3.3. Aldehydes

MacEwan[15] prepared a one-layer acetaldehyde complex with halloysite by treating the mineral with a large excess of the (cold) liquid. Using a similar technique followed by baking the clay-organic liquid mixture at $333°K$, Larson and Sherman[114] obtained evidence from infra-red measurements for complex formation between an untreated bentonite (montmorillonite) sample and a number of aldehydes.

The infra-red spectra of the complexes pressed in potassium bromide disks showed that the $C=O$ stretching vibration of the adsorbed molecules, with the exception of acrolein, shifted to a lower frequency compared with the carbonyl band in the spectra of the pure liquid (Table 17). The peaks for aldol, benzaldehyde, and crotonaldehyde were diffuse, thus obscuring similar frequency shifts that might have occurred. This observation, together with a change in relative absorbance values of the $3,597$ cm^{-1} (*free hydroxyl*) and $3,378$ cm^{-1} (*bonded hydroxyl*) bands of the parent clay attending adsorption, led Larson and Sherman to suggest that the carbonyl group of the aldehyde is hydrogen bonded to hydroxyl of the silicate layer. However, in the absence of adsorption data and those relating to the behaviour of the $C—C$ stretching and water deformation bands, it is difficult to postulate any particular mechanism of bonding beyond saying that the carbonyl group is involved. By analogy with ketones, it seems likely that aldehydes interact with the exchangeable cations either directly or through a bridging water molecule.

Larson and Sherman[114] did not examine the basal spacing of their complexes, nor was the amount adsorbed specified, so that it could not be ascertained

73

TABLE 17

Carbonyl stretching frequencies (cm^{-1}) of some aldehydes, acetic acid and acetone in the unadsorbed (liquid) phase and when adsorbed on montmorillonite.[114]

Compound	Unadsorbed	Adsorbed	Remarks
Acetic acid	1712	1692	
Acetone	1715	1698	see also Table 14
Acetaldehyde	1727	1709	
Propionaldehyde	1730	1715	
n-Butyraldehyde	1724	1706	
iso-Butyraldehyde	1733	1715	
3-Hydroxybutyraldehyde (aldol)	—	—	Obscured, diffuse peaks
2-Propenal (acrolein)	1692	1718	Polymerization (?)
Crotonaldehyde	—	—	Obscured, diffuse peaks; enolization.
Benzaldehyde	—	—	Obscured, diffuse peaks.
Formaldehyde	—	—	C = O band absent; polymerization (?)

whether or not interlayer penetration had occurred. It is interesting to note that the infra-red spectrum of adsorbed formaldehyde did not show the characteristic C=O stretching band near 1,700 cm^{-1}, but a strong hydroxyl contribution was observed. They suggested that this might be due to polymerization. Although it was not stated, formation of the hydrate $(CH_2(OH)_2)$ which may or may not be followed by polymer formation to give a polyoxymethylene glycol, can conceivably occur. A similar observation with crotonaldehyde was ascribed by Larson and Sherman to enolization. Acrolein is also known to polymerize readily and this may account for the unusual behaviour of the carbonyl stretching band on adsorption, that is, it occurs at a higher frequency than that for the free liquid (Table 17).

Glaeser and Méring[28] have claimed that acetaldehyde converted to acetic acid when adsorbed by calcium-montmorillonite. How this occurs is not clear, although the ability of layer silicates to catalyse the oxidation of adsorbed organic molecules is now well established.[120]

3.3.4. Ethers and Other Compounds

Bradley[35] first reported the formation of two-layer complexes of montmorillonite with some ethers and polyethers. Not long after, MacEwan[15] prepared one-layer complexes of halloysite with some alcohol ethers. The basal spacings of complexes with the glycol ethers, for example, were similar to those given by complexes with the corresponding diols. However, 1,4-dioxane yielded a montmorillonite complex with a basal spacing of 1·47–1·50 nm, too low a value for a double-layer of intercalated molecules but comparable with that obtained for a single-layer pyridine complex with the plane of the ring lying normal to the silicate surface.

The adsorption of some ethers, ether-alcohols, and ether-esters from dilute aqueous solutions by calcium-montmorillonite has been examined by Hoffmann and Brindley[2] (cf. Fig. 27). Both single-layer and double-layer complexes were formed giving basal spacings of 1·31–1·34 and 1·57–1·76 nm, respectively. The shortening of contact distances and the probable orientation in the interlayer space of the organic molecules in the single-layer complexes may be deduced from the respective Δ values. The X-ray data on these complexes, together with those reported by MacEwan[15] and Bradley[35], are shown in Table 18.

Following up this work, Brindley *et al.*[121] have compared the adsorption of acetoaceticethylester, $\beta:\beta'$-oxydipropionitrile (as well as 2,5-hexanedione and triethyleneglycoldiacetate) from aqueous solutions by gibbsite, kaolinite, and montmorillonite. As might be expected from the large surface area of mont-morillonite, the amount adsorbed (expressed as moles per unit weight of clay) is greatest for this mineral; but gibbsite shows the highest adsorption when the results are expressed per unit surface or as an equivalent number of packed organic layers. In the absence of any exchangeable cations in the gibbsite structure, cation-dipole interactions cannot be responsible for the uptake of polar organic molecules. Because the gibbsite surface is composed of hydroxyl ions it would seem that O . . . H—O bonding is important. Displacement of the water molecules from the gibbsite surface would also be easier than in the case of montmorillonite, where this water exists in coordination with the interlayer cations.

The interaction of benzonitrile with montmorillonite containing different interlayer cations (Mg^{2+}, Ca^{2+}, and Ba^{2+}) has been examined by Serratosa[113] using infra-red spectroscopy. Complexes were prepared by immersing the clay film in the organic liquid, removing the excess by evacuation and determining the amount adsorbed by extraction with acetone.

Figure 31 shows the spectra of the magnesium-montmorillonite film at ambient air humidity (A), after evacuation (B), and containing intercalated benzonitrile (C). Of interest here are the changes observed in the vibrational bands of adsorbed water with maxima at 3,395, 3,250, and 1,637 cm^{-1}. The residual water (after evacuation) gave rise to a deformation band at about 1,623 cm^{-1}, while the O—H stretching bands shifted upwards and became obscured by the strong band due to structural O—H of the silicate layer. As pointed out in Chapter 2 (e.g. Farmer and Russell[122]) these observations indicate less hydrogen bonding between water molecules and (probably) between water molecules and surface oxygens, and may be interpreted in terms of direct water-to-cation coordination. Intercalation of benzonitrile, on the other hand, caused a decrease of the O—H stretching frequencies of water from those of the original (before evacuation) sample and an increase in the deformation frequency. This observation led Serratosa to conclude that the benzonitrile molecule is hydrogen-bonded through its —C≡N group to the magnesium ion by way of a water bridge:

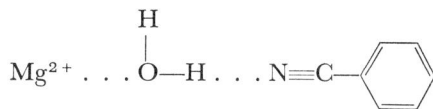

$$Mg^{2+} . . . \overset{\overset{\textstyle H}{|}}{O}—H . . . N≡C—\hexagon$$

TABLE 18

Basal spacings (nm), and Δ value for complexes formed between calcium-montmorillonite and some polar organic compounds together with calculated thickness and probable orientation of the intercalated molecules.[2]

Compound	Basal Spacing		Δ value (one-layer spacing —0·95)	Calculated thickness (nm) of molecules oriented*		Shortening of contact distance (nm)		Probable orientation
	One-layer	Two-layer		⊥	//	⊥	//	
Bis-(2-ethoxyethyl)-ether	1·34	1·67	3·9	4·5	4·1	0·6	0·2	⊥
Bis-(2-methoxyethyl)-ether	1·32	1·70	3·7	4·5	4·1	0·8	0·4	⊥
Ethyleneglycoldiglycidether	1·34	1·76	3·9	?	?	—	—	—
Triethylene glycol	1·33	1·73	3·8	4·5	4·1	0·7	0·3	⊥
Diethyleneglycol	1·33	1·57	3·8	4·5	4·1	0·7	0·3	⊥
Ethoxy-trimethyleneglycol (a)		1·67						
Carbitol (a)		1·69						
Methoxy-triethyleneglycol (a)		1·74						
Dimethylether of ethylene glycol (a)		1·70						
Dimethylether of triethylene glycol (a)		1·75						
Dimethylether of tetraethylene glycol (a)		1·77						
1,4-Dioxane (a,b)	1·47—1·50		0·52—0·55	5·9	—	0·4	0·3	⊥(?)
Triethyleneglycoldiacetate	1·33	1·65	3·8	4·5	4·1	0·7	0·3	⊥
Diethyleneglycoldiacetate	1·31	1·65	3·6	4·5	4·1	0·9	0·5	//(?)
β:β′-Oxydipropionitrile	1·31	1·57	3·6	4·8	4·1	1·2	0·5	//
β-Ethoxypropionitrile	1·32	—	3·7	4·8	4·1	1·1	0·4	//

* Calculated using Pauling's van der Waals radii.

// Plane of aliphatic chain parallel to the silicate surface (α_{II})

⊥ Plane of aliphatic chain perpendicular to the silicate surface ($\alpha_{\perp}$).

a = ref. 35; b = ref. 15.

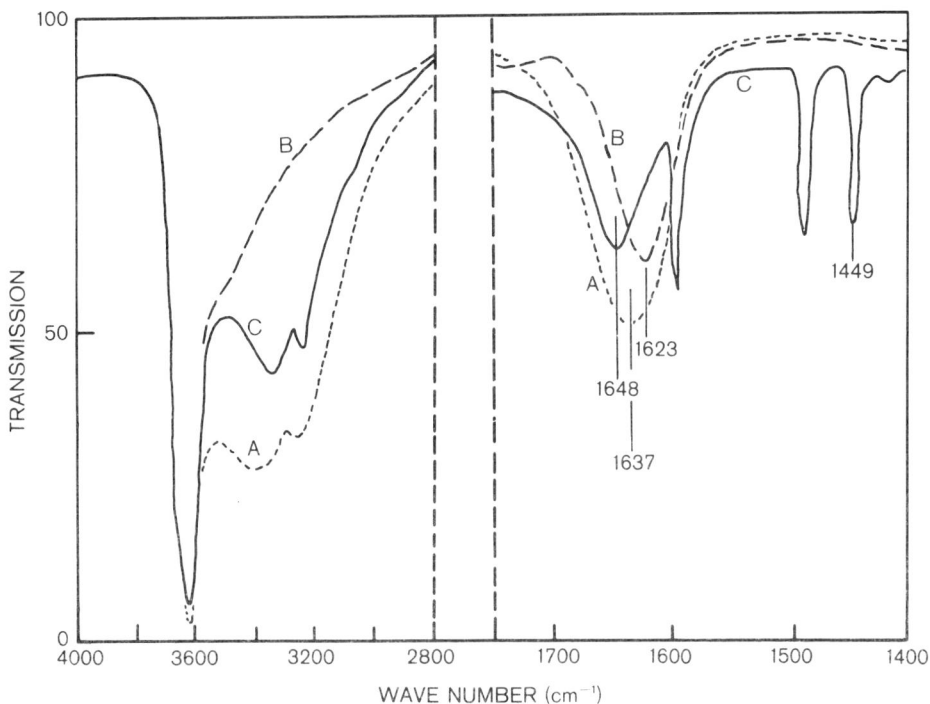

Fig. 31   Infra-red spectra of magnesium-montmorillonite. (A) at ambient air humidity;
(B) after evacuation; (C) containing benzonitrile, after Serratosa[113]

This conclusion was further supported by the increase in the —C≡N stretching frequency from 2,228 cm^{-1} (in the liquid) to 2,240 cm^{-1} (in the adsorbed state).

On evacuating the system, some of this cation-coordinated water was removed, allowing the benzonitrile molecule to be directly linked to the interlayer cation. This was shown by the development of the 2,261 cm^{-1} band (—C≡N stretching vibration) which intensified as the 2,240 cm^{-1} band weakened since on coordination, the —C≡N stretching frequency in the spectra of molecules having —CN and —NC groups is known to increase.[118] As might be expected, the amount of displacement of the —C≡N stretching frequency from that shown by the free liquid is influenced by the nature of the cation, being greater for the more highly polarizing type: 12 cm^{-1} for Ba^{2+}, 21 cm^{-1} for Ca^{2+}, and 33 cm^{-1} for Mg^{2+}.

Prolonged evacuation at ∼353°K of the magnesium system failed to reduce the water content below that corresponding to the presence of 2 water molecules (and 4 organic molecules) per Mg^{2+} ion. Nor could this residual water be displaced by further exposure of the film to benzonitrile.

Besides providing evidence for the reality of the cation-water-organic and cation-organic type of interaction, infra-red studies also yielded information

77

on the arrangement and disposition of the adsorbed species in the interlayer space. Using the same technique as described earlier for the pyridine- and pyridinium-montmorillonite systems (Section 2.4), Serratosa deduced a model of the structure of the magnesium-water-benzonitrile coordination complex. His argument was based on the observed change in intensity of certain absorption bands as the angle of incidence of the infra-red beam with respect to the normal to the clay film was altered by rotating the clay film.

Like pyridine, benzonitrile and water belong to the point group C_{2v} (see Fig. 15). As the angle of incidence was changed from 0 to $\sim$0·698 rad ($\sim$40°), only those modes corresponding to vibrations of the B_1 class (1,293, 1,337, and 1,449 cm^{-1}) of benzonitrile[123] and the 1,650 cm^{-1} of water (class A_1, dipole moment change along the symmetry axis) showed a significant ($\sim$100 per cent) increase in intensity. These observations unambiguously demonstrate that the benzonitrile molecules are intercalated with the aromatic ring inclined at a high angle and the —C≡N bond parallel to the silicate layers. Similarly, the water molecules associated with the magnesium ion are oriented so that their molecular axes make a high angle with the clay surface and their oxygen vertices point inwards towards the magnesium ion positioned midway between two opposing oxygen planes (Fig. 32). The basal spacing of this complex was 1·51 nm, indicating slight keying and/or tilting of the organic molecule. Serratosa further showed that it was feasible to accommodate the full complement of magnesium ions ($\frac{1}{3}$ per unit cell), each of which is octahedrally coordinated to 4 benzonitrile and 2 water molecules in the interlayer space (Fig. 33).

Calcium- and barium-montmorillonite gave rise to similar complexes with benzonitrile but the cation-coordinated water molecules were more easily

Fig. 32 Proposed arrangement of benzonitrile molecules in the interlayer space of magnesium-montmorillonite containing some interlayer water; each Mg^{2+} ion, lying midway between two opposing silicate layers, is coordinated to 4 benzonitrile and 2 water molecules. After Serratosa[113]

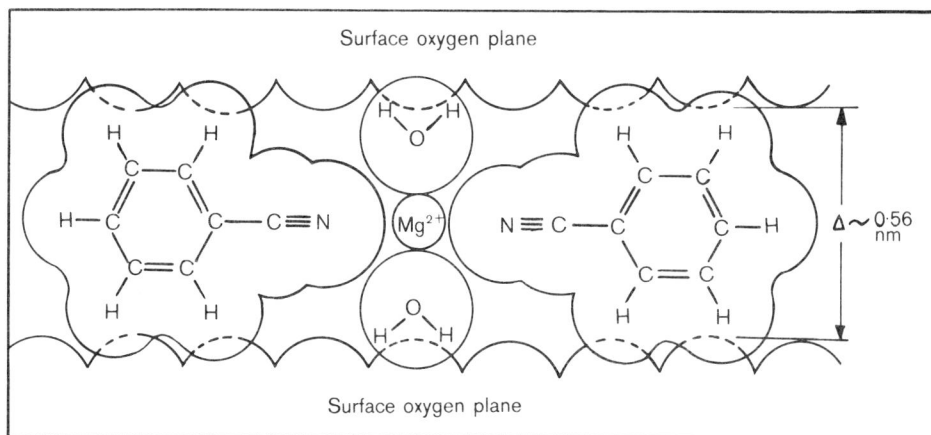

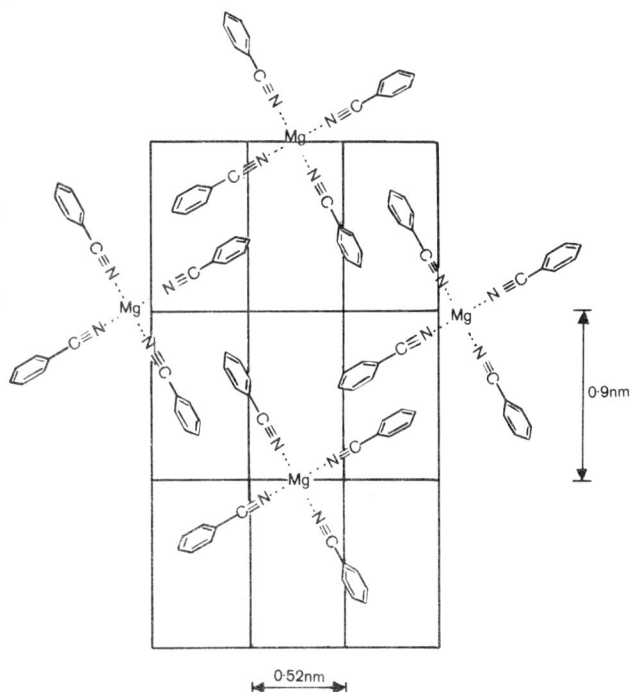

Fig. 33 Idealized arrangement of interlayer benzonitrile molecules in magnesium-montmorillonite projected on the *a, b* plane. Each rectangle represents a unit cell and six unit cells ($\sim$1.56 $\times$ 1.8 nm^2) contain two (Mg^{2+} . 4 C$_6$H$_5$CN . 2H$_2$O) assemblages. The population of Mg^{2+} ions is consistent with the montmorillonite layer charge ($\sim$0.33 per formula unit. After Serratosa[113]

lost as compared with the magnesium system. Thus, the complex with the calcium and the barium clays was completely dehydrated by evacuating the system at $\sim$353°K and at ambient temperature, respectively. Differences in the polarizing power of the cations examined were also reflected in the position of the O—H vibrational bands of water coordinated to the respective cation. The stretching and bending frequencies observed were 3,390 and 1,635 cm^{-1} for Ca^{2+}, 3,405 and 1,630 cm^{-1} for Ba^{2+}, and 3,348 and 1,648 cm^{-1} for Mg^{2+}.

Tarasevich et al.[124] have examined the infra-red spectra of acetonitrile adsorbed by montmorillonite films containing Li$^+$, Na$^+$, K$^+$, Ca^{2+}, Co^{2+}, and Cu^{2+} ions previously evacuated at 383 to 423°K. They showed that acetonitrile interacted with the interlayer cation through the —C≡N group as indicated by the behaviour of the C—C (stretching) +C—H (deformation) combination and C≡N stretching bands. Table 19 shows the dependence of the C≡N vibration on the nature of the cation. Little or no interaction between acetonitrile and interlayer potassium seemed to occur since the C≡N frequency of the adsorbed organic was virtually coincident with that of liquid acetonitrile (2,254 cm^{-1}) and independent of the relative pressure of the organic vapour to which the clay film was exposed. On the other hand, $\nu_{(C≡N)}$ in the spectra of the sodium, calcium, and copper complexes exposed to acetonitrile vapour at P/P$_0$ = 0·1 showed an appreciable increase. The shift in frequency follows the order Na$^+$ (13 cm^{-1}) < Ca^{2+} (24 cm^{-1}) < Co^{2+} (64 cm^{-1}) < Cu^{2+} (76 cm^{-1}). For the copper system the magnitude of this displacement is comparable with that observed ($\sim$80 cm^{-1}) for the interaction of acetonitrile with AlCl$_3$ and SnCl$_4$ both of which are strong electron acceptors (Lewis acids). The increase in the

79

TABLE 19

$C \equiv N$ Stretching frequency (cm^{-1}) of acetonitrile adsorbed by
montmorillonite containing different interlayer cations.[124]

Relative pressure of acetonitrile	Interlayer cation					
	K^+	Na^+	Ca^{2+}	Li^+	Co^{2+}	Cu^{2+}
0·1	2255	2267	2278	2281	2293	2330
0·20—0·25	2253	2266 2255 (Sh)	2274 2253 (Sh)	2279 2255 (Sh)	2290 2252 (Sh)	2332 2271 (W) 2262 (W)
0·45—0·50	2253	2264 2252	2271 2253	2272 2255	2290 2252	2332 2271 (W) 2256 (W)
1·0	2253	2264 2255	2271 2252	2272 2255	2290 2252	2328 2262 2253

(W) = weak; (Sh) = shoulder.

$C \equiv N$ stretching vibration frequency may therefore be attributed principally to the increase in the force constant of the $C \equiv N$ bond when the nitrile group coordinates directly to the interlayer cation. At high relative pressures, a band characteristic of liquid acetonitrile appeared. On desorption, this band disappeared before those which were due to the cation-coordinated organic molecule.

The magnitude of the frequency shift of the $C \equiv N$ vibration indicates the strength of the respective cation-acetonitrile interaction. Figure 34 shows the

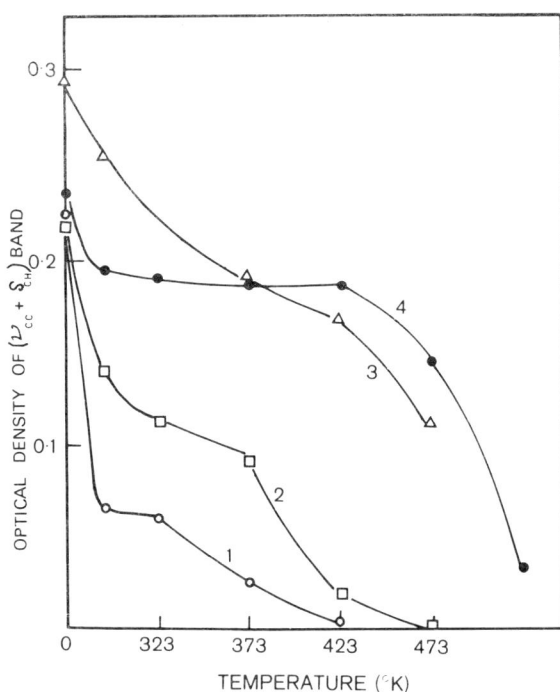

Fig. 34 Variation in optical density of $(\nu_{C-C} + \delta_{C-H})$ vibrational band of adsorbed acetonitrile with temperature of evacuation for complexes of acetonitrile with montmorillonite containing interlayer sodium (1), calcium (2), lithium (3), and cobalt (4) ions, after Tarasevich et al[124]

decrease in optical density of the $\nu_{(C-C)} + \delta_{(C-H)}$ band as the temperature at which the film is evacuated is increased. On this basis, the stability of the different complexes increases in the order $Na^+ < Ca^{2+} < Li^+ < Co^{2+}$. The position of lithium with respect to calcium is unexpected and may reflect a difference in the structure and geometry of the lithium-acetonitrile coordination complex compared with that involving the other cations of the series.

Entirely parallel observations were reported by Yariv *et al.*[112] for nitrobenzene and benzoic acid adsorbed by montmorillonite (Table 20). These compounds can penetrate the interlayer space of the clay crystals, where they displace the water in the outer coordination spheres of the interlayer cation and so form hydrogen bonds with water molecules (directly) coordinated to the cation. When the cations are weakly polarizing, the organic molecules can displace this cation-coordinated water completely, and hence, link directly to the interlayer cation. On the other hand, the more highly polarizing cations tend to retain their primary hydration shell. Bonding of the organic molecule to the cation is mediated by a bridging water molecule, although even here direct cation-organic interaction is possible with special treatment, such as preheating the clay to dehydrate the cation.

Thus, nitrobenzene coordinates directly through its nitro group to interlayer potassium and ammonium ions. However, in the hydrated complexes of

TABLE 20

Infra-red band frequencies (cm^{-1}) and assignments for nitrobenzene and benzoic acid in the free state and when intercalated by calcium-montmorillonite.[112]

Vibration		Nitrobenzene		Vibration	Benzoic acid	
Type	Class (species)	Liquid	Adsorbed	Type	Dimer	Adsorbed
NO$_2$	$\left\{ \begin{array}{l} B_1 \\ A_1 \end{array} \right.$	1527	1523	C $=$ O stretching	1690	1684
(stretching)		1350	1353	C—OH	$\left\{ \begin{array}{l} 1425 \\ 1294 \end{array} \right.$	1415
						1275
NO$_2$	A$_1$	853·5	856		—	
(deformation)	A$_1$	—	—		1605	1604
	B$_1$	—	—		1585	1584
Ring	A$_1$	1480	1480		1499	1495
vibrations	B$_1$	1456	1458		1453	1455
	B$_1$	1318	1320		1326	1320
	A$_1$	1175	1182		$\left. \begin{array}{l} 1181 \\ 1189 \end{array} \right\}$	1182
	B$_2$	795	796			
	B$_2$	704	709		708	717
	A$_1$	682·5	685		668	—
	B$_2$	677	675		684	685

copper, magnesium, and calcium ions it forms hydrogen bonds with the cation-coordinated water. This is indicated by the depression and broadening of the 856 and 685 cm^{-1} bands compared with the corresponding vibrations in the anhydrous potassium and ammonium complexes, the former band involving an in-plane vibration of the nitro group[125] and the latter being due to substituent-sensitive vibration in mono-substituted benzene derivatives.[80,81] On heating under vacuum (323–373°K) new absorption bands (symmetrical stretching of NO_2) were observed at 1,300 cm^{-1} (Cu^{2+}), at 1,315 cm^{-1} (Mg^{2+}), and at 1,340 cm^{-1} (Ca^{2+}), while the antisymmetrical stretching mode at 1,523 cm^{-1} broadened and decreased in frequency.

The infra-red and X-ray data indicate that in freshly prepared complexes, the nitrobenzene molecule is intercalated in an orientation similar to that shown by benzonitrile (Fig. 32), that is, with the plane of the aromatic ring at a high angle to the silicate surface. Displacement of much of the interlayer nitro-benzene by water on exposing the complex to water vapour caused a change in molecular conformation, from an upright orientation to one in which the benzene ring lies parallel to the silicate layer. This change was indicated by a doubling in the intensity of the B_2 vibrational mode (Fig. 15) on altering the angle of incidence of the infra-red beam from 0 to 0·785 rad (45°), whereas the intensity of the A_1 and B_1 vibrations remained virtually the same. A similar change in orientation was earlier reported for pyridine by Greene-Kelly[22] who noted that this change in the opposite sense (that is, from a flat to an up-right orientation) is of general occurrence in montmorillonite complexes with aromatic compounds as the concentration of the intercalated organic species is increased.

Similarly, benzoic acid interacts through its carboxyl group either with the interlayer cation or forms a hydrogen bond with the water molecule coordinated to the cation, depending on the polarity of the cation and the hydration status of the clay:

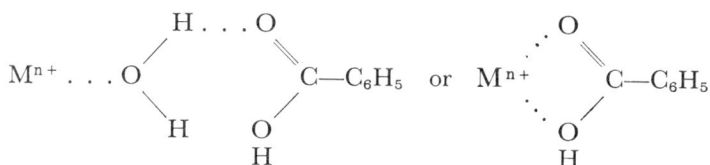

$$M^{n+} \ldots O \underset{H}{\overset{H \ldots O}{<}} \overset{\big\|}{C}-C_6H_5 \quad \text{or} \quad M^{n+} \; O \; C-C_6H_5$$

Evidence for the formation of the above structures is provided by the changes in the position of the —C=O stretching as a function of the saturating cation and hydration state of the complex (Table 21). Thus in the absence of interlayer water (vacuum, 323–373°K) the carbonyl frequency decreased in the order $Na > Li > NH_4 > Ca > Mg > Cu > Al$. At ambient temperature and humidity, this frequency was independent of the cation for Ca, Mg, Cu, and Al and was higher than that observed *in vacuo*. On the other hand, the —C=O frequency in complexes with monovalent cations was independent of the state of hydration. The decrease in carbonyl stretching frequency of acetic acid adsorbed

82

TABLE 21

Carbonyl stretching frequencies (cm^{-1}) and adsorption/retention data for complexes of benzoic acid with montmorillonite containing different interlayer cations.[112]

Conditions	Interlayer cations							
	Na$^+$	K$^+$	Li$^+$	NH$_4{}^+$	Ca^{2+}	Mg^{2+}	Cu^{2+}	Al^{3+}
ir	1706	$\begin{cases}1706\\1689\end{cases}$	1695	1689	1684	1684	1684	1684
in vacuo at 323—373°K	1706	1706	$\begin{cases}1695\\1724(W)\end{cases}$	1689	1689	1664	1639	1625
in vacuo at 423°K	(VW)	(VW)	(VW)	lost	$\begin{cases}1739\\1667\end{cases}$	$\begin{cases}1740(W)\\1712\\1661\end{cases}$	1650	$\begin{cases}1660(d)\\1620(d)\end{cases}$
benzoic acid on clay (per cent) A	6·7	10	13·4	12	20	14·6	17	16—17
benzoic acid on clay (per cent) B	5·3	0·6	6	7·7	8·6	9·2	9·8	8·7

A. after superficial washing with CCl$_4$.
B. after evacuation at 323°K.
(W) = weak; (VW) = very weak; (d) = diffuse.

by montmorillonite observed by Larson and Sherman[114] and Kohl and Taylor[115] can be explained in similar terms.

The X-ray diffractograms of montmorillonite-benzoic acid complexes were diffuse and only the calcium and magnesium systems showed sharp peaks and nearly integral series of reflections. Basal spacings of 1·24–1·28 nm were observed for complexes with copper and monovalent cations, indicating a flat orientation of the intercalated benzoic acid. The complex with calcium-montmorillonite, however, gave a basal spacing of 1·49 nm corresponding to an upright orientation of the organic molecule.

The presence of a hydroxyl group in benzoic acid offers scope for further reactions in the interlayer space. Thus, heating and evacuation gave rise to the formation of the anhydride, particularly in divalent cation systems. In complexes with polyvalent cations the benzoate anion may form as indicated by the presence of an absorption band near 1,550 cm^{-1} in the infra-red spectra, for which Yariv et al.[112] proposed the following reaction:

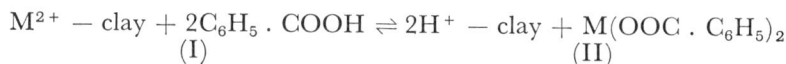

$$M^{2+} - \text{clay} + 2C_6H_5 \cdot COOH \rightleftharpoons 2H^+ - \text{clay} + M(OOC \cdot C_6H_5)_2$$
$$\text{(I)} \qquad\qquad\qquad\qquad\qquad\qquad \text{(II)}$$

In which direction the reaction will largely proceed clearly depends on which of the phases I and II is the more stable under a given set of experimental conditions, as well as on the nature of M. Divalent cations would be expected to behave in the manner shown since a single cation may not compensate for two negative charges in the montmorillonite surface to the same extent. In other cation-clay systems such as aluminium-montmorillonite, basic species of the type $(\text{Al OH})^{2+}$ may obtain which would be capable of reacting with benzoic acid to yield $(\text{Al OOC} \cdot C_6H_5)^{2+}$ without the concurrent formation of protons (or hydronium ions). These effects can explain, at least qualitatively, the differences in the amount of benzoate ion formed between clay samples.

It ought to be pointed out in this connection that organic compounds presented in the anionic form are usually repelled by, rather than attracted to, the negatively charged clay surface. This effect is referred to as *negative adsorption*. However, (positive) adsorption of anions can occur under conditions in which such compounds exist in the molecular (undissociated) form as exemplified by benzoic acid.

3.4. Complexes with Amines

Outside the alcohols, complex formation between expanding 2:1 type minerals and amines of both the aliphatic and aromatic type has perhaps received most attention. Like the alcohols, homologous series of alkylamines (RNH_2) from R = 1 to R = 18 are available or may be readily synthesized. An advantage of studying the adsorption of a homologous series of a given compound is that the properties of the complexes formed might be expected to vary regularly along the series. Amines, of course, can exist in the cationic form as the corresponding alkylammonium ions which can replace the inorganic cations occupying exchange sites at the clay surface. The interactions of layer silicates with organic cations are treated in a separate chapter. Protonation of amino groups at the clay surface is also a well-known process.

3.4.1. Aliphatic Monoamines

One of the many compounds used by Bradley[35] in his study on clay-organic systems was methylamine which formed an interlayer complex with montmorillonite giving a basal spacing of about 1·27 nm. Subsequently, using methylamine and *n*-heptylamine and sodium-montmorillonite, Haxaire and Bloch[126] reported basal spacings of about 1·28 nm. In this instance, however, the amines were added to the clay as their respective aqueous salt solution so that they were, in fact, intercalated as the corresponding *n*-alkylammonium ions. This was further supported by the observation that adsorption released a stoichiometric amount of sodium ions into solution.

For methylammonium, Rowland and Weiss[127] have shown that a basal spacing of ~1·27 nm is compatible with an orientation in which the organic ion is positioned over a ditrigonal hole of the silicate layer with each hydrogen touching two adjacent oxygen ions. This geometry permits keying of the cation to the extent of 0·07 nm. A similar interlayer arrangement may be assumed for the uncharged methylamine molecule. The available evidence indicates that neutral primary *n*-amines with less than about 5 carbon atoms in their molecule prefer an orientation in which the hydrocarbon chain lies parallel to the silicate layer. In this arrangement the plane of the carbon zig-zag may either be perpendicular (α_I) or parallel (α_{II}) to the clay surface. Using X-ray diffraction and infra-red spectroscopic techniques, Fripiat and co-workers[128-130] have shown that both types of disposition can be realized. Which of the two alternative orientations is adopted appears to depend on the amount of amine present.

As we have noted, amines presented to montmorillonite as their aqueous salt solution enter the interlayer space as the corresponding alkylammonium ions, replacing the inorganic cations initially present. Uncharged amines may also be retained by clays in their cationic form when a ready source of protons is available at the silicate surface. An example of a surface protonation reaction is the formation of the respective alkylammonium and alkyldiammonium ions when n-propylamine, n-butylamine, ethylenediamine, and propylenediamine are intercalated by acid-treated montmorillonite.[128] With sodium- and calcium-montmorillonite, protonation does not obtain or if it occurs (under certain conditions of low moisture status) its extent is limited. Here the adsorption process is apparently determined by ion-dipole interactions, the nature of the interlayer cation being of prime importance. Thus, at a given concentration, calcium-montmorillonite takes up a greater quantity of the above amines than the sodium-saturated material.[128]

Sieskind and Wey[131,132] have reported that the adsorption of some primary n-amines by hydrogen-montmorillonite was influenced by the pH of the system and by the size (chain length) of the organic molecule. With n-butylamine, for example, the initial slope of the isotherm (below an added amount of about 1 mmol g^{-1}) at pH 11 where the amine was presented as the free base, was much steeper than that at pH 3 where the organic was added as the hydrochloride salt. However, in both instances the maximum amount adsorbed (at the plateau) was similar, being close to the exchange capacity of the clay. Evidently, at pH 11 the organic molecule enters the interlayer space as an uncharged species which then accepts a proton, as demonstrated by Fripiat and co-workers.[128-130] On the other hand, at pH 3 the amine is intercalated as the corresponding n-alkylammonium ion replacing the (inorganic) cations initially present at the clay surface. Because of the inherent instability of acid-treated montmorillonites,[133] an appreciable proportion of these cations would consist of Al^{3+} which have to diffuse out of the interlayers as they are being replaced. This difference in uptake mechanism between pH 11 and pH 3 would explain why the process is apparently less energetic under acid conditions, at least in the initial stages of adsorption. These workers further observed that below pH 5 and for a given concentration, adsorption of monoamines by montmorillonite increased with an increase in the number of carbon atoms in the molecule. With long-chain molecules, such as n-decylamine, the amount taken up may actually exceed the cation exchange capacity of the clay. Cowan and White[134] have reported similarly for some n-alkylammonium ions adsorbed from aqueous solutions by sodium-montmorillonite, the excess being present as the free amine. This behaviour is generally attributed to the increased contribution of van der Waals forces to the adsorption energy[131,134-137] although H bonding[138] and ion hydration effects[139] may also be important in these systems.

The kinetics of the adsorption of monomethylamine (Me_1), dimethylamine (Me_2), trimethylamine (Me_3), and monoethylamine (Et_1) from the vapour phase by a previously degassed sample of sodium-montmorillonite, have been examined by Palmer and Bauer.[140] Straight line plots were obtained when the ratio of the amount of amine taken up at time t, Q_t, to that present at 'infinite' time (24 hours), Q_∞, was plotted against $t^{1/2}$ with the possible exception of Me_3. Conformity to the $t^{1/2}$ law indicated that the adsorption process was diffusion-controlled. From the slope of these lines, the apparent rate constants, K, may be derived. Values of K at different temperatures, the apparent activation energy, E, and Q_∞ are set out in Table 22. The decrease in K with molecular size may be ascribed to steric hindrance, since the individual silicate

TABLE 22

Kinetic data for the adsorption of some amines from the vapour phase by sodium-montmorillonite.[140]

	Apparent rate constant, K			Activation energy	Q_∞ (293°K)
Compound	273·5°K	298°K	313°K	E (J mol^{-1})	mg g^{-1}
Monomethylamine	0·29	0·38	0·40	5857·6 ± 2092	85
Monoethylamine	—	0·30	0·33	5439·2 ± 2092	67
Dimethylamine	0·22	0·29	—	5857·6 ± 2092	64
Trimethylamine	—	0·08	0·13	20920 ± 8368	29

layers must be forced apart before the interlayers can be penetrated. Curiously enough, the activation energy for diffusion did not vary between the amines used, although the amount adsorbed at t_∞ decreased in the order $Me_1 > Me_2 > Me_3$. This observation may be explained, at least partly, in terms of the orientation and space requirement of the organic molecules in the clay inter-layers. Using models of the corresponding methylammonium ions and the silicate surface, Rowland and Weiss[127] have shown that by positioning the Me_1^+ ion over a ditrigonal hole, the exchange capacity of the clay ($\sim$1 mmol g^{-1}) is satisfied when one hole out of three is occupied by the amine. Hence, in a single-layer complex an amount of methylammonium ions up to at least three times the exchange capacity can feasibly be accommodated in the interlayer space. In (single-layer) complexes with Me_2^+ and Me_3^+ ions, only two and one out of three holes, respectively, can be filled by the amine cations. The proportion of 3:2:1 for Me_1^+, Me_2^+, and Me_3^+ agrees reasonably well with that of the observed Q_∞ values for the corresponding methylamines (Table 22).

In the system investigated by Palmer and Bauer,[140] little interaction is expected between the exchangeable sodium ions and the intercalated amine molecules. This may, in part, account for the observed decrease in Q_∞ from

Me_1 to Me_3 and from Me_1 to Et_1. This behaviour may be compared and contrasted with that of the copper-alkylamine complexes prepared by Bodenheimer *et al.*[141] by titrating copper-montmorillonite with Me_1, Et_1, Et_2, Et_3, and *n*-hexylamine (Hx_1) in an aqueous medium. These workers noted that as the amine was added, the pH of the system slowly increased until at the end-point the pH rose abruptly as the amine passed into solution; beyond the end-point, the pH slowly rose again. The slope of the first part of the titration curve indicates the buffer capacity of the system, and hence may be regarded as a measure of the relative stability of the respective copper-amine complex. Since this slope decreased in going from the primary (e.g. Et_1) to the tertiary (e.g. Et_3) amine and with increasing chain length (from Me_1 to Hx_1), the stability of the complexes formed increased in the same sense. This contrasts with the corresponding copper-amine complexes in aqueous solution where on both counts there is a decrease in stability. Formation of $[Cu\,(amine)]^{2+}$ and $[Cu\,(amine)_2]^{2+}$ complexes in the interlayer space was inferred from X-ray diffraction and differential thermal analysis measurements.

Using infra-red spectroscopy, Farmer and Mortland[138] have subsequently confirmed that in the absence of water, amines are capable of coordinating directly to interlayer copper ions in montmorillonite. They prepared their complex by exposing the preheated clay to ethylamine vapour. The blue colour and the basal spacing ($\sim 1 \cdot 29$ nm) of the complex together with the $4:1$ molar proportion of amine to copper indicated the presence in the interlayer space of a square planar coordination complex, $[Cu(Et_1)_4]^{2+}$, with the plane of the molecule lying parallel to the silicate surface. More convincingly, coordination of the amine to copper caused marked changes in the position of certain bands in the infra-red spectrum compared with that shown by the amine dissolved in carbon tetrachloride. Thus, the $1,626$ cm^{-1} band (NH_2 scissoring) was displaced to $1,590$ cm^{-1} and the $1,470$ cm^{-1} vibration (C—H deformation) was split into two bands at $1,472$ and $1,456$ cm^{-1}. Complexes of this type involving transition metal ions and diamines in montmorillonite and vermiculite have been reported by Bodenheimer and co-workers[73,142-145] and more recently by Laura and Cloos,[146] and Santos *et al.*[147,148] Their formation and properties are discussed in the following section. What is important to note here, is that the formation and stability of these complexes are clearly dependent on the propensity of the cation to accept electrons from suitable electron-donating groups in the organic molecule. This would account, for example, for the strong interaction between transition metal cations having unfilled *d* orbitals and organic bases, such as amines, which are capable of acting as electron donors.

In connection with the ethylamine-montmorillonite system studied by Farmer and Mortland,[138] we must mention their evidence for the formation of a 'hemisalt' complex. Such a complex may arise when the amount adsorbed of a base, B, exceeds that of hydrogen ions capable of protonating the base. The protonated species, BH^+, then 'shares' its proton with the neutral (excess) base to yield a cation of the type $(B \ldots H \ldots B)^+$. When, for example, ethylamine is brought into contact with montmorillonite saturated with ethylammonium ions, a complex containing interlayer

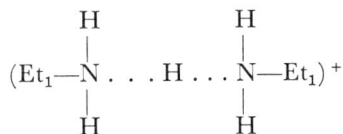

$$(Et_1 \overset{\overset{\displaystyle H}{|}}{\underset{\underset{\displaystyle H}{|}}{N}} \ldots H \ldots \overset{\overset{\displaystyle H}{|}}{\underset{\underset{\displaystyle H}{|}}{N}} Et_1)^+$$

ions is formed, as indicated by the changes in the infra-red spectrum of the ethylammonium-montmorillonite sample. Thus, the spectrum of the material saturated with Et_1^+ ions shows a strong band at 1,510 cm^{-1} attributable to the symmetric deformation vibration of $-NH_3^+$. After adsorption of ethylamine, this band disappears and the spectrum is characterized by the presence of strong, broad, absorption bands between 3,500 and 1,200 cm^{-1}. The ethylammonium-ethylamine hemisalt complex is stable against evacuation, but on exposure to air at ambient humidity ethylamine is lost and the infra-red spectrum of the (exposed) complex then resembles that of the (parent) ethylammonium-montmorillonite.

The formation of hemisalt complexes in which two molecules of an organic base are joined by a symmetrical hydrogen bond has also been observed in montmorillonite complexes with ethylenediammonium-ethylenediamine,[130] pyridinium-pyridine,[149] urea,[150] and amides.[151]

Somewhat related to this process is the transfer of a proton from the cationic form of one organic base (B_1H^+) acting as a Brønsted acid (proton donor), to another adsorbed base (B_2) according to the general equilibrium:

$$B_1H^+ + B_2 \rightleftharpoons B_2H^+ + B_1 \qquad (4)$$

In which direction reaction (4) will move for any given system clearly depends on the relative basicity of B_1 and B_2 as well as on the concentration of the reactants and products. Russell et al.[152] for example, have shown by infra-red spectrosocopy that treatment of ammonium-montmorillonite with 3-aminotriazole gives rise to the formation of 3-aminotriazolium cations to the extent of the exchange capacity. Besides confirming the validity of this reaction, Raman and Mortland[153] have been able to observe a similar proton interchange between a variety of organic compounds intercalated by montmorillonite and nontronite. Some of these systems are listed in Table 23.

By analogy with the corresponding primary n-alcohols[13] (Section 3.1) long-chain primary n-amines (containing five or more carbon atoms per molecule) tend to be intercalated as a double-layer with the alkyl chain inclined at a high angle to the silicate surface. Such complexes, first observed with graphitic acid by Cano Ruiz and MacEwan[154] and referred to as β-type interlayer complexes, also form with montmorillonite and vermiculite.

Using primary n-amines from C_4 to C_{18} and sodium-montmorillonite, Aragon et al.,[155] for example, have shown by X-ray diffractometry that, with the exception of C_{18}, the basal spacing of the complexes increased linearly with chain length (Fig. 35). The slope of this line corresponded to about 0·26 nm

TABLE 23
Proton interchange between some adsorbed compounds in
montmorillonite.[153]

Compound acting as proton donors (B_1H^+)	Clay minerals studied	Compound actings as proton acceptors (B_2)
Ammonium	Bentonite, nontronite	Pyridine, methylamine, 3-aminotriazole
Pyridinium	Bentonite, nontronite	Ammonia, methylamine, 3-aminotriazole
Ethylammonium	Bentonite, nontronite	Ammonia, pyridine, 3-aminotriazole
Methylammonium	Bentonite, nontronite	Ammonia, pyridine, 3-aminotriazole
Urea $(NH_2)_2COH^+$	Bentonite	Ammonia, pyridine

Fig. 35 Variation in basal spacing with number of carbon atoms in alkyl chain for complexes with primary n-alcohols and n-amines, after Brindley.[158] Data from Brindley and Ray[13] for alcohol-montmorillonite (cf. Fig. 19); Aragon et al.[155] for amine-montmorillonite; and, Sutherland and Mac-Ewan[156] for amine complexes with two vermiculites, Irish, and West Chester. Line 1, $d(001) = 1.42 + 0.248n$ (n even); line 2, $d(001) = 1.363 + 0.2288n$ (n even); line 3, $d(001) = 1.333 + 0.2288n$ (n odd)

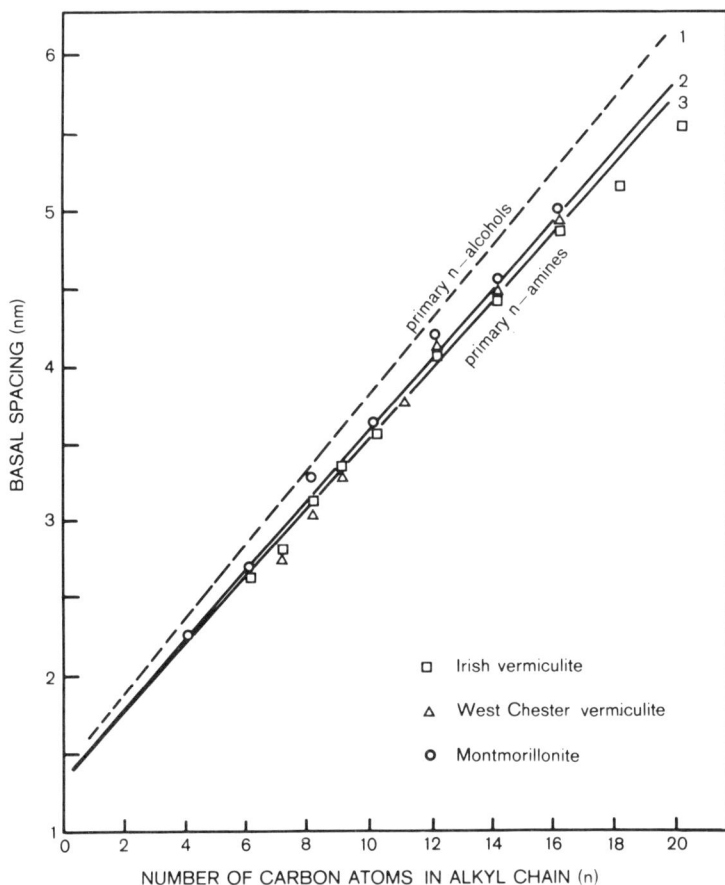

per carbon atom on which basis they have suggested that the amines were adsorbed as a double layer of extended molecules (the projected length of two C—C bonds being near to 0·254 nm).

Sutherland and MacEwan[156] later extended their work to vermiculite samples from two localities (Irish and West Chester) both of which yielded β-type complexes. Here the slope of the line relating the number of C atoms to basal spacing was slightly less (∼0·23 nm per C atom) than that with montmorillonite.

Since the respective Δ values (interlayer distances) of the complexes were actually too small to accommodate two layers of non-overlapping organic molecules with their alkyl chain normal to the clay surface, Aragon et al.[155] proposed that either the free ends of the molecules overlapped to the same extent in each complex, or the first two or three methylene groups next to the —NH$_2$ group were in contact with the surface, while the rest of the molecule, through bond rotation, was perpendicular to the mineral surface. A similar folding-down of the terminal parts of the amine on the surface of a number of planar mineral substrates has been suggested by MacEwan[157] to account for the observed spacing deficit.

Analysing the data afresh and comparing them with those obtained for n-alcohol-montmorillonite complexes (Fig. 19), Brindley[158] has proposed another plausible alternative to account for the incremental increase of ∼0·23 nm per C-atom and for the actual values of basal spacing reported by MacEwan and co-workers. Using the same approach as described for the alcohol complexes, the amine chain inclination, φ, to the plane of the silicate layer may be estimated from

$$\sin \phi = 0·23/0·254 \qquad (5)$$
$$= 1·134 \pm 0·087 \text{ rad } (65 \pm 5°)$$

The figure of 0·254 nm corresponds to the increase in aliphatic chain length per two carbon atoms, or the increase in basal spacing per carbon atom for a double layer of molecules oriented perpendicularly to the silicate layer.

By fitting a line to the experimental points of C$_6$ through C$_{16}$ amines using the method of least squares, the following equations were obtained for the two vermiculites:

$$d(001) = 1·283 + 0·232 \text{ n } (\phi = 1·152 \text{ rad}) \qquad (6)$$

and

$$d(001) = 1·256 + 0·236 \text{ n } (\phi = 1·187 \text{ rad}) \qquad (7)$$

respectively, where n is the number of carbon atoms in the amine molecule. The question arises whether the observed basal spacings agree with the corresponding calculated values. In order to estimate the latter, however, certain assumptions must be made for the geometrical arrangement of the terminal amino group in relation to the surface oxygen network of the silicate layer and of the terminal methyl group at the mid-plane for the interlayer space. The model adopted by Brindley is depicted in Fig. 36 which is arrived at by taking the

90

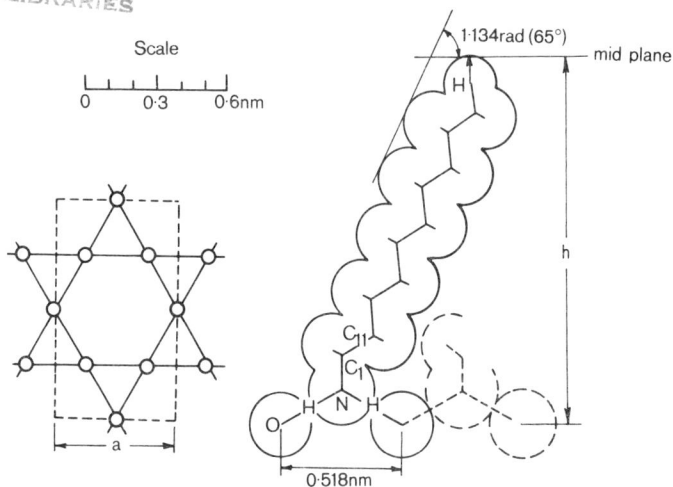

Fig. 36 Brindley's[158] model for the arrangement of straight-chain aliphatic amine molecules in a double-layer complex with montmorillonite and vermiculite. Left: idealized structure of surface oxygens of the silicate layer; Right: orientation of amine molecule with two NH — O contacts to surface oxygens and angle of inclination $\phi = 1.134$ rad (65°). Basal spacing is given by 2h + 0.66 nm. The possible position of an adjacent molecule is indicated by broken lines

O–O distance across the ditrigonal depression of the surface as 0·518 nm, the N—H . . . O distance as 0·29 nm, the NC_I bond as normal to the surface, the C—C bond as 0·154 nm, and bond angles as 1·91 rad (109° 28′). Further, the terminal CH_3 groups of both layers are assumed to make van der Waals contact at the centre of the interlayer space, the radius of the hydrogen atom being 0·12 nm. On the basis of the above model the nitrogen atom is found to be at a distance of 0·13 nm from the line of O–O. If now the NC_IO_{II} angle is taken as 1·91 rad, $\phi = 0·96$ rad (55°); this angle has to be increased to 2·09 rad (120°) if the value of 1·134 rad (65°) for ϕ is to be accepted.

If the NC_IC_{II} angle is taken as 2·09 rad, however, the following equations

$$d(001) = 1·363 + 0·2288\, n \ (n = \text{even}) \tag{8}$$

and

$$d(001) = 1·333 + 0·2288\, n \ (n = \text{odd}) \tag{9}$$

adequately describe the experimental data (Fig. 35). Deviation from the calculated lines for C_{18} and C_{20} could be attributed to the presence in the samples of these compounds of lower molecular weight fractions.

It is of interest to note that the 'constant' terms in Eqs. (8) and (9), that is, the ordinate intercept values, are appreciably greater than those given using Eqs. (6) and (7). As Brindley pointed out, however, the latter are extremely sensitive to the magnitude of the chain inclination angle, ϕ. Moreover, Eqs. (6) and (7) represent lines which are extrapolatable to n = 0, while by analogy with the corresponding n-alcohols (Fig. 19) β-type complexes would not be expected to form for amines with n < 6. The above model, together with Eqs. (6) and (7) which are derived on the basis that the organic molecules are inclined at a high angle to the clay surface, would therefore be only strictly applicable to amines with n > 8. These observations indicate that for n > 8, van der Waals attractive forces between adjacent alkyl chains rather than ion-dipole interactions exert a determining influence on the adsorption process. As a consequence the

91

tilted (or extended chain) rather than the flat conformation is favoured. On the other hand, for shorter chain amines (n < 6) adsorption would primarily be influenced by the tendency of the interlayer cation to form the appropriate coordination complex with the respective organic molecule. Here the nature of the cations occupying exchange positions is decisive and an orientation in which the alkyl chain lies parallel to the silicate surface would be favoured. Recent work by Hach-Ali and Martin Vivaldi[159] has confirmed the above predictions.

The orientation and packing of amines in β complexes with montmorillonite and vermiculite (Fig. 36) clearly offers scope for hydrogen bonding between the terminal amino groups and the oxygen ions as well as keying of these groups into the ditrigonal holes of the silicate surface. Infra-red examination of the corresponding alkylammonium-vermiculite systems[21] has shown that N—H . . . O hydrogen bonding, although weak, does exist. Complementary X-ray diffraction measurements indicate that the basal spacing of these complexes can be rationalized by making assumptions regarding bond lengths, bond angles, and chain inclinations similar to those adopted in constructing the model depicted in Fig. 36, which allows a keying-in of the terminal ammonium (or amine) groups to the extent of about 0·15 nm below the imaginary plane placed over the surface oxygens of the silicate layer.

Brindley[158] has drawn attention to the fact that about 1/6 of the ditrigonal sites are associated with monovalent cations in montmorillonite and with divalent cations in vermiculite. In a double-layer complex of extended amine molecules, 5/6 of the total sites are therefore available on each oxygen surface. This means that the amount of organic molecules which can be accommodated in a complex of this type far exceeds that which is present in the corresponding one-layer complex. This space requirement effect must be responsible, at least in part, for the tendency of long chain polar organic compounds to be intercalated as double layers of molecules. Single-layer complexes of n-alkylammonium ions with vermiculites may be obtained by washing and drying of the corresponding two-layer systems.

3.4.2. Aliphatic Polyamines and Cyclic Amines

Following the early work of Bradley,[35] MacEwan,[15] and Aragon et al.[155] the formation and properties of montmorillonite complexes with diamines have been investigated by Bodenheimer and co-workers.[73,142–145]

Using a titration technique[142] these workers first showed that, besides being capable of extracting metal-diamine complexes from aqueous solutions, montmorillonite readily allowed such complexes to be formed in its interlayer space. This complexing ability of montmorillonite was particularly evident when the exchange positions of the clay were occupied by transition metal cations such as copper. The copper-ethylenediamine and copper-1,2-propylenediamine systems, for example, gave titration curves showing a sharp end-point which corresponded to an amine:copper ratio of 2:1. Subsequently, they examined the sorption of ethylenediamine (en), 1,2-propylenediamine (1,2 pn), 1,3-propylenediamine (1,3 pn), and 1,5-pentamethylenediamine (1,5 pen) from their aqueous solutions by montmorillonite as influenced by copper ions and *vice versa*, and characterized the complexes so obtained by X-ray diffractometry.[143] In the absence of copper, sorption of diamines depended on the type and concentration of the amine used, while the presence of copper in solution markedly enhanced amine uptake. En and 1,2 pn behaved differently from 1,3 pn and 1,5 pen. The first two diamines, presumably because they were capable of chelating copper ions, did not cause a precipitation of $Cu(OH)_2$ in dilute solutions containing copper and amine in a 1:2 proportion. Likewise, the type and concentration of the amine used influenced the amount of copper ions sorbed by montmorillonite. Thus, an increase in amine concentration gave rise to a greater uptake of Cu^{2+} up to a maximum and then decreased. The basal spacing (after washing) of the copper-diamine-clay complexes did not vary greatly with the type of diamine used, ranging from 1·26 to 1·30 nm which indicates a flat conformation of the intercalated species with a certain amount of keying into the ditrigonal depressions of the silicate surface.

Bodenheimer *et al.*[143,144] extended this investigation to complexes of montmorillonite containing copper, nickel, and mercury ions with diethylenetriamine (dien), triethylenetetramine (trien), and tetraethylenepentamine (tepa). Two well-defined complexes of dien and one each of trien and tepa with copper-montmorillonite were identified, dien and tepa apparently forming planar copper complexes in the interlayer space. Similar, stable complexes also obtain in aqueous solution although, in the absence of clay, the structure of these compounds may not be identical with that of the complexes formed in the montmorillonite interlayers. Thus, unlike the complexes formed in aqueous solutions, the interlayer counterparts show non-integral amine:copper ratios. The data suggest that below the end-point the added amine participates in complex formation with the interlayer cations; beyond this point, an additional phase could be distinguished, ascribable to the entry into the interlayer space of ionic amine.

Similarly, dien, trien, and tepa readily complex with interlayer nickel and mercury ions[144] by replacement of the cation-coordinated water molecules. Both trien and tepa give rise to a single type of complex with Ni^{2+} whereas two definite complexes are formed between Ni^{2+} and dien, [Ni (dien)]$^{2+}$, and [Ni (dien)$_2$]$^{2+}$. These complexes with integral metal:amine ratios are stable against displacement by ionic amines.

Although X-ray diffraction, differential thermal analysis (DTA), and thermogravimetric analysis (TGA) have proved useful in characterizing metal-diamine complexes[143-145] the application, in more recent years, of infra-red spectroscopy (in conjunction with the above techniques) to the study of these systems has provided structural information about the metal-organic link. This combined approach was used by Laura and Cloos[146] to examine the coordination of ethylenediamine (en) to copper(II) ions in the montmorillonite interlayers. For this reason we discuss their data and conclusions at some length.

TABLE 24

Basal spacings (nm) of copper-montmorillonite and its complexes with ethylenediamine (en) under the specified conditions of sample treatment.[146]

Sample	En/Cu ratio (molecule/ion)	Treatment						
		Ambient temp.	Outgassed	353°K	393°K	433°K	493°K	Rehydrated overnight
Cu-en-clay								
1	1	1·25	1·24	—	1·24	1·04(17)	1·03(ir)	1·25
2	2	1·26	1·25	1·25	1·25	1·12(ir)	1·05(ir)	1·08(ir)
3	2	1·27	1·27	—	1·27	1·13(ir)	1·09(ir)	1·05(ir)
Cu-clay	—	1·25	1·21	0·99	0·99	0·98	0·97	0·97

Except where indicated, a rational (or nearly) rational series of basal reflections were observed. The stated values are an average of the first, second, and fourth order.
ir = irrational series of reflections.

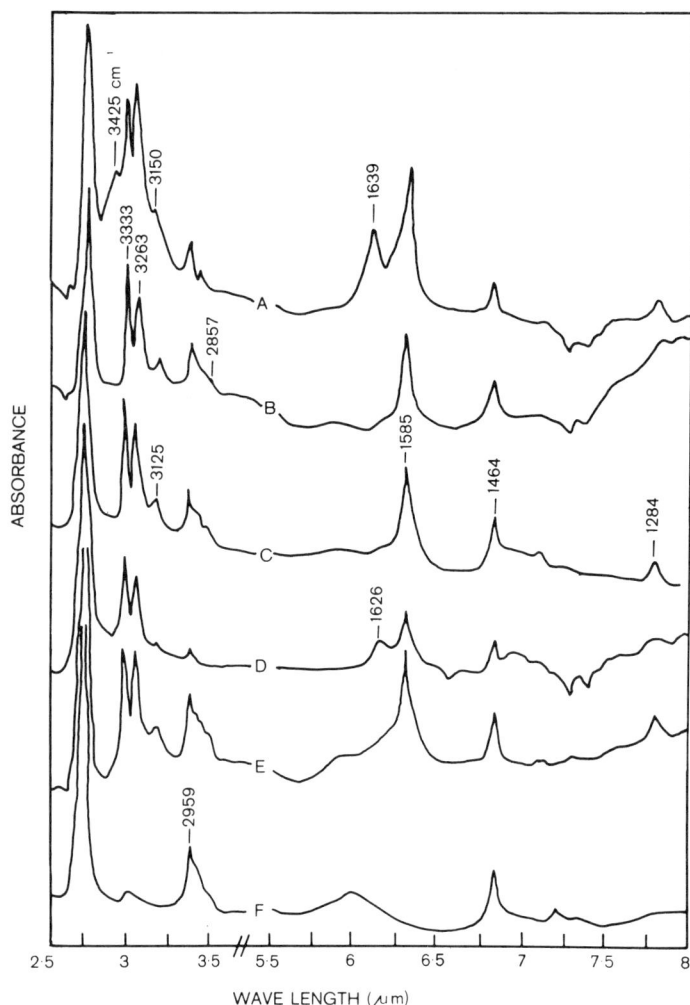

Fig. 37 Infra-red spectra of ethylenediamine adsorbed by copper-montmorillonite taken under the specified conditions. (A) sample 3 (see text) at ambient temperature; (B) outgassed for 3 hours at ambient temperature; (C) sample 2 (see text); (D) sample 1 (see text); (E) copper-montmorillonite film immersed in liquid ethylenediamine for 16 hours and outgassed at 393°K for 1 hour before deuteration; (F) same as (E) but after deuteration. Data from Laura and Cloos[146]

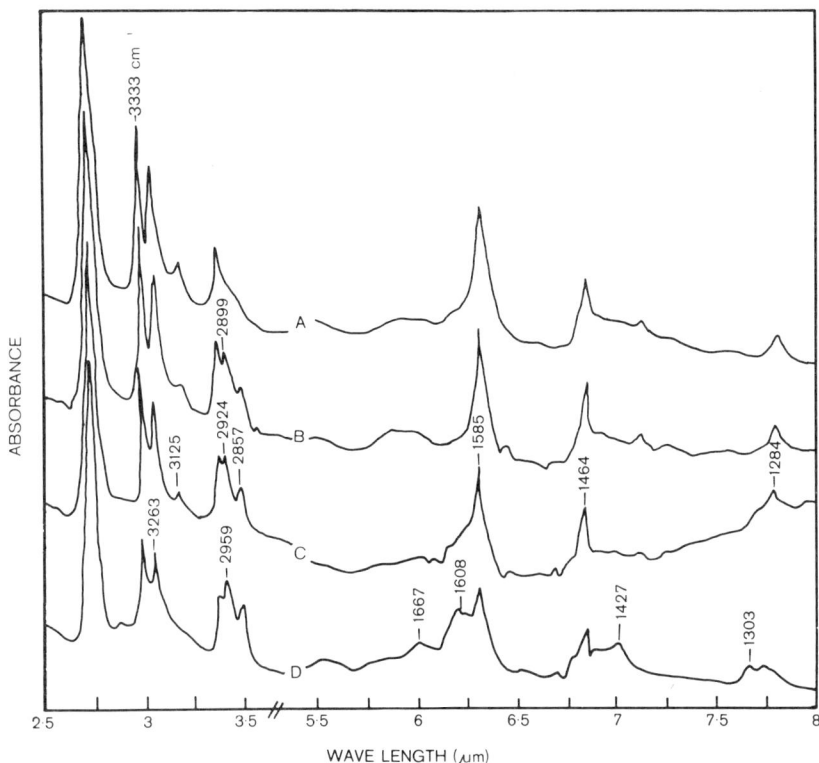

Fig. 38 Changes in the infra-red spectra of ethylenediamine complexes with copper-montmorillonite on heating at different temperatures. (A) outgassed for 3 hours at ambient temperature; (B) heated at 393°K under vacuum for 3 hours; (C) same as (B) but at 473°K; (D) heated at 473°K for 16 hours. After Laura and Cloos[146]

The interlayer copper-ethylenediamine complexes in montmorillonite were prepared by titrating a dilute (1 per cent) suspension of the copper clay with an aqueous solution of ethylenediamine. Samples 1, 2, and 3 correspond to points below, coincident with, and above the inflection point of the titration curve, respectively. Using thin, oriented films the basal spacing and the infra-red spectrum of the respective complexes (samples) were measured. The X-ray difffraction data are summarized in Table 24 and the infra-red spectra, recorded under the specified conditions, are shown in Figs. 37 and 38. By analogy with the spectrum of [Cu (en)$_2$] PtCl$_4$ studied by Powell and Sheppard,[160] some of the absorption bands may be identified with specific vibrational modes (Table 25).

The X-ray data showed that a profound change occurred between 393 and 433°K possibly owing to some decomposition of the complexes giving rise to interstratified layers within a crystal. However, infra-red measurements (Fig.

95

TABLE 25

Frequencies and assignments of the principal bands in the infra-red spectra
of [Cu(en)$_2$] PtCl$_4$ and Cu-en-montmorillonite complex.[146]

Frequencies (cm^{-1})				Assignment
[Cu(en)$_2$]PtCl$_4$		Cu-en-mont.		
Normal	Deuterated	Normal	Deuterated	
—	—	3425 (Sh)	—	ν_{OH} (stretching mode of hydration water)
3320	2475	3333 (VS)	2494 (VS)	ν_{NH_2} (asymmetric stretching)*
3245	2380	3263 (VS)	2410 (VS)	ν_{NH_2} (symmetric stretching)*
		3125 (MB)	—	NH . . . o or δ_{NH_2} overtone
		2959 (S)	2959 (S)	ν_{CH_2} (asymmetric and symmetric stretching)
		2924 (M)	2924 (M)	
		2899 (M)	2899 (M)	
		2857 (MB)	2857 (MB)	
		1639 (S)	—	δ_{OH} (deformation of hydration water)
1571 (VS)	1178 (VS)	1585 (VS)	—	δ_{NH_2} (scissoring mode)*
1478 (W)	1479 (W)	—	—	δ_{CH_2} (scissoring mode)*
1463 (W)	1463 (W)	1464 (S)	1464 (S)	
1376 (M)	1358	1370 (W)	—	δ_{CH_2} (wagging mode)*
1321	1033	1325 (W)	—	δ_{NH_2} (wagging mode)*
1282	—	1284 (MB)	—	δ_{CH_2} (twisting mode)*

W = Weak; Sh = Shoulder; S = Strong; VS = Very Strong; M = Medium; B = Broad.
* Powell and Sheppard [160], measured in liquid paraffin and hexachlorobutadiene.

38) indicate that the nature of the complex was essentially unaltered by heat treatment. It is interesting to note that neither the presence of some interlayer water (vibrations near 3,425 and 1,639 cm^{-1}, Fig. 37) nor the amount of diamine adsorbed seemed to affect the position of the NH$_2$ bands (at 3,333 and 3,263 cm^{-1}, Fig. 37). However, as might be expected, the intensity of the NH$_2$ and CH$_2$ vibrations increased with the amount of en present. On deuteration, the NH$_2$ stretching bands were shifted to lower frequencies, the ratio ν_{N-H}/ν_{N-D} being 1·35. The 3,125 cm^{-1} band cannot be assigned with certainty but is undoubtedly associated with the presence of an NH$_2$ group.

The similarity in X-ray patterns and infra-red spectra between samples 1, 2, and 3 before and after heating to different temperatures indicates that a complex of essentially identical structure is formed between Cu^{2+} ions and ethylenediamine. This finding contrasts with the earlier suggestion of Bodenheimer et al.[143] that two types of Cu/en complexes may form in the interlayer space of montmorillonite. Laura and Cloos suggested that the 1:1 Cu/en ratio in sample 1 merely reflected the fact that only about half of the interlayer copper was involved in complex formation. In samples 2 and 3, where this ratio attains 1:2 the formation of a stable [Cu (en)$_2$]$^{2+}$ complex is indicated in which four NH$_2$ groups coordinate to one Cu^{2+} ion, probably in a square planar structure.

As Laura and Cloos pointed out, two possible arrangements of en with respect to Cu^{2+} might be envisaged: (1) in which the diamine molecule, in the *trans* form, is linked to two copper ions; and (2) in which the two NH_2 groups of the diamine, either in the *cis* or *gauche* form, are linked to the same copper ion (Fig. 39). Structure 1 is unlikely to exist on the ground that the length of the diamine molecule is incompatible with the intercharge distance in the mont-

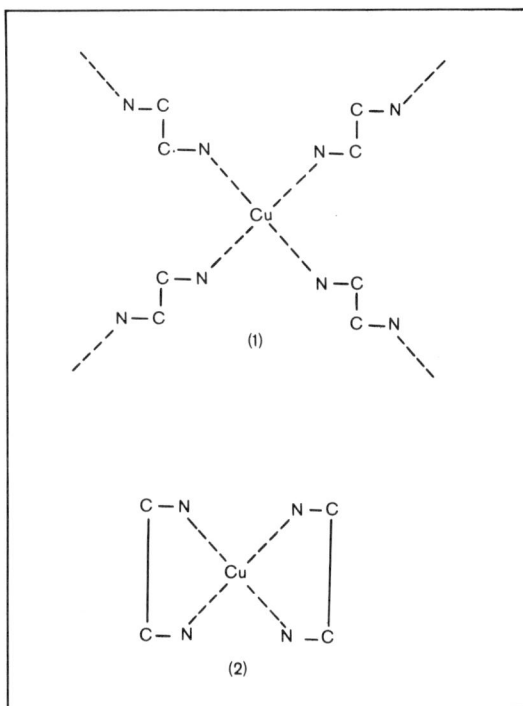

Fig. 39   Possible arrangements of the copper-ethylenediamine complexes. (1) each ethylenediamine molecule in the *trans* form is linked to two copper ions; (2) each ethylenediamine molecule, either in the *cis* or *gauche* form, has two of its NH_2 groups linked to the same copper ion. After Laura and Cloos[146]

morillonite layer. Further, both the (heat) stability of the complexes (Fig. 38) and the close correspondence of the infra-red band frequencies between the Cu/en/clay complex and the [Cu (en)$_2$] PtCl$_4$ compound of Powell and Sheppard[160] strongly point to the presence of a square planar chelate compound in the interlayer of montmorillonite.

The d(001) spacing at about 1·26 nm indicates that the plane of the complex is parallel to the silicate layers. In the *cis* form the C and N atoms are coplanar with the Cu^{2+} ion, whereas in the *gauche* form one C atom lies above and one below the Cu/N plane. Using scale models of the ethylenediamine molecule, Laura and Cloos estimate thicknesses of 0·27 and 0·33 nm for the *cis* and *gauche* forms, respectively. Taking a value of 0·95 nm for the thickness of an individual montmorillonite layer, the corresponding basal spacing is calculated as 1·22 nm (*cis*) and 1·28 nm (*gauche*). On this basis the *gauche* form appears to be intercalated since a certain amount of keying, and hence some contraction in contact distances, is to be expected (cf. Chapter 2). Recent quantum conformational analyses[161] have shown that ethylenediamine is predominantly in the *gauche* form since this is energetically the most stable conformation, the *cis* conformer being the least stable. Indeed, *cis*-chelated complexes are not known. The analysis of Hadjiliadis *et al.*[161] suggests that the ethylenediamine molecule prefers the *gauche* form for normal chelates and the *trans* form for bridge complexes.

On decreasing the incidence angle of the infra-red beam from 1·57 rad (90°) to 0·87 rad (50°) the asymmetric NH_2 stretching vibration (ν_{asym}) increased by 47 per cent. This observation, however, does not allow a distinction to be made between the *cis* and the *gauche* conformers since both forms would give rise to dipole moment changes in a direction perpendicular to the silicate layers. However, there was evidence to suggest that the hydrogen atoms in the diamine-copper complex existed in closely similar environments so that little interaction seemed to occur between N—H groups and the surface oxygen ions. In addition, protonation of ethylenediamine was indicated on heating the complexes above 393°K; the ethylenediammonium ion, in turn, probably decomposed to give ammonium and hydrocarbon fractions. Heat treatment, however, did not cause a reduction in the amount of nitrogen held by the clay. Such a process would account for the marked change observed in the X-ray diffractograms (the decrease in basal spacing between 393 and 433°K remarked on earlier) while maintaining an invariant nitrogen content.

A parallel study involving *n*-propylamine, *n*-pentylamine, ethylenediamine, 1,2-propylenediamine, and 1,2-butylenediamine and montmorillonite containing cobalt, nickel, copper, and magnesium ions together with copper saturated vermiculite was carried out by Santos *et al.*[148] The infra-red spectra showed that both protonation reactions (indicated by the presence of a band near 1,510 cm^{-1} due to the symmetric deformation vibration of NH_3^+) and hemisalt formation[138] occurred in these systems. Magnetic susceptibility measurements on the nickel-ethylenediamine montmorillonite sample indicated that the [Ni (en)$_2$]$^{2+}$ complex is diamagnetic. This observation led Santos *et al.* to conclude that this complex has a planar structure. By analogy with Laura and Cloos' data, the ethylenediamine is most likely to be chelated in the *gauche* form.

The interaction of cyclohexylamine with montmorillonite from two localities differing in the source of their layer charge has been investigated by Yariv and Heller.[162] In the Wyoming sample (W) the layer charge is about equally divided between the tetrahedral and the octahedral sheet, whereas in the sample from Camp Berteau (CB) this charge resides principally in the octahedral sheet. The complexes prepared by immersing thin, self-supporting films of the clay containing different interlayer cations in liquid cyclohexylamine were examined by X-ray diffractometry and infra-red spectroscopy.

Table 26 lists the observed frequencies for the NH_2 and NH_3^+ stretching and bending modes of cyclohexylamine in the liquid state, and when adsorbed as either the neutral molecule or the cation. The values for cyclohexylammonium-

TABLE 26

NH$_2$ and NH$_3^+$ stretching and bending frequencies (cm^{-1}) for liquid cyclohexylamine (A), cyclohexylammonium-montmorillonite (AH$^+$—mont) and montmorillonite containing different interlayer cations immediately after immersion in liquid cyclohexylamine.[162]

Vibration	Liquid A	AH$^+$—mont	Montmorillonite immersed in liquid A												
			Interlayer cation												
			AH$^+$	NH$_4$	Cs	Na	Li	Ca	Mg	Al	Co	Ni	Cu	Zn	Cd
NH$_2$, NH$_3^+$ stretching	3358 3278	3240 —	3340 3220	3338 3228	3346(VW) 3282(VW)	3358(S) 3281(W)	3356(S) 3285(VW)	3340(W) 3232	3346(W) 3226	3346(W) 3225(SP)	3344*(W) 3226	3344(VW) 3244	3315(SP) 3238(S)	3325 3255 3225*	3307(SP) 3262(SB)
NH$_3^+$ bending		1620 1510	1518— 1630*	1608 1530	— —	— —	— 1522	1610(B) 1526	1608 1526	1604 1525	1605(SP) 1515	— 1515	1610(SP) 1528	— 1518	1608(SP) 1528
NH$_2$ bending	1610	1595	1590(SP)	1590(SP)	1585(B)	1582(S)	1580	1590(SP)	—	—	—	—	1583(VS)	1588(SP)	1588

* only after 24 hour immersion.

SP = Sharp; V = Very; W = Weak; S = Strong; B = Broad.

montmorillonite are also given; the band near $1{,}510$ cm^{-1} due to the NH$_3{}^+$ deformation vibration serves as a diagnostic criterion for the presence of the ionic form. A band at about $1{,}450$ cm^{-1} (CH$_2$ deformation) may likewise be used as an index for the total amount of amine adsorbed since both the ionic and molecular forms of the molecule show absorption at this frequency. The intensity ratio of the $1{,}510$ cm^{-1} to the $1{,}450$ cm^{-1} band thus represents the fraction of the total present as the cation or protonated species (Table 27).

With the exception of the sodium- and cesium-saturated samples, protonation occurs immediately after immersion of the clay in the liquid. Prolonged exposure to the liquid gives rise to an increase in the amount of neutral cyclohexylamine present. The total (ionic and molecular amine) amount taken up after 24-hour exposure does not very greatly between samples, although the rate of sorption is much influenced by the nature of the saturating cation (Table 27).

The NH$_3{}^+$ bending frequencies (after a short period of exposure) are coincident with those shown by cyclohexylammonium-montmorillonite. The NH$_3{}^+$ deformation bands, initially weak, intensify on prolonged exposure, and concomitantly the asymmetric and symmetric bending modes shift to lower and higher frequencies, respectively.

To account for these changes, Yariv and Heller suggested that only the hydrated form of cyclohexylammonium ions was present at low concentrations of the ions. When the concentration is raised, inorganic cation hydroxides are formed and these interact with the organic ions. For example, washing with water of calcium-W which has been immersed in the amine for 24 hours, results in a spectrum which is identical with that of cyclohexylammonium-montmorillonite. Evidently, this treatment removes both the neutral amine and Ca(OH)$_2$ from the interlayers. On the other hand, treatment with CCl$_4$ does not cause a change in NH$_3{}^+$ frequencies since all the hydroxides are insoluble in this liquid.

TABLE 27

Intensity ratio of ~ 1510 cm^{-1} band (ionic cyclohexylamine) to ~ 1450 cm^{-1} band (ionic and molecular cyclohexylamine) in the spectra of montmorillonite containing different interlayer cations after different periods of exposure to liquid cyclohexylamine.[162]

Duration of exposure	Interlayer cation									
	NH$_4$	Li	Ca	Mg	Al	Co	Ni	Cu	Zn	Cd
1 min	0·62(vab)	1·30(ap)	1·30(ab)	ab	1·7(ab)	1·6(ap)	1·9(ap)	0·31(ab)	0·6(ma)	1·0(ab)
15 min		0·80	0·33	0·42	0·45	1·0	—	0·2	—	0·2
24 hr	0·25	0·54	0·22	0·33	0·42	0·2	0·2	0·15	0·46	0·18

Amount of cyclohexylamine taken up initially is indicated in brackets after the appropriate figure. Thus: ma = minor amounts; ap = appreciable; ab = abundant; vab = very abundant.

100

X-ray diffraction measurements of the W and CB samples after different periods of immersion in liquid cyclohexylamine show a tendency for most to expand to a basal spacing of ~2·05 nm. This tendency, like that for the 'instant' formation of the protonated form on immersion (Table 26), is more pronounced with CB than with W montmorillonite. The Mg, Al, Co, Ni, and Cu samples of W show less interlayer swelling (d(001) ~ 1·5 nm) after 20 hours immersion but these also eventually attain the fully expanded state. Washing with CCl_4 causes a collapse of the basal spacing from ~2·05 to ~1·47 or 1·40 nm indicating that interlayer expansion is influenced by the presence and maintenance of the cation-water-amine structural combinations in the interlayer space. The structure, proposed by Yariv and Heller, compatible with a spacing of ~2·05 nm is shown in Fig. 40. This is analogous to that of the 2·3 nm pyridine complex of Farmer and Russell.[122] The nitrogen of

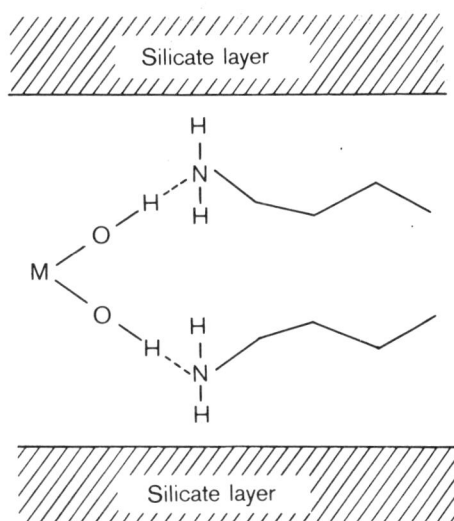

Fig. 40 Suggested arrangement of molecular cyclohexylamine in the interlayer space of montmorillonite (d(001) = 2.05 nm); M represents the exchangeable cation. After Yariv and Heller[162]

the amino group forms a hydrogen bond with water or with hydroxide groups coordinated to interlayer cations (M), while the hydrogens may be similarly bonded to surface oxygens. The ~14·7 nm and ~1·40 nm spacings (observed after CCl_4 washing) are thought to be determined by the presence of the hydrated inorganic cations and the hydroxide, respectively. The former situation tends to obtain when hydroxide/cyclohexylammonium interactions are weak, whereas the latter prevails when these interactions are pronounced.

Yariv and Heller[162] have compared and contrasted the behaviour of cyclohexylamine with that of aniline, which is discussed below. Despite the fact that cyclohexylamine (pK_a = 10·6) is a much stronger base than aniline (pK_a = 4·6) the amount adsorbed, under comparable conditions, is similar. Cyclohexylamine, however, accepts protons much more readily (in accord with its stronger basic character) giving rise to interactions involving metal hydroxides, hydrated

101

cations, cyclohexylammonium, and cyclohexylamine. On the other hand, only limited protonation occurs with aniline and the sorption products are correspondingly less varied, consisting of aniline-water-cation type structures. Of great significance, however, is their suggestion that, all things being equal, hydrogen bonding of interlayer water to the silicate surface may be an important, if not a decisive factor in the protonation of adsorbed organic bases. For the same total charge the strength of the water-to-surface hydrogen bonding would clearly be influenced by the source of the layer charge. On this premise, hydrogen bonding is relatively weak for the CB material where most of the charge resides in the octahedral sheet, and this is reflected by the observation that both interlayer expansion with and protonation of cyclohexylamine occur more readily in the CB than in the W samples. Conversely, complex formation between adsorbed amines and transition metal cations is more easily attained with W than with CB montmorillonite although Swoboda and Kunze[163] have previously argued for the opposite case, that is, that protonation reactions are facilitated in layer silicates where most of the negative charge originates in the tetrahedral sheet.

3.4.3. Aromatic Amines

Considerable attention has been paid to the interaction of aniline with montmorillonite. Firstly, this compound serves as a model for the sorption of aromatic bases by expanding layer silicates, and secondly, many amine herbicides consist of substituted anilines.

As we noted above, aniline is a relatively weak base ($pK_a \sim 4.6$) so that under comparable conditions of pH and concentration, less would be adsorbed than the aliphatic amines. Haxaire and Bloch,[126] for example, have reported that whereas the aliphatics, presented to sodium-montmorillonite as their aqueous salt solution at pH ~ 7, were taken up to the exchange capacity of the clay, appreciably less aniline and other aromatic amines (e.g. N,N-dimethylaniline, diphenylamine) were intercalated by an ion-exchange process. This difference in sorption behaviour also applies within the aromatic amine group itself. Thus, Hendricks[164] had earlier observed that bases such as benzidine, p-aminodimethylaniline, p-phenylenediamine, and α-naphtylamine were capable of neutralizing most of the exchangeable hydrogen ions in montmorillonite. On the other hand, o- and m-nitroanilines being extremely weak bases (in comparison) failed to form salts with the clay.

Using sodium- and hydrogen-montmorillonite, Bailey and White[165] have observed that the amount of aniline adsorbed is greater for the acid-treated clay than for the sodium material. This they ascribed to the ability of H,Al-montmorillonite to protonate the intercalated organic molecule which is then retained as the anilinium ion. In the sodium system, on the other hand, the neutral amine molecules compete with water for sorption sites (around the interlayer cation). Since the dipole moment of aniline is smaller than that of water, less aniline is adsorbed as compared with the acid-treated clay.

That aniline is protonated when brought into contact with acid-treated montmorillonite has been demonstrated by Harter and Ahlrichs[166] who noted that the infra-red spectrum of H,Al-montmorillonite containing aniline showed features identical with those of the complex obtained by adding a solution of aniline hydrochloride to the clay at pH 2·50.

The interaction of aniline with montmorillonite containing different interlayer cations has been studied in some detail by Yariv et al.[167] By immersing thin, self-supporting films of the clay in liquid aniline and examining the complexes so formed using infra-red spectroscopic, X-ray diffraction, and differential thermal analysis (DTA) methods, the profound influence of the interlayer cation on the amount adsorbed and the type of bonding is clearly demonstrated.

The infra-red spectrum of anilinium-montmorillonite shows bands typical of the amine cation. Vibrations at 1,553, 1,533, and particularly at about 1,520 cm^{-1} are assigned to NH_3^+ bending modes. The double bands at 2,625 and 2,650 cm^{-1} are thought to be combination bands. On exposure of this sample to liquid aniline, marked changes occur in the spectrum. Of the vibrations due to anilinium ions, only the combination bands are visible but these occur at a lower frequency range (2,500–2,600 cm^{-1}). Suppression of the 1,520 cm^{-1} band and an appreciable enhancement of absorption between 2,500 and 2,600 cm^{-1} have also been reported by Mortland et al.[168] who also observed that the CN vibration of aniline occurring at ~1,278 cm^{-1} in the pure liquid was shifted to 1,250 cm^{-1} in the anilinium-aniline complex. These changes are indicative of strong hydrogen bonding consistent with the formation of a $ØNH_3^+ - NH_2Ø$ (hemisalt) complex. Hydrogen- and aluminium-montmorillonites likewise form anilinium-aniline ions when an excess of aniline is present, while the ammonium saturated clay gives rise to ammonium-aniline ions. Because of the presence of strong hydrogen bonding interactions, these structures show considerable stability. Exposure of the anilinium-aniline-day complexes to air for 10 days and to liquid water, for example, did not materially alter their infra-red spectra.

Lithium-, sodium-, potassium-, magnesium-, and calcium-montmorillonites, on the other hand, retain aniline principally by means of cation-water-organic bonding. Aniline is capable of displacing all the interlayer water from cesium-montmorillonite, giving rise to an anhydrous system. As a consequence, water bridges do not occur, nor are the NH_2 vibrations perturbed. With montmorillonites containing transition metal ions, aniline may be bonded directly to the cation (type I) or linked to the ion through a water bridge (type II). Structural variations are possible in each type and these are discussed below. The various types of bonding established between aniline and different cations in the montmorillonite interlayers observed by Yariv et al.[167] from infra-red measurements are summarized in Table 28.

Besides determining the type of complex formed, the interlayer cation also influences the amount of aniline taken up. Thus, adsorption is enhanced in the presence of ammonium and transition metal ions, while alkaline earth samples intercalate greater amounts of aniline than either sodium- or potassium-montmorillonites.

103

TABLE 28

Principal interlayer aniline complexes formed on immersion in liquid aniline
of montmorillonite containing different interlayer cations.[167]

Interlayer cation	Interlayer aniline compound
$\emptyset$—NH_3^+, H^+, Al^{3+}	$\left[\begin{array}{c} H \quad\quad H \\ \| \quad\quad \| \\ \emptyset\text{—N—H}\ldots\text{N—}\emptyset \\ \| \quad\quad \| \\ H \quad\quad H \end{array}\right]^+$
NH_4^+	$\left[\begin{array}{c} H \quad\quad H \\ \| \quad\quad \| \\ \text{H—N—H}\ldots\text{N—}\emptyset \\ \| \quad\quad \| \\ H \quad\quad H \end{array}\right]^+$
Li^+, Na^+, K^+, Mg^{2+}, Ca^{2+}	$\left[\begin{array}{c} \quad\quad\quad\quad H \\ \quad\quad\quad\quad \| \\ \text{M}\ldots\text{O—H}\ldots\text{N—}\emptyset \\ \quad\quad \| \quad\quad \| \\ \quad\quad H \quad\quad H \end{array}\right]^{+\,(or\ 2+)}$
Cs^+	$\begin{array}{c} H \\ \| \\ \text{M}^+ + \text{N—}\emptyset \\ \| \\ H \end{array}$
Cr^{2+}, Mn^{2+}, Co^{2+}, Ni^{2+}, Cu^{2+}, Zn^{2+}, Cd^{2+}, Hg^{2+}	$\left[\begin{array}{c} H \quad\quad\quad\quad\quad H \\ \| \quad\quad\quad\quad\quad \| \\ \emptyset\text{—N}\ldots\text{H—O}\ldots\text{M}\ldots\text{N—}\emptyset \\ \| \quad\quad \| \quad\quad\quad \| \\ H \quad\quad H \quad\quad\quad H \end{array}\right]^{2+}$

M = exchangeable (interlayer) cation; $\emptyset$ = phenyl ring.

Aniline uptake appears to require the presence of interlayer water and, with the possible exception of the cesium-saturated sample, residual amounts of water are invariably retained even after boiling the specimens with aniline. The basal spacing of air-dry samples immersed in aniline is about 1·5 nm, compatible with an orientation in which the plane of the benzene ring is either perpendicular[22] to or slightly tilted making a high angle[111] with the silicate surface. With the copper saturated clay, and to a less pronounced extent with the lithium and ammonium samples, a basal spacing of 1·26 nm obtains after short immersion. This spacing increases to 1·5 nm after prolonged exposure to aniline, indicating that there is a conformational change from one in which the aromatic ring is parallel to one in which it is more or less perpendicular to the clay surface.

The DTA patterns show endothermic peaks between 373 and 473°K due to water, and possibly loosely held aniline. As the period of immersion is increased the peak near $373°K$ is reduced in accord with the infra-red data, showing that water is being largely but not completely displaced by aniline. The relative magnitudes of the endothermic and exothermic peaks reflect the amounts of organic material which are weakly and strongly retained, respectively. In this regard, the DTA patterns show a similar dependence on the interlayer cation as did the infra-red spectra. Thus, the pattern for the sodium clay resembles that for the potassium material, both of which hold aniline weakly. On the other hand, the divalent cation (Ca^{2+} and Mg^{2+}) systems show features similar to those observed for complexes with transition metal ions which bind aniline more firmly.

In a subsequent paper, Heller and Yariv[169] described the interaction of aniline and some of its derivatives with montmorillonite containing interlayer transition metal ions. The presence in the interlayer space of two basic structural types is indicated, that is, the organic molecule may either be directly coordinated to the cation (type I), or it may be linked to the cation through a water bridge (type II). Variants are possible in each type.

Type I complexes may be distinguished from those of type II on the basis of the behaviour of the NH_2 and CN stretching bands whose frequencies are lower than those of the corresponding liquids (Table 29). Thus, either two or three bands occur between 3,220 and 3,420 cm^{-1}. Vibrations in the range of 3,220 to 3,270 cm^{-1} and between 3,375 and 3,420 cm^{-1} can be assigned to symmetric NH_2 stretching of type I and to asymmetric NH_2 stretching of type II, respectively. In addition both types of complex give rise to a band at 3,305 to 3,340 cm^{-1}. For example, when the spectrum shows only two bands, vibrations in the 3,220 to 3,270 cm^{-1} range being absent, a type I complex does not obtain.

By this criterion, cadmium- and copper-montmorillonite form both type I and type II complexes with all the compounds used. Which of these is preferentially formed is indicated by the relative intensity of the bands. For the cadmium and copper systems, type I appears to predominate in all instances with the possible exception of o-chloroaniline. For a given compound, the tendency to form a type I complex depends on the interlayer cation decreasing in the order $Cd = Cu > Zn > Ni > Co > Mn$. Aniline, m- and p-toluidine and m-chloroaniline always form both types of complex, while only type II obtains with some other ligands.

Transition between types may occur on heating the samples in vacuo, type I tending to form at the expense of type II as shown by changes in the CN stretching frequency. On heating the m-chloroaniline-nickel-montmorillonite sample, for example, the band at 1,263 cm^{-1} (due to type II) is removed while the 1,244 cm^{-1} band (attributed to type I) remains intact.

The infra-red spectra of the o-chloroaniline complex taken before and after heating the sample suggest that this compound behaves differently from the other aniline derivatives examined. Although o-chloroaniline is the weakest base in this series and tends to form a type II complex, the N—M bond (M is

105

TABLE 29

NH$_2$ and CN stretching frequencies (cm^{-1}) in the infra-red spectra of aniline and some aniline derivatives intercalated by montmorillonite containing different interlayer cations.[169]

Compound	Group	Interlayer cation						Pure liquid
		Mn	Co	Ni	Cu	Zn	Cd	
aniline	NH$_2$	3270, 3320, 3385	3265[1a], 3315, 3380	3270, 3320, 3385	3240(w), 3280(sh), 3315, 3360	3245,3315, 3385	3260, 3325, 3385	1260
	CN	1228(2), 1250	1225(sh), 1250	1224(sh), 1252		1248	1220, 1250	
o-toluidine	NH$_2$	3325, 3380	3315, 3390	3260(vw), 3320, 3395, 1245[4], 1238(sh)[5], 1242[5]	3295, 3350, 3390(sh)	3245[3] 3315, 3390	3265, 3320, 3390	1268
	CN	1242	1246		1234	1234(sh), 1243	1235(w), 1246	
m-toluidine	NH$_2$	3255(sh)[6], 3310, 3380	3255(sh)[6], 3310, 3380	3260, 3310, 3380	3225, 3290, 3380(w)	3240(w)[7], 3330[7], 3420[7]	3260, 3310, 3380	1290
	CN	1256(w)[6], 1276	1245(w)[6], 1262(w)[6], 1275	1260, 1275	1260, 1280	1275, 1243(w)[5], 1260(w)[5]	1258, 1274	
f-toluidine (CCl$_4$ solution)	NH$_2$	3310(br), 3375	3270, 3320, 3380	3270, 3330, 3390(vw)	3230, 3295, 3380	3225, 3300, 3380	3260, 3320, 3385	
	CN	1248	1246	1247		1245	1240	
2,5-dimethyl-aniline	NH$_2$	3305(br), 3385	3305(br), 3385	3315, 3385	3290, 3330	3250[8], 3300, 3385	3260(br), 3315, 3385 1247(sh), 1263[4] 1244(sh)[5], 1254(sh)[5]	1282
	CN	1263	1264[4], 1256[5]	1264[4], 1252[5]	1240	1263[4], 1263[4]		
2,6-dimethyl-aniline	NH$_2$	3325, 3395	3325, 3395	3330, 3395	3245(sh), 3320, 3395	3305(br), 3385	3260, 3325, 3395	1273
	CN	1260	1264[4], 1257[5]	1260		1260	1264[4], 1258[5]	
2,4,6-trimethyl-aniline	NH$_2$	3240, 3400	3320, 3390	3330, 3385	3295(w), 3340, 3390	3335, 3390	3255(br), 3315, 3395	1251
	CN	1228	1230	1230	1230	1228	1230	
o-chloroaniline	NH$_2$	3330, 3395	3320, 3390	3270(vw)[9], 3325, 3390	3280, 3340, 3385(sh)	3220, 3315, 3390	3250, 3320, 3390	1312
	CN	1292	1292	1280	1283	1278(sh), 1290	1290[4], 1282[5]	
m-chloroaniline	NH$_2$	3270(vw), 3325, 3390[4], 1250[4], 1260(w)[4], 1240[5]	3260[1b], 3320, 3385[9] 1245, 1260(w)[9]	3265, 3320, 3385 1244, 1263(sh)[9]	3270, 3330, 3385	3235, 3320, 3385 1244, 1263(w)[9]	3265, 3325, 3390 1245–1263(br)	1293
	CN							

1 Becomes very broad after vacuum treatment at 373°K and shifts to (a) 3255 cm^{-1}, and (b) 3195–3255 cm^{-1}
2 Shoulder detected only after vacuum treatment at 373°K
3 After vacuum treatment at 373°K changes to 3240 cm^{-1} and becomes very broad
4 Before vacuum treatment at 373°K
5 After vacuum treatment at 373°K

6 Detected only after vacuum treatment at 373°K
7 After vacuum treatment at 373°K bands change to 3230(br), 3305, 3385 cm^{-1} respectively
8 Reduced to shoulder after vacuum treatment at 373°K
9 Band disappears after vacuum treatment at 373°K
sh = shoulder; w = weak; vw = very weak; b = broad.

metal cation) once formed is the strongest. These observations suggest chelate formation either directly to the cation:

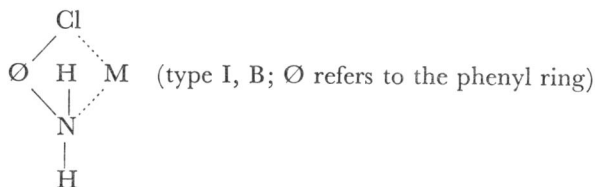

$$
\begin{array}{c}
\text{Cl} \\
\diagup \quad \diagdown \\
\text{Ø} \quad \text{H} \quad \text{M} \qquad \text{(type I, B; Ø refers to the phenyl ring)} \\
\diagdown \quad | \quad \diagup \\
\text{N} \\
| \\
\text{H}
\end{array}
$$

or, through a water molecule:

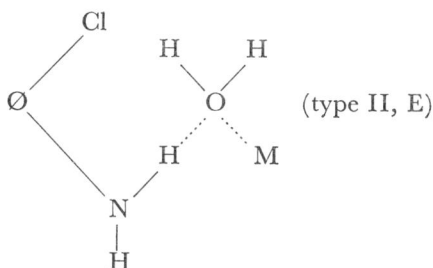

$$
\begin{array}{c}
\text{Cl} \\
\diagup \qquad\qquad \text{H} \qquad \text{H} \\
\text{Ø} \qquad\qquad\qquad \diagdown \quad \diagup \\
\diagdown \qquad\qquad\qquad \text{O} \qquad \text{(type II, E)} \\
\diagdown \qquad\qquad \text{H} \qquad \diagdown \text{M} \\
\text{N} \diagup \\
| \\
\text{H}
\end{array}
$$

Two other structural arrangements of type II variety were postulated on the basis of the infra-red data.

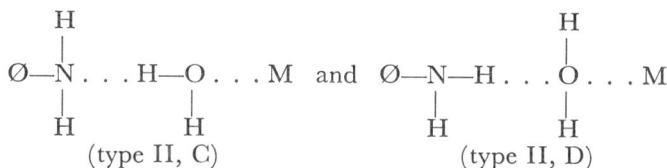

$$
\begin{array}{cc}
\qquad\text{H} \qquad\qquad\qquad\qquad\qquad\qquad\text{H} \\
\qquad | \qquad\qquad\qquad\qquad\qquad\qquad | \\
\text{Ø—N} \ldots \text{H—O} \ldots \text{M} \quad \text{and} \quad \text{Ø—N—H} \ldots \text{O} \ldots \text{M} \\
\qquad | \qquad\quad | \qquad\qquad\qquad\quad | \qquad\quad | \\
\qquad\text{H} \qquad\quad\text{H} \qquad\qquad\qquad\quad\text{H} \qquad\quad\text{H} \\
\qquad\text{(type II, C)} \qquad\qquad\qquad \text{(type II, D)}
\end{array}
$$

Steric hindrance such as would be expected to occur in *ortho*-substituted anilines would reduce accessibility of the electron pair of nitrogen to the hydrogen of water, and hence favour the formation of type II, D.

Anilines intercalated by montmorillonite containing transition metal cations may thus form five possible structural types, type I, A being

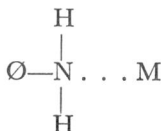

$$
\begin{array}{c}
\text{H} \\
| \\
\text{Ø—N} \ldots \text{M} \\
| \\
\text{H}
\end{array}
$$

The X-ray data of the different complexes are summarized in Table 30. On heating the samples there is a decrease in basal spacing indicating the loss of ('free') amine molecules from the interlayer space. These molecules are thought to be sorbed by van der Waals forces and not coordinated to the interlayer cations. According to their space requirement the compounds may be divided into three groups: (1) aniline itself, (2) the toluidines and chloroanilines, and (3) the di- and trimethylanilines. Increased sorption and hence closer packing and larger basal spacings would be expected as the basicity and polarity of the

TABLE 30

Basal spacing (nm) of montmorillonite saturated with different cations after immersion in the respective liquid amine for 48 hours (a) and then evacuated at 373 °K for 15 minutes (b).[169]

Compound		Interlayer Cation					
		Mn	Co	Ni	Cu	Zn	Cd
aniline	a	1·50	1·50	1·51	1·50	1·51	1·51
	b	1·50*	1·47	1·47		1·49*	1·50*
o-toluidine	a	1·62	1·64	1·60	1·61[1]	1·63	1·63
	b	1·58**	1·62*	1·52	1·56**	1·59**	1·63*
m-toluidine	a	1·67	1·63	1·58	1·64[1]	1·67	1·59
	b	1·52*	1·45	1·46	1·54*	1·47	1·52*
2,5-dimethylaniline	a	1·72	1·77	1·71	1·73	1·71	1·72
	b	1·59(d)	1.61	1·58**		1·60	1·55*
2,6-dimethylaniline	a	1·65	1·63	1·65		1·65	1·65
	b	1·62(d)	1·63*	1·61**		1·62*	1·62**
2,4,6-trimethylaniline	a		1·67	1·68	1·66[1](1·26)[1]	1·67	1·71
	b	1·59	1·71*	1·62	1·52	1·63	1·68
o-chloroaniline	a	1·62	1·60	1·68	1·64[1]	1·65	1·65
	b	1·55*	1·53	1·51	1·53	1·55	1·52**
m-chloroaniline	a	1·50	1·50	1·53		1·55	1·56
	b	1·53**	1·51*	1·51*		1·51*	1·51*

[1] Sample immersed for 4 hours only.
d = diffuse;　* integral series of reflections;　** almost integral series.

organic compound increase. This holds true for the di- and trimethylanilines, the spacing of the wet complexes decreasing with a lowering in the pK_a value of the compounds.

The same compound, however, may give different basal spacings depending on the interlayer cation, and this is interpreted in terms of the extent to which the various structural types influence the amount and orientation of the free amine with which they are associated. This effect is characteristically absent in the heated samples from which the excess amine has been removed, so that the basal spacing in this instance reflects the arrangement of the cation-organic assemblage. However, even after prolonged immersion in the organic liquid followed by heating *in vacuo* at 373°K, some interlayer water is still present and the basal spacing of ~1·5 nm of the various complexes after heat treatment may indicate that the interlayer separation is determined more by the thickness of two molecules of water than by any specific conformation of the adsorbed amines.

This fact, together with the probable presence of different structural types and the fact that the interlayer cations may have varying numbers of nearest neighbours within a single layer of a clay crystallite, makes it difficult to deduce a particular orientation of the intercalated molecule from basal spacing data. The frequently observed non-integral basal reflections must in part be due to the above effects although some mixed layering also seems probable.

Yariv et al.[170] have extended the above study to montmorillonite samples containing interlayer cations of varying 'acidity', such as Cs^+, K^+, Na^+, Mg^{2+}, Al^{3+}, and H^+. Their aim was to separate the part played by the oxygen surfaces of the silicate layers from that played by intercalated weak organic bases, both of which may accept protons donated by the interlayer water.

Water associated (or coordinated) to interlayer cations is known to be acidic since it can dissociate to yield protons under the polarizing influence of the cation.[171,172] Hence, the acidity of this type of water increases with an increase in the polarizing power of the exchangeable cation, that is, in the order of $K < Na < Mg < Al$. In cesium-montmorillonite, however, the interlayer water is associated more with the negatively charged silicate surface than with the exchangeable cation, the cesium ion being virtually unhydrated. Some of the molecules of this type of water will be oriented with the positive ends of their dipoles directed towards the surface oxygens, and hence this water has a 'basic' character. Because acidic water in many situations predominates over its basic counterpart, the latter's presence is frequently obscured.

As we noted previously, in potassium-, sodium, magnesium-, and aluminium-montmorillonites, aniline and its derivatives are linked to the cations through water bridges. Being weak bases, these compounds would have to compete with the surface oxygens for protons derived from the acidic water molecules. Thus, anilines may act either as proton donors (type II, D) or proton acceptors (type II, C). The former structure is favoured by an increase in the polarizing power of the cation and also by steric hindrance due to *ortho*-substitutions (e.g. in *o*-chloroaniline). In the cesium sample, however, anilines are thought to be primarily bonded to the surface oxygens either directly or through a water bridge.

The postulate that aniline may form a structure in which the NH_2 group is hydrogen-bonded to the oxygen atoms of a water molecule (type II, D) has been questioned by Farmer,[173] who suggests that the data from which the presence of this structure is deduced may well be interpreted in terms of the formation of anilinium ions.

The term *surface acidity* is widely used to refer to the ability of layer silicates to act as Brønsted acids, that is, to donate protons to adsorbed bases (e.g. Swoboda and Kunze[163]). As Yariv and Heller[162] have remarked, this term can be misleading since, strictly speaking, the oxygen surface of the silicate layers has a basic character. The acid properties of layer silicates are in fact derived from the ionization of the adsorbed water molecules associated with the exchangeable cations or possibly, but to a much lesser extent, with the silicate surface. Swoboda and Kunze and Yariv and Heller all recognized the importance of the source of (negative) charge in the silicate structure in protonation reactions.

Using calcium- and magnesium-saturated montmorillonites of different surface charge density and tetrahedral/octahedral layer charge ratios, Swoboda and Kunze reported that adsorbed weak organic bases such as pyridine and aniline were protonated. This reaction appears to occur more readily with those minerals in which the tetrahedral layer is the principal source of the negative charge (e.g. nontronite) than with, say, Wyoming montmorillonite where the total charge is distributed in about equal proportion between the tetrahedral and octahedral layers. To account for this behaviour they proposed that the stronger the hydrogen bond between water and surface oxygens the weaker the O—H bond in water, and hence the greater the extent to which it can dissociate to produce protons.

On the other hand, Yariv and Heller presented evidence to indicate that montmorillonite in which the negative charge chiefly resides in the octahedral layer (e.g. Camp Berteau) displayed a greater tendency to protonate adsorbed organic bases than the Wyoming sample. In accord with Swoboda and Kunze's idea, they also suggest that the O—H . . . O—Si hydrogen bond between water and surface oxygens is weaker in the Camp Berteau sample than in the Wyoming specimen, but then say that the breaking of this bond must precede protonation. This would account for the observation that the Camp Berteau montmorillonite is a better proton donor compared with the Wyoming clay. This view is further supported by the measurements of Mortland and Raman[174] who found that under comparable conditions, calcium-Wyoming montmorillonite protonated a greater amount of adsorbed ammonia molecules than did calcium-nontronite. Furthermore, weak bases such as pyridine[111] and aniline tend to link to interlayer calcium (or magnesium) ions through a water bridge rather than be adsorbed as the corresponding protonated species. Indeed, the detection by infra-red spectroscopy of pyridinium and anilinium ions in divalent cation clay systems, if present, may prove difficult because of the formation of hemisalt complexes when excess base is intercalated.[168]

In connection with the question of acidity of clay surfaces *per se*, Yariv et al.[170] have stated that although hydrogen bonding interactions between aniline and the silicate surface occurred in both cesium-montmorillonite and allophane, the infra-red spectra of the respective complexes were quite dissimilar. They ascribed this difference to the fact that the allophane surface, unlike that of montmorillonite, is acidic. The behaviour of uncharged organic bases in soils derived from volcanic ash where allophane is the dominant clay present would therefore be different from those soils whose clay fraction consists chiefly of crystalline layer silicate minerals.

Many aromatic amines are known to convert to their coloured derivatives when brought into contact with clays. This interaction involves charge transfer between the organic molecule and the mineral surface as well as adsorption. The literature on this topic has recently been summarized by Theng.[120] A separate chapter is devoted to charge-transfer complexes of layer silicates with aromatic amines and organic monomers.

110

3.5. Complexes with Amides

In acidic media, amides may *a priori* accept a proton on either the oxygen or the nitrogen atom. However, the weight of evidence from both spectroscopic and solution studies[175-180] lies on the side of the former alternative. Using infra-red spectroscopy, Tahoun and Mortland[151] have confirmed that amides predominantly protonate on the oxygen atom in acidic montmorillonite systems.

The infra-red spectra of hydrogen-montmorillonite containing different amounts of acetamide are shown in Fig. 41. The main absorption bands and

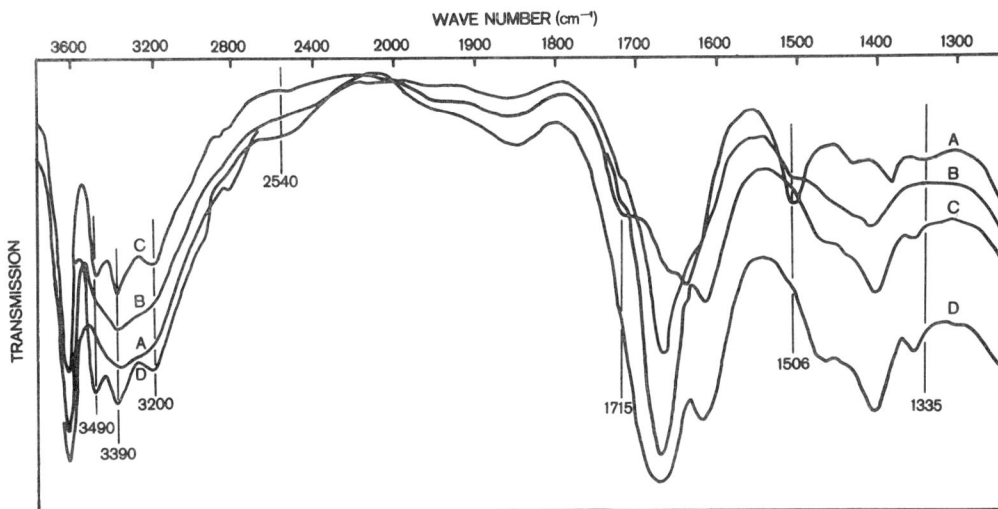

Fig. 41   Infra-red spectra of hydrogen-montmorillonite complexes with acetamide for different amounts of acetamide adsorbed per g clay. (A) 0.45 mmol; (B) 0.90 mmol; (C) 1.80 mmol; (D) 3.60 mmol. All curves have the same base line at 2000 cm^{-1}. After Tahoun and Mortland[151]

their assignments are listed in Table 31. Of importance is the occurrence of two bands near 1,700 and 1,500 cm^{-1} which provide evidence for O-protonation as well as for the existence of the resonance structures

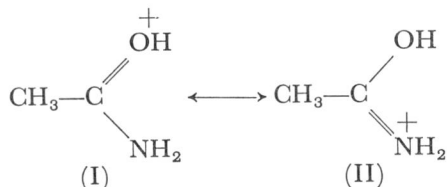

$$CH_3-C \overset{\overset{+}{O}H}{\underset{NH_2}{}} \longleftrightarrow CH_3-C \overset{OH}{\underset{\overset{+}{N}H_2}{}}$$

$$\text{(I)} \qquad\qquad \text{(II)}$$

Thus, O-protonation decreases the double-bond character of the C=O group while that of the C—N group is increased, causing a reduction and an increase in the respective absorption frequency. Further, the acquisition of a positive charge on the nitrogen atom (II) would reduce the NH force constant and

111

TABLE 31

Frequencies (cm^{-1}) of principal absorption bands and their assignments in the infra-red spectra of acetamide and its protonated form.[151]

Solid acetamide	Acetamide cation of			Assignment
	Chloride	H-mont*	Al-mont*	
3330 } 3160 }	3310 } 3125 }	3240(b)	3240(b)	NH stretching
	2530	2540(b)	2500(b)	OH stretching
	1718	1713	1700	C=N stretching
1681				Amide I band (C=O stretching)
1610	1664	1668	1652	NH deformation
	1497	1507	1502	$\overset{\mid}{-}C\overset{+}{=}OH \leftrightarrow \overset{\parallel}{-}C-OH$ stretching
	1345	1335(b)	1325(b)	OH deformation

* containing 0·45 mmol g^{-1}.
b = broad.

concomitantly increase hydrogen bonding interactions leading to a decrease in the NH stretching frequency relative to that shown by the neutral molecule.

Similar conclusions emerge from the spectra of N-ethylacetamide and N,N-diethylacetamide adsorbed by hydrogen- and aluminium-montmorillonite. Being stronger bases than acetamide because of the inductive effect of the ethyl group(s), these compounds are protonated to a greater extent than acetamide. The presence of absorption near 1,700 cm^{-1} (C=N stretching) is again evidence for protonation to have occurred (presumably on the oxygen) although the 1,500 cm^{-1} counterpart of acetamide is absent. A broad band near 2,500 cm^{-1} may be assigned to O—H stretching vibration. The substantial displacement of this band from its 'normal' frequency near 3,500 cm^{-1} is attributed to extensive and strong hydrogen bonding and to the polar character of the bond in the protonated species.[177]

In the acid montmorillonite system, hemisalt formation is observed when excess amide is present, that is, two amide molecules share a proton through a symmetrical hydrogen bond. The existence of acetamidehemihydrochloride has been reported by Fraenkel and Franconi[176] and by Kutzelnigg and Mecke.[178]

The cationic form of the amides is unstable against heat and evacuation, the neutral species being obtained when the protonated amide-clay complex is subjected to such treatments. This behaviour may be explained in terms of the removal of water from the complex, since exposure of the heat- and vacuum-treated sample to air at ambient humidity restores the protonated form. The presence of water at the clay surface is clearly essential for protonation of adsorbed amides. For the acetamide systems, for example, Tahoun and Mortland[151] have proposed the following reaction schemes:

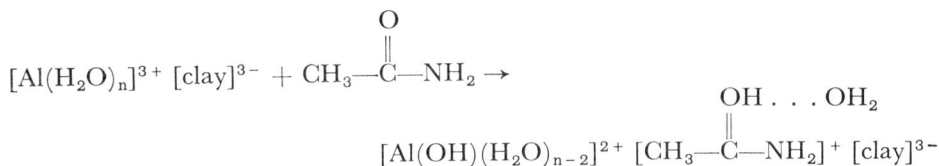

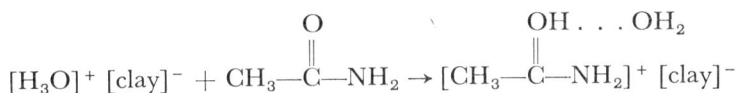

$$[Al(H_2O)_n]^{3+} [clay]^{3-} + CH_3-\overset{\overset{\textstyle O}{\|}}{C}-NH_2 \rightarrow$$

$$[Al(OH)(H_2O)_{n-2}]^{2+} [CH_3-\overset{\overset{\textstyle OH...OH_2}{\|}}{C}-NH_2]^+ [clay]^{3-}$$

$$[H_3O]^+ [clay]^- + CH_3-\overset{\overset{\textstyle O}{\|}}{C}-NH_2 \rightarrow [CH_3-\overset{\overset{\textstyle OH...OH_2}{\|}}{C}-NH_2]^+ [clay]^-$$

and

$$[CH_3-\overset{\overset{\textstyle OH...OH_2}{\|}}{C}-NH_2]^+ [clay]^- \underset{H_2O}{\overset{heat\ and\ vacuum}{\rightleftharpoons}} CH_3-\overset{\overset{\textstyle O}{\|}}{C}-NH_2 + H\text{-clay} + H_2O \uparrow$$

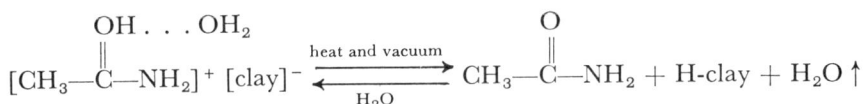

The infra-red spectra of the complexes with aluminium-montmorillonite indicate that besides undergoing protonation reactions, some of the adsorbed amides may coordinate to the exchangeable cation. In order to verify the formation of cation-amide coordination complexes, Tahoun and Mortland[181] have examined the infra-red spectra of acetamide, N-ethylacetamide, and N,N-diethylacetamide adsorbed by copper-, calcium-, and sodium-montmorillonite. Their data are summarized in Tables 32 and 33.

Acetamide is clearly coordinated to the metal ions at the clay surface as evidenced by the decrease in the frequency of the amide I band and the displacement of the NH and CN stretching bands to higher wavenumbers. Coordination apparently occurs through the oxygen atom of the amide, since the infra-red data indicate that there is an increase in the relative contribution of the resonance structure II. The possibility that the shift to a lower frequency of the amide I band is due to hydrogen bonding can be discounted for several reasons. The magnitude of this shift, for example, is greatest for the copper and least for the sodium system as would be expected from the coordinating properties of these cations. Furthermore, this shift is accompanied by a displacement of the CN and NH stretching bands to higher frequencies, and a splitting of the amide II and amide III bands in the N-ethylacetamide complex (Table 33).

The spectra of the N-ethylacetamide-copper clay system under different treatments are shown in Fig. 42. Unlike the protonated form, N-ethylacetamide coordinated to a cation is stable against heating, a property which is shared by all three amides examined. The amide I band in the dehydrated sample is shifted to $1,605$ cm^{-1}, that is, some 20 wavenumbers lower than for the hydrated and rehydrated specimens. Tahoun and Mortland ascribed this to the fact that at an amount present of 0.45 mmol g^{-1} there was one amide for each copper ion.

113

TABLE 32

Frequencies (cm^{-1}) of principal absorption bands in the infra-red spectra of acetamide and its complexes.[181]

Solid acetamide	Acetamide complexes				Assignment
	Cu^{2+} (*)	Cu-mont*	Ca-mont*	Na-mont*	
3330	3373	3490	3500	3500	
3160	3260	3452	3382	3400	NH stretching
		3368		3260	
1681	1664	1659	1661	1667	Amide I band (C=O stretching)
1610	1590	1576	1608	1607	Amide II band (NH deformation)
1458	?	1471	1468	?	CH asym. deformation
1399	1412	1416	1407	1395	CN stretching

(*) ref. 178. * dehydrated complex containing 0·45 mmol g^{-1}.

TABLE 33

Frequencies (cm^{-1}) of principal absorption bands in the infra-red spectra of N-ethylacetamide and its complexes.[181]

Liquid N-ethylacetamide	N-ethylacetamide complexes*			Assignment
	Cu-mont	Ca-mont	Na-mont	
3300	3455	3410	3465	NH stretching
3090	3285 3130	3315 3120		
1650	1621	1637	1635	Amide I band (C=O stretching)
1555	1564 1495(b)	1578 1548	1545 1490 }	Amide II band (CN stretching and NH deformation)
1474	1477	1476	1475	CH$_3$(C)=O asym. bending
1467	1462	1461	1460	CH$_2$ scissoring
1445	1437	1440	1425(b)	CH$_3$(N) bending
1373	1387	1379	1378	CH$_3$(C) bending
1357	1356	1356	1358	CH$_3$(C)=O sym. bending
1300	1339 1292	1336 1289	1335 1288 }	Amide III band (CN stretching and NH deformation)

* hydrated, containing 0·45 mmol. g^{-1}. b = broad.

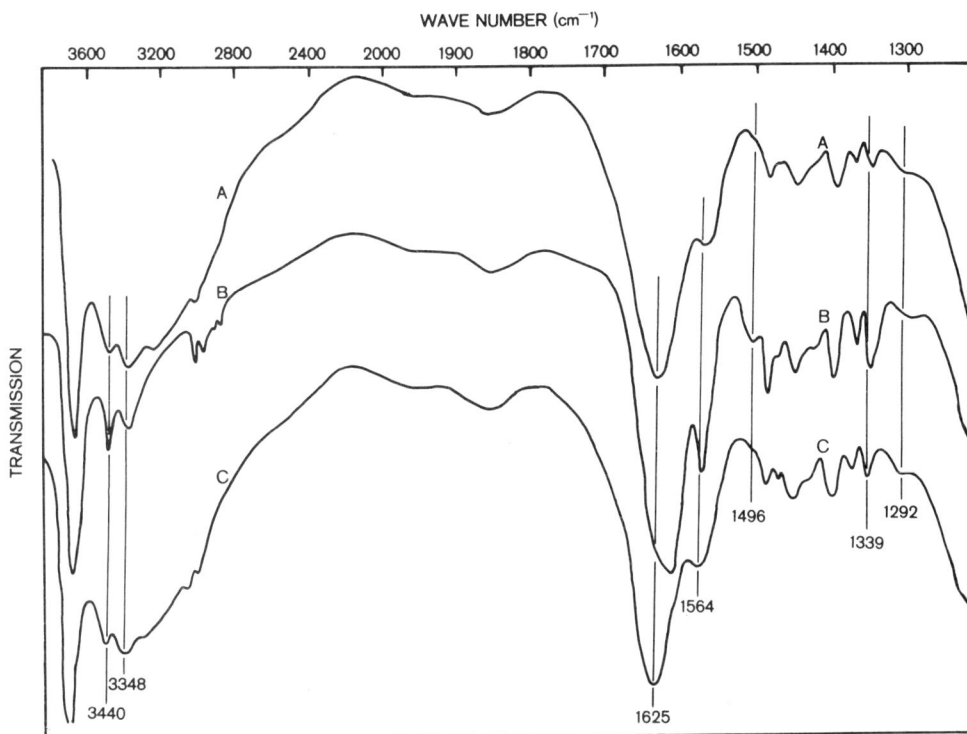

Fig. 42   Infra-red spectra of *N*-ethylacetamide, adsorbed in copper-montmorillonite. (A) the hydrated complex; (B) the dehydrated complex (heated under vacuum at 353°K for 1 hour); (C) the rehydrated complex (exposed to air for 3 weeks). In all cases the amount of amide adsorbed was 0.45 mmol.g^{-1}. After Tahoun and Mortland[181]

Since the coordination number of copper is 4, only a fourth of the coordination requirement is satisfied in the dry sample. In the wet samples, water molecules presumably complete this requirement and the copper-amide coordination bond is evidently weakened. It would also seem that coordination through oxygen induced the splitting of the amide II band at 1,555 cm^{-1} into two vibrations at 1,564 and 1,496 cm^{-1} corresponding to NH deformation and CN stretching, respectively. Similarly, the amide III band is split into two from 1,300 cm^{-1} into one at 1,339 cm^{-1} (CN stretching) and one at 1,292 cm^{-1} (NH deformation) for reasons which are as yet obscure.

Changes in features between 3,500 and 3,000 cm^{-1} in the N-ethylacetamide spectra, as the amount adsorbed was increased, could be explained in terms of the formation of intermolecular N—H . . . O hydrogen bonds and of the relative abundance of the *trans* and *cis* isomers of the molecule. The spectra of N,N-diethylacetamide (not shown) can be similarly interpreted. In addition, there is evidence to show that N,N-diethylacetamide links to the cation either directly or through a water bridge.

Spectra obtained at different inclinations of the clay film to the infra-red beam indicate that the amide molecules adopt a flat orientation in the interlayer

115

space, at least for low amounts adsorbed. On the basis of the infra-red data, Tahoun and Mortland have proposed three different structural types, representing three stages of amide sorption. Structure I obtains when the amount present equals or is less than that corresponding to the coordination requirement of the cation; structure II is formed for amounts exceeding the coordination requirement; and structure III for still greater amounts of amide sorbed.

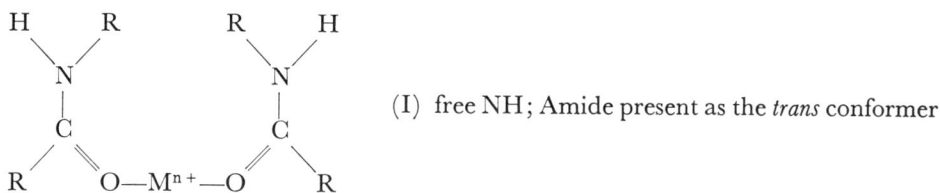

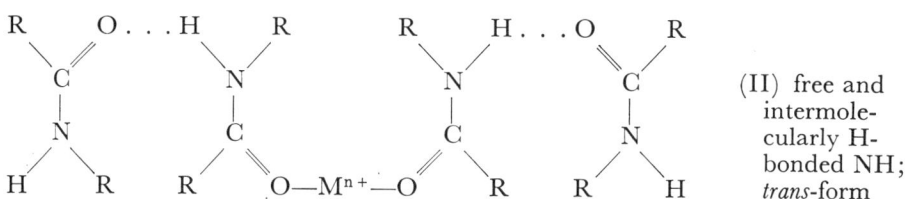

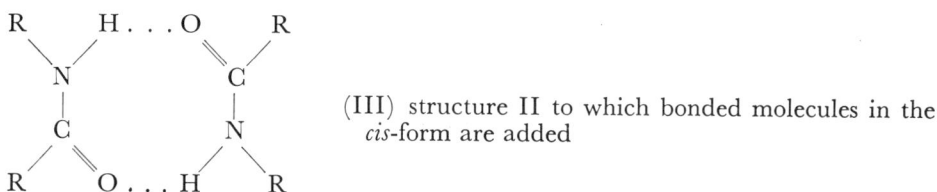

$$
\begin{array}{cc}
\text{H} \quad \text{R} & \text{R} \quad \text{H} \\
\backslash \diagup \text{N} & \text{N} \diagup \\
| & | \\
\text{C} & \text{C} \\
\diagup \; \| & \| \; \backslash \\
\text{R} \quad \text{O}-\text{M}^{n+}-\text{O} \quad \text{R}
\end{array}
$$

(I) free NH; Amide present as the *trans* conformer

$$
\begin{array}{cccc}
\text{R} \quad \text{O}\ldots\text{H} \quad \text{R} & & \text{R} \quad \text{H}\ldots\text{O} \quad \text{R} \\
\backslash \| \diagup \; \backslash \diagup & & \backslash \diagup \; \backslash \| \diagup \\
\text{C} \quad \text{N} & & \text{N} \quad \text{C} \\
| \quad\quad | & & | \quad\quad | \\
\text{N} \quad \text{C} & & \text{C} \quad \text{N} \\
\diagup \; \backslash \; \diagup \| & & \| \diagup \; \backslash \\
\text{H} \quad \text{R} \quad \text{R} \quad \text{O}-\text{M}^{n+}-\text{O} & & \text{R} \quad \text{R} \quad \text{H}
\end{array}
$$

(II) free and intermolecularly H-bonded NH; *trans*-form

$$
\begin{array}{c}
\text{R} \quad \text{H}\ldots\text{O} \quad \text{R} \\
\backslash \diagup \; \backslash \| \diagup \\
\text{N} \quad \text{C} \\
| \quad\quad | \\
\text{C} \quad \text{N} \\
\diagup \| \; \diagup \; \backslash \\
\text{R} \quad \text{O}\ldots\text{H} \quad \text{R}
\end{array}
$$

(III) structure II to which bonded molecules in the *cis*-form are added

In conclusion, it can be said that amides form coordination bonds to the exchangeable cations in montmorillonite through their oxygen atoms. For a given amide order, this bond is strongest with transition metal ions, weakest with alkali ions, and intermediate with alkaline earth ions. For a given cation, bond strength falls in the order tertiary > secondary > primary.

The question arises as to how amides and compounds containing a carbonyl group are retained by montmorillonite of which the exchange positions are occupied by organic cations. In an attempt to answer this, Doner and Mortland[182] examined the interactions between some dialkylamides and montmorillonite saturated with trimethylammonium and tetramethylammonium ions using infra-red spectroscopy. The basicity of a molecule of the type

$$
\begin{array}{c}
\text{O} \\
\| \\
\text{X}-\text{C}-\text{Y}
\end{array}
$$

is known to be influenced by the nature of the X and Y substituents. Indeed, it has been proposed[183] that the carbonyl group may show different bond characters from a triple to a single bond depending on the relative electron-with-drawing or electron-releasing power of X and Y, as shown below

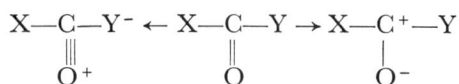

$$X—C≡Y^- \leftarrow X—C=Y \rightarrow X—C^+—Y$$
$$\underset{\overset{|||}{O^+}}{} \qquad \underset{\overset{||}{O}}{} \qquad \underset{\overset{|}{O^-}}{}$$

Taft[184] has provided a method of estimating this power for various sub-stituents and a summation of the Taft polar substituent constants $\Sigma\sigma*$ can therefore serve as a useful index of the basicity of

$$\underset{\overset{||}{X—C—Y}}{\overset{O}{}}$$

type compounds. For the dialkylamides chosen by Doner and Mortland, the fol-lowing $\Sigma\sigma*$ values are applicable[185]: N,N-dimethylformamide (DMF, +0·49), N,N-diethylformamide (DEF, +0·41), N,N-dimethylacetamide (DMA, 0), N,N-diethylacetamide (DEA, −0·20), and N,N-dipropylacetamide (DPA, −0·23).

The infra-red spectra of trimethylammonium-montmorillonite (Me_3NH-mont), liquid DEA, and of DEA adsorbed by Me_3NH-mont are shown in Fig. 43. The NH stretching frequency of Me_3NH-mont and of solid Me_3NHCl occurs at 3,200 and 2,725 cm^{-1}, respectively. Treatment of the clay with dialkylamides such as DEA lowers this frequency to 2,725 cm^{-1}. Concomi-tantly, the CO stretching vibration of liquid DEA at 1,645 cm^{-1} is shifted to 1,613 cm^{-1} when adsorbed. The depression in frequency for the NH ($\Delta\nu_{NH}$) and CO ($\Delta\nu_{CO}$) stretching bands of the dialkylamides when adsorbed by Me_3NH-mont are given in Fig. 44, showing the linear relationship between

Fig. 43 Infra-red spectra of trimethylammonium-montmorillonite, A; of liquid, N,N-diethy-lacetamide (DEA), B; and of the DEA complex with trimethylammonium-montmorillonite, C; comparing the N–H and C–O stretching frequencies. After Doner and Mortland[182]

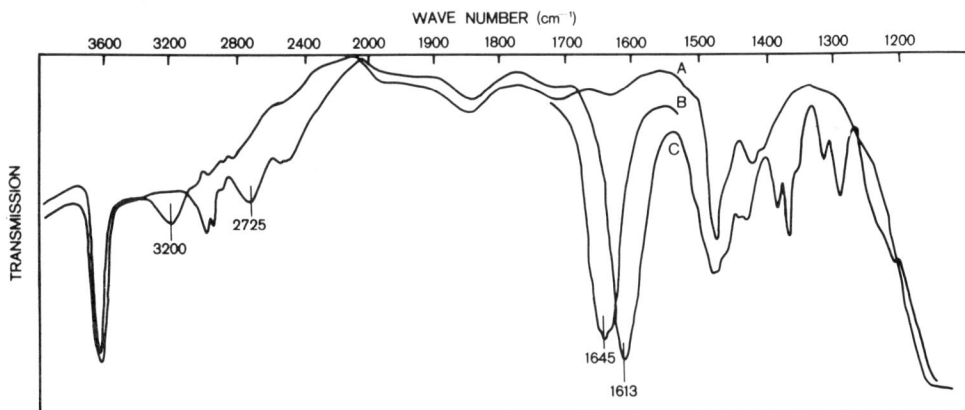

117

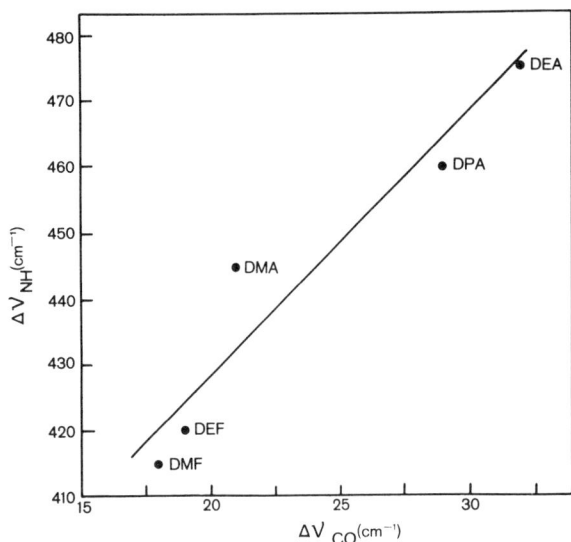

Fig. 44 Relationship between the frequency depression for the N—H stretching (Δv_{NH}) and that for the C—O stretching (Δv_{CO}) band in complexes of some dialkyl amides with trimethyl-ammonium-montmorillonite. Δv_{NH} and Δv_{CO} are measured with respect to the parent trimethylammonium-montmorillonite and to the corresponding amides, respectively. After Doner and Mortland[182]

these two parameters. These results clearly indicate that amides interact with trimethylammonium ions by a hydrogen-bonding mechanism of the type

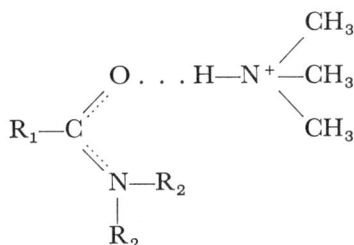

$$R_1—C \begin{array}{c} O \ldots H—N^+—CH_3 \\ \\ N—R_2 \\ | \\ R_2 \end{array}$$

R_1 is hydrogen (as in the formamides) or methyl (as in the acetamides); R_2 is either methyl or ethyl.

The strength of this interaction evidently increases in the order DMF < DEF < DMA < DPA < DEA as might be expected from the respective $\Sigma\sigma^*$ values. In other words, as the electron-withdrawing ability of R_1 and R_2 increases, the basicity of the amide and hence the displacement of the CO and NH frequencies decreases.

Adsorption of the dialkylamides by tetramethylammonium-montmorillonite (Me_4N-mont) also gives rise to a decrease in the CO stretching frequency of the amides, but the extent of the frequency reduction is less than for the trimethylammonium system. The reason for this behaviour is not clearly evident since the Me_4N^+ ion has no active hydrogen available for hydrogen-bond formation. Nevertheless, the greater depression in the C—O stretching frequency in complexes with Me_3NH-mont as compared with that for Me_4N-mont seems to be of general applicability to the adsorption of carbonyl compounds (Table 34).

118

TABLE 34

C=O stretching frequency (cm^{-1}) of some carbonyl compounds adsorbed by trimethylammonium—(Me$_3$NH$^+$) and tetramethylammonium—(Me$_4$N$^+$) montmorillonite.[182]

Compound	Interlayer cation	
	Me$_3$NH$^+$	Me$_4$N$^+$
Acetone	1710	1712
Propionaldehyde	1695	1717
o-Chlorobenzaldehyde	1667	1712
p-Chlorobenzaldehyde	1666	1695
Benzaldehyde	1665—1670	1663
N,N-diphenylacetamide	1638	1679
N,N-diphenylformamide	1650	1637
Ethyl N,N-di-n-propylthiolcarbamate (EPTC)	1611	

X-ray diffraction measurements on the amide-alkylammonium-clay complexes indicate little or no change in basal spacing of Me$_3$NH- and Me$_4$N-montmorillonite on amide adsorption. The basal spacing of the parent alkylammonium clays reported by Doner and Mortland is comparable with that obtained by Theng *et al.*[136] The latter workers observed a slightly higher value for the Me$_3$NH saturated sample, presumably because their sample contained some water. In any case, the interlayer separation is sufficiently large for the dialkylamide molecules to be intercalated without causing any further or appreciable expansion (Table 35).

The compound ethyl N,N-di-n-propylthiolcarbamate (EPTC) mentioned in Table 34 has earlier been shown by Mortland[186] to form a stable complex

TABLE 35

Basal spacing (nm) of trimethylammonium—(Me$_3$NH$^+$) and tetramethylammonium—(Me$_4$N$^+$) montmorillonite before and after intercalation of some dialkylamide compounds.[182]

Compound	Basal spacing	
	Me$_3$NH$^+$	Me$_4$N$^+$
Parent clay	1·29 (1·33)	1·38 (1·38)
N,N-dimethylformamide (DMF)	1·28	1·38
N,N-diethylformamide (DEF)	1·30	1·38
N,N-dimethylacetamide (DMA)	1·29	1·38
N,N-diethylacetamide (DEA)	1·35	1·38
N,N-dipropylacetamide (DPA)	1·34	1·38

Figures in brackets from ref. 136.

with pyridinium-montmorillonite. On the basis of the changes in the NH and CO vibrational bands observed in the infra-red spectrum of the complex, hydrogen bonding between EPTC and pyridinium ion of the type

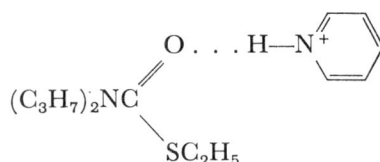

$$(C_3H_7)_2NC \begin{array}{c} O \ldots H—N^+ \\ \diagdown \\ SC_2H_5 \end{array}$$

can be postulated for this system. It seems probable, therefore, that EPTC interacts with Me_3NH^+ ions by a similar mechanism when intercalated by Me_3NH-mont. Hydrogen bonding is also implicated in the adsorption of the other compounds listed in Table 34, although direct comparison with the dialkylamides cannot be made because of differences in structure, size, and nature of the functional groups. What is clear, however, is that intermolecular hydrogen bonding is an important mechanism in the adsorption and retention of neutral organic compounds by clays containing interlayer organic cations. If these cations are the protonated form of the adsorbed uncharged species, hemisalt complexes tend to form as exemplified by the ethylammonium-ethylamine and anilinium-aniline systems. Indeed, the formation of such complexes may be regarded as a special case of intermolecular hydrogen bonding between two organic species.

Because of their properties, notably their polar nature and high dielectric constant, amides which are liquids at ambient temperature lend themselves to being used as a swelling agent of clay mineral systems.

More recently, Weismiller[187] has examined the interlayer expansion of montmorillonite and vermiculite crystals in N-ethylacetamide. Irrespective of the kind of interlayer cation present, montmorillonite apparently expands to a basal spacing of about 2·05 nm which is interpreted in terms of the intercalation of three layers of N-ethylacetamide molecules with the exchangeable cations occupying positions midway between two opposing silicate layers. The plane of the carbon atoms and the C=O group of the organic molecule lies parallel to the 001 plane, thus confirming Tahoun and Mortland's[181] earlier suggestion.

In the initial stages of adsorption, the amide molecule coordinates to the interlayer cation through the oxygen atom, but as the amount present increases the 'excess' (presumably corresponding to the amount over and above the coordination requirement of the cation in question) amide may hydrogen bond to the cation-linked species and/or to the oxygen ion of the silicate surface.

With vermiculite, interlayer swelling is usually less pronounced, probably because the silicate layers are more strongly held together compared with montmorillonite. The amide molecules tend to link to the cation by means of water bridges. Some vermiculite samples, however, may expand to a basal spacing of 6·3 nm after prolonged immersion in liquid N-ethylacetamide. This behaviour is reminiscent of that shown by n-butylammonium-vermiculite in water producing large interlayer separations.[188] Extensive crystalline swelling

120

of this type is attributed to the operation of osmotic repulsive forces giving rise to the formation of diffuse double layers. Similar effects are also produced when vermiculite crystals are immersed in strong amino acid solutions.[16,189]

Among the amides, urea occupies a special position because it is an important nitrogen fertilizer and because some of its derivatives possess herbicidal properties. In this section we shall confine ourselves to discussing the interactions of urea compounds with 2:1 layer silicates, notably the montmorillonites. The intercalation of urea and other amides by kaolinite minerals is dealt with in a later chapter.

It might be expected that the behaviour of urea would parallel that of, say, formamide and acetamide discussed previously in that it becomes protonated in acidic (H—, Al—) montmorillonite systems but is coordinated to the interlayer cation in clays containing other metal cations. The infra-red study of Mortland[150] confirms this expectation. Table 36 summarizes his data for complexes with hydrogen-, aluminium- and iron-montmorillonites. The broad bands with a maximum near 2,600 cm^{-1} and near 1,360 cm^{-1} assigned to OH stretching and deformation vibration of the protonated carbonyl group, respectively,

TABLE 36

Frequencies (cm^{-1}) and assignments of bands in the infra-red spectra of urea adsorbed by hydrogen-, aluminium-, and iron-montmorillonite.[150]

	Interlayer cation			
Urea-HCl*	H	Al	Fe	Assignment
3360	3470 (m)	3475 (m)	3490 (m)	NH stretching
3165	3400 (m) 3235 (m)	3385 (m) 3220 (m)	3380 (m) 3230 (m)	
2772 2570	2600 (m,br)	2600 (w,br)	2600 (w,br)	OH stretching
1700	1710 (s)	1708 (s)	1708 (s)	CN stretching
1642 1625	1634 (s)	1640 (s)	1640 (s)	NH_2 bending
1550	1565 (s)	1565 (s)	1565 (s)	CO stretching
1475				
1407				
1317	1360 (m,br)	1360 (w,br)	1360 (w,br)	OH deformation

* taken from Spinner[192].
s = strong;   m = medium;   w = weak;   br = broad.

disappear on dehydration of the sample and reappear when the protonated species forms on rehydration.

When the amount present is increased from 0·5 to 2·0 mmol g^{-1} the spectrum of the hydrogen-urea-clay complex is markedly altered. Strong, broad absorption bands appear between 1,700 and 1,200 cm^{-1} attributed to the formation of a hemisalt structure containing a strong, symmetrical hydrogen bond

$$
\begin{array}{ccc}
H_2N & & NH_2 \\
\diagdown & & \diagup \\
& C=O \ldots H^+ \ldots O=C & \\
\diagup & & \diagdown \\
H_2N & & NH_2
\end{array}
$$

The dehydration/rehydration process follows that suggested for acetamide and may be represented by one or both of the following equilibria:

$$[\text{urea-H}_3O]^+ [\text{clay}]^- \underset{H_2O}{\overset{\text{heat and vacuum}}{\rightleftharpoons}} \text{urea} + \text{H-clay} + H_2O \uparrow$$

$$2[\text{urea-H}_3O]^+ [\text{clay}]^- \underset{H_2O}{\overset{\text{heat and vacuum}}{\rightleftharpoons}} [(\text{urea})_2H_3O]^+ [\text{clay}]^- + \text{H-clay} + H_2O \uparrow$$

The similarity in spectral features between the hydrogen, aluminium, and iron saturated samples containing urea indicate that the latter two systems were sufficiently acidic to cause protonation of adsorbed area. Harter and Ahlrichs[166] also noted that the infra-red spectra of urea and urea hydrochloride adsorbed by hydrogen-montmorillonite were identical. On the other hand, protonation does not appear to be important in samples containing other metal ions in which cation-urea coordination is the dominant process.

Coordination through the nitrogen would cause the CO stretching frequency to increase and the CN stretching frequency to decrease. On the other hand, if coordination occurs through the oxygen, the CO and CN stretching bands would be shifted to a lower and a higher frequency, respectively. These deductions are based on the existence of three resonance structures for urea,

$$
\begin{array}{ccc}
NH_2 & \overset{+}{N}H_2 & NH_2 \\
\diagup & \diagup & \diagup \\
O=C & \overset{-}{O}-C & \overset{-}{O}-C \\
\diagdown & \diagdown & \diagdown \\
NH_2 & NH_2 & NH_2 \\
& & \overset{+}{} \\
(I) & (II) & (III)
\end{array}
$$

thus making it possible for the CO and CN bonds to have either a greater or a smaller double-bond character.

The infra-red results for the magnesium-, calcium-, lithium-, sodium-, and potassium-montmorillonite systems listed in Table 37 are interpreted on the basis of coordination of urea to the cation through nitrogen, because the spectra bear some resemblance to those of Pd(II)- and Pt(II)-urea complexes studied by Penland et al.[190] who suggested that coordination occurred

TABLE 37

Frequencies (cm⁻¹) and assignment of bands in the infra-red spectra of urea adsorbed by montmorillonite saturated with alkali and alkaline-earth cations.[150]

Pd(NH₂CONH₂)₂Cl₂*	Interlayer cation					Assignment
	Mg	Ca	Li	Na	K	
3390	3510(m)	3515(m)	3510(s)	3510(s)	3510(s)	ν_{NH} stretching
3290	3390(m)	3390(m)	3400(s)	3390(s)	3385(s)	
3140	3230(m,sh)	3240(m,sh)	3250(w,sh)	3240(w,sh)	3200(w)	
3030						
1725	1736(nm)	1731(m,sh)	1722(m)	1720(w,sh)	1718(w,sh)	ν_{CO} stretching (coordinated)
1615	1668(s)	1668(s)	1670(s)	1674(s)	1670(s)	CO stretching (uncoordinated) and
	1637(s)	1634(2s)	1633(s)	1630(s)	1625(s)	NH bending (uncoordinated)
1585	1588(w,sh)	1600(m,sh)	1598(m,sh)	1595(m,sh)	1590(m,sh)	NH bending (coordinated)
1400	1480	1474	1460(s)	1450(s)	1460(s)	ν_{CN} stretching (uncoordinated)
	1350(w)	1350(w)	1340(w)	1350(w)	1330(w)	ν_{CN} stretching (coordinated)

* taken from Penland et al.[190]

s = strong; m = medium; w = weak; sh = shoulder.

through the nitrogen. There are, however, some features in the spectra of the clay-urea complexes which do not accord with this interpretation. For example, the frequencies of the bands at 1,330–1,350 cm^{-1} ascribed to CN stretching (coordinated) are appreciably lower than those for the Pd(II) complex; the CN stretching and NH$_2$ bending vibrations in the potassium clay are more intense than in other cation systems; ν_{CN} (uncoordinated) varies widely between 1,450 and 1,480 cm^{-1}. It is possible, of course, that the N—H bond may ionize the metal ion then substituting for the hydrogen ion to form the respective metal salts of urea. If so, the 1,350 cm^{-1} band might be assigned to NH deformation vibration.

This ambiguity as to the mode of coordination of urea to Mg^{2+}, Ca^{2+}, Li$^+$, Na$^+$, and K$^+$ ions does not arise with montmorillonite saturated with transition metal ions, such as copper (II), manganese (II), and nickel (II). In these instances, coordination to the metal ions through oxygen is clearly indicated by the infra-red spectra (Table 38) which resemble those of the Cu [OC(NH$_2$)$_2$]Cl$_2$ complex of Penland et al.[190] The NH stretching frequencies at 3,500–3,400 cm^{-1} are higher than those of urea in KBr although the NH bending vibrations near 1,640 cm^{-1} are comparable with those of the unadsorbed molecule. This suggested a greater freedom of the NH$_2$ group in the montmorillonite interlayers compared with that in the crystalline solid. The position of the CO and CN stretching bands occurring at about 1,590 and 1,490 cm^{-1}, respectively, is consistent with coordination through oxygen since this would enhance the contribution of (resonance) structures II and III with respect to I.

Earlier, Mitsui and Takatoh[191] had suggested that urea might be adsorbed by montmorillonite through hydrogen bonding to the silicate surface on the ground that the NH stretching frequencies were slightly decreased in the clay complex. Their evidence, however, lacks conviction since a frequency reduction of similar, if not greater, magnitude is also observed in the spectrum of solid urea suspended in KBr or in a paraffin mull.[192]

Using deuterated urea and urea derivatives, and montmorillonite saturated with calcium, nickel, and aluminium ions, Farmer and Ahlrichs[193] have attempted to verify Mortland's conclusions regarding the different types of bonding mechanisms. Their data for urea-d$_4$, methylurea-d$_3$, and 1,1-dimethylurea-d$_2$ are set out in Table 39. In all instances, there is an increase in the ND stretching frequency over that of crystalline (ND$_2$)$_2$CO confirming Mortland's[151] suggestion that the ND group in the intercalated molecule is freer than in the solid, where it is probably hydrogen-bonded to oxygen of an adjacent molecule. At the same time, the frequency of the CO and CN stretching vibrations decreases and increases, respectively. These observations accord with urea being coordinated to the exchangeable cations through oxygen, the strength of coordination increasing in the order Ca^{2+} < Ni^{2+} < Al^{3+}. However, the shift in frequency of these bands for the calcium system is inappreciable, if at all detectable. For example, ν_{CO} of solid urea-d$_4$ is coincident with that of the adsorbed species (Table 39). This does not demonstrate unequivocally that a

124

TABLE 38

Frequencies (cm^{-1}) and assignments of bands in the infra-red spectra of urea adsorbed by montmorillonite saturated with transition metal cations.[150]

Urea[a]	Urea[b]	Cu[OC(NH_2)$_2$]$_2$I$_2$Cl$_2$[c]	Interlayer cation			Assignment
			Cu(II)	Ni(II)	Mn(II)	
3450(s)	3440(s)	3425	3500(s)	3505(s)	3510(s)	ν_{NH} stretching
3355(s)	3350(s)	3340	3390(s)	3400(s)	3395(s)	
3265(w)	3260(w)	3275	3265(w,sh)	3260(w,sh)	3260(w,sh)	
	3210(w)					
1682(s)	1670(s)	1655	1640(s,br)	1658(s)	1660(s,br)	Mostly NH bending
		1640		1635(s,sh)	1637(s,br)	NH bending and CO stretching for uncoordinated urea in clay systems
1630(s)	1622(s)	1620		1590(s)	1595(s)	NH_2 scissoring
1605(s)	1605(sh)		1584(s)		1490(s)	Mostly CO stretching
1468(s)	1447(s)	1580	1498(s)	1483(m)		ν_{CO} stretching
		1485				ν_{CN} stretching
		1470				
1158(w)	1163(w)	1160				NH_2 rocking

a   KBr pellet.

b   KBr pellet, impure sample; spectrum somewhat modified.

c   Taken from Penland et al.[190]

s = strong;   m = medium;   w = weak;   sh = shoulder;   br = broad.

TABLE 39

Frequencies (cm^{-1}) of principal absorption bands and their assignments in the infra-red spectra of urea-d$_4$, methyl urea-d$_3$, and 1,1,-dimethylurea-d$_2$ in the solid and when adsorbed on calcium-, nickel-, and aluminium-montmorillonite.[193]

Compound	Interlayer cation			Assignment
urea-d$_4$	Ca^{2+}	Ni^{2+}	Al^{3+}	
2590	2627	2630	2617	ν_{ND}
2500				ν_{OD}
2430	2490	2480	2473	ν_{ND}
1620	1620	1600	1560	ν_{CO}
1490	1495	1522	1595	ν_{CN}
methylurea-d$_3$				
2950	2950	2950	2945	ν_{CH}
2880	2890	2890	2890	ν_{CH}
2570	2620	2625	2615	ν_{ND}
2460	2485	2490	2485	ν_{ND}
1595	1605	1590	1595	ν_{CO}
1510	1525	1550	1560	ν_{CN}
1410	1415	1420	1425	δ_{CH}
1,1-dimethylurea-d$_2$				
2935	2932	2940	2945	ν_{CH}
2875	2880	2890	2890	ν_{CH}
2555	2620	2630	2625	ν_{ND}
2375	2475	2485	2480	ν_{ND}
2345				ν_{ND}
		1668	1665	ν_{CN}
1610	1600	1595	1612	ν_{CO}
1515	1525	1535	1540	ν_{CN}
1425	1415	1423	1425	δ_{CH}
1410				δ_{CH}

coordination complex is formed between calcium and the organic molecule, although it seems reasonable to propose that the carbonyl group is involved in binding the molecule to the clay. The fact that little or no protonation appears to occur with the aluminium saturated sample is also a departure from Mortland's data. The infra-red spectrum of the aluminium clay-urea complex indicated that a small amount of heavy water was present in the system. Evidently, this residual water failed to provide enough hydrogen ions for appreciable protonation to occur.

The infra-red results for methylurea-d$_3$ and 1,1,dimethylurea-d$_2$ (Table 39) can be interpreted in similar terms, that is, these molecules tend to coordinate to the exchangeable cations through oxygen of the carbonyl group. This tendency is much less pronounced with 1,1,dimethylurea-d$_2$ because of steric hindrance due to the presence of the methyl groups in the molecule.

As might be expected, the interlayer swelling of montmorillonite immersed in aqueous urea solutions is influenced by the nature of the saturating cation. Using a range of concentrated solutions (1–10 mol l^{-1}) Shiga[194] has noted that samples containing sodium, magnesium, barium, and calcium ions may swell to a basal spacing of 1·7–2·1 nm, whereas the potassium- and ammonium-saturated samples did not expand under similar treatment. Similar results were reported by Libor and co-workers[195–197] using X-ray diffractometry, differential thermal, and thermogravimetric analyses. The basal spacing of complexes containing urea and cations in a 4:1 mole ratio has been measured by Mortland[151] who obtained a value of 1·38 nm for the Cu(II), Ni(II), and Mn(II) systems, and 1·43 for the Ca and Mg systems.

The clay-urea interaction has found practical application in the manufacture of nitrogenous fertilizers based on urea. Since the amount intercalated by and the strength of bonding to the clay are influenced by the exchangeable cation, it is possible to vary the rate of urea release into the soil solution by using montmorillonite containing different interlayer cations.[197] It is clearly important for some systems that the nitrogen requirement of the crop is supplied in small amounts over an extended period of its growth rather than all at once when much may be lost by leaching.

Similarly, the mode of bonding will largely influence the bioactivity and persistence of herbicidal urea derivatives added to soil. Farmer[198] did not observe measurable interlayer sorption of phenylurea, fenuron (3-phenyl-1,1-dimethylurea), and monuron (3-(p-chlorophenyl)-1,1-dimethylurea by montmorillonite. Some adsorption on external and crystal edge surfaces of the clay may, of course, occur but this is unlikely to be large. At least, the infra-red spectra of the adsorbed species are similar to those of the corresponding pure compounds.

Failure of the phenylureas to enter the interlayer space may partly be due to steric and concentration factors. Using application rates of 0·1 to 2 mmol g^{-1}, Kim[199] has recently reported that mono- and double-layer intercalation complexes of fenuron, OMU(3 -cyclo octyl-1,1-dimethylurea), herban (3-(hexahydro-4,7-methano indan-5-yl)-1,1-dimethylurea), and monuron were formed with sodium-, calcium, and aluminium-montmorillonite. The average number of molecules adsorbed per cation in a monolayer was about 1·4, 3·0, and 3·8 for the sodium, calcium, and aluminium samples, respectively. The infra-red spectra of the complexes indicated that these compounds were bonded to the interlayer cations (M^{n+}) chiefly through water bridges of the type

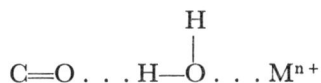

$$C{=}O \ldots H{-}\overset{\overset{\displaystyle H}{\displaystyle |}}{O} \ldots M^{n+}$$

no evidence being found for $N \ldots M^{n+}$ type linkages. Some $N{-}H \ldots O{-}Si$ bonding may occur. Direct organic-to-cation bonding only occurred after dehydrating the samples.

The interactions of clays with herbicides and pesticides are more fully discussed in a later chapter.

3.6. Complexes with Aliphatic and Aromatic Hydrocarbons

Strictly, these compounds should not be included in a chapter dealing with complexes between clays and polar organic species because hydrocarbons are non-polar or, at best, only very weakly polar. However, compared with organic compounds possessing permanent dipoles, very little is known about the behaviour of paraffinic molecules at the clay surface to warrant a separate chapter on their interactions with clays.

The lack of information on this segment of clay-organic reactions has also produced seemingly conflicting reports about the ability of hydrocarbons to penetrate the interlayer space of expanding layer silicates. We have already stressed that ion-dipole interactions have a determining influence on the uptake of uncharged polar molecules by clay minerals. Indeed, the intercalation of such molecules is essentially a process in which all or part of the interlayer water associated with the exchangeable cations is replaced by the organic species. In other words, polar organics occupy similar sites to water at the silicate surface to satisfy the coordination requirement of the interlayer cation. On the other hand, ion-dipole effects are not expected to play an important part in the interactions of clays with non-polar compounds, which are thought to be adsorbed by relatively weak, non-specific London-van der Waals (dispersion) forces alone.

In dehydrated systems where the silicate layers of a clay crystal are fully collapsed, intercalation of non-polar organic liquids and gases is either absent or proceeds only with difficulty. This is attributed to the fact that the compounds, being weakly adsorbed, are incapable of forcing the silicate layers apart. Adsorption then takes place predominantly on external crystal surfaces.[200] In partly dehydrated or air-dry systems where the layers are already slightly separated, intercalation may be limited by the inability of non-polar molecules to replace the interlayer water or to establish links to the exchangeable cation through water bridges.

Between these two states of hydration, optimum conditions may be found under which interlayer sorption can occur. X-ray diffraction measurements by Barshad,[16] for example, have indicated that liquid benzene and n-hexane enter the interlayer space of some montmorillonite and vermiculite samples when these have previously been dehydrated at $293°K$. However, the same specimens, dried at $523°K$, failed to intercalate these hydrocarbon liquids. By boiling air-dry samples of montmorillonite (saturated with Ca^{2+} or NH_4^+ ions) with the organic liquid, MacEwan[15] observed no interlayer penetration by n-hexane and n-heptane but benzene, naphthalene, and tetrahydronaphthalene were intercalated. Benzene formed a double-layer complex as Bradley[35] had earlier shown. On the other hand, using a similar method to MacEwan but with montmorillonite dried at 353°K, Greene-Kelly[22] failed to achieve interlayer sorption of benzene.

However, there is no doubt about the interlayer sorption of hydrocarbons in systems where the silicate layers are held 'permanently' apart by exchanging the inorganic cations, initially present, for alkylammonium ions such as monomethylammonium ($Me_1NH_3^+$), tetramethylammonium (Me_4N^+), and tetraethylammonium (Et_4N^+). These organic cations fill only a fraction of the interlayer space of montmorillonite, leaving ample room between them to accommodate non-polar gases and vapours.[200-204] Furthermore, 2:1 expanding-type layer silicates containing alkylammonium and long-chain quaternary ammonium ions (e.g. dimethyldioctadecylammonium) show different selectivities for paraffins and aromatic hydrocarbons, a property which forms the basis of their wide use as gas-chromatographic separation media.[205-209]

Both saturated and unsaturated hydrocarbons may also be present in the interlayer space of montmorillonite by the decomposition *in situ* of intercalated organic compounds. Chaussidon and Calvet,[210] for example, have shown by infra-red spectroscopy that a mixture of *n*-alkanes and *n*-alkenes were produced when alkylammonium-montmorillonite was heated at temperatures up to 523°K. Similar results have been reported by a number of workers using montmorillonite containing alcohols[211,212], fatty acids,[213,214] and various alkylammonium cations.[215] The transformation of adsorbed organic compounds catalysed by clay minerals is discussed more fully later.

The recent report by Eltantawy and Arnold[216] presenting evidence for the interlayer sorption of *n*-hexane and *n*-dodecane by calcium-Wyoming montmorillonite is, if somewhat unexpected, illuminating in that it provides a clue as to the likely factor limiting the intercalation by montmorillonite of non-polar gases and liquids. These workers were able to form a single-layer complex with *n*-hexane having a sharp basal spacing of 1·41 nm on exposing the air-dry clay to the hydrocarbon vapour for about 3 hours at room temperature. The complex so formed contained 110 mg of *n*-hexane per g of oven-dry (383°K) clay. Increasing the exposure period beyond 3 hours yielded a mixed single- and double-layer complex which could also be obtained by immersing the air-dry mineral in liquid *n*-hexane (d(001) spacings 1·403 and 1·796 nm). Infra-red examination of the *n*-hexane complex showed that a considerable amount of water was retained in the interlayer space. This suggests that ion-organic interactions were not significant and, as noted in Chapter 2, C—H . . . O—Si hydrogen bonding was also unlikely to be important. Yet, as indicated by the results of differential thermal microanalysis the complex, once formed, was markedly stable.

It would seem that water influences the rate of intercalation. Thus, immediate X-ray examination of the material previously heated at 493°K for 24 hours and then immersed in water-free liquid *n*-hexane failed to show any evidence of interlayer complex formation. Intercalation did occur, however, when this sample was left immersed for 2 weeks. Similar results were obtained for *n*-dodecane. These observations strongly suggest that *n*-alkanes can penetrate the interlayer space of montmorillonite but at a very slow rate. This would largely account for the failure on the part of some workers to observe interlayer sorption of non-polar organic molecules when insufficient time was allowed for significant intercalation.

Following De Boer and Zwikker,[217] Eltantawy and Arnold[216] explained their data in terms of dipole-induced dipole interactions, that is, the electrostatic field at the clay surface induces the formation of dipoles in the hydrocarbon molecules. This postulate is by no means unlikely since, as the analysis of Fripiat[218] has shown, the electric field arising from the exchangeable cations can be quite substantial ($\sim 10^8$–10^9 volt cm^{-1}), capable of inducing a dipole of appreciable magnitude in the adsorbed species. Why the intercalation rate is apparently so slow is an open question because, in theory, van der Waals adsorption occurs instantaneously. An important factor likely to slow down intercalation is the rate at which the *n*-alkanes diffuse into the montmorillonite interlayers. In addition, the intercalated molecules must be distributed and ordered in a certain way and this would cause a further delay in complex formation.

On the other hand, the nature of the interlayer cation present, so crucial in the adsorption of polar organic compounds, is apparently of no great influence in the intercalation of *n*-alkanes (Table 40). Slight variations in basal

TABLE 40

Basal spacing (nm) of interlayer *n*-alkane complexes with montmorillonite containing different interlayer cations.[216]

Compound	Interlayer cation			
	Ca^{2+}	Mg^{2+}	Na^+	K^+
n-Hexane	1·1418	1·473	1·237	1·228
n-Dodecane	1·405	1·409	1·288	1·177

spacing between samples containing different interlayer cations were ascribed by Eltantawy and Arnold to differences in the packing and orientation of the molecules in the interlayer space.

Using infra-red spectroscopy, Mortland and co-workers[219-221] have demonstrated that benzene and some methyl substituted benzenes, presented as their respective vapour, are readily intercalated by montmorillonite saturated with copper (II) ions. Benzene itself yields two distinct complexes:[219,220] type I which is green and type II having a red colour. Which type will form appears to be critically dependent on the degree of hydration of the montmorillonite sample. Type I obtains at a higher water content than type II, both of which

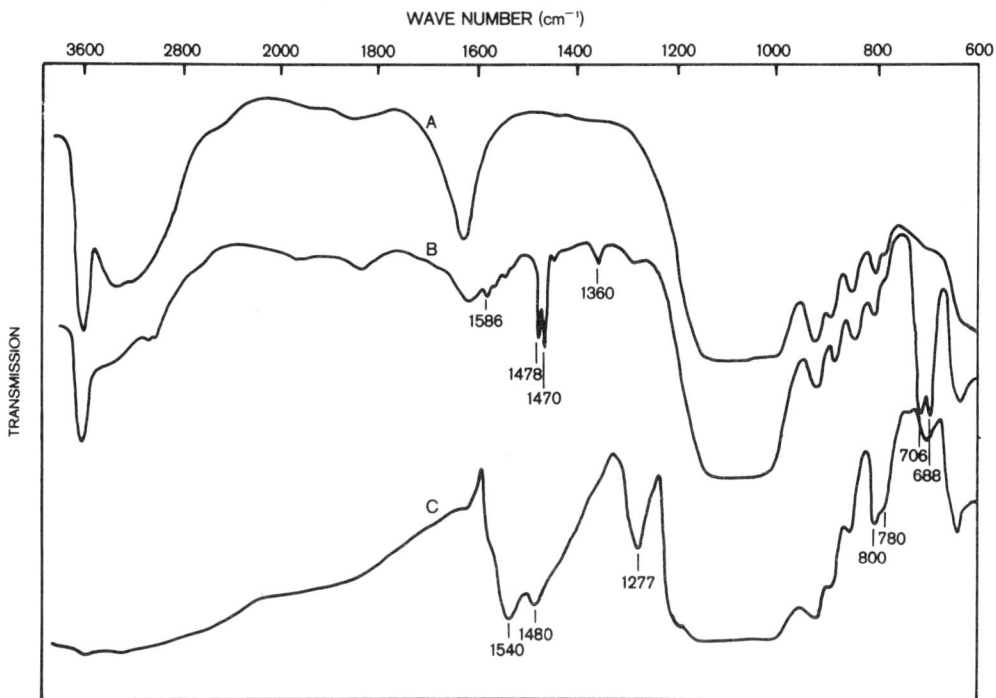

Fig. 45 Infra-red spectra of copper (II)-montmorillonite equilibrated at 40 per cent relative humidity and 298°K (A); the green form of coordinated benzene physically adsorbed on copper-montmorillonite (B); the red form of coordinated benzene on copper-montmorillonite (C). Spectrum B was obtained after the clay was degassed at 298°K over P_2O_5 for 4 hours and then exposed to benzene vapour for 24 hours. Spectrum C was taken after more exhaustive degassing (24 hours over P_2O_5 at 298°K) and then exposed to benzene vapour for 24 hours. After Mortland and Pinnavaia[220]

are interconvertible by simply changing the hydration status of the system within a given, narrow limit.

The infra-red spectrum of the parent copper clay taken at ambient temperature and relative humidity (40 per cent) is shown in Fig. 45(A). The strong, sharp band near 3,600 cm⁻¹ is the stretching mode of OH groups of the silicate structure. The broad bands centring near 3,250 and 1,630 cm⁻¹ are identified with the stretching and deformation modes, respectively, of both physically adsorbed and coordinated water. Other bands are due to vibrations of the silicate layer. When benzene vapour was introduced into the system two new bands appeared in the spectrum. Absorption at 1,478 cm⁻¹ can be ascribed to C—C stretching (ν_{19}) and at 688 cm⁻¹ to C—H out-of-plane (ν_{11}) vibrations, which in liquid benzene occurs at 1,478 and 675 cm⁻¹, respectively. This observation suggests that the intercalated benzene is physically adsorbed, that is, it interacts with the silicate surface rather than with the copper ion. This postulate is further supported by the fact that the position of the ν_{19} and ν_{11} bands is independent of the nature of the exchangeable cation, a behaviour which parallels that of n-alkanes described earlier.

131

The spectrum of a partially dehydrated (degassed at 298°K) copper (II)-montmorillonite after being exposed to benzene vapour shows bands at 1,586, 1,470, 1,360, and 706 cm^{-1} in addition to those due to physically adsorbed benzene (Fig. 45(B)). The sharp intense absorption at 1,470 and 706 cm^{-1} may be attributed to the ν_{19} and ν_{11} modes, respectively, of benzene coordinated to a partially hydrated copper ion.

The spectrum of benzene adsorbed by an exhaustively degassed sample is characterized by the presence of two broad, intense bands in the C—C stretching region (at 1,540 and 1,480 cm^{-1}) and two in the C—H out-of-plane region (at 800 and 780 cm^{-1}) together with broad, strong absorption from 1,600 and extending to beyond 4,000 cm^{-1} ascribed to low energy electronic transitions (Fig. 45(C)). A second type of benzene-copper coordination complex is therefore indicated under these conditions.

Mortland and co-workers suggested that coordination is effected by donation of the π electrons of benzene to the copper (II) ion acting as an electron acceptor. In the type I (green) complex, the coordinated benzene molecule evidently has essentially the same D_{6h} local symmetry as that of the liquid and physically adsorbed forms. The increase and decrease in frequency of ν_{11} and ν_{19}, respectively, from their position in liquid benzene are consistent with the formation of a coordination complex. Less than D_{6h} local symmetry for coordinated benzene in type II (red) complex may be inferred from the complex nature of the spectrum. This could occur if, for example, the copper ion bonds to the edge of the benzene ring.

It may therefore be concluded that the intercalated benzene molecule in the green complex is planar, essentially retaining its aromatic character, whereas in the red complex the ring of the molecule appears to be distorted and the π electrons are more localized.

It ought to be mentioned in this connection that the contribution of π electrons to the adsorption of some polyaromatic organic bases has earlier been referred to by Haxaire and Bloch[126] who found that the amount taken up in excess of the cation exchange capacity of montmorillonite, Q, was more or less linearly related to the ratio of Δ/π (Δ is the Δ value of the appropriate complex and π is the number of π electrons in the organic molecule). This is shown in Fig. 46, indicating that Q decreases with an increase in Δ/π, and for $\Delta/\pi > 0.34$, Q becomes negligibly small and the adsorption process is then solely one of cation exchange.

Pinnavaia and Mortland[221] have subsequently extended this study to some methyl-substituted benzenes, such as toluene, the xylenes, and mesitylene. Toluene is both physically adsorbed and coordinated to the exchangeable cation when intercalated by copper (II) montmorillonite which has previously been dried over P_2O_5 at room temperature (d(001) $= 1.58$ nm). This is indicated by the infra-red spectrum of its intercalation complex (Fig. 47), the bands and assignments[222] of which are listed in Table 41.

The two forms of intercalated toluene are easily distinguished. The frequencies of the various bands of physically adsorbed toluene are essentially

Fig. 46 Relationship between the amount adsorbed in excess of the cation exchange capacity, Q, and the ratio of Δ value of the complex to the number of π electrons in the molecule, Δ/π, for complexes of montmorillonite with some aromatic organic bases. (1) o-phenylenediamine: (2) p-aminodiphenyl; (3) solid green 2B (same basic structure as malachite green but containing two additional chlorine substituent groups); (4) α-napthylamine; (5) malachite green; (6) phenergan (dimethylamino-2' propyl-1')-N-dibenzoparathiazine); (7) p-phenylenediamine; (8) hexamethyl violet; (9) methylene blue; (10) benzidine; (11) neutral red. After Haxaire and Bloch[120]

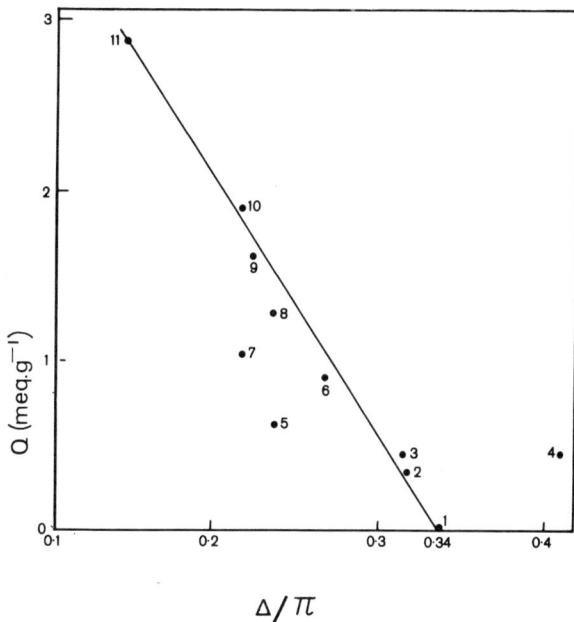

Fig. 47 Infra-red spectra of liquid toluene (A); copper (II)-montmorillonite (B); and the toluene-copper (II)-montmorillonite complex (C). Bands designated 'P' and 'L' refer to physically adsorbed and ligand coordinated toluene, respectively. Data from Pinnavaia and Mortland[221]

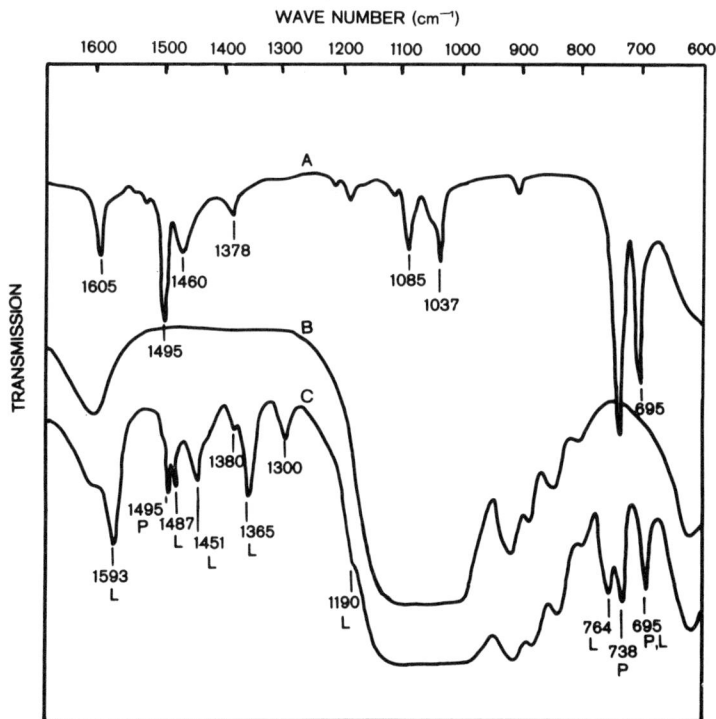

TABLE 41

Frequencies (cm^{-1}), relative optical densities, and assignments of the principal absorption bands of toluene in the liquid state and when adsorbed by copper (II)-montmorillonite.[221]

Liquid		Cu (II)—montmorillonite complex		Assignments*
Frequency	Relative OD	Frequency	Relative OD	
1605	100	1593	100	C—C stretching (ν_{8a})
1495	300	1487	36	C—C stretching (ν_{19a})
1460	80	1451	36	C—C stretching (ν_{19b})
1460		1435 (sh)		Asymmetric CH_3 deformation
1378	37	1365	56	Symmetric CH_3 deformation
		1300	25	C—H in-plane deformation (ν_3) ?
1212	17	1190		C—H in-plane deformation (ν_{13}) ?
728	780	764	59	C—H out-of-plane deformation (ν_{11})

* The assignments for liquid toluene were taken from ref. 222.

coincident with those shown by the liquid form, with the exception of ν_{11} which is shifted to a higher frequency by about 10 cm^{-1}. On the other hand, the bands ascribed to the coordinated form of toluene display appreciable frequency shifts as well as changes in relative intensity compared with the corresponding vibrations of liquid toluene (Table 41).

Similarly, the xylenes and mesitylene are intercalated by copper (II)-montmorillonite by physical adsorption as well as by coordination to the cation. Changes in frequency of the C—C stretching and C—H out-of-plane vibrations attending interlayer sorption of the methyl substituted benzenes examined are summarized in Table 42.

TABLE 42

Frequencies (cm^{-1}) of C—C stretching and C—H out-of-plane vibrations of some methyl-substituted benzenes in the liquid state and when adsorbed by copper (II)-montmorillonite.[221]

Compound	C—C stretching (ν_{19})			C—H out-of-plane deformation (ν_{11})		
	Liquid	Physically adsorbed	Cu(II)-complex	Liquid	Physically adsorbed	Cu(II)-complex
Toluene	1495	1495	1487	728	738	764
o-Xylene	1495	1495	1483	743	750	773
p-Xylene	1517	1516	1503	795	800	817
m-Xylene	1481	1480	1473	769	777	803
Mesitylene	1473	1473	1463	837	844	890*
Benzene	1478	1478	1470**	675	688	706**

* Obscured by vibration of the silicate structure.
** These figures refer to the green (type I) complex of benzene.

134

Comparison of the spectra with those given by benzene indicates that the methyl-substituted aromatics show a behaviour analogous to that of benzene in the green (type I) complex. The spectra give no evidence for the formation of the type II complex. The reason for this difference in coordination property between benzene and its methyl-substituted derivatives is as yet obscure, but steric factors would be expected to influence the extent to which the ring can be distorted to enable greater localization of the π electrons.

As with benzene, the basal spacing of the complexes with methyl-substituted benzenes is compatible with the formation of a single-layer intercalation complex in which the aromatic ring of the molecule is more nearly parallel than perpendicular to the silicate layer.

4

Interactions with Organic Compounds of Biological Importance

4.1. Complexes with Organic Pesticides

4.1.1. Introduction

The term 'pesticides' is used here to denote the wide range of synthetically prepared organic compounds which are used to control and eradicate weeds and insects—that is, as herbicides and insecticides. The pesticides may be grouped into three broad classes depending on their predominant charge characteristics: cationic; anionic; and non-ionic (polar). Because most compounds possessing pesticidal activity have a complex chemical structure, they are frequently referred to by their common or registered trade names. Some of the more important members in each class, together with their corresponding chemical names and properties, are listed in Table 43.

Compounds in the first class, such as the bipyridilium halides, are generally applied as their aqueous salt solution. Being completely ionized, they are adsorbed by clays (and soils) through an ion-exchange process, replacing the inorganic cations initially present at the silicate surface. As might be expected, the anionic pesticides tend to be negatively adsorbed, that is, they are repelled from rather than attracted to the negatively charged clay surfaces. At low pH, however, the acidic functional groups in the molecule may accept protons, giving rise to an uncharged species. Positive adsorption can then take place. In the pH range commonly encountered in soils, the non-ionic pesticides are taken up and retained by clays predominantly in molecular form. Their adsorption is therefore determined chiefly by ion-dipole interactions as outlined in the preceding chapter. In acidic media these compounds may acquire a positive charge by accepting a proton and so behave as cations. Some may even be doubly or triply protonated as Hirt and Schmitt[1] have suggested for melamine, for example.

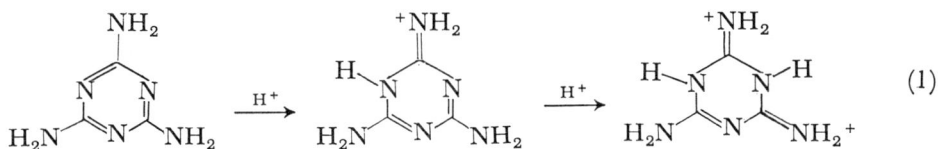

(1)

basic and neutral

Under conditions of high pH, dissociation may occur with some compounds, such as the hydroxy-s-triazines, and the molecules then display anionic properties.

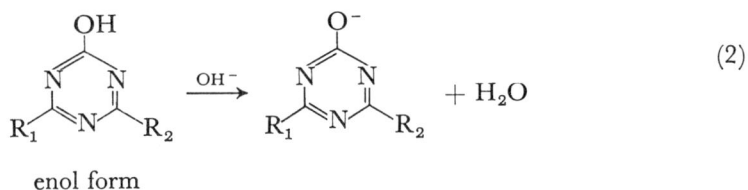

enol form

(2)

and

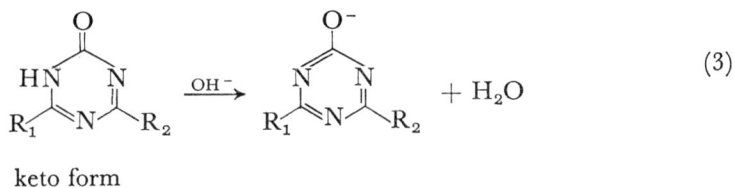

keto form

(3)

Clearly, the basicity (pK_a) of the compound as well as the pH of the system would exert a profound influence on the adsorption process. The interactions of all three classes of pesticides with clay minerals and other soil colloids have been discussed at some length by Bailey and White[2] while Weber[3] has dealt more specifically with the s-triazine-clay mineral system.

4.1.2. Non-ionic Pesticides
Among the non-ionics, the substituted s(ymmetric) triazines are perhaps the most widely used soil-applied pesticides. A considerable amount of attention

TABLE 43

Common (trade) names, chemical nomenclature, and physico-chemical properties of some organic pesticides and herbicides.

Class	Family	Common Name	Chemical Name	Water Solubility g dl⁻¹ (293°K)	pKₐ
Non-ionic	Substituted s-triazines	Atratone	2-methoxy-4-ethylamino-6-isopropylamino-s-triazine	0·1800	4·20
		Prometone	2-methoxy-4, 6-bis (isopropylamino)-s-triazine	0·0750	4·28
		Simetone	2-methoxy-4, 6-bis(ethylamino)-s-triazine	0·3200	4·15
		Ipatryne	2-methylthio-4-isopropylamino-6-diethylamino-s-triazine	—	4·43
		Prometryne	2-methylthio-4, 6-bis(isopropylamino)-s-triazine	0·0048	4·05
	symmetric triazine ring	Atrazine	2-chloro-4-ethylamino-6-isopropylamino-s-triazine	0·07	1·68
		Ipazine	2-chloro-4-isopropylamino-6-diethylamino-s-triazine	0·0040	1·85
		Propazine	2-chloro-4, 6-bis(isopropylamino)-s-triazine	0·0009	1·85
		Simazine	2-chloro-4, 6-bis(ethylamino)-s-triazine	0·0005	1·65
		Trietazine	2-chloro-4-diethylamino-6-ethylamino-s-triazine	0·0020	1·88
		Hydroxypropazine	2-hydroxy-4, 6-bis(isopropylamino)-s-triazine	—	5·20, ~11 } keto-enol
		Hydroxyipazine	2-hydroxy-4-isopropylamino-6-diethylamino-s-triazine	—	5·32, ~11 } keto-enol
	Triazole	Amitrole, Aminotriazole	3-amino-1,2,4-triazole	28	4·17
	Substituted ureas	Monuron	3-(p-chlorophenyl)-1,1-dimethylurea	0·0230	-1 to -2
		Diuron	3-(3,4-dichlorophenyl)-1,1-dimethylurea	0·0020	-1 to -2
		Linuron	3-(3,4-dichlorophenyl)-1-methoxy-1-methylurea	—	—
	Esters and Organophosphates	CIPC	m-chloroisopropyl carbanilate	0·0108	—
		EPTC	ethyl N,N-di-n-propylthiolcarbamate	—	—
		IPC(propham)	isopropyl carbanilate	0·0032	—
		Diazinon	O,O-diethyl O-(2-isopropyl-4-	0·004	—

Class	Family	Common Name	Chemical Name	Water Solubility g dl⁻¹ (293°K)	pKa
			methyl-6-pyrimidyl) phosphorothioate		
		Dursban	O,O-diethyl O-3,5,6-trichloro-2-pyridyl phosphorothioate	—	—
		Malathion	O,O-dimethyl S-bis(carbethoxy) ethyl phosphorodithioate	0·0145	—
		Ronnel	O,O-dimethyl O-2,4,5-trichlorophenyl phosphorothioate	0·004	—
		Zytron (DMPA)	O-(2,4-dichlorophenyl)O-methyl isopropyl-phosphoramidothioate	0·0005	—
	Chlorinated hydrocarbons	DDE	2,2-bis (p-chlorophenyl)-1,1-dichloroethylene	sparing	—
		DDT	2,2-bis(-chlorophenyl)-1,1,1-trichloroethane (p,p' isomer)	sparing	—
		Dieldrin	1,2,3,4,10,10-hexachloro-6,7-epoxy-1,4,4a,5,6,7,8,8a-octahydro-1,4-endo, exo-5,8-dimethanonaphtalene	sparing	—
		Endrin	1,2,3,4,10,10-hexachloro-6,7-epoxy-1,4,4a,5,6,7,8,8a-octahydro-1,4-endo, endo-5,8-dimethanonaphtalene	sparing	—
		Heptachlor	1,4,5,6,7,8,8-heptachloro-3a,4,7,7a-tetrahydro-4,7-methanoindene	sparing	—
Anionic	Anilides	Dicryl	3',4'-dichloro-2-methylacrylanilide	0·0009	—
		Propanil	3,4'-dichloro-propionanilide	0·0500	—
	Amide	Solan	3'-chloro-2-methyl-p-valerotoluidide	0·0008	—
	Benzoic acid	Amiben	3-amino-2,5-dichloro-benzoic acid	0·0700	—
		Benzoic acid	benzoic acid	0·270	4·12
	Picolinic acid	Picloram	4-amino-3,5,6-trichloro-picolinic acid	0·040	—
	Phenylalkanoic acid	2,4-D	2,4-dichlorophenoxy-acetic acid	0·050	2·80
		2,4,5,T	2,4,5-trichlorophenoxy-acetic acid	0·024	2·65
		MCPA	4-chloro-2-methylphenoxy-acetic acid	—	—
Cationic	Dipyridilium halides	Diquat	1,1'-ethylene-2,2'-dipyridilium dibromide	70	completely ionized
		Paraquat	1,1'-dimethyl-4,4'-dipyridilium dichloride	100	completely ionized

has therefore been paid to their interactions with layer silicates. The s-triazine compounds share a common ring structure differing only in the type of substituent at the 2-, 4-, and 6-positions (Table 43). Because of the presence of 3 nitrogen atoms in the ring, s-triazine shows little aromatic character. The chemical properties and reactions of the s-triazines, and hence the behaviour of these compounds at the clay surface, are determined more by the nature and location of the substituent groups than by the heterocyclic ring.

One of the early attempts to elucidate the mechanisms underlying the clay-pesticide interaction was due to Frissel[4] whose findings and conclusions were subsequently summarized by Frissel and Bolt[5]. These workers studied the uptake from an aqueous environment of some 14 organic herbicides by montmorillonite, illite (a soil mica), and kaolinite as a function of pH and electrolyte concentration. Among the compounds examined were the chloro-s-triazines, such as simazine and trietazine. Their observation indicated that these compounds were adsorbed as the uncharged species in the neutral and basic range of pH and as the corresponding cations under acidic conditions. The presence of added electrolytes affected the adsorption process in so far as it modified the hydration properties of the clay and those of the organic herbicide. In addition, high electrolyte concentrations (>0.1 M) may enhance adsorption by a 'salting-out' effect.

Frissel and Bolt[5] have also attempted to separate the relative contributions of coulombic and van der Waals forces to the adsorption of herbicides, such as trietazine, by (sodium-) montmorillonite. They assumed that van der Waals forces alone were responsible for uptake between pH 8 and 10 but that in the region of pH 2.9 to 8 both types of interactions were operative. However, the success achieved using this approach was only limited. The discrepancy between the observed and the calculated amount adsorbed was particularly marked below pH ~ 4 where adsorption was considerably less than predicted. Comparing the capacity of different clays to take up non-ionic herbicides, adsorption was observed to decrease in the order montmorillonite > illite > kaolinite.

That adsorption of s-triazine compounds by clays increases as the suspension pH is lowered has been repeatedly confirmed and substantiated by many workers.[6-9] Similarly, the observation that montmorillonite shows greater adsorption than either illite or kaolinite is of general applicability in these systems.[7,10,11] Indeed, in some instances, the amount of s-triazines taken up by kaolinite is so small as to be undetectable.[5,8] Under pH conditions where the compounds exist as the uncharged form, adsorption must predominantly occur by replacement of water molecules from the clay surface. The bonding between pesticide and clay is therefore not a strong one[5] and the adsorbed molecule may be readily desorbed by adding water to the system.[6,11]

The greater uptake by montmorillonite as compared with illite and kaolinite is partly due to its larger hydrateable surface. Its capacity to take up pesticides as either the uncharged or the cationic form is, of course, greatly enhanced when the organic species can penetrate the interlayer space of the mineral.

However, only limited interlayer penetration would be expected with the non-ionics unless the silicate layers have already been separated as is the case, for example, with sodium-montmorillonite in water or dilute salt solutions. On the other hand, if the compound is protonated either before or during adsorption such as would occur at low suspension pH or when hydrogen-montmorillonite is used, intercalation is readily effected by replacement of the interlayer inorganic cations.

Thus, Weber et al.[7] have observed that prometone was intercalated by both sodium- and hydrogen-montmorillonite. The basal spacing of the complexes so formed rose from ~1·30 to ~1·85 nm as the amount adsorbed was increased. This indicates that as more prometone enters the montmorillonite interlayers the organic molecules tend to rearrange from a position in which the s-triazine ring is parallel (flat orientation) to one in which the ring is tilted at an increasingly higher angle to the silicate layer. Basal spacings near 1·8 nm would indicate either the formation of a single-layer complex in which the plane of the ring is perpendicular to the clay surface or a double layer of flatly oriented molecules is intercalated.[12] We have already cited a number of conformational changes of this type in earlier chapters.

The influence of molecular structure and pH on the interaction of 13 s-triazine compounds with sodium-montmorillonite has been examined by Weber.[8] As expected, adsorption increases with a decrease in the pH of the suspension, reaching a maximum when the pH approaches the pK_a of the appropriate compound and then decreases. For a given concentration of pesticide the height of this maximum is related to molecular structure. In general, the maximum amount adsorbed decreases as the nature of the substituent group in the 2-position is changed in the order $-SCH_3 > -OCH_3 > -OH > -Cl$ as represented, for example, by the series prometryne > prometone > hydroxypropazine > propazine (Fig. 48) and ipatrine > ipatone > hydroxyi-

Fig. 48   Effect of pH on the adsorption of four related s-triazines (4,6-bis (isopropyl-amino) series) by montmorillonite; 25×10^{-6} M of each compound was used with 2·5 mg of clay and a contact time of 20 minutes was allowed for each pH change. (A) prometryne; (B) prometone; (C) hydroxypropazine; (D) propazine. After Weber[3]

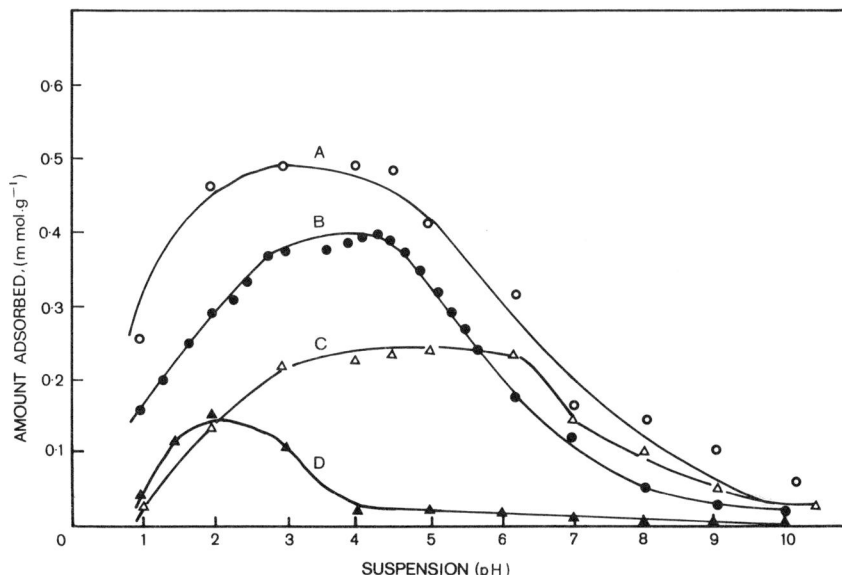

pazine > ipazine and by ametryne > atrazine in calcium-montmorillonite.[9] This maximum is also lowered as the number and length of the alkyl chain of substituents in the 4- and 6-positions are increased. For example, with a series of ethyl-substituted derivatives adsorption is reduced in the order $(C_2H_5)_4$ > $(C_2H_5)_3$ > $(C_2H_5)_2$. These observations have led Weber to propose that the adsorption process is primarily controlled by the nature of the 2-substituent. Since the basicity of the compounds is also influenced by the alkyl groups in the 4- and 6-positions, the latter substituents determined the amount adsorbed of a given series of 2-substituted s-triazines.

Similar substituent effects have been observed by Lambert[13] for 2,6-dinitro-4-alkylsulphoryl-N, N-diallylanilines and by Briggs[14] for substituted phenylureas and alkyl-N-phenylcarbamates in soil systems. Thus, a linear relationship is obtained between the logarithm of the partition coefficient, K_m, and the parachor or the Hammett constant for substituents in the phenyl ring.

Another important parameter controlling the adsorption of s-triazines is water solubility. Using six s-triazine compounds and sodium-montmorillonite, Bailey et al.[15] have observed that adsorption conformed to the Freundlich equation,

$$x/m = KC^{1/n} \tag{4}$$

where x is the amount adsorbed by a unit weight m of clay; K and n are constants. K may therefore serve as an index of adsorbability and on this basis the order of simetone ≫ atratone > prometone > trietazine > propazine > atrazine is found. With the exception of atrazine, this is also the order of their respective solubility in water (Table 43).

The interactions of s-triazines with montmorillonite in an aqueous environment have been summarized by Weber[8] in terms of the following equilibria:

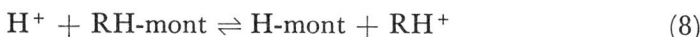

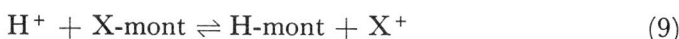

$$R + H^+ \rightleftharpoons RH^+ \tag{5}$$

$$R + X\text{-mont} \rightleftharpoons RX\text{-mont} \tag{6}$$

$$RH^+ + X\text{-mont} \rightleftharpoons RH\text{-mont} + X^+ \tag{7}$$

$$H^+ + RH\text{-mont} \rightleftharpoons H\text{-mont} + RH^+ \tag{8}$$

$$H^+ + X\text{-mont} \rightleftharpoons H\text{-mont} + X^+ \tag{9}$$

$$R + H\text{-mont} \rightleftharpoons RH\text{-mont} \tag{10}$$

R is the s-triazine compound; RH^+ is the cationic (protonated) form of R; X is the exchangeable cation on montmorillonite (mont) and H^+ refers, in reality, to the hydronium ion (H_3O^+). Eq. (5) is controlled by the pK_a of the compound according to

$$pK_a = \log [RH^+]/[R] + pH \tag{11}$$

Since R may have more than one pK_a as depicted by Eqs. (1), (2), and (3), Weber has restricted the discussion to the range of pH 1–8 in which the pK_a

142

values listed in Table 43 are applicable. In an aqueous suspension of, for example, sodium-montmorillonite at pH 7 containing an s-triazine, the dominant reaction may be represented by Eq. (6) where the pesticide may interact with the cation either directly by replacing the water in the primary shell of the cation, or indirectly through a water bridge. If now the suspension is acidified, the organic molecule becomes protonated (Eq. (5)) and the cationic form of the pesticide is adsorbed by exchanging for X (Eq. (7)) giving rise to increased uptake (Fig. 48). Addition of acid would also result in Eq. (9) being established and some neutral s-triazine molecules may adsorb by Eq. (10). The decrease in adsorption beyond pH $\sim$ pK$_a$ may in part be due to reaction (8) taking place while the depressive effect of electrolytes may partly be ascribed to the fact that Eq. (7) is driven to the left.

So far, we have only considered interactions in aqueous media and as we have noted, protonation of s-triazines may occur in situations represented by Eqs. (5) and (10). In a 'water-free' environment, such as when the organic is added as a solution in chloroform or ethanol to air-dry montmorillonite, some uncharged s-triazine derivatives may also become protonated, the active protons being derived from the dissociation of residual water molecules coordinated to the exchangeable cations. The protonation and hydrolysis of some adsorbed s-triazines have been demonstrated by Cruz et al.[12] using infrared spectroscopy.

The infra-red spectra of propazine and of a film of ammonium-montmorillonite exposed to a chloroform solution of the pesticide under the specified conditions are shown in Fig. 49 (A)–(D), while those of the ammonium- and hydrogen-saturated clays reacting with propazine in ethanol are illustrated in Fig. 49 (E) and (F). Adsorbed hydroxypropazine, and prometone yielded similar spectra, being essentially identical with the spectrum of montmorillonite treated with the corresponding s-triazine cations. Hence the uncharged organic molecule must convert to the protonated species and it is this form which is present at the clay surface. Of particular interest is the appearance of an absorption band near 1,750 cm^{-1}, indicative of a carbonyl group probably arising from (acid) hydrolysis at the 2-position followed by keto-enol tautomerism (cf. Eqs. (2) and (3)). Both propazine and prometone, for example, would yield hydroxypropazine, according to the following reactions:

$$P—Cl \xrightarrow{H_2O/H^+} P—OH + HCl \qquad (12)$$
$$P—O—CH_3 \xrightarrow{} P—OH + CH_3OH \qquad (13)$$

where P is the 4,6-bis (isopropylamino-s-triazine) moeity.

Except for prometryne, the phases adsorbed from chloroform initially failed to show a 1,746 cm^{-1} band, but weak absorption at this frequency appeared on exposing the sample to water vapour.

The s-triazines propazine, prometone, and hydroxypropazine adsorbed by montmorillonite containing other interlayer cations were not hydrolysed at very low moisture levels, but were so upon exposure of the samples to water

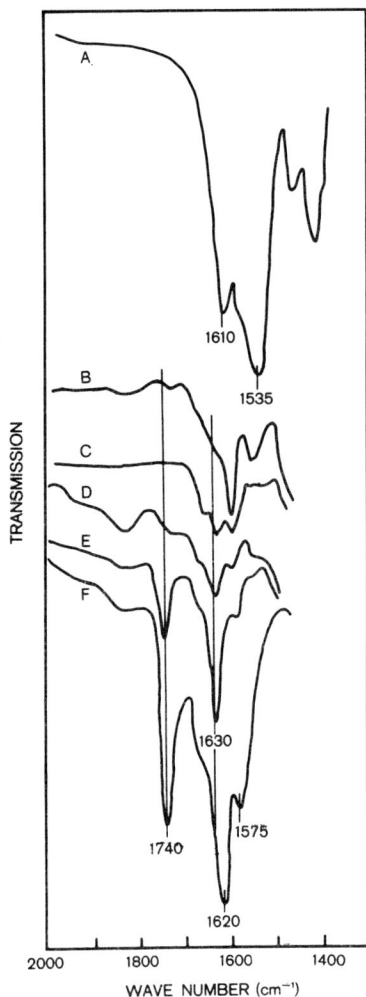

Fig. 49 Infra-red spectra of propazine and propazine-montmorillonite systems. (A) propazine in AgCl disk; (B) ammonium-montmorillonite film treated with a chloroform solution of propazine for 3 days; (C) same as (B) for 19 days; (D) same as (C) for 55 days after exposure to water vapour; (E) ammonium-montmorillonite film treated with an ethanol solution of propazine for 3 days; (F) hydrogen-montmorillonite film treated with an ethanol solution of propazine. After Cruz *et al*[12]

vapour. The changes in the position and intensities of the bands between 1,500 and 1,600 cm^{-1} prior to the appearance of the 1,740 cm^{-1} band after hydrolysis suggest that protonation may precede hydrolysis. It ought to be mentioned in this context that neither hydroxypropazine nor its protonated derivative is bio-active. Hydrolysis and protonation reactions may therefore play an important part in the degradation (deactivation) of *s*-triazine pesticides in natural soil systems.

Atrazine is likewise converted to its protonated hydroxy derivative which is then adsorbed by the clay.[16] The extent of hydrolysis of the chloro-*s*-triazines follows the order of trietazine > chlorazine > propazine > ipazine > simazine = atrazine. Compared with the chloro-substituted compounds, the corresponding methoxy and methylthio derivatives appear to undergo less hydrolysis

144

under similar conditions.[17] This is consistent with the fact that the 2-sub-stituted chloro-compounds are less basic, that is, more acidic than either the 2-methoxy- or the 2-methylthio-*s*-triazines.

The reactions of 3-amino-1,2,4-triazole (aminotriazole or amitrole) with montmorillonite have been studied by Russell *et al.*[18] By examining the infra-red spectra of dry montmorillonite saturated with different cations and con-taining varying levels of amitrole, they were able to deduce that much of the organic pesticide retained by the clay is in the protonated (cationic) form. The extent of protonation depends on the nature of the exchangeable cation. For example, the reaction is maximal for the aluminium clay and decreases in roughly the same order as the polarizing power of the cation (Table 44).

The strong band at 1,696 cm^{-1} is attributed to the C=N stretching vibration of the exocyclic

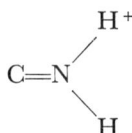

$$C=N\diagup\diagdown\begin{matrix}H^+\\H\end{matrix}$$

group. Support comes from the observed shift of this band to 1,683 and 1,666 cm^{-1} upon dehydration and deuteration, respectively. The infra-red spectra of the complexes with nickel- and copper-montmorillonite indicate that besides becoming protonated, the amitrole molecule may coordinate to the exchange-able cation.

The importance of protonation reactions in amitrole adsorption by mont-morillonite extends to aqueous systems.[19] Indeed, at pH > 5 of the suspension

TABLE 44

Extent of protonation of 3-aminotriazole by montmorillonite saturated with different cations.[18]

Amount of 3-aminotriazole added	Amount of 3-aminotriazolium ion formed (% exchange capacity)*						
	Interlayer cation						
	Na$^+$	NH$_4$$^+$	Ca^{2+}	Mg^{2+}	Al^{3+}	Cu^{2+}	Ni^{2+}
25	6	17			23	3	22
50	6	28	18	29	47	12	46
100	6	48	23	45	100	32	42
200	6	49	21	42			

* Based on the optical density of the 1696 cm^{-1} band of 3-aminotriazolium ion using montmoril-lonite saturated with the organic cation as standard. Film weights were standardized either by weighing or by measuring the optical density at 800 cm^{-1} due to the silicate structure.

or the equilibrium solution when amitrole is mostly present as the uncharged species, there is very little adsorption by calcium-montmorillonite. Under acid conditions and with (H, Al)-montmorillonite samples, however, appreciable uptake is noted. The process is fully reversible and no molecular (physical adsorption) takes place.

The adsorption of substituted ureas (3-phenylurea, fenuron, monuron, and diuron) by sodium- and hydrogen-montmorillonite in aqueous suspensions (pH 6·80 and 3·35, respectively) has been reported by Bailey et al.[15] As might be expected, the hydrogen clay system takes up greater amounts than sodium-montmorillonite. As with the s-triazines, the 'affinity' of the urea derivatives for the clay as measured by the magnitude of the Freundlich K-value (Eq. (4)) is related to their respective water solubility, although others working with soil systems have found no such relationship[20] or an inverse one[21] between these two parameters. Extrapolation to soils, under field conditions, of measurements on 'pure' clays must, of course, be made with caution since soil systems contain a variety of actively sorbing constituents besides clays. Evidence is accumulating that soil organic matter, for example, is capable of adsorbing appreciable amounts of applied non-ionic pesticides.[10,22-25]

The profound influence of the pH of the system on adsorption is also evident in the interactions of aniline, the phenylcarbamates, the anilides, and the amides with montmorillonite. Thus, IPC, dicryl, and solan failed to adsorb on sodium-montmorillorite (pH $\sim$ 6·80) but did so on the hydrogen clay (pH $\sim$ 3·35). As might be expected for such structurally varied compounds the extent of adsorption, as measured by the Freundlich K-value, is not related to water solubility.

The data of Bailey et al.[15] indicates that, within a given chemical family or an analogous series of compounds, adsorption is a function of and controlled by the solubility of the pesticide in water, whereas between families the basicity (pK_a) of the respective compound is the principal factor determining uptake. Although ion-dipole interactions and protonation reactions are undoubtedly involved in the systems investigated by Bailey et al., coordination of the pesticides to the exchangeable cations would not be of great importance. On the other hand, when the exchange positions are occupied by cations of the polyvalent and transition metal type, organic-to-cation coordination either directly or indirectly through a water bridge is perhaps the most important mechanism in the adsorption of non-ionic pesticides both in clay mineral and soil systems.[26-31]

The interaction between ethyl N,N-di-n-propylthiolcarbamate (EPTC) and montmorillonite saturated with different cations has been reported by Mortland and Meggitt[30]. The infra-red spectra of the complexes so formed showed that the C=O and C—N stretching frequencies of the adsorbed pesticide decreased and increased, respectively, as compared with the free liquid. The extent of these shifts was determined by the nature of the exchangeable metal ion on the clay (Table 45). These observations were explained by coordination of EPTC to the cation through the oxygen of the carbonyl group rather than through the

146

TABLE 45

C=O and C—N stretching frequencies (cm^{-1}) of EPTC adsorbed by
montmorillonite saturated with different cations.[30]

	Saturating Cation	C=O	C—N
Free liquid	—	1655	1222
Montmorillonite complex	Li^+	1593	1227
	Na^+	1590	1222
	Ca^{2+}	1594	1232
	Mg^{2+}	1587	1232
	Al^{3+}	1570	1234
	Cu^{2+}	1566	1232
	Co^{2+}	1583	1231

nitrogen, which was sterically hindered by the presence of the bulky propyl groups attached to it. We have already noted (Section 3.5) that EPTC may also be held to montmorillonite by hydrogen-bonding interactions when the exchange positions at the clay surface are occupied by organic cations such as trimethylammonium, tetramethylammonium[31] (Table 34), and pyridinium ions.[32]

Coordination of the carbonyl group to the exchangeable cation either directly or through a water bridge has also been proposed by Bowman et al.[33] as the mode of adsorption of malathion by montmorillonite. Their evidence comes from the observed shift in the frequency of the C=O stretching band from about 1,740 cm^{-1} (for liquid malathion) to lower levels for the adsorbed species. The amount of this frequency depression is again influenced by the kind of interlayer cation present (Table 46).

TABLE 46

C=O stretching frequency (cm^{-1}) of malathion adsorbed by montmorillonite as influenced by the nature of the exchangeable cation and the hydration status of the sample.[31]

Hydration status of complex	Saturating cation				
	Na^+	Ca^{2+}	Cu^{2+}	Fe^{3+}	Al^{3+}
Hydrated	1730	1714 1695 (sh)	1725 1715	1723 (sh) 1714 1695 (sh) 1684 (sh)	1740
Dehydrated	1722 1714	1695 1683 (sh)	1740 1734 1715 (sh) 1695 (sh) 1683 (sh)	1740 1664—1678	1740 1678 1635

sh = shoulder.

For a given cation the decrease in frequency is greater in the dehydrated complex than when some interlayer water is present. This indicates that adsorption takes place through the oxygen atom of the carbonyl group of malathion which coordinates to the interlayer metal ion directly (in the absence of water) or through a bridging water molecule. Water apparently needs to be present for intercalation to occur, and at relative humidities >40 per cent a double layer of organic molecules is present in the montmorillonite interlayers.

4.1.3. Anionic Pesticides

Among the compounds investigated by Bailey et al.[15] were picloram, phenylalkanoic, and benzoic acids and their derivatives, all of which possess a carbonyl group (Table 43). None of these acidic pesticides were adsorbed by sodium-montmorillorite, while the hydrogen clay showed both positive and negative adsorption.

Frissel and Bolt[5], using sodium-montmorillonite, have earlier reported that 2,4-D and 2,4,5-T were negatively adsorbed between pH 10 and 4 but that positive uptake occurred at pH < 4. This transition probably coincided with the formation of the uncharged species, the anionic form rapidly disappearing below pH 3–4. Some uptake occurred with illite at pH > 4, possibly due to adsorption on positive sites at crystal edges. Adsorption may also be enhanced by high electrolyte concentrations owing to the salting-out effect. Other acidic herbicides, such as MCPA (2-methyl-4 chlorophenoxyacetic acid), DNC (3,5-dinitro-o-cresol), DNBP (4,6-dinitro-o-sec-butylphenol), picric and picolinic acids behave in a similar manner.[4]

Negative adsorption of 2,4-D by sodium-saturated montmorillonite and kaolinite at pH 6 was also observed by Weber et al.[7] while Schwartz[34], using radio-carbon-labelled 2,4-D with montmorillonite, illite, and kaolinite at pH 4,8–9·3, reported very little uptake by any of the clay sample examined.

Haque and Sexton[35] have carried out kinetic and equilibrium measurements of the adsorption (at pH 6) of 2,4-D by montmorillonite, illite, sand, silica gel, alumina, and humic acid. The equilibrium data can be described by the Freundlich equation, the heats of adsorption (which generally were small) changing exponentially with the amount adsorbed. Montmorillonite showed negative adsorption but there was some (positive) uptake by illite. The rate data indicate that physical adsorption is the dominant mechanism, possibly supplemented by weak hydrogen-bonding interactions in some systems.

4.1.4. Cationic Pesticides

Of the compounds in this class, the bipyridilium halides diquat and paraquat are perhaps the best known and most widely studied members (Table 43). In both compounds the two pyridilium rings are coplanar, and the ability to adopt a planar configuration in these and related structures appears to be necessary for herbicidal activity.

In an early reference to the interaction of diquat and paraquat with clay minerals, Weber et al.[7] reported that these compounds, added as their aqueous solutions, were taken up by montmorillonite and kaolinite in amounts approaching the exchange capacity of the respective layer silicate. They inferred that diquat and paraquat were adsorbed by an ion-exchange process replacing an equivalent amount of the inorganic cations (sodium) initially present at the clay surface. Although cation exchange is undoubtedly the principal mechanism underlying the uptake of these bipyridilium ions by clays (and soils), recent ultra-violet and infra-red spectroscopic measurements by Haque et al.[36,37] have indicated that charge transfer between the organic cation and the anionic silicate framework may also be involved.

On both montmorillonite and kaolinite, the adsorption isotherms for diquat and paraquat, besides being essentially coincident, were insensitive to variations in temperature (283 and 328°K)[7]. The isotherms belong to the H or high-affinity class,[38] characterized by a region in which all the added herbicide is adsorbed, followed by a short shoulder in which the extent of adsorption is related to the herbicide concentration in solution, finally reaching a plateau approaching, in value, the exchange capacity of the adsorbent and for which the amount taken up is little influenced by the solution concentration.

These observations indicate that diquat and paraquat are adsorbed strongly and with similar affinity (with paraquat perhaps slightly preferred). Further, these herbicide cations are evidently intercalated by montmorillonite. X-ray diffraction examination[7] of their complexes with montmorillonite suggests that the plane of the rings lies parallel to the silicate layers, thus affording close van der Waals contact between molecule and clay surface. No interlayer penetration occurs with kaolinite, adsorption taking place only on external basal and edge surfaces of the crystals. This difference in the location of adsorption sites partly accounts for the observation that montmorillonite retains adsorbed diquat and paraquat much more tenaciously than kaolinite. Repeated extraction using 1M barium chloride solutions, for example, removed about 5 per cent and 80 per cent of the herbicide cations held by montmorillonite and kaolinite, respectively.[11] Another important factor influencing retention is geometrical 'fit' and this, as we shall see later, depends on the surface charge density of the clay and the size/shape of the organic molecule.

The general features of the uptake and retention of diquat and paraquat by clays described above have been confirmed by the work of Tomlinson and co-workers[39,40]. In an attempt to relate the rapid and virtually complete loss of activity of these herbicides on contact with soils, these workers have introduced the term Strong Adsorption Capacity (SAC). This is the capacity of the adsorbent to reduce the amount of herbicide in solution to a level which is chemically undetectable (below point B in Fig. 50). Its practical usefulness is shown by the finding that the SAC corresponds to the capacity for herbicidal inactivation to within at least 50 per cent.[39]

Since the rate of field application is normally well below the SAC limit, an understanding of the behaviour of paraquat (as an example and model of the bipyridiliums) in this region of the isotherm is of prime importance. To this end, Tomlinson et al.[40] have used radio-carbon-labelled paraquat and demon-

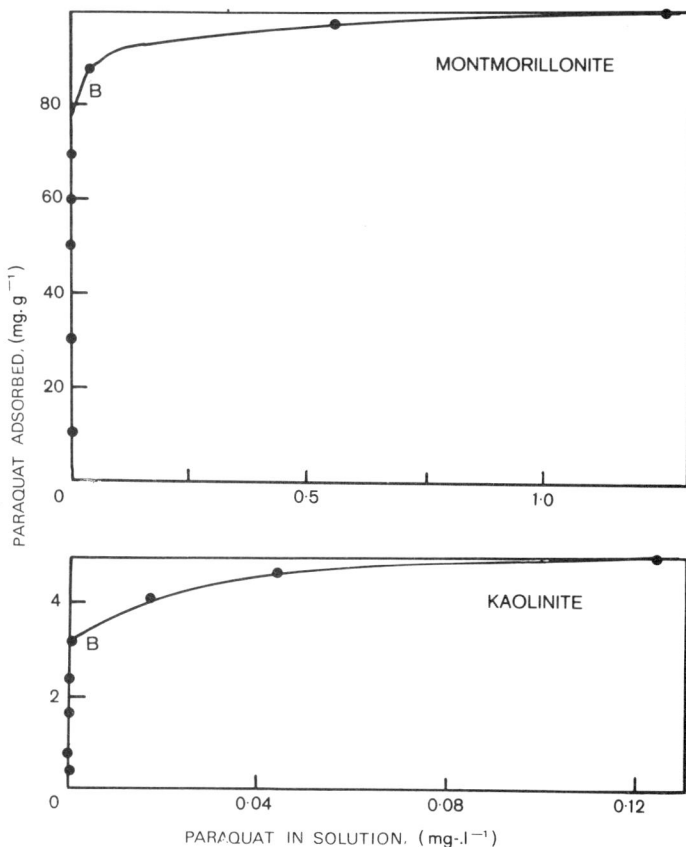

Fig. 50 Isotherms for the adsorption of paraquat (1,1'-dimethyl-4,4'-dipyridilium dichloride) on montmorillonite (top) and kaolinite (bottom). The amount adsorbed below point B in the isotherms represents the *Strong Adsorption Capacity*, that is, the capacity of the clay to reduce the solution concentration to a chemically undetectable level. After Tomlinson *et al*[40]

strated that all the adsorbed paraquat is in equilibrium with the cation in solution. This indicates that even the more strongly held species is readily accessible and isotopically exchangeable. These workers then examined the replaceability of the adsorbed labelled paraquat (in kaolinite) for other bipyridilium cations. Displacement occurred most readily with 4,4-diquaternary salts in which the rings are coplanar, and was further enhanced by the presence of large alkyl substituents.

Tomlinson *et al*.[40] also measured the relative ability of montmorillonite, illite, and kaolinite to retain adsorbed paraquat against treatment with a solution of ammonium nitrate. This salt (rather than barium chloride) was chosen because of its high water solubility. In a closed system, that is, without leaching, little desorption of paraquat was observed even at high concentrations (up to 6 M) of ammonium nitrate. For a given ratio of ammonium/paraquat

150

ion in solution, the order of displacement increases in the order montmoril-lonite < illite < kaolinite. As desorption progresses, this ratio increases and the solubility of ammonium nitrate is exceeded before paraquat is completely displaced from the exchange complex. We may conclude that very little, if any, desorption of adsorbed bipyridilium cations should occur under field condi-tions, and hence the phytotoxicity of soil-applied diquat and paraquat would only be of very short duration.

The phytotoxic effect and availability to plants and microorganisms of diquat and paraquat adsorbed on different clay mineral types have been ex-amined by Weber and co-workers[41-44]. As might be expected from their desorption behaviour, these organic cations when complexed with montmoril-lonite cannot significantly inhibit the growth of young plants such as cucumber seedlings[41,42] or be decomposed by microorganisms.[43] But diquat and paraquat adsorbed on kaolinite and vermiculite minerals can be slowly released to the surrounding medium and so affect plant growth. This is because these herbicide cations are relatively weakly held by kaolinite and adsorption is restricted to external surfaces.[11,40] Although interlayer penetration apparently occurs with vermiculite, the extent of intercalation is limited since complete exchange for the inorganic ions initially present is not attained. Diquat present at mont-morillonite surfaces may, however, be made available to plants on addition to the system of (large) organic cations capable of displacing the adsorbed herbi-cide.[40] Thus, when the non-phytotoxic N-(4-pyridyl) pyridinium cations are incorporated into the montmorillonite-diquat/cucumber seedling system, the phytotoxic effect of diquat is observed.[44]

The effect of adsorbent charge and the kind of exchangeable cation present on the adsorption and retention of diquat and paraquat by 2:1 type layer sili-cates has been investigated by Weed and Weber[45]. Samples of montmorillonite, vermiculite, and K^+-depleted mica were used covering a range of exchange capacities from 1·03 to 1·62 meq g^{-1} and containing different charge-balancing cations. As expected, maximum uptake by montmorillonite was close to the exchange capacity of the sample, being little influenced by the nature of the saturating organic cations. Under similar conditions, vermiculite adsorbed appreciably less than its exchange capacity while only 13 per cent of the ex-change capacity of the muscovite sample was satisfied by the organic cations. Furthermore, with vermiculite the extent of adsorption depended on the saturat-ing inorganic cation decreasing in the order $Na^+ > Ca^{2+} \geqslant Mg^{2+}$.

The desorption characteristics of the various samples also showed marked differences. Thus, a single equilibrium with 0·005 N chloride solutions of Al^{3+}, Ca^{2+}, Mg^{2+}, and K^+ greatly in excess of the amount present, displaced less than 15 per cent of adsorbed diquat or paraquat from its respective complex with montmorillonite. On the other hand, up to 70 per cent was desorbed from the corresponding vermiculite complexes. The actual proportion displaced depends on the kind of electrolyte used, decreasing in the order $Al^{3+} > Ca^{2+} > Mg^{2+} > K^+$ (for diquat) and $Ca^{2+} > Al^{3+} > Mg^{2+} > K^+$ (for paraquat). Differences in charge density and hence in intercharge separation on interlayer surfaces, together with variations in stability of cation-water assemblages in the interlayer space, were invoked by Weed and Weber to account for these observations. More recently, Dixon et al.[46] have reported similar differences between montmorillonite and vermiculite complexes with diquat in their response to treatment with 1 M potassium chloride solutions.

The influence of surface charge density of the adsorbent on the uptake of diquat and paraquat when these organic cations are both present has been studied by Weed and co-workers[47-49]. They determined the exchange isotherms for the herbicide cations on a series of 2:1 layer silicates with varying surface density of charge, σ, in the reasonable expectation that the competitive adsorption of diquat and paraquat is related to σ, which is a measure of the distance of separation between negatively charged sites at the clay surface. This approach, of course, assumes that the positive charges on the diquat and paraquat molecules are localized, and that the negative charge on the silicate framework arising from isomorphous substitution exists as discrete entities rather than smeared out over the surface.

The dimensions of diquat and paraquat estimated from their respective molecular models are given in Fig. 51, showing that the intercharge distance in paraquat is more than twice that of diquat. There is also a difference in the

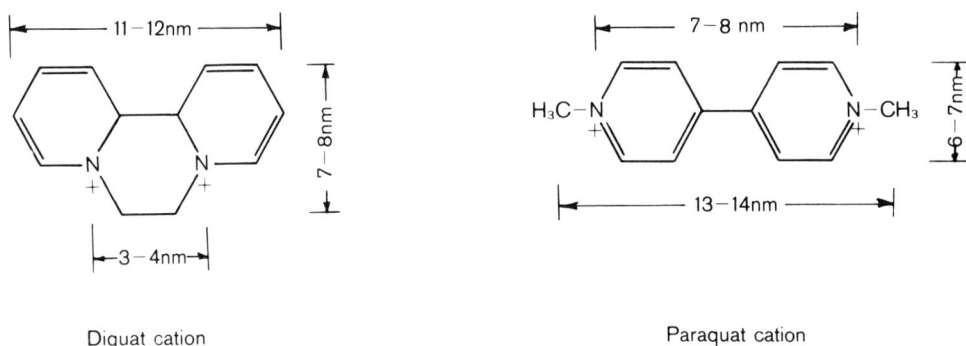

Diquat cation

Paraquat cation

Fig. 51 Molecular dimensions given in nm units of diquat (1,1'-ethylene-2,2'-dipyridilium) and paraquat (1,1'dimethyl-4,4'-dipyridilium) cations, according to Weed and Weber[47]

overall geometry and shape of these organic cations; paraquat is a more flexible cation and can establish closer van der Waals contact with the clay surface compared with diquat.

From the exchange isotherms, obtained by plotting the equivalent fraction of diquat in solution (c/c_0) against that of diquat adsorbed (q/q_0), Weed and Weber[47] showed that paraquat was preferred to diquat by all montmorillonites. This is to be expected on structural grounds, if the underlying assumptions are valid. Similarly, with vermiculites of higher σ this preference was apparently restricted to the external crystal surface, a reversal being observed

152

for the interlayer surface where diquat was preferentially taken up. Non-expanding mica minerals also preferred diquat to paraquat. Evidently, the larger intercharge distance in paraquat (Fig. 51) favours its uptake in clays of relatively low σ (montmorillonites). As σ increases—going from montmorillonite through vermiculite to mica—the distance separating pairs of negative charges at the silicate surface decreases. Now diquat can more readily be fitted on the negatively charged clay surface by virtue of its smaller positive charge separation as compared with paraquat. The data also support the discreteness-of-charge concept proposed earlier by Edwards et al.[50]

Philen et al.[48,49] have extended the above analysis to a greater number of clay minerals and to soil clays ranging in σ from ~0·1 to ~0·34 C m^{-2}. Where adsorption is restricted to external crystal surfaces, such as in collapsed vermiculites and micas, there is a simple linear relationship between σ and the extent of (preferential) adsorption of the organic cations measured at some arbitrary point in the appropriate isotherms. The derivation of this point is illustrated in Fig. 52, which gives the exchange isotherms for the competitive adsorption of diquat and paraquat by some vermiculites (V_1, V_2, V_3) and micas (M_1, M_2) of

Fig. 52 Competitive exchange isotherms of diquat^{2+} versus paraquat^{2+} on external surfaces of two mica (M) and three vermiculite (V) samples of different surface charge density (σ). Solid diagonal represents line of no preference for either organic cation. Broken diagonal shows method for relating competitive adsorption to σ as explained in the text. After Philen et al[48]

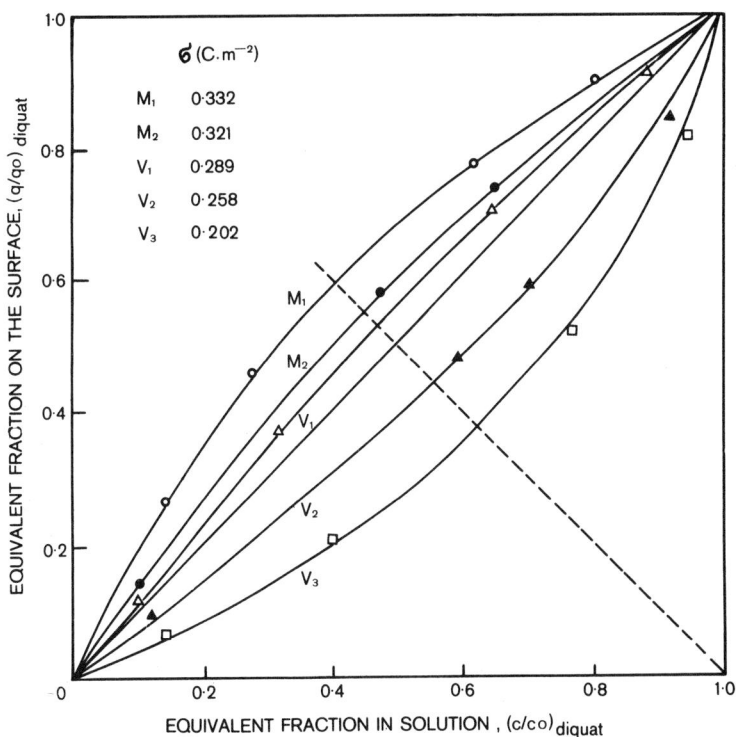

varying surface charge density. The values of q/q_0 at points where the curves intersect the diagonal drawn from $c/c_0 = 1$; $q/q_0 = 0$ to $c/c_0 = 0$; $q/q_0 = 1$, are taken as an index of the relative preference of the adsorbent for the organic cations. These intercept values (χ) plotted against the corresponding values of σ yield the relationships shown in Fig. 53, line 1.

If adsorption occurs on both external and interlayer surfaces as in montmorillonites, a different line is obtained (Fig. 53, line 2). Interlayer adsorption

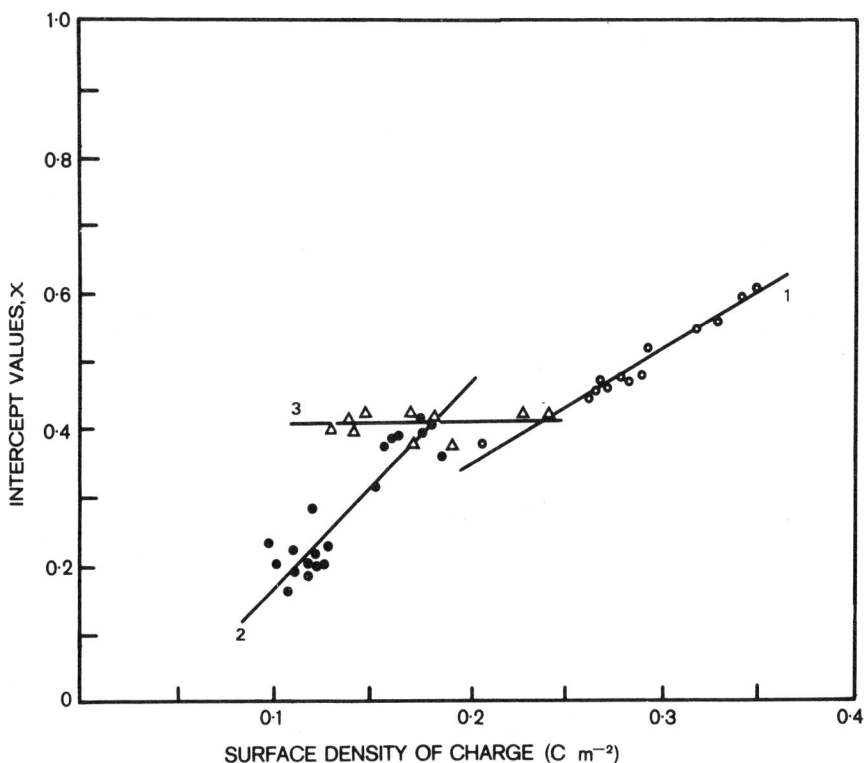

Fig. 53 Relationship between equivalent fraction χ of diquat^{2+} adsorbed and surface charge density for micas (line 1), montmorillonites or smectites (line 2), and kaolinites (line 3). Values of χ are obtained from the respective exchange isotherms (see Fig. 52) and represent the points where the broken diagonal intersects the isotherms. After Philen et al[49]

was characterized by a strong preference for paraquat in low-charge montmorillonites, a relative decreasing preference for paraquat with high-charge montmorillonites, and a marked preference for diquat on high-charge expanding vermiculites. Preferential adsorption for paraquat was also observed with kaolinite minerals, resembling the uptake in external vermiculite surfaces. However, no apparent relationship between competitive adsorption and σ is observable for the kaolinites, that is, the χ values remain sensibly invariant as σ increases (Fig. 53, line 3).

154

These analyses provide yet more evidence that low-charge montmorillonites can be readily distinguished from high-charge vermiculites, but that the distinction between the high-charge members of the smectite group and the low-charge vermiculite samples (Table 1) is ill-defined. However, if the different types of minerals present in a given clay system need not be known, the above procedure may be profitably employed to estimate the surface charge density of the clay fraction of soils.[48]

Clay minerals may detoxify adsorbed pesticides by catalysing their decomposition. In most instances, this degradation is brought about by hydrolysis of the organic compounds. The detoxification of propazine and prometone cited earlier (cf Eqs. (12) and (13)) is an example of a clay-catalysed decomposition reaction. Decomposition may also occur by molecular rearrangement. This property of clays to convert pesticidal agents to their corresponding non-toxic forms is of prime importance in the practical formulation of pesticides. Finely ground clays are frequently used as 'inert' diluents of toxic organic substances to reduce the pesticide concentration to a level suitable for agricultural use as well as to improve handling properties.

The catalytic properties of clays can be ascribed to the ability of clay minerals to either donate protons and so act as Brønsted acids, or accept electrons and behave as Lewis acids. One or the other type of acidity is normally predominent for a given set of conditions and sample preparation. Both Brønsted and Lewis acidities may be simultaneously involved in some catalytic reactions of clays, and it is often difficult to distinguish between them clearly.

As we pointed out, Brønsted acidity derives essentially from the dissociation of water molecules coordinated to the exchangeable cations giving rise to active protons. This type of acidity is therefore strongly influenced by the hydration status of the clay and by the kind of cation occupying exchange sites at the silicate surface.[51-53] Low water contents, as achieved by mild heat treatment ($\sim373°K$) or evacuation of the clay sample, and highly polarizing cations promote Brønsted acidity. Lewis acidity, on the other hand, is little influenced by the exchangeable cations present and has been identified with certain ions making up the silicate structure, such as aluminium ions exposed at the edges of the clay crystals. Under conditions when most or all of the adsorbed water is removed by preheating the clay to elevated temperatures ($\sim600°K$) and when the reaction is carried out in an anhydrous non-polar organic liquid, Lewis acidity will predominate. This type of acidity persists even after destruction or collapse of the clay structure as may be caused, for example, by heating the solid at $\sim1,200°K$.

The conversion of DDT to DDE catalysed by kaolinite and montmorillonite samples perheated at $\sim400°K$, observed very early on by Fleck and Haller[54], might be attributed to Brønsted-type acidity which was probably also responsible for the degradation of heptachlor by palygorskite (attapulgite)[55]. The effect of heating the clay to various temperatures on the extent of decomposition of the insecticide ronnel has been examined by Rosenfield and van Valkenburg[56]. Extensive degradation was noted in samples preheated at 573°K. As the clay was heated to increasingly higher temperatures, dehydroxylation occurred and the structure gradually collapsed, leading to a diminution of Lewis acidity and catalytic activity. On samples preheated at $1\,223°K$, ronnel was stable against hydrolysis although a small amount of decomposition occurred owing to molecular rearrangement catalysed by Lewis-acid-free Al^{3+}:

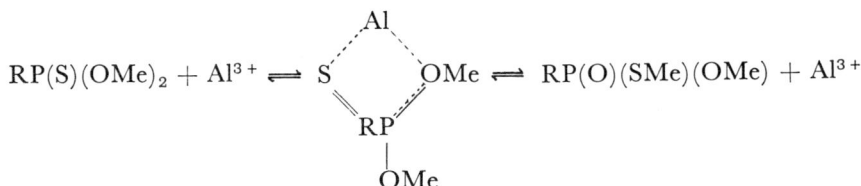

$$RP(S)(OMe)_2 + Al^{3+} \rightleftharpoons S \overset{Al}{\underset{RP}{\diamond}} OMe \rightleftharpoons RP(O)(SMe)(OMe) + Al^{3+}$$

$$\underset{OMe}{|}$$

where $R = 2,4,5\text{-}Cl_3C_6H_5O$.

The difficulty of separating Brønsted acidity from that of the Lewis type has led to attempts at deriving an index of overall acid strength which may be correlated with catalytic activity. This is done by titrating the solid adsorbent with an organic base such as n-butylamine in an inert hydrocarbon liquid (e.g. benzene) using Hammett indicators.[57-59] Some of the most common indicators are listed in Table 47. The range of acidities of some clay materials, measured in

TABLE 47

Some commonly used Hammett indicators and their properties.[58]

Indicator	Basic colour	Acid colour	pK_a
Phenylazonaphtylamine	Yellow	Red	+4·0
Butter yellow	Yellow	Red	+3·3
Benzeneazodiphenylamine	Yellow	Purple	+1·5
Dicinnamalacetone	Yellow	Red	−3·0
Benzalacetophenone	Colourless	Yellow	−5·6
Anthraquinone	Colourless	Yellow	−8·2

terms of Hammett and Deyrup's[60] H_0 scale, are summarized in Table 48. This scale may be considered an extension of the pH scale into media where the activity of the hydrogen ion ceases to be equivalent to its stoichiometric concentration. The limits of the H_0 are obtained by noting the colour of the adsorbed indicator. Thus, an adsorbent having H_0 of −5·6 to −8·2 gives a yellow colour with benzalacetophenone but produces no colour with anthra-quinone. On the other hand, an adsorbent sample having $H_0 < -8·2$ gives acid colours with all of the indicators in Table 47.

In examining the acid strength of representative clay mineral types by this method supplemented by infra-red measurements, Solomon et al.[61] have shown that Brønsted acidity predominates on clays as normally supplied (moisture

156

TABLE 48
Acid strength of some clay minerals as measured using Hammett indicators.

Mineral	H_o (pK$_a$ range of strongest sites)		
	I	II	III
Kaolinite	$-3\cdot0$ to $-5\cdot6$	$-5\cdot6$ to $-8\cdot2$	$< -8\cdot2$
Montmorillonite	$-3\cdot0$ to $-5\cdot6$	$-3\cdot0$ to $-5\cdot6$	$-5\cdot6$ to $-8\cdot2$
Palygorskite (attapulgite)	$+1\cdot5$ to -3	$-5\cdot6$ (tr)	$-8\cdot2$
Talc	$+3\cdot3$ (tr)		$+4\cdot0$ to $+3\cdot3$
Calgon-treated kaolin*			$+1\cdot5$ to $-3\cdot0$

I and II from Benesi[58] and Fowker et al.[62] Columns I and II refer to samples as received and after drying at 393°K for 16 hours, respectively.
tr = transition colour.
III from Solomon et al.[61] measured on dry samples (water content $\sim$0%).
* Calgon is the trade name for sodium hexametaphosphate which specifically adsorbs on crystal edge surfaces.

content 0·5 to 1 per cent). At still lower levels of hydration, the acid strength increases and concomitantly the materials behave more like Lewis rather than Brønsted acids.

Fowker et al.[62] have earlier shown that the decomposition rates of the organo-chlorines dieldrin and endrin rose with an increase in the acid strength of the clay samples used as given by H_0 (Table 48). Treatment of the clays with organic bases, such as urea and hexamethylene diamine, reduces the acidity of the silicates causing the reaction rates to diminish. Such additives or stabilizers apparently raise the activation energy, indicating that these compounds separate the pesticide from the active site at the clay surface. This accords with the recent data of Solomon et al.[61] who noted that the efficiency of amines and various other polar organic molecules in neutralizing the clay depends on the chain length, basicity, and chain branching of the additive.

The influence of the exchangeable cation on acid strength and, consequently, on the rate of decomposition of dieldrin by kaolinite is shown in Table 49. Similarly, the insecticide ronnel was found to be hydrolysed to 2,4,5-trichloro-phenol when incubated with montmorillomite at 323°K.[56] The extent of hydrolysis increased by saturating the exchange sites on the clay with Al^{3+} or Fe^{2+} ions, or when the mineral had been preheated to 573°K, reminiscent of the malathion-montmorillonite system (Table 46). Decomposition may further be enhanced when the organic molecule forms a coordination-type complex with the exchangeable cation. Some organophosphate insecticides adsorbed by copper-montmorillonite may be inactivated by a coordination-hydrolysis

TABLE 49

Decomposition rates of dieldrin by kaolinites containing
different exchangeable cations.[62]

Cation	Amount of Cation on clay $\mu M\ g^{-1}$	Total acidity* $\mu M\ g^{-1}$	H_0	$t_{\frac{1}{2}}$ at 338°K** (min)
Natural	—	25	−3 to −5·6	400
H⁺	—	24—31	−5·6 to −8	30
NH₄⁺	24	—	−3 to −5·6	150
Na⁺	26	19—22	−3 to −5·6	400
Ca²⁺	15	10—20	+1·5 to −3	330
Al³⁺	8	10—12	−5·6 to −8	30

* determined by titration with *n*-butylamine.
** $t_{\frac{1}{2}}$ is the half-time of the reaction.

mechanism.[63] The reaction for Dursban, for example, may follow the following
scheme:

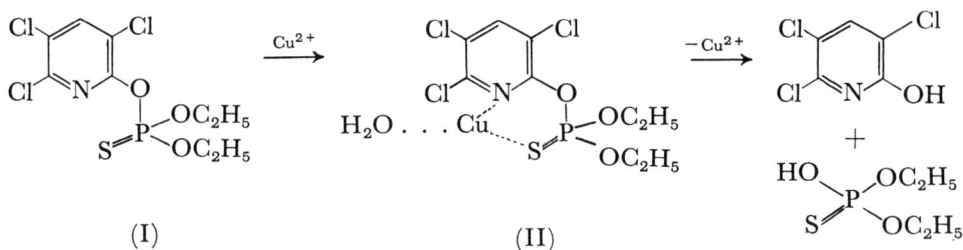

(I)   (II)

Electron shifts in the closed, cyclic resonance system of structure II weaken the
bonding of the side chain to pyridine (or pyrimidine), thereby promoting
hydrolysis.

 The ease with which copper-montmorillonite hydrolyses compounds of this
type decreases in the series Dursban > diazinon > ronnel ⩾ Zytron (DMPA).
Both the total and the source of the negative charge at the silicate surface
influence the reaction. Copper-saturated beidellite, nontronite, and vermiculite, for example, were less active than the corresponding montmorillonite
sample, while a sample of an organic soil similarly treated showed no hydrolytic
activity. The reason for this difference in reactivity between the materials
examined lies in the packing and space requirement of the adsorbed insecticides.

4.2. Complexes with Amino Acids and Peptides

Because amino acids (and peptides) play a vital role in biochemical processes
and organic matter transformation in soil, great interest attaches to the interactions of these compounds with clays. Acid (carboxyl) and basic (amino)

functional groups in the same structure allow amino acids to exist in the form of their corresponding cations, zwitterions (or dipolar ions), or anions. Which species is predominatly present in a given system is clearly determined by the prevailing pH conditions. For the monoamino monocarboxylic acids, the relationship between the different forms may be described by the following equilibria:

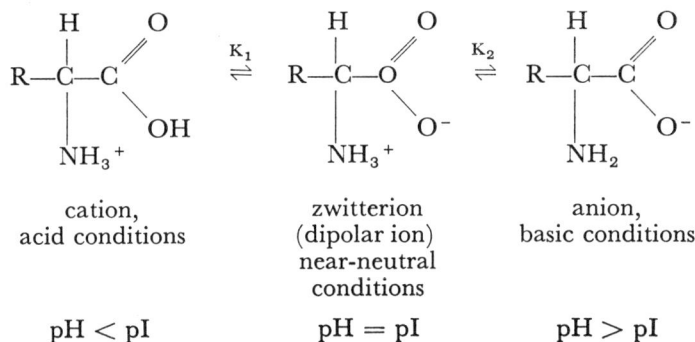

$$
\begin{array}{ccc}
\underset{\underset{\displaystyle NH_3{}^+}{|}}{\overset{\overset{\displaystyle H}{|}}{R-C-C}}\diagup\!\!\!\!\diagdown\!\!\overset{O}{_{OH}} & \underset{K_1}{\rightleftharpoons} & \underset{\underset{\displaystyle NH_3{}^+}{|}}{\overset{\overset{\displaystyle H}{|}}{R-C-O}}\diagup\!\!\!\!\diagdown\!\!\overset{O}{_{O^-}} & \underset{K_2}{\rightleftharpoons} & \underset{\underset{\displaystyle NH_2}{|}}{\overset{\overset{\displaystyle H}{|}}{R-C-C}}\diagup\!\!\!\!\diagdown\!\!\overset{O}{_{O^-}}
\end{array}
$$

cation, acid conditions	zwitterion (dipolar ion) near-neutral conditions	anion, basic conditions
$pH < pI$	$pH = pI$	$pH > pI$

R is hydrogen or alkyl; pI is the 'isoelectric' pH which for most amino acids of interest lies in the range of 5 to 6. For the monoamino dicarboxylic acids (e.g. aspartic acid) the first two dissociation constants (pK_1 and pK_2) refer to the first and second carboxyl group, respectively, while pK_3 denotes the constant for the $NH_3{}^+$ group. Similarly, for diamino monocarboxylic acids (e.g. arginine), pK_1 is the constant for the carboxyl group while pK_2 and pK_3 are those for the first and second amino groups, respectively.

The adsorption of amino acids by clay and soil systems is extremely sensitive to variations in the pH of the medium. By analogy with the anionic pesticides, the anionic form of amino acids is relatively unreactive towards the negatively charged clay surfaces. Our prime attention is therefore directed to the reactions of the cationic and zwitterionic forms. The latter are expected to be dominant in soils under natural field conditions since the aqueous suspension pH of many soil systems does not greatly deviate from neutrality. However, as we have frequently stressed, the pH in the vicinity of clay surfaces may be depressed by several units from that in the surrounding (bulk) solution, particularly when the water content of the system falls below 5 per cent. Cation exchange and proton transfer are therefore expected to control the adsorption process at the clay/solution interface and in the interlayer space of expanding-layer silicates. In addition, stable coordination complexes may form at the clay surface if strongly chelating cations such as copper, cobalt, and zinc are present. Besides basicity, the size and shape of the organic molecule would influence uptake since these factors determine both the extent of physical (or van der Waals) adsorption and the ease with which the amino acid may gain entry into the clay interlayers.

The systematic investigation into the influence of basicity and molecular size on the adsorption of some amino acids and proteins by montmorillonite under acid conditions was initiated by Talibudeen[64]. On the basis of the amount of sodium ions replaced by the organic cations he concluded that, of these factors, basicity was the more important. Thus, for a given ratio of amino acid (protein) cations in solution to sodium ions on the clay, arginine ($pK_3 = 12\cdot48$ guanidium) could displace more sodium than glycine ($pK_2 = 9\cdot78$). However, the protein salmine with 40 arginine residues per molecule was less efficient in displacing the exchangeable sodium ions compared with arginine itself. Talibudeen ascribed this to the fact that the large salmine cation (mol. wt. 7,870) experienced greater difficulty, for steric reasons, in exchanging for the inorganic cations initially present. However, as Hendricks[65] had earlier pointed out, the amount of inorganic ions desorbed (or hydrogen ions neutralized) by bulky organic cations (or bases) did not reflect the extent to which such organic species could be adsorbed. Indeed, many high-molecular-weight organic compounds can be taken up in excess of the exchange capacity of the clay, although the amount of inorganic cations displaced may be less than that observed during the uptake of small organic bases. To account for this behaviour Hendricks proposed that a bulky organic molecule can cover more than the area per exchange site. It seems probable therefore that, in addition to steric factors, this 'cover-up' effect also operates in the salmine-montmorillonite system.

Talibudeen's conclusions were confirmed by Sieskind[66] who determined the isotherms for the adsorption by montmorillonite at pH2 of a number of amino acids. Two types of amino acid can be distinguished on the basis of their respective isotherms: those which rapidly reach a plateau adsorption approaching the exchange capacity of the clay ($\sim$1 meq g^{-1}); and those which do not level off to a discernible plateau, the amount adsorbed at the limit of the concentration used ($\sim$6 $\times$ 10^2 eq g^{-1} added) being appreciably less than the exchange capacity of the sample. Compounds of type I are characterized by having their functional groups at the extremities of the chain, the carboxyl and amino groups being separated by at least two carbon atoms. They include such amino acids as β-amino butyric acid, γ-amino butyric acid, and ε-amino caproic acid, all of which have a pK_2 value greater than 10·5. The compounds of type II have a pK_2 near 9·8 and are represented by the α-amino acids, such as glycine, α-alanine, α-amino butyric acid, and norvaline. The importance of basicity is illustrated by the observation that at the limit of solubility, no more than 0·47 meq g^{-1} of α-alanine was taken up at pH 2. Contrast this with the earlier finding of McLaren et al.[67] who reported extensive uptake (up to $\sim$3 meq g^{-1}) of alanine under similar conditions. The discrepancy probably arises because Sieskind's figure was derived from the amount retained by montmorillonite against washing with water which would have removed the physically adsorbed fraction, whereas the value reported by McLaren et al. represented the difference between the quantity added and that found at equilibrium with the mineral. That these two methods of assessing uptake can yield markedly different results has, more recently, been demonstrated for glycine, diglycylglycine, and β-alanine by Cloos et al.[68]

Although basicity is clearly important, the influence of molecular weight on adsorption is by no means inconsequential. But uptake is evidently not simply related to cation size. This is illustrated by the data of Sieskind and Wey[69] who measured uptake by montmorillonite at a fixed concentration corresponding to the initial part of the isotherms as a function of pH. They obtained a family of curves (Fig. 54) which, as expected, showed that adsorption decreases with a rise in suspension pH. Chassin[70] has reported similarly for glycine. Maximum uptake occurred at about pH 2, and above pH $\sim$ 3 there was a steep decline;

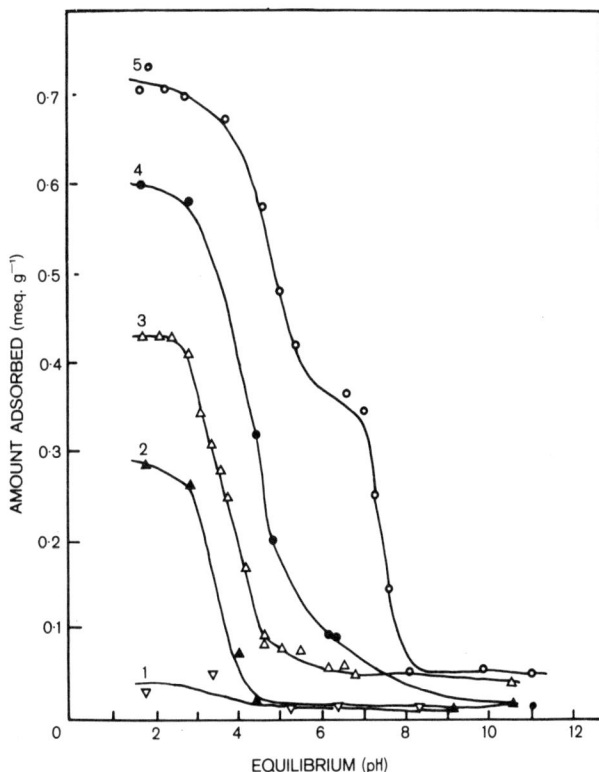

Fig. 54 Variation of the amount adsorbed (actually retained against washing with water) by montmorillonite, with suspension pH for some amino acids. (1) α-amino acetic acid (glycine): (2) β-amino propionic acid (β-alanine); (3) γ-amino butyric acid; (4) δ-amino valeric acid; (5) ε-amino caproic acid. After Sieskind and Wey[69]

the point at which the slope changed abruptly was close to pK_1. A similar but less marked change might be expected in the alkaline range near pK_2 such as McLaren et al.[67] had observed for α-alanine. Figure 55 shows that the amount retained (against washing) at pH 2 is linearly related to the number of carbon atoms, n, separating the terminal functional groups, rather than to molecular weight per se. For example, the points for glycine (n = 1; mol. wt. 75), α-amino butyric acid (n = 1; mol. wt. 103), and α-amino caproic acid (n = 1; mol. wt. 131) are coincident, as are those for β-alanine (n = 2; mol. wt. 89) and β-amino butyric acid. This observation suggests that proximity of the COOH and NH_3^+ groups in the amino acid cation is more important than the length

161

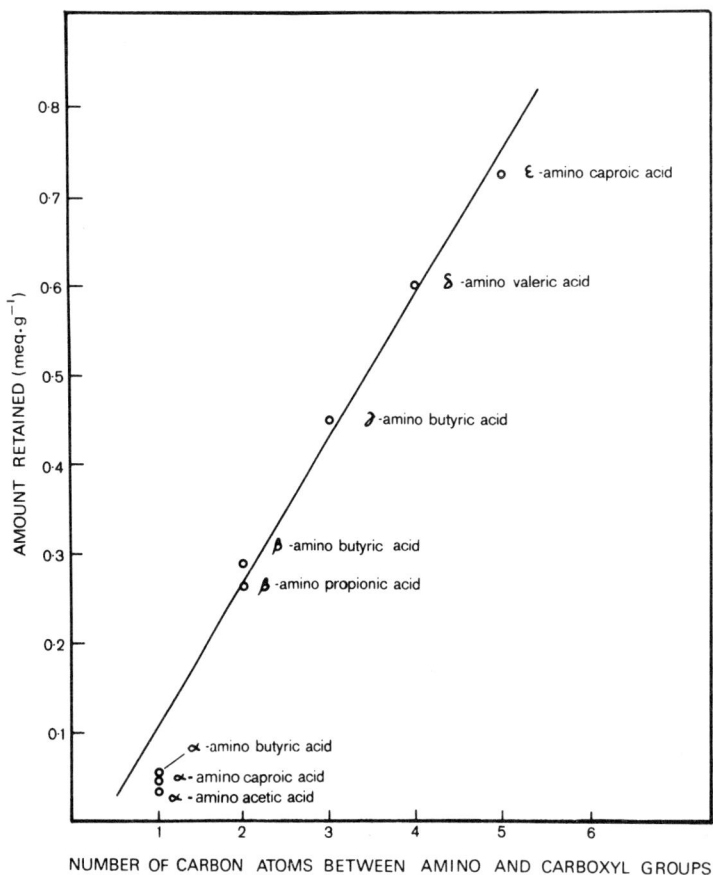

Fig. 55 Amount retained (against water washing) by montmorillonite at pH 2 as a function of the number of carbon atoms separating the amino and carboxyl group for some amino acids, according to Sieskind and Wey[69]

of the alkyl substituent in the α position. Greenland *et al.*[71-73] have determined the isotherms for the adsorption of glycine and its peptides together with a range of naturally occurring amino acids by montmorillonite and illite saturated with calcium, hydrogen, and sodium ions.

The isotherms on the *calcium* clays are of the C class,[38] the amount adsorbed increasing linearly with solution concentration (Figs. 56 and 57). This indicates that the number of energetically equivalent sites at the clay surface does not decrease as adsorption progresses.[38] The creation of fresh sites is initiated by the ability of the solute (rather than the solvent) to act as a swelling agent for the clay. This suggestion is further supported by the observed increase in basal spacing from 1·9 nm for the moist parent calcium-montmorillonite to 2·2–2·4 nm for the complexes with glycine and its peptides. Interlayer expansion was probably brought about by an increase in the dielectric constant of the interlayer solution as the amino acid molecules entered the clay interlayers (Table 50). The increase in interlayer volume can be shown to be more than

162

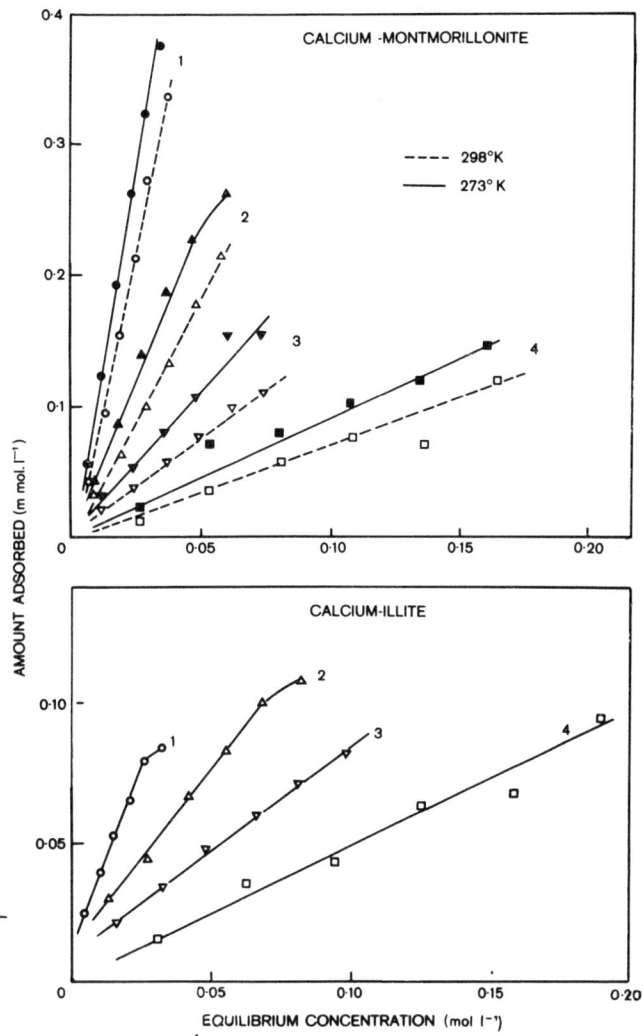

Fig. 56 Adsorption isotherms for glycine (4), glycyl glycine (3), diglycyl glycine (2), and triglycyl glycine (1) on calcium montmorillonite (top) and on calcium-illite (bottom). Open symbols: at 273°K; filled symbols: at 298°K. Data from Greenland *et al*[73]

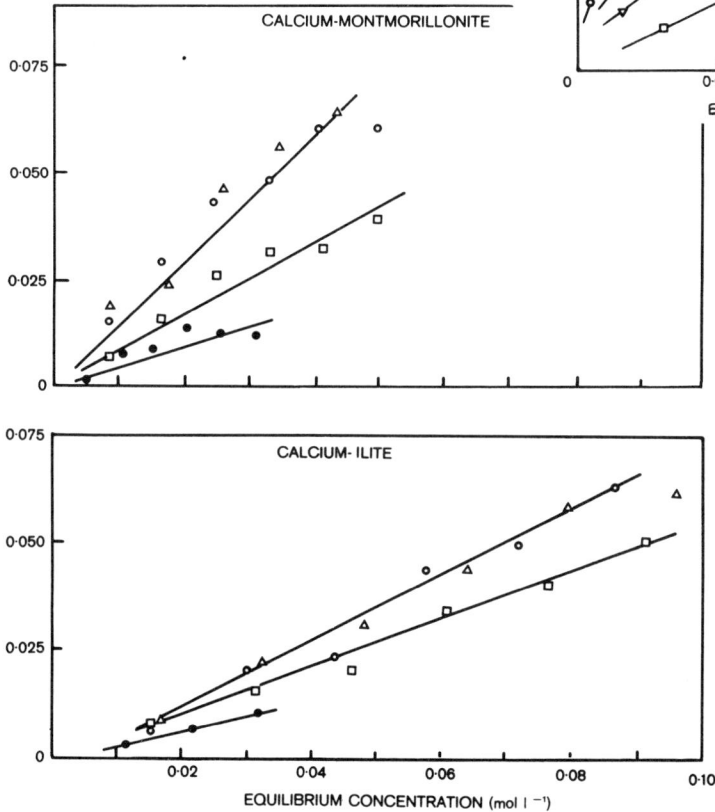

Fig. 57 Adsorption isotherms for α-alanine (○), β-alanine (Δ), leucine (●), and serine (□) on calcium-montmorillinite (top) and on calcium-illite (bottom). All measurements were carried out at 298°K. Data from Greenland *et al*[73]

TABLE 50

Selectivity coefficients (K_m) and non-standard free energy changes ($-\Delta G^m$) for physical adsorption of some amino acids and peptides by calcium-montmorillonite and calcium-illite, together with dielectric increments (δ) of the respective compounds.[73]

Compound	δ	Calcium-montmorillonite				Calcium-illite	
		K_m		$-\Delta G^m$ (J mol^{-1}10^3)		K_m	$-\Delta G^m$ (J mol^{-1}10^3)
		273°K	298°K	273°K	298°K	298°K	298°K
Glycine	23	2·28	1·77	1870·2	1418·4	6·6	4686·1
Glycyl glycine	71	5·42	3·61	3836·7	3179·8	10·2	5773·9
Diglycyl glycine	113	14·5	10·5	6067	5815·8	17·2	7029·1
Triglycyl glycine	159	34·3	27·9	8033·3	8242·5	34·3	8744·6
α-alanine	23	—	4·1	—	3502·0	10·1	5690·2
β-alanine	35	—	4·1	—	3502·0	10·1	5690·2
Leucine	25	—	1·8	—	1430·9	4·9	3933·0
Serine	23	—	2·4	—	2129·7	7·3	4937·1

adequate to fit adsorbed glycine, glycyl glycine, and diglycyl glycine so that no net loss of interlayer water need occur. For tetra-glycine, however, a net transport of water away from the adsorbed phase must take place beyond 50 per cent of the solute adsorbed shown in Fig. 56.

That the amino acids and peptides were taken up principally as their respective zwitterions ($RH^\pm$) was indicated by the observation that very few, if any, calcium ions were displaced from the clay on adsorption of these compounds. The uptake mechanism is therefore one of physical adsorption which, by reason of the linear nature of the isotherms, can be expressed in terms of a constant partition of solute between the bulk water and the water adhering to the clay surface (Stern-layer water):

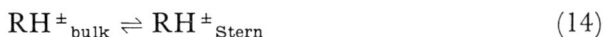

$$RH^\pm{}_{bulk} \rightleftharpoons RH^\pm{}_{Stern} \tag{14}$$

for which the equilibrium constant, K_a, is given by

$$K_a = C_{RH^\pm Stern}/C_{RH^\pm bulk} \cdot \gamma^\pm{}_{Stern}/\gamma^\pm{}_{bulk} \tag{15}$$

$$= K_m \cdot K_\gamma \tag{16}$$

C and γ refer to the concentration and activity coefficient of the appropriate species, respectively; K_m is the selectivity (partition) coefficient and is derived by multiplying the assumed Stern-layer thickness of 0.5 nm by the specific surface of the clay sample (750×10^3 and 150×10^3 m^2 kg^{-1} for montmorillonite and illite, respectively). A similar treatment of the data has recently been used by Theng[74] for the adsorption of n-alkylammonium chlorides by soil allophane.

The non-standard free energy of the partition reaction, $-\Delta G^m$, was estimated from

$$-\Delta G^m = RT \ln K_m \tag{17}$$

Values of K_m and $-\Delta G^m$ together with those of the dielectric increment, δ, for the various amino acids and peptides, are listed in Table 50.

For glycine and its peptides, $-\Delta G^m$ was found to be linearly related to both molecular weight (Fig. 58) and to δ. The reason is that δ (which is positive for the zwitterionic form of amino acids and peptides) is theoretically proportional

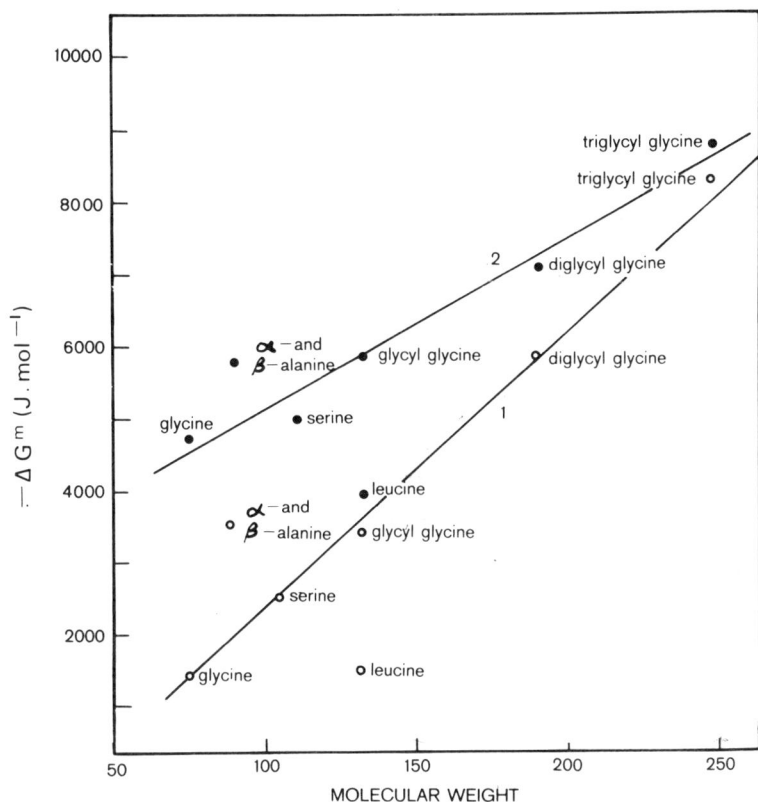

Fig. 58 Variation of $-\Delta G^m$ with molecular weight for the adsorption of some amino acids and peptides by (1) calcium-montmorillonite (O) and (2) calcium-illite (●). After Greenland et al[73]

to the negative logarithm of the activity coefficient, γ_n, of the non-electrolyte (in this instance, the zwitterion), defined as

$$-\ln \gamma_n = \ln (S/S^0) \qquad (18)$$

where S and S^0 are the solubility of the non-electrolyte in the salt solution (in the Stern layer) and in pure water, respectively. Since, to a first approximation,

$$K_m = S/S^0 \qquad (19)$$

165

$$-\Delta G^m = - RT \ln \gamma_n \tag{20}$$

S and hence γ_n reflect the combined operation of ion-dipole interactions, which tend to enhance S (salting-in effect), and of the salting-out effect due to the presence of alkyl side chains in the organic molecule. The deviation of leucine, for example, from the line drawn through glycine and its peptides (Fig. 58) was ascribed by Greenland et al.[73] to the presence of the bulky isobutyl group in the molecule tending to shift Eq. (14) to the left. The reason for the enhanced uptake of α- and β-alanine with respect to the glycine derivatives is not clear, but it is interesting to note that the points for the alanine compounds coincide. On the other hand, where proton transfer is the dominant process, β-alanine is apparently adsorbed in preference to its α-isomer.[69]

For compounds with a positive δ, K_m and $-\Delta G^m$ would increase with a rise in electrolyte concentration according to Eqs. (18) to (20). Because of its higher surface charge density (about 1·6 times that of montmorillonite) the Stern concentration of illite (4·0 N) is proportionately higher than that of montmorillonite (2·5 N). It might therefore be expected that, for a given amino acid concentration in solution, illite would adsorb a greater amount of amino acid zwitterions than montmorillonite; this is borne out by the experimental data (Fig. 57; Table 50).

A comparison of the entropy change for the adsorption of glycine and its peptides by calcium-montmorillonite shows that the entropy becomes more favourable to the process as the molecular weight increases. However, for all but the tetra-peptide, the entropy change was positive. The negative entropy found with tetra-glycine is consistent with the X-ray diffraction data which, as noted earlier, indicates that its uptake was accompanied by a concomitant displacement of water from the interlayer space.

The isotherms for the adsorption of the zwitterions of glycine and its peptides by the sodium clays tended to be of the L or Langmuir type[38] rather than linear. Partition of the compounds between the bulk water and the Stern layer water, if it did occur, must therefore have varied with the amount adsorbed. Greenland et al.[72] have suggested that this might be due to the development of extensive diffuse double layers on the surfaces of these samples. The uptake of amino acid zwitterions by the systems would greatly modify the distribution of sodium ions and, hence, the balance of attractive and repulsive forces in the double layer.

Adsorption of neutral and acidic amino acid and peptide zwitterions by hydrogen-montmorillonite, and of basic amino acid and peptide cations by sodium- and calcium-saturated montmorillonite and illite, gave rise to L-type isotherms[71,72] (Figs. 59 and 60). This was ascribed to the fact that uptake occurred either by proton transfer (as in the H-clay) or by cation exchange for the inorganic cations present (Na- and Ca-clays).

For the hydrogen-montmorillonite system, the reaction may be written as

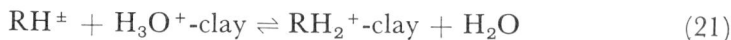

$$RH^\pm + H_3O^+\text{-clay} \rightleftharpoons RH_2^+\text{-clay} + H_2O \tag{21}$$

166

Fig. 59 (A) Adsorption isotherms for aspartic acid (□), glutamic acid (Δ), phenylalanine (●), and p-amino benzoic acid (○) at 298°K on hydrogen-montmorillinte. Dashed line represents amount of titratable H_3O^+ ions. (B) Adsorption isotherms for DL-arginine (○), L-arginine (Δ), and glycine (□), all in the cationic form, at 293°K on calcium-montmorillonite. Filled symbols represent amount of Ca^{2+} ions liberated by the exchange process. Data from Greenland et al[72]

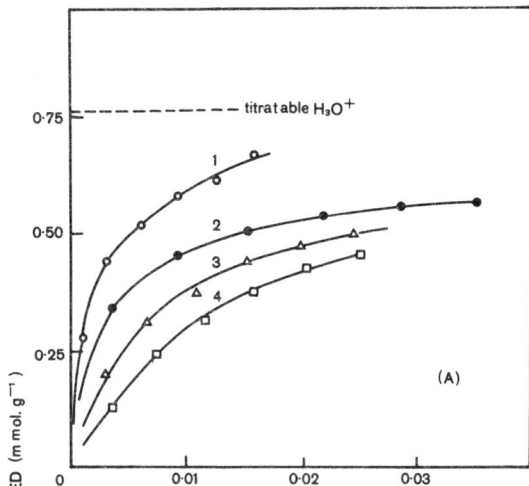

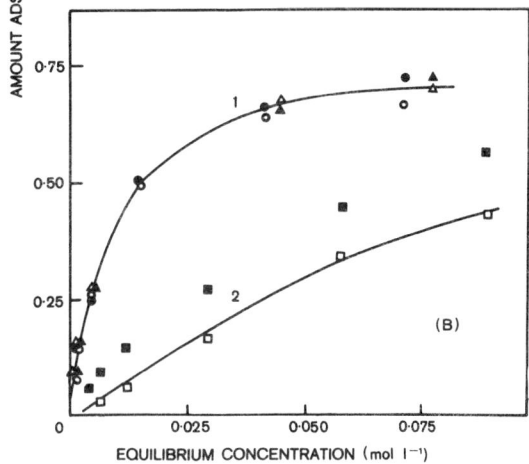

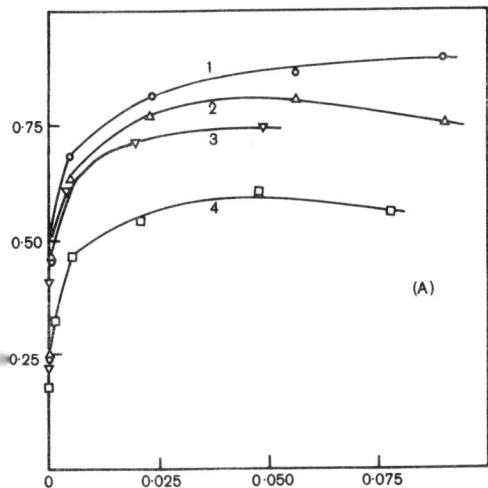

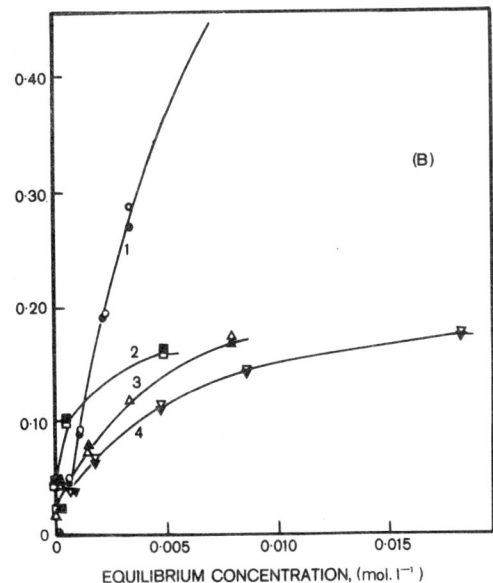

Fig. 60 (A) Adsorption isotherms for L-arginine (Δ), histidine (○), and lysine (□), at 293°K, and for carnosine (∇), at 298°K on sodium-montmorillonite. (B) Adsorption isotherms for carnosine (□) on sodium-illite, and for histidine on calcium-montmorillonite (○), on sodium-illite (Δ), and on calcium-illite (∇), all at 298°K. The filled symbols represent the Na^+ or Ca^{2+} ions displaced on adsorption of the amino acid cations. Data from Greenland et al[72]

167

so that the equilibrium constant, K_t, is given by.

$$K_t = \frac{a^*_{RH_2^+}}{a^*_{H_3O^+} \cdot a_{RH^\pm}} = \frac{C^*_{RH_2^+}}{C^*_{H_3O^+} \cdot C_{RH^\pm}} \cdot \frac{\gamma^*_{RH_2^+}}{\gamma^*_{H_3O_+} \cdot \gamma_{RH^\pm}} \tag{22}$$

$$= K_n K_\gamma \tag{23}$$

where a, C, and γ refer to the activity, concentration, and activity coefficient, respectively. Asterisks apply to the adsorbed species. Equilibrium (21) may be compared with the reverse of the first dissociation of the amino acid or peptide cation:

$$RH_2^+ + H_2O \rightleftharpoons RH^\pm + H_3O^+ \tag{24}$$

The equilibrium constant K_1 for this equilibrium is available from the literature. If physical adsorption forces contribute to the process, $K_t > 1/K_1$ and since K_t principally reflects the effect of basicity, the product $K_t K_1$ (or $K_n K_\gamma K_1$) is a useful index of the relative contribution of basicity to physical adsorption forces. Table 51 lists the values of K_n, K_1 and the product $K_n K_1$ for the different amino acids and peptides. K_n is derived from the slope of the linear plots relating $1/C_{RH^\pm}$ to $C^*_{H_3O^+}/C^*_{RH_2^\pm}$ but K_γ cannot be evaluated. However, since a plot of $K_n K_1$ against molecular weight yields a reasonably good straight line (Fig. 61), K_γ is apparently similar for the compounds examined. Deviations from the linear plot may be due to differences in K_γ values but may also arise from effects associated with molecular structure and orientation, as we noted earlier for the adsorption of cationic herbicides.

The large value of $K_n K_1$ for p-amino benzoic acid, for example, is attributable to the interaction of the planar aromatic ring with the clay surface. The X-ray data (Table 53) suggest that, like benzoic acid,[75] the p-amino derivative is

TABLE 51
Mass-action quotients (K_n) for the adsorption of some amino acids and peptides by hydrogen-montmorillonite, first dissociation constants of the organic cations (K_1), and the product $K_n K_1$.[72]

Compound	K_n (M^{-1})	K_1 $(M\ 10^3)$	$K_n K_1$
Glycine	34	4·60	0·16
Glycyl glycine	730	0·85	0·62
Digylcyl glycine	2100	0·55	1·15
Trigylcyl glycine	1800	0·89	1·60
α-alaniné	123	4·48	0·55
β-alaniné	312	0·25	0·08
Leucine	179	4·70	0·84
Serine	111	6·17	0·68
Aspartic acid	60	7·95	0·48
Glutamic acid	114	7·95	0·90
Phenylalanine	272	2·63	0·72
p-aminobenzoic acid	415	4·38	1·82

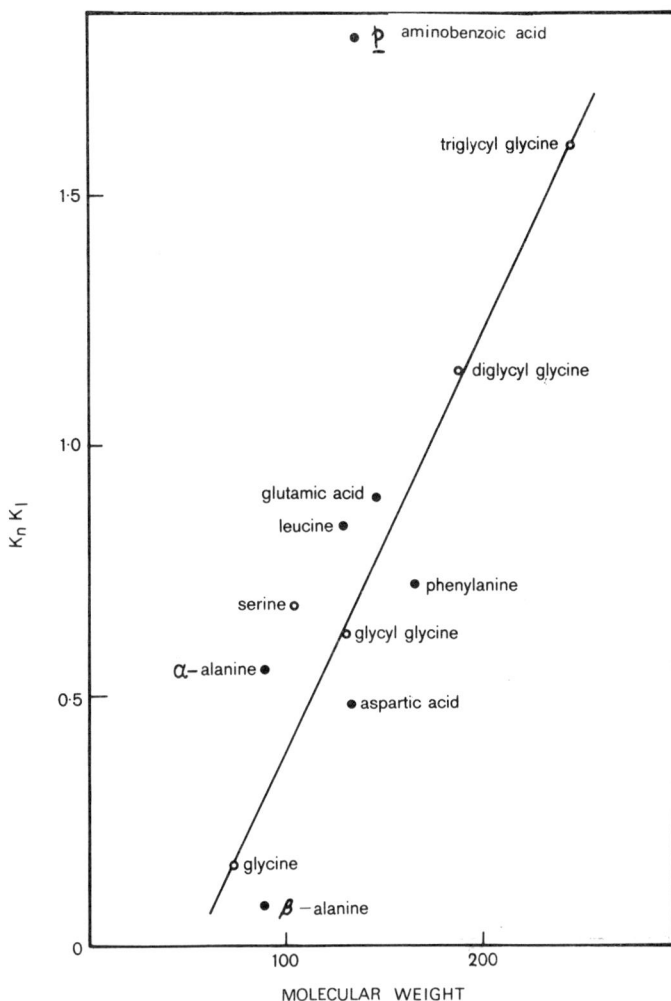

Fig. 61 Dependence of the product $K_n K_1$ on molecular weight for the adsorption of some amino acids by hydrogen-montmorillonite. $K_n K_1$ values for glycine and its peptides were taken from a previous study.[71] Data from Greenland *et al*[72]

intercalated with its benzene ring parallel to the silicate surface (d(001) = 1·30 nm). The basal spacing of the phenylalanine complex is 1·385 nm which is apparently too large to allow intimate van der Waals contact between the amino acid cation and the clay surface. Consequently the value of $K_n K_1$ for phenylalanine is appreciably less than that for *p*-aminobenzoic acid. On the basis of basicity alone (K_n values, Table 51), β-alanine is preferred to α-alanine, in accord with Sieskind and Wey's[66] observation. The effect of physical adsorption forces, however, evidently outweighs that of basicity so that the point for the α isomer lies above that for the β compound (Fig. 61).

The adsorption of basic amino acids and peptide cations by sodium- and calcium-montmorillonite may be treated in a similar manner. For the weakly basic amino acids, considerable hydrolysis occurs in the aqueous solution of their hydrochloride salts and an additional parameter has to be introduced into the general equilibrium equation. The values of K_m and $-\Delta G^m$ for the

169

TABLE 52

Mass action quotients (K_m) and non-standard free energy change ($-\Delta G^m$) for the adsorption of some amino acid and peptide cations by montmorillonite and illite.[72]

Adsorbent	Amino acid or peptide									
	Arginine		Histidine		Lysine		Carnosine		Glycine	
	K_m	$-\Delta G^m$	K_m	$-\Delta G^m$	K_m	$-\Delta G^m$	K_m	$-\Delta G^m$	K_m	$-\Delta G^m$
Na-montmorillonite	8·1	5188·2	8·2	5230	3·4	3012·5	12·0	6150·5	—	—
Ca-montmorillonite	5350	21254·7	10^4	22844·6	—	—	—	—	83	10878·4
Na-illite	—	—	2·3	2050·2	—	—	—	—	—	—
Ca-illite	—	—	2860	19664·8	—	—	—	—	—	—

Free energy values are expressed in kJ per g ion.

adsorption by montmorillonite and illite of the various compounds used are summarized in Table 52.

The free energy change of cation exchange is usually larger for montmorillonite than for illite, presumably because the organic cation interacts with two opposing surfaces in the interlayer space of montmorillonite. Intercalation also results in stronger retention of the adsorbed organic species against extraction with electrolyte solutions (1 M KCl for 20 hours), with the possible exception of lysine. On the other hand, the amino acid cations held at the external surfaces of illite are readily desorbed by such a treatment.

The X-ray results are presented in Table 53. In general, the compounds are intercalated as a single layer, agreeing with the results of other workers[64,70,76] although for high amounts adsorbed, both interstratified single- and double-layer, and double-layer complexes may also form. For all the compounds examined the observed Δ values are smaller than the minimum molecular thickness. Using models of the montmorillonite-oxygen surface and the appropriate amino acid, this apparent shortening of contact distances can be accounted for by keying of the organic species into the silicate surface.

TABLE 53

X-ray diffraction data and Δ values of montmorillonite complexes with some amino acids and peptides after drying over P_2O_5 at room temperature.[71,72]

Compound	Amount adsorbed* (mmol g^{-1})	d(001) (nm)	d(004) (nm)	Δ value (d(001)− 0·95)	Type of complex
Glycine	0·63	1·250	0·312	0·30	single layer
Glycyl glycine	0·53	1·300	0·317	0·35	single layer
Diglycyl glycine	0·62	1·320	0·320	0·37	single layer
	0·74	1·414	0·323	—	single and double layer**
Triglycyl glycine	0·54	1·330	0·321	0·38	single layer
	0·66	1·525	0·323	—	single and double layer**
	0·90	1·675	—	0·725	double lcyer
α-alanine	0·44	1·370	0·320	0·42	single layer
β-alanine	0·60	1·300	0·322	0·35	single layer
Leucine	0·39	1·480	0·364	0·53	single layer
Serine	0·42	1·320	0·324	0·37	single layer
Aspartic acid	0·31	1·320	0·323	0·37	single layer
p-aminobenzoic acid	0·66	1·300	0·318	0·35	single layer
Phenylalanine	0·54	1·385	0·339	0·44	single layer
Arginine	0·76	1·360	0·340	0·41	single layer
Carnosine	0·75	1·352	0·332	0·40	single layer
Histidine	0·91	1·352	0·332	0·40	single layer
Lysine	0·61	1·360	0·334	0·41	single layer

* slightly underestimated because these complexes were obtained by drying the corresponding centrifuged materials containing some entrained solute.
** interstratified structure.

In examining further the cation exchange and proton transfer reactions of amino acids (glycine, glycyl glycine, and β-alanine) with montmorillonite, Cloos et al.[68] have arrived at similar conclusions. However, the replacement of the inorganic cations (Na^+ and Ca^{2+}) by the amino acid cations, presented as their respective acidic hydrochloride solution, was only approximately stoichiometric since appreciable amounts of Al^{3+} ions, liberated by acid attack of the silicate structure, entered the exchange positions. This is because the adsorption process is accompanied by proton uptake as indicated by a rise in the pH of the equilibrium solution. The intercalated amino acid cations were effectively retained against washing but could be completely displaced by treatment with $M/2$ $BaCl_2$.

When neutral solutions of the amino acids were added to hydrogen-montmorillonite, two types of adsorbed species were apparently formed. The first type, probably produced by proton transfer, is strongly held against repeated washing; the second type is more loosely bound, being removed by a washing treatment, and probably consists of associated zwitterions. Cloos et al. proposed that the NH_3^+—R—COO^- ions initially adsorbed accept protons, yielding NH_3^+—R—COOH cations. As proton transfer progresses, the amount of available protons decreases and dissociation of the carboxyl group occurs, the extent of which depends on K_1 as described earlier. The negatively charged carboxylate group may then associate with the —NH_3^+ of other zwitterions. On this basis the proportion of zwitterions present will increase with a rise in K_1 and this accords with experimental results, that is, glycine ($pK_1 = 2\cdot35$) > glycyl glycine ($pK_1 = 3\cdot12$) > β-alanine ($pK_1 = 3\cdot60$).

Using infra-red spectroscopy, Fripiat et al.[77] have shown that this dissociation (of COOH) process was enhanced by moderate heating (below $400°K$). As the complex was further dehydrated by heat treatment to about $430°K$ a secondary amide linkage —R—CO—NH—R was formed for glycine and β-alanine. This was indicated by the development and intensification of the amide I band ($\sim 1,640$ cm^{-1}), and to a lesser extent, of the amide II band ($\sim 1,550$ cm^{-1}). At the same time the band due to the un-ionized carboxyl group was strongly reduced, that is, the ratio of $RH_2^+/RH^\pm$ decreased. The demonstration that clay minerals can catalyse the formation of peptide bonds has important implications for the abiogenic synthesis of proteins from simple precursors. The use of clay minerals as substrate and template in the synthesis of amino acids, peptides, and proteins is discussed later.

Similar changes in the infra-red spectrum of lysine cations adsorbed on sodium-montmorillonite have been reported by Ovcharenko et al.[78] Thus, on heating the complex, the band near $1,750$ cm^{-1} ascribable to un-ionized COOH group weakened and, concomitantly, there appeared a band near $1,417$ cm^{-1} due to the symmetric stretching mode of COO^-. As water was removed, part of the amino acid cation converted to the zwitterionic form, the excess of negative charge in the clay being compensated by the proton released during the process. No evidence for the formation of an amide-type band was found on heating the complex to $488°K$. Rather, the infra-red spectrum showed

172

weak, broad adsorption at 1,642 cm^{-1} and a sharp band at 1,419 cm^{-1} attributed to ammonium ions.

Ovcharenko *et al.* also investigated the influence of pH and the nature of the exchangeable cation on adsorption of lysine and norleucine by montmorillonite. In confirmation of the findings by McLaren *et al.*[67] and Sieskind,[66] the amount adsorbed decreased with a rise in pH. The highest uptake occurred with the sodium-saturated sample at pH ~ 2 being slightly less than the exchange capacity. At pH 2 and for a given initial solution concentration (0·16 meq g^{-1}), adsorption decreased in the order Na$^+$ > K$^+$ > Mn^{2+} > Zn^{2+} > Cr^{3+} > Ca^{2+} > Fe^{3+}. No explanation was given for this result although selectivity and steric factors are likely causes. Extensive interlayer expansion of the sodium sample, for example, would facilitate the exchange process, while for the samples containing transition metal cations formation of a co-ordination complex between the amino acid and the cation might be expected.

Using samples of natural (non-homoionic) kaolinite, illite, and montmorillonite, Siegel[79] has shown, for example, that uptake of zinc in the presence of glycine was higher than would have been predicted from stability constant measurements. This was explained on the basis that the [zinc-glycine]$^+$ complex formed was adsorbed by the clays. Estimation of the binding parameters for this complex and the clays indicated that, for some of the minerals used, the binding to the clay of the monovalent zinc-glycinate was almost as strong as that of the divalent zinc ion. This contrasts with the behaviour of inorganic cations of which divalent ions are more strongly bonded than the monovalent species. Similarly, Bodenheimer and Heller[80] have reported that uptake by montmorillonite (Wyoming) of amphoteric and basic amino acids such as glycine, α-alanine, methionine, and lysine from dilute solutions (< 2 mM g^{-1} added) at pH > 5 was enhanced in the presence of copper.

Formation of copper-glycine and copper-alanine complexes apparently occurs both in aqueous solution and in the interlayer space of montmorillonite. The data suggest that complex formation between the copper clay and amphoteric amino acids is favoured at near-neutral pH values and in dilute solutions. Under similar conditions, neither glycine nor α-alanine were found to be adsorbed by natural Wyoming bentonite (that is, in the absence of copper).

The strong adsorption of lysine by copper-montmorillonite was attributed to the presence in the lysine zwitterion of a basic (amino) group which could coordinate to copper. Two configurations of lysine exist:

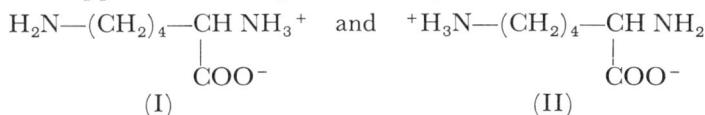

$$H_2N-(CH_2)_4-CH\ NH_3^+ \quad \text{and} \quad {}^+H_3N-(CH_2)_4-CH\ NH_2$$
$$\underset{\text{(I)}}{|\ COO^-} \qquad\qquad\qquad \underset{\text{(II)}}{|\ COO^-}$$

Zwitterion (II) predominates in aqueous solution, but which of the two forms is preferentially coordinated to copper in the montmorillonite interlayers cannot be deduced from Bodenheimer and Heller's data. However, recent infra-red measurements by Jang and Condrate[81], discussed fully below, indicate that the species coordinated to interlayer transition metal ions in montmorillonite has

173

the type (II) configuration. Zwitterion association involving COO^- and NH_3^+ groups of different lysine molecules also seems possible, as suggested by Cloos et al.[68]

The sulphur-containing amino acid methionine shows a behaviour intermediate between the amphoteric and the basic compounds. But glutamic acid, although it forms a complex with copper in aqueous solution, does not appear to complex with interlayer copper ions in montmorillonite. Failure to yield an interlayer copper complex is attributed to the fact that an aqueous solution of glutamic acid is acidic, containing either the $HOOC—CHNH_2—$ $(CH_2)_2—COO^-$ or the $^-OOC—CH\,NH_2—(CH_2)_2—COOH$ anionic form, both of which are presumably repelled by the negatively charged montmorillonite surface.

That amphoteric and basic amino acids can complex with copper and other transition metal ions in montmorillonite has been unequivocally demonstrated by Jang and Condrate[81,82] using infra-red spectroscopy.

The band frequencies and their assignments for valine adsorbed by hydrogen-montmorillonite and by copper-montmorillonite at different pH values are shown in Table 54[81]. From a comparison of the frequencies for the asymmetric and symmetric carboxyl stretching vibrations of valine intercalated at $pH \leqslant 5\cdot6$ with those of the unadsorbed zwitterion and of the crystalline copper-valine compound[83], Jang and Condrate were able to conclude that the carboxyl group is free and ionized. The NH stretching frequencies for the valine-copper-clay complex prepared at pH 5·6, on the other hand, indicate that the amino group of the molecule is coordinated to copper. The copper-valine complex formed under acid conditions in the montmorillonite interlayers must therefore be of the monodentate type, having the following configuration:

$$\left[\begin{array}{c} H_2O \qquad NH_2—CH—CH(CH_3)_2 \\ \diagdown \diagup \qquad \qquad | \\ Cu \qquad \qquad O \\ \diagup \diagdown \qquad \qquad \big| \diagup \\ H_2O \qquad OH_2 \quad C \\ \qquad \qquad \diagdown \diagdown \\ \qquad \qquad \quad O \end{array}\right]^+$$

At pH 8·2, the carboxyl asymmetric and symmetric vibrations occur at 1,630 and 1,376 cm^{-1}, respectively. These frequencies are comparable with those given by solid copper-valine chelate compounds having a bidendate structure:

$$\left[\begin{array}{c} H_2O \qquad NH_2—CH—CH(CH_3)_2 \\ \diagdown \diagup \qquad \qquad | \\ Cu \qquad \qquad | \\ \diagup \diagdown \qquad \qquad | \\ H_2O \qquad O——C{=}O \end{array}\right]^+$$

In the hydrogen clay system, proton transfer occurs giving rise to the formation of the corresponding valine cation as indicated by the appearance in the

174

TABLE 54

Band frequencies (cm⁻¹) and assignments for valine in the zwitterionic form, in the copper complex, and when adsorbed by copper- and hydrogen-montmorillonite at the indicated pH.[81]

Zwitterion*	Cu-complex*	Cu-clay complex pH 2·9	Cu-clay complex pH 5·6	Cu-clay complex pH 8·2	H-clay complex pH 2·9	Assignments
	3297		3345		3220(m,br)	NH_2 asymmetric stretching
	3250		3275		3180(sh)	NH_2 symmetric stretching
3132						NH_3^+ stretching
2989	2986		2960(br)		2970(w)	CH stretching
2884	2876					
		1730(vw)			1742(s)	COOH stretching
1627(sh)	1610(s)	1615(s,br)	1615(sh)	1630(s)	1610(s)	NH_3^+ degenerate deformation
1596(s)	1583(sh)					COO^- asymmetric stretching
	1572(sh)	1570(sh)	1570(s)	1590(sh)		
1500						NH_2 bending
1469	1466	1465	1465	1465	1492(s)	NH_3 symmetric deformation
	1455				1465(sh)	CH_3 degenerate deformation
					1410	CO stretching plus OH bending
1411	1381(s)	1417(m)	1417(m)	1376		COO^- symmetric stretching
1389						CH_3 symmetric deformation
1364	1370(sh)	1375(vw)	1375(w)		1380(w)	CH bending coupled with
1356	1352	1340	1340		1350	NH_2 wagging for Cu
1324	1326	1310	1310	1310		
1317	1309					
1269	1277					

* taken from Nakagawa et al.[83]

s = strong;   m = medium;   w = weak;   v = very;   br = broad;   sh = shoulder.

175

spectrum of absorption near 3,220 and 3,180 cm^{-1}, both of which can be assigned to NH_3^+ stretching modes.

Jang and Condrate[82] have extended this approach to the examination of lysine complexes with montmorillonite containing different interlayer cations. The results are summarized in Table 55. In the spectrum of the copper-clay-lysine complex, bands are observed at 3,355, 3,268, 3,150, and 3,080 cm^{-1}. By analogy with the valine complex, absorption near 3,350 cm^{-1} is probably due

TABLE 55

Band frequencies (cm^{-1}) and assignments for lysine adsorbed by montmorinollite saturated with hydrogen and some transition metal ions.[82]

	Exchangeable cation on montmorillonite				Assignments
H$^+$	Co^{2+}	Ni^{2+}	Cu^{2+}	Zn^{2+}	
	3310(w,sh)	3350(w)	3355(w)	3345(w)	NH$_2$ stretching
	3260(s,br)	3270(s,br)	3268(s)		
3276(s,br)	3100(br)	3100(br)	3150(w)	3137(sh)	NH$_3^+$ stretching
3040(br)			3080(w)		
2937	2930	2960	2944	2930	CH stretching
2867	2860	2870	2870	2862	
1625(vs,br)	1630(vs,br)	1640(vs,br)	1630(vs,br)	1640(vs,br) ⎫	COO$^-$ asymmetric stretching
1695(sh)	1590(sh)	1590(sh)	1590(sh)	1580(sh) ⎬	and NH$_3^+$ deformation
1497(s,br)	1515(m,br)	1510(m,sh)	1515(m)	1512(m,sh)	NH$_3^+$ symmetric stretching
1475(w,sh)	1470(vw,sh)	1470(w,sh)	1475(vw,sh)	1470(w,sh)	
1460(m)	1459	1465(m)	1460(w)	1458(w)	CH$_2$ deformation
	1440(sh)	1445(sh)	1450(w,sh)	1445(sh)	
1395(s)	1382(s)	1380(s)	1378(s)	1389(vs)	COO$^-$ symmetric stretching
1340	1340	1340	1340		
1305	1311	1310	1305		CH deformation
1275	1280	1275	1275	1280	
		1220	1220		

s = strong;   m = medium;   w = weak;   v = very;   br = broad;   sh = shoulder

to NH_2 groups coordinated to the metal cation. Similar modes occur in the cobalt-, nickel-, and zinc-saturated samples indicating similar types of bonding. The remaining bands occur at frequencies close to those observed for the hydrogen and calcium systems and may therefore be assigned to NH stretching of an NH_3^+ group. No absorption occurs near 1,700 cm^{-1}, indicating the absence of interlayer species containing un-ionized COOH groups. This may be compared with the results of Ovcharenko et al.[78] who observed a band at 1,750 cm^{-1} in the spectrum of lysine adsorbed by sodium-montmorillonite at pH 2. Strong, broad absorption near 1,600 cm^{-1} in the hydrogen sample is assigned to NH_3^+ deformation and COO$^-$ symmetric stretching modes. In the calcium and natural clays this broad band occurs at 1,615 cm^{-1} but is less intense. For the copper-, cobalt-, nickel-, and zinc-montmorillonite complexes absorption in this region is even broader, suggesting that the carboxyl stretching vibration

176

is shifted to a higher frequency by coordination to the metal through oxygen while one of the NH deformation modes shifts to a lower frequency for the same reason. The 1,497 cm^{-1} band on the spectra of the hydrogen, calcium, and natural clay samples is undoubtedly due to the NH$_3{}^+$ symmetric deformation mode which occurs at about 1,515 cm^{-1} in the other cation-clay systems.

Strong, broad absorption at 1,395 cm^{-1} for the hydrogen- and natural-montmorillonite complexes, and at 1,416 cm^{-1} for the calcium sample, is assigned to COO$^-$ symmetric stretching. This mode occurs at lower frequencies for complexes with transition metal ions, the order of wavenumber being Zn^{2+} < Co^{2+} < Ni^{2+} < Cu^{2+}. Nakamoto et al.[84] have made similar observations for a number of α-amino acid-metal chelate compounds. These observations are explicable in terms of an increase in the covalent character of the metal-to-oxygen bond and provide further evidence for the O-coordination bond between the carboxyl group of lysine and transition metal cations.

On the basis of these data and on the assumption that the α-amino rather than the ε-amino group of lysine is involved in coordination to the transition metal cation M^{2+} (since a 5-membered chelate ring is more stable than either an 8- or 9-membered structure), Jang and Condrate proposed the following model for the intercalated complex:

$$\left[\begin{array}{c} \overset{NH_2-CH-(CH_2)_4\,NH_3}{} \\ M^{2+} \\ O-C=O \end{array} \right]^{2+}$$

which may either be in the cis or trans configuration.

For the hydrogen-, calcium-, and natural-montmorillonite systems where chelation is absent, the lysine cations are probably intercalated in the form of

$$^+H_3N-(CH_2)_4-CH-COO^-$$
$$NH_3{}^+$$

The foregoing discussion on complex formation between amino acids and clays is almost wholly confined to the montmorillonite-type minerals. As might be expected, vermiculites can also intercalate various amino acids. The relative lack of experimental data on vermiculite-amino acid complexes is perhaps explained by the fact that interlayer sorption of organic species is usually less readily achieved with vermiculite than with montmorillonite because of the greater charge and, hence, greater electrostatic attraction between the vermiculite layers.

As early as 1952, Barshad[85] demonstrated by X-ray diffraction techniques that vermiculites could intercalate a number of amino acids from their respective aqueous solution, causing the mineral layers to expand. He was able to relate the extent of this expansion (or intracrystalline swelling) to the nature of the interlayer cation present, and to the magnitude of the dipole moment and dielectric constant of the interlayer solution. The work of Greenland et al.[71,73], discussed earlier, has established the quantitative basis for the dependence of interlayer adsorption on dielectric constant. Barshad further noted that after treatment of the vermiculite crystals with strong aqueous amino acid solutions, some of the samples became gel-like so that he could no longer measure their basal spacing by X-ray diffraction.

Walker and Garrett[86] have subsequently re-examined this effect using single crystals of vermiculite immersed in strong solutions of amino acids present in their zwitterionic form. Swelling may take place at room temperature in a matter of minutes, and its extent increases directly with concentration of the amino acid and eventually gives rise to interlayer separations of several tens of nanometres. Interlayer expansion here is clearly controlled by the dielectric constant, D, of the medium since D is usually a linear function of the solute concentration:

$$D = D_w + \delta c \qquad (25)$$

where D_w is the dielectric constant of water; δ, as before, is the dielectric increment, and c is the solute concentration. Possibly the charge on the interlayer cations present is masked, reducing the effective electrostatic (coulombic) attraction between the cations and the silicate layers. Intracrystalline swelling of this type occurs with amphoteric amino acids such as glycine, β-alanine, γ-amino butyric and, and ε-amino caproic acid.

Extensive interlayer expansion can also occur in vermiculite complexes with amino acid cations.[86] Thus, single crystals of vermiculite of which the inorganic cations initially present have been replaced by ornithine, lysine, and γ-amino butyric acid cations, swell in the respective amino acid solution when its concentration is below a critical value. Interlayer expansion increases as c is further decreased, reaching a maximum for $c \to 0$, that is, for an infinitely dilute solution. Following Norrish[87], this type of swelling is initiated by the hydration of the interlayer amino acid cations which dissociate from the clay surface. This is followed by the development of diffuse double layers on interlayer surfaces so that subsequent interlayer expansion is controlled by osmotic repulsive interactions.

Following Walker and Garrett's work, Rausell-Colom and Salvador[88,89] have recently examined the nature of the interlayer complexes formed between amino acids and vermiculite single crystals, paying particular attention to the factors affecting the formation and swelling of these complexes. They used two samples of vermiculite, from Macon County (V_3) and from Malawi (V_6) having a cation exchange capacity of 1·64 and 1·22 meq g^{-1}, respectively, and two amino acids (γ-amino butyric and ε-amino caproic). These workers measured the basal spacing of the complexes formed on immersing the clays in solution of the respective amino acid together with the amount of inorganic cations displaced by the amino acid. Solute concentrations up to 4 M were employed, the pH of which was adjusted to give a concentration of the cationic form (RH_2^+) of about $1·5 \times 10^{-3}$ M.

∂ – aminobutyric acid

BASAL SPACING, (nm)

2·2   2·0   1·8   1·6   1·4   1·2

Mg²⁺ – V₃
1·43nm inflexion
still present for c>2·5M

1   2   3   4

Mg²⁺–V₆
2·03nm inflexion
absent for c>2·5M

2·0   1·8   1·6   1·4   1·2

1   2   3   4

ε – aminocaproic acid

1·7   1·6   1·5   1·4   1·3

Mg²⁺ – V₃
1·43nm inflexion
still present for c>3M

1   2   3   4

Mg²⁺ – V₆
1·49nm inflexion
still present for c>3M

1·7   1·6   1·5   1·4   1·3

1   2   3   4

AMINO ACID CONCENTRATION (mol. l⁻¹)

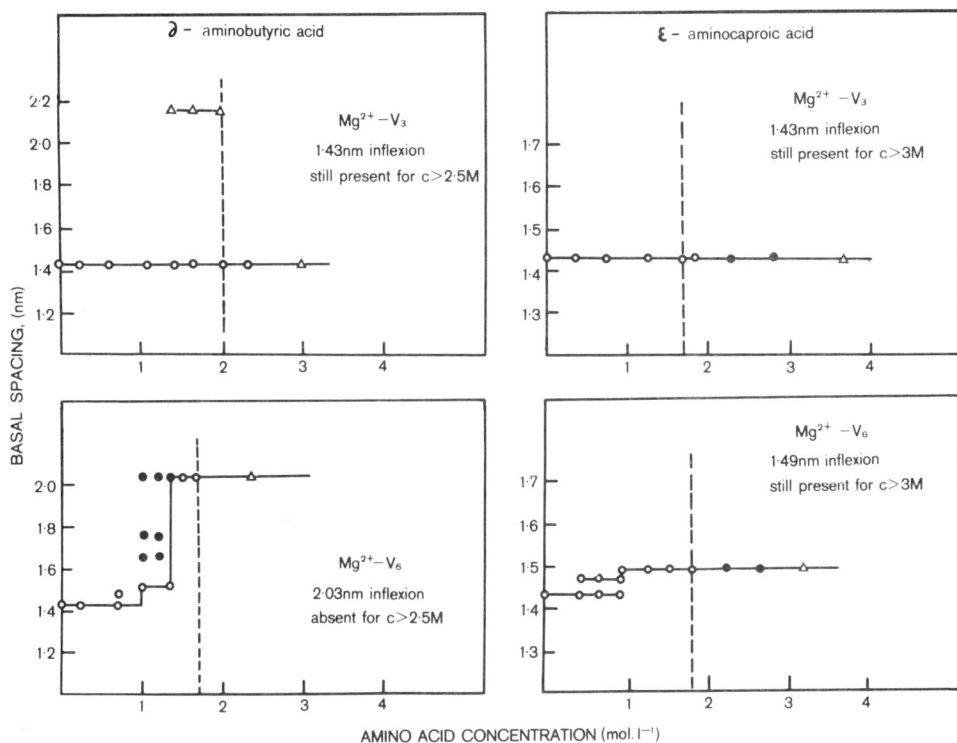

Fig. 62   Basal spacing (nm) variation of complexes of two magnesium-vermiculite samples (V_3 and V_6) with γ-amino butyric acid and ε-amino caproic acid as a function of amino acid concentration, *c*. After Rausell-Colom and Salvador[88]

○ several orders of basal reflections observed
● higher orders of basal reflections absent
△ diffuse

The basal spacing of the complexes as a function of amino acid concentration (c) is shown in Fig. 62. The increase in d(001) spacing (from an initial value of ~1·43 nm for the magnesium-saturated sample) with an increase in c is apparently discontinuous. Beyond a given critical value of c, which varies with the type of vermiculite and amino acid used, extensive intracrystalline (or 'macroscopic') swelling occurs giving interlayer separations greater than 20 nm, as previously reported by Walker and Garrett[86]. This type of swelling does not appear to take place uniformly throughout the crystal, since some of the layers (within a crystal) maintain their collapsed (unexpanded) spacing, even in saturated solutions. An exception is the V_6 complex with γ-amino butyric acid which shows no residual small spacings for $c > 2·5$ M. This complex was therefore chosen as material for subsequent study.

The amount of strontium retained by a strontium-saturated sample of V_6 on adsorption of γ-amino butyric acid under the specified conditions is shown in Table 56. It appears that a proportion of the amino acid is taken up in the cationic form, this proportion increasing with an increase in c and with prolonged immersion. Although these data are seemingly at variance with those

TABLE 56

Percentage of strontium retained by strontium-saturated samples of vermiculite (V_6) on adsorption of γ-aminobutyric acid ($[RH_2{}^+] = 1.5 \times 10^{-3}$ mol l⁻¹) under the specified conditions.[88]

Duration of treatment (hours)	Amino acid concentration (mol l⁻¹)				
	0·25 (pH 6·44)	0·75 (pH 6·91)	1·25 (pH 7·14)	1·5 (pH 7·23)	2·5 (pH 7·5)
9	77 (1·53 nm)				56 extensive swelling
24		83 (1·53 nm, 1·65 nm)	75 (1·53 nm, 1·65 nm, 1·85 nm)	72 (1·53 nm, 1·65 nm, 2·03 nm)	30 extensive swelling
56			26 (2·03 nm, 1·53 nm, 1·74 nm)	50 (2·03 nm, 1·53 nm)	15 extensive swelling
80			20 (2·03 nm)	36 (2·03 nm)	
150	28 (1·50 nm)	35 (1·49 nm, 1·65 nm, 2·03 nm)	12 (2·03 nm)	28 (2·03 nm)	8 extensive swelling
200	6 (1·50 nm)	7 (1·50 nm)	3 (2·03 nm)	5 (2·03 nm)	2 extensive swelling

reported by Greenland *et al.*[71,73] for montmorillonite, it must be remembered that the earlier workers used concentrations less than 0·20 M and relatively short equilibration periods (24 hours) so that uptake of the cationic form by an ion exchange process would have been inappreciable.

Rausell-Colom and Salvador[88] have explained the basal spacings below the point of extensive swelling (<2·1 nm) as the intercalation of a single and a double layer of organic species adopting different orientations, from one in which the molecular chain is nearly parallel to one in which this chain is inclined at a high angle to the silicate layer (Fig. 63). The extent of interlayer expansion can be reconciled by making assumptions similar to those previously

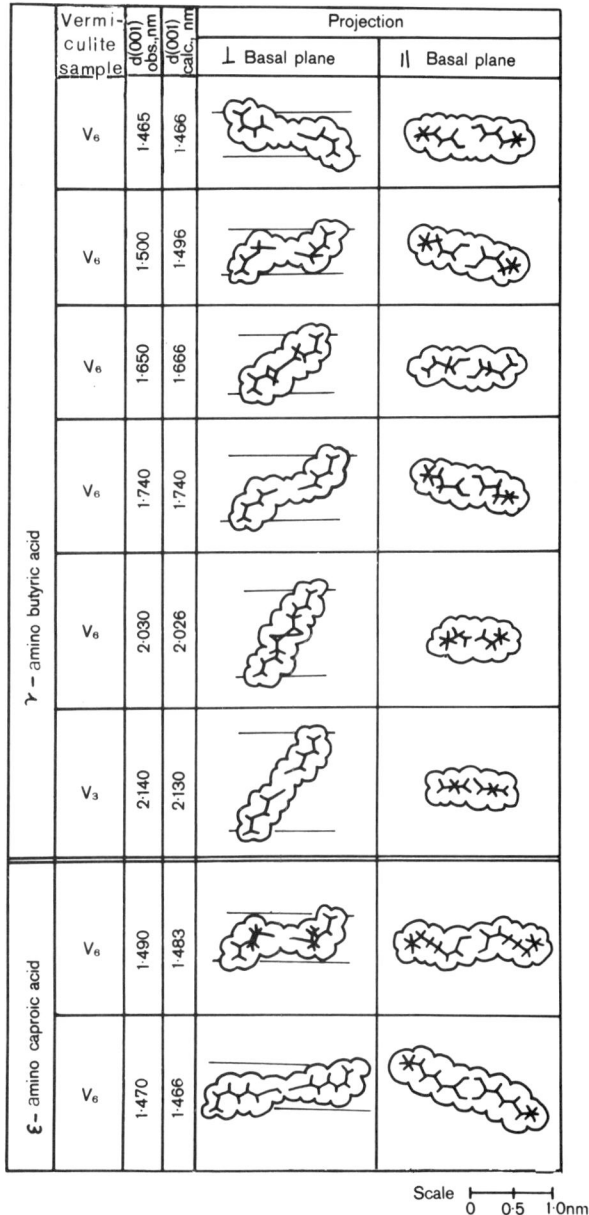

Fig. 63 Diagram showing possible interlayer arrangement of γ-amino-butyric acid and ε-amino caproic acid in double-layer complexes with vermiculite, after Rausell-Colom and Salvador[88]

outlined for *n*-alkylamines in montmorillonite (cf. Fig. 36). We further assume that the carboxyl groups of amino acid cations adsorbed on opposite surfaces and terminating at the mid-plane between two opposing silicate layers form a double hydrogen bond of the type

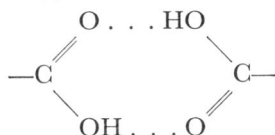

$$
\begin{array}{ccc}
 & O \ldots HO & \\
 & \diagup \qquad \diagdown & \\
-C & & C- \\
 & \diagdown \qquad \diagup & \\
 & OH \ldots O &
\end{array}
$$

That the type of complex formed (that is, single- or double-layer) and the orientation of the amino acid chains in the interlayer region are influenced by the method of preparation have also been demonstrated by Kanamaru and Vand[90]. Thus, immersion of a sodium-vermiculite flake in 1 N solution of ε-amino caproic acid (6-amino hexanoic acid) at pH 5·5 and at room temperature gave rise to a complex showing two sets of basal reflections with d(001) of 1·42 and 1·732 nm, respectively. When this complex was treated with the same solution at 323°K for a week, a new phase with d(001) = 2·1 nm was formed. This phase, however, was unstable in air, giving way to one with d(001) = 1·732 nm. From a one-dimensional Fourier synthesis of the stable 1·732 nm complex, these workers showed that a double layer of ε-amino caproic acid cations was intercalated with the zig-zag chain lying nearly parallel to the *b* axis, while the plane of the carbon zig-zag was tilted at an angle of about 0·873 rad (50°) to the silicate layer. This is a configuration intermediate between an α_I and an α_{II} arrangement (Chapter 2). The O—H . . . O and N—H . . . O distances of 0·282 and 0·286 nm, respectively, are consistent with a certain amount of hydrogen-bond interaction between the terminal functional groups of the molecule and the silicate oxygen surface.

Rausell-Colom and Salvador have suggested that, under their experimental conditions, the progressive increase in basal spacing with concentration (Fig. 62) was controlled by the zwitterion component $(RH^\pm)$ rather than by the cationic form of the intercalated amino acid. It can be shown that, in all instances, the maximum number of $RH^\pm$ ions required to give complete surface coverage (this varies with the orientation adopted), divided by the interlayer volume of the complex, remains essentially constant at about 6·5–5·5 ions per 1 cm^3 which is equivalent to a 9 molar solution. This large difference in concentration between the external (bulk) and the interlayer solutions would partly account for both the limited interlayer expansion (prior to the commencement of macroscopic swelling and gelation) and the observed increase in basal spacing with a rise in concentration. For longer-chain amino acids, intermolecular van der Waals interactions become important and hence the corresponding complexes with ε-amino caproic acid show little tendency to expand and are stable at high solute concentrations. These complexes give way directly to the gel state.

Extensive intracrystalline swelling and gel formation may qualitatively be explained in terms of the accumulation in the interlayer space of —COO$^-$

groups belonging to the intercalated amino acid zwitterions adsorbed on opposite surfaces. Interactions between such groups give rise to electrostatic repulsion and when this exceeds the cohesive forces holding the silicate layers together, the layers abruptly separate.

In a subsequent paper, Rausell-Colom and Salvador[89] have attempted to obtain quantitative data on the gelation process using vermiculite single crystals treated with concentrated solutions of γ-amino butyric acid. By measuring the basal spacing using low-angle single-crystal X-ray diffraction techniques when an external pressure is being applied in a direction normal to the basal planes of the crystal, the swelling pressure, and hence the magnitude of the repulsive forces, may be estimated. The relationship between swelling pressure and equilibrium interlayer separation is found to depend on both the concentration of the amino acid cations, $[RH_2^+]$, and the dielectric constant of the zwitterion solution in the interlayer space. Since $[RH_2^+]$ is a function of pH, small variations in pH conditions can bring about marked changes in the extent of macroscopic swelling (Fig. 64). Similarly, the solute concentration in the interlayer space, through its effect on the dielectric constant (Eq. (25)), determines the extent of (interlayer) expansion (Fig. 65).

These observations have been explained in terms of the interaction of diffuse double layers of amino acid cations developed on interlayer surfaces.

Fig. 64 Influence of solution pH on swelling pressure. In all cases the concentration of γ-amino butyric acid zwitterions is 2·5 mol. l^{-1} while that of the cationic form $[RH_2^+]$ of the amino acid varies with pH as indicated. Data from Rausell-Colom and Salvador[89]

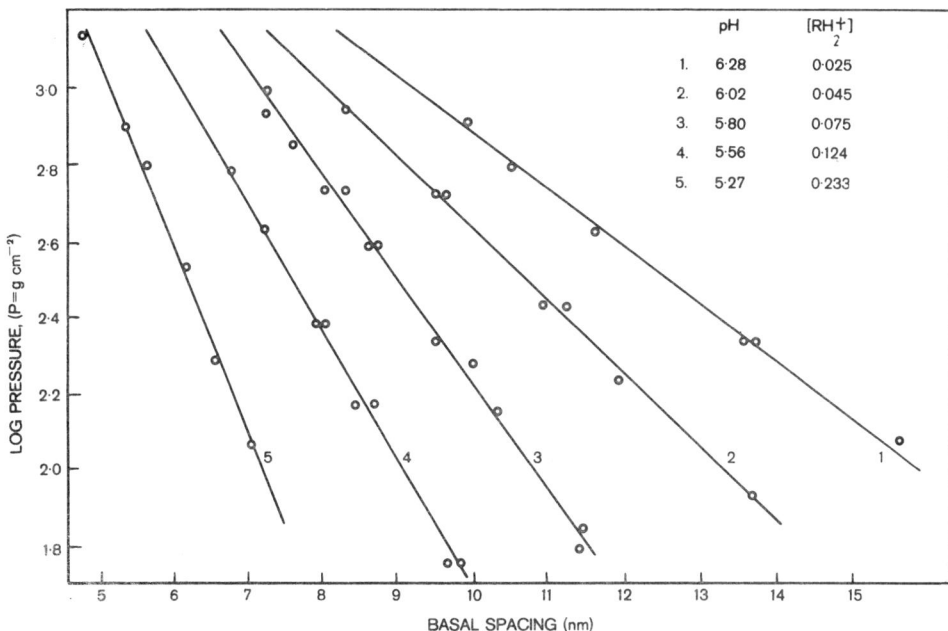

	pH	$[RH_2^+]$
1.	6·28	0·025
2.	6·02	0·045
3.	5·80	0·075
4.	5·56	0·124
5.	5·27	0·233

LOG PRESSURE, ($P = g\ cm^{-2}$)

BASAL SPACING (nm)

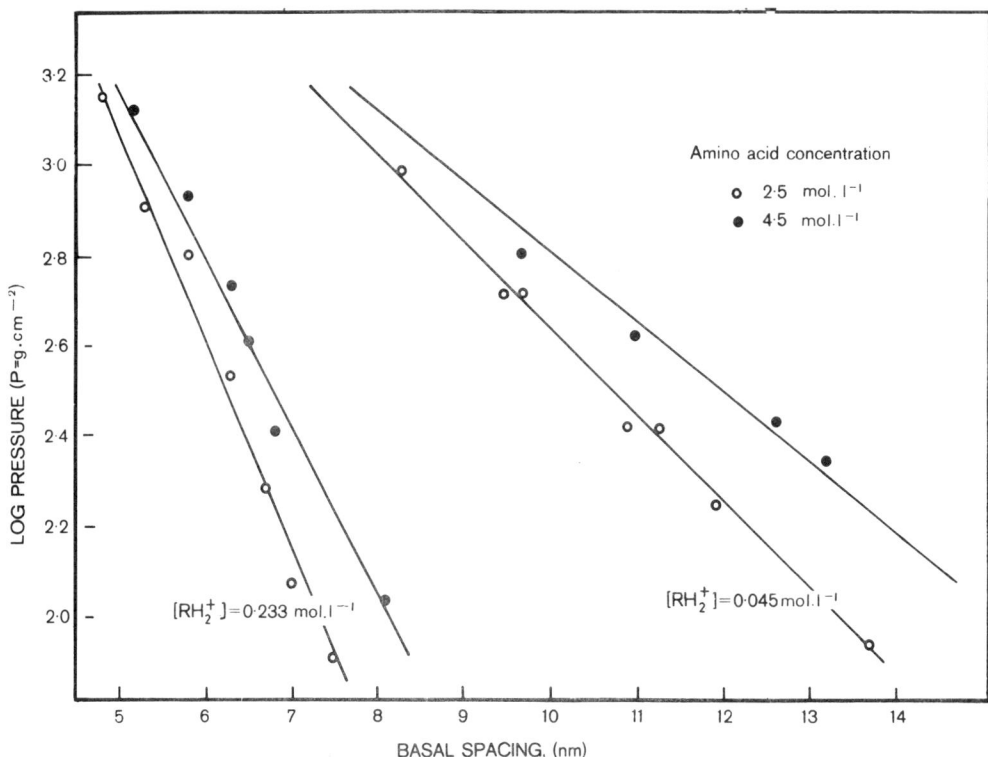

Fig. 65 Influence of amino acid (γ-amino butyric) concentration, in its zwitterionic form (2·5 and 4·5 mol. l^{-1}) and in its cationic form [RH$_2$$^+$], on swelling pressure. After Rausell-Colom and Salvador[89]

At first sight, it might appear that the double-layer theory[91] would not be applicable in view of the high amino acid concentrations used. However, this interaction is controlled by [RH$_2$$^+$] which, under the conditions employed, never exceeds 0·3 M. The concentrated solution in the interlayer space refers to that of the zwitterionic form (non-electrolyte) and therefore only serves as a dielectric medium. Indeed, Eq. (25) predicts that the dielectric constant, D, will increase as the zwitterion concentration, c, increases.

Reasonably good agreement is found between the observed and the calculated values for swelling pressure by assuming the presence of a Stern layer, 0·5 nm thick, in which the amino acid cations are specifically adsorbed on the silicate surface with an interaction energy of $13,807 \pm 837$ J mol^{-1}.

The kinetics of the exchange reaction between ornithine cations,

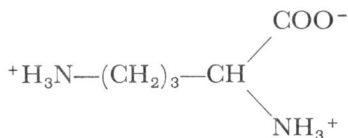

$$^+H_3N\!\!-\!\!(CH_2)_3\!\!-\!\!CH\!\!\begin{array}{c} COO^- \\ \\ NH_3{}^+ \end{array}$$

and interlayer strontium (or magnesium) ions in vermiculite single crystals (from Malawi) has been investigated by Mifsud *et al.*[92] The reaction which begins at the edges of the vermiculite flake and gradually proceeds towards the crystal centre can be readily followed under the light microscope. While the vermiculite crystal is immersed in a (0·5 M) aqueous solution of ornithine hydrochloride, the basal spacing of the reacted portion is about 4·22 nm. Removal of the crystal from this solution followed by exposure to air at ambient temperature gives rise to a sequence of phase changes, each phase being characterized by a definite basal spacing with 4 to 5 integral orders. However, some of these phases are only transitory (a matter of minutes) until a basal spacing of 2·02 nm obtains, marking the formation of a relatively stable (~3 hours) phase. If this complex is now dehydrated at 333°K for 14 hr, a new stable crystalline phase appears with a basal spacing at about 1·63 nm. Further heating at 493°K for 10 hr produces yet another complex (basal spacing = 1·45 nm) in which peptide bond formation is indicated by infra-red spectroscopy, confirming the earlier observations by Fripiat et al.[77] for glycine and β-alanine.

The rate of advance of the sharp boundary (front) between the expanded (d(001) = 4·22 nm) and the original, unexpanded strontium-saturated (d(001) = 1·52 nm) regions, measured at different temperatures, enables the

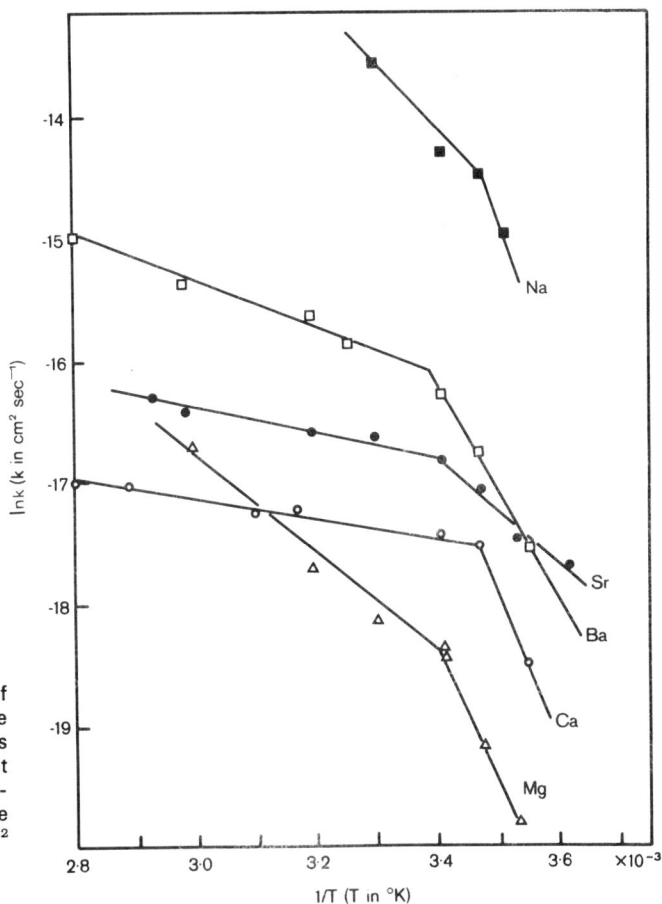

Fig. 66 Variation of the logarithm of the diffusion coefficient, k with the reciprocal of temperature (Arrhenius plot) for the exchange between different interlayer inorganic cations in vermiculite single crystals and ornithine cations in solution, from Mifsud et al[92]

activation energy (E) of the diffusion-controlled exchange process to be evaluated from the Arrhenius equation:

$$\frac{\mathrm{d}\ln k}{\mathrm{d}\,T} = \frac{E}{RT^2} \tag{26}$$

The plots of $\ln k$ (here k is the diffusion coefficient) against $1/T$ for vermiculite samples containing different interlayer cations are shown in Fig. 66. Table 57 lists the corresponding values of E and k derived from these measurements. The reason for the abrupt break in the Arrhenius plots between 288 and 293°K is as

TABLE 57

Activation energies, E, and diffusion (Arrhenius) coefficients, k, for the exchange reaction between ornithine cations and different inorganic cations in vermiculite.[92]

Interlayer cation	E ($J\ mol^{-1} \times 10^3$)		k ($cm^2\ s^{-1}$)	
	(1)	(2)	(1)	(2)
Mg^{2+}	98·7	31·8	$4\cdot6 \times 10^9$	$5\cdot4 \times 10^{-3}$
Ca^{2+}	97·5	6·7	$1\cdot2 \times 10^{10}$	$4\cdot3 \times 10^{-7}$
Sr^{2+}	38·1	8·4	$8\cdot8 \times 10^{-2}$	$1\cdot8 \times 10^{-6}$
Ba^{2+}	67·8	15·9	$1\cdot0 \times 10^5$	$7\cdot1 \times 10^{-5}$
Na^+	113·0	44·8	$1\cdot5 \times 10^{14}$	$7\cdot5 \times 10$

(1) for temperatures below 293°K.
(2) for temperatures above 293°K.

yet obscure, but it seems probable that a change in equilibrium occurs near 293°K between two hydration states of the ornithine cations, causing an alteration in the rate of exchange. Further, as exchange progresses the concentration of strontium ions in solution increases. The displaced strontium ions would compete with the ornithine cations for exchange sites in the vermiculite interlayers. At some critical temperature this may be a factor contributing to the observed decrease in the rate of exchange diffusion.

The interaction of amino acids by kaolinite-type minerals has recently attracted much interest in view of the observation that kaolinite seems capable of preferentially adsorbing and polymerizing the L-optical amino acids over the corresponding D-optical isomers. The role of clay minerals in transforming and polymerizing adsorbed organic compounds is discussed in Chapter 7.

4.3. Complexes with Antibiotics and Alkaloids

Like the pesticides, antibiotic compounds may be classified into the basic (cationic), the uncharged (neutral), and the acidic (anionic) groups. In addition, some antibiotics are amphoteric; in this respect they resemble amino acids in that the molecules contain both acidic and basic functional groups. The interactions of a particular group of antibiotics with clays (and soils) might therefore be expected to be analogous to those shown by the corresponding group of organic pesticides (or amino acids). Although relatively little work has been carried out on clay-antibiotic complexes, the available data indicate that this is so.

Working with streptomycin, Siminoff and Gottlieb,[93] for example, have noted that this basic antibiotic was readily taken up by samples of bentonite, illite, and soil whereas little, if any, adsorption occurred for compounds of the neutral and anionic type.[94]

From the practical point of view, the availability to microorganisms and hence the release from the clay (or soil) complex of adsorbed antibiotics is of prime importance. Although the retention of these compounds by clays is clearly influenced by and dependent on the mechanisms underlying their adsorption, it is frequently difficult to accurately predict whether antibiotics, once adsorbed, maintain their antimicrobial activity. The early work by Siminoff and Gottlieb and by Jeffreys[95] indicated that streptomycin became inactivated by adsorption on montmorillonite and soil whereas Pramer and Starkey[96] found that this antibiotic could be released from the mineral complex by the assay procedure, so that inactivation by this means should not be regarded as permanent.

Pinck and his co-workers[97-101] have attempted to clarify the role of clay minerals in relation to bacterial populations in soils by studying the adsorption and desorption of different types of antibiotics by various clay minerals. The main conclusions about the reactions of these compounds with montmorillonite are summarized by Pinck.[102]

In the first of a series of papers, Pinck et al.[97] established that appreciable uptake occurred with the basic compounds such as streptomycin, dihydro-streptomycin, neomycin, and kanamycin. On the other hand, adsorption of the anionic and uncharged types such as penicillin, chloramphenicol, and cyclo-heximide, was either low or undetectable while the extent to which the amphoteric species (bacitracin, aureomycin, and terramycin) were taken up by the clays was, as expected, dependent on the pH conditions. For the clay samples examined, adsorption decreased in the order montmorillonite > vermiculite > illite > kaolinite. X-ray diffraction measurements indicated that the basic and amphoteric antibiotics were intercalated by montmorillonite since there was an increase in interlayer separation of the order of 0·44 and 0·76 nm, respectively, indicative of the formation of single- and double-layer complexes. A limited amount of interlayer penetration was also observed with vermiculite. Although no data were given as to the amount of inorganic cations displaced, it was inferred that the adsorption process is principally one of ion exchange.

The results of subsequent desorption measurements using bioassay techniques[98] agreed with the general conclusions derived from adsorption studies, in that the basic antibiotics were strongly retained by montmorillonite, vermiculite, and illite although streptomycin and dihydrostreptomycin were released to varying extents from kaolinite. The amphoteric compounds, on the other hand, were more readily desorbed from all the clay samples used. Apparently, the adsorbed compound was displaced from the mineral complex by exchanging for the cation of the buffer solution and then diffused through the agar. The clay-antibiotic complex itself, however, was incapable of diffusing through the supporting medium. Of the buffers tested, phosphate and citrate were found to be most efficient in releasing the basic and the amphoteric antibiotics, respectively, from their clay complexes; the amphoterics were desorbed at an appreciably higher rate than the basic compounds.[99]

These investigations were later extended to the reactions of basic polypeptides (polymixin B sulphate and viomycin sulphate) and 'macrolides' (erythromycin and carbomycin), a term denoting the presence of a large lactone ring in their molecules.[100] Again, montmorillonite took up more than either illite or kaolinite, presumably because these compounds were readily intercalated by mont-morillonite giving an increase of about 0·57 nm in interlayer separations. Poly-myxin and viomycin were shown to be strongly retained compared with the other two antiobiotics.

The ease by which adsorbed antibiotics were released from their complexes with soils is reflected by changes in the bacterial population of the soils in question. Even in this situation the type of clay dominantly present has a marked influence on the extent to which the adsorbed organic species can become available to the microbes. A soil which is predominantly montmorillonitic, for example, shows a much greater growth-suppressing activity than one in which the clay minerals present are chiefly of the kaolinite type. This indicates that physico-chemical processes are at least as important as biological action in controlling antibiotic activity in soil systems.[101] Stotzky et al.[103] studied the relationship between soil mineralogy and *Fusarium* wilt infestation of banana trees in Panama. Among the many soil factors examined such as texture, pH, soluble salts, cation exchange capacity, drainage, organic matter, and available phosphorus levels as well as clay minerals content, only the presence or absence of montmorillonite appeared to be directly correlated with the spread of the pathogen. Soils containing this clay mineral species were better suited for growing bananas than those in which montmorillonite was absent, reflecting the ability of this type of silicate structure to effectively inactivate either the patho-gen itself, or the toxins produced by it, or both.

Because of the ability of clay minerals to take up toxic agents and bacteria causing enteric infections they are widely used as active ingredients in intes-tinal-adsorbent preparations. Kaolin and structurally related minerals have long been used to this end, presumably because of their abundance and rela-tively low cost. However, as indicated above, the adsorptive capacity of this type of structure is inferior to that of the expanding 2:1 type later silicates, such as montmorillonite. On the other hand, if the mineral is used as a carrier of pharmaceutical compounds, such as antibiotics and alkaloids, it is clearly desirable that activity be maintained, that is, the adsorbed substance must be capable of being slowly, but readily, released into the medium.

Barr and co-workers[104-107] have compared the efficiency of kaolin, halloysite, and palygorskite (attapulgite) as adsorbents of various alkaloids, bacteria, and toxins. Of these minerals, palygorskite shows the highest adsorption. The adsorptive capacity may be further enhanced by prior acid washing and heat treatment of the silicate. Such 'activated' palygorskite is therefore finding increasing pharmaceutical application. Palygorskite is superior to kaolinite and halloysite because of its open structure, enclosing channels into which the organic compounds or bacteria can be accommodated.

The isotherms for the adsoption of different alkaloids, such as strychnine, atropine, and quinine, can generally be described by the Langmuir equation[104,105] which, in its linear form, may be written as

$$\frac{C}{x/m} = \frac{1}{kV_m} + \frac{C}{V_m} \tag{27}$$

where C is the equilibrium concentration; x is the amount adsorbed per unit weight m of clay; k and V_m are constants, the latter being identifiable with the monolayer capacity of the adsorbent. Conformity of the data to Eq. (27) is indicated by the straight line observed when $C/(x/m)$ is plotted against C. However, this in itself does not provide information on the mechanisms of adsorption.

Hendricks[65] had earlier demonstrated that the alkaloids brucine and codeine, being strong bases, were intercalated by hydrogen-montmorillonite through proton transfer and physical adsorption (van der Waals) forces. Because of their large size, the cover-up effect operates: that is, some of the hydrogen ions are not available for neutralizing other molecules. On this basis, strychnine, atropine, and quinine would be expected to show a similar effect because their molecular weights (in the order of 300) are comparable with those of brucine and codeine. Nicotine (mol. wt. = 162), on the other hand, can neutralize stoichiometric amounts of hydrogen ions held by various clay minerals[108] because the molecule occupies less than the equivalent area per exchange site.

In extending to higher concentrations the work of Evcim and Barr[104] on the adsorption of atropine by different clays, Ridout[109] has observed that uptake of this alkaloid by kaolinite could be described in terms of a stepped Langmuir-type isotherm. Applying Eq. (27) to the adsorption data yields two straight lines of different slopes indicating the presence of two energetically distinct sites. This might be expected if sorption occurs on both the planar basal and the edge surfaces of the kaolinite crystals. Stepped isotherms of this type are not uncommon and have even been observed for anion-kaolin systems.[110]

4.4. Complexes with Pyrimidines, Purines, and Nucleosides

Some purines, pyrimidines, and nucleosides were among the variety of organic bases examined by Hendricks[65] in his classic study as complex-forming compounds with montmorillonite. He was able to show that purines and nucleosides were intercalated by hydrogen-montmorillonite through a proton transfer reaction and that the basal spacing of the complexes so formed was correlated with molecular configuration. For example, adenine and guanine complexes containing about 0.7 meq g^{-1} gave basal spacings of 1.24 nm, indicating the presence in the clay interlayers of a single layer of organic cations, the atoms of which are coplanarly arranged. On this basis, we can also infer that guanine is intercalated as the *enol* rather than as the *keto* form. On the other hand, 3-methylcytosine must exist and be intercalated as the *keto* tautomer. A coplanar arrangement is therefore no longer realizable and the basal spacing of the complex with this pyrimidine base is accordingly larger (d(001) = 1.29 nm). Similarly, complexes with the nucleosides adenosine and guanosine, in which a ribofuranose group substitutes at the 9-position in the purine ring, give still greater interlayer expansion (d(001) = 1.35 nm) although this is only 0.10 nm larger than would be required for a strictly coplanar atomic arrangement. The intercalated species apparently tend to adopt the flattest possible configuration since this allows maximum van der Waals interaction between the adsorbate and the clay surface.

In view of the biological importance of the title compounds it is perhaps surprising that there was little follow-up work between 1941 when Hendricks' paper appeared and the late 1960s when Brindley and co-workers[111-114] examined the interactions of purines, pyrimidines, and nucleosides with montmorillonite and illite as a function of pH, exchangeable cation, and molecular constitution.

Firstly, these workers looked into the uptake by lithium-, sodium-, magnesium-, and calcium-montmorillonites using initial organic concentrations of about 1 mM and clay-to-organic ratios such that no more than about 25 per cent of the exchange positions could be satisfied by the organic cations (under acidic conditions).[111] Their results for the sodium and calcium clays are reproduced in Figs. 67 to 70. The corresponding data for lithium- and magnesium-montmorillonite are not shown since they are similar to the sodium and calcium systems, respectively.

The pH-uptake curves show that adsorption of these compounds, besides being dependent on pH conditions, is also influenced by the basicity and constitution of the organic molecules and by the nature of the interlayer cation present. Although these factors are clearly interdependent in any given situation, we discuss each separately in order to avoid confusion.

The effect of pH is perhaps self-evident and mechanisms similar to those suggested by Weber[8] for the uptake of *s*-triazines by montmorillonite are likely to operate in this instance. Indeed, a series of equilibrium situations represented by Eqs. (5) to (10) (Section 4.1.2) has been proposed by Lailach *et al.*[111] to account for the observed isotherms. Thus, at pH < pK$_a$ Eqs. (5) to (10) would apply. The amount of RH$^+$ increases as the pH is lowered and RH-montmorillonite forms. Some interlayer cations are displaced by protons to give H-montmorillonite, which may take up the uncharged R molecules yielding RH-montmorillonite by proton transfer. Figure 70 for example, shows that at low pH, the amount of inorganic cations released far exceeds the quantity of

190

Fig. 67 Adsorption by sodium- and calcium-montmorillonite as a function of pH of adenine (○), 7-methyladenine (△), 9-methyladenine (▲), cytosine (x), and pyrimidine (+), after Lailach et al[111]

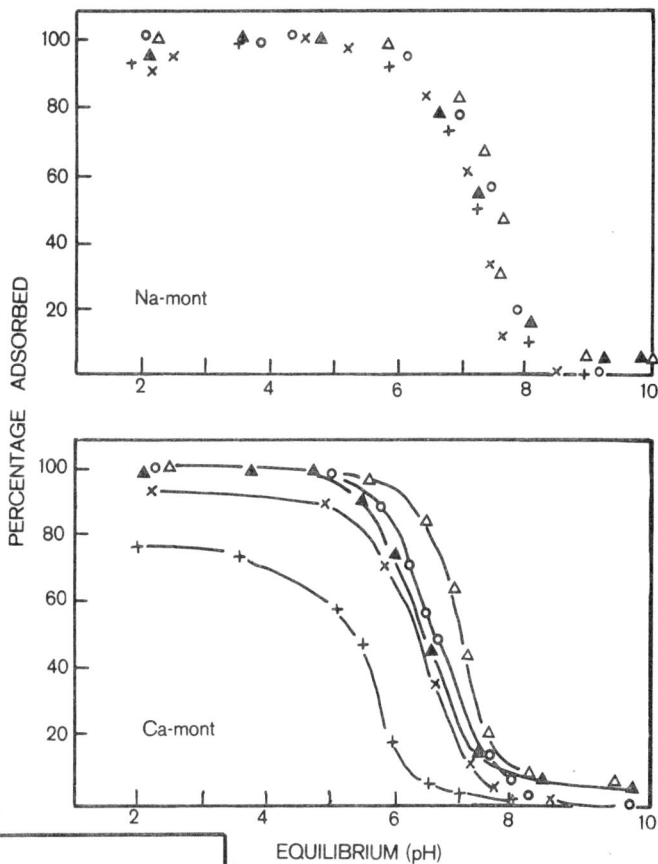

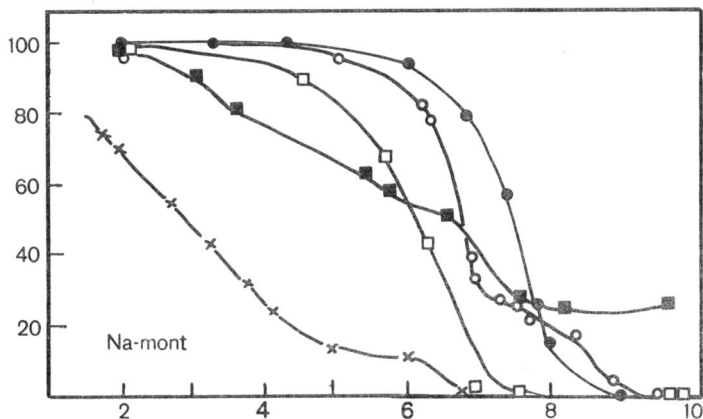

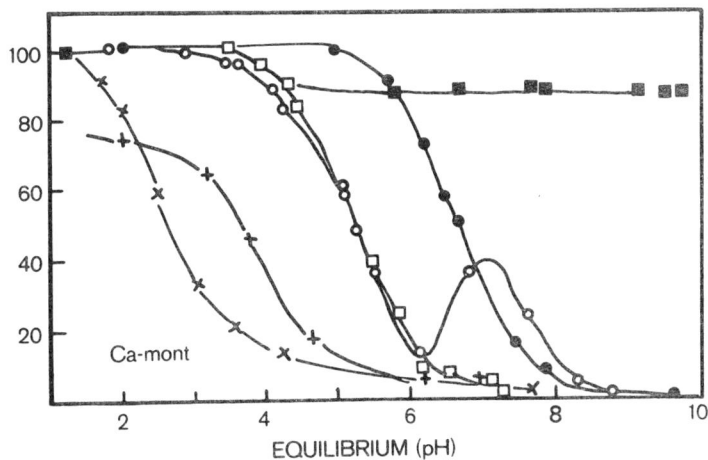

PERCENTAGE ADSORBED

Na-mont

Ca-mont

EQUILIBRIUM (pH)

Na-mont

Ca-mont

EQUILIBRIUM (pH)

Fig. 68 Adsorption by sodium- and calcium-montmorillonite as a function of pH of adenine (●), hypoxanthine (○), purine (□), 6-chloropurine (x), 5-amino-6 methyl-uracil (+), and caffeine (▦), from Lailach et al[111]

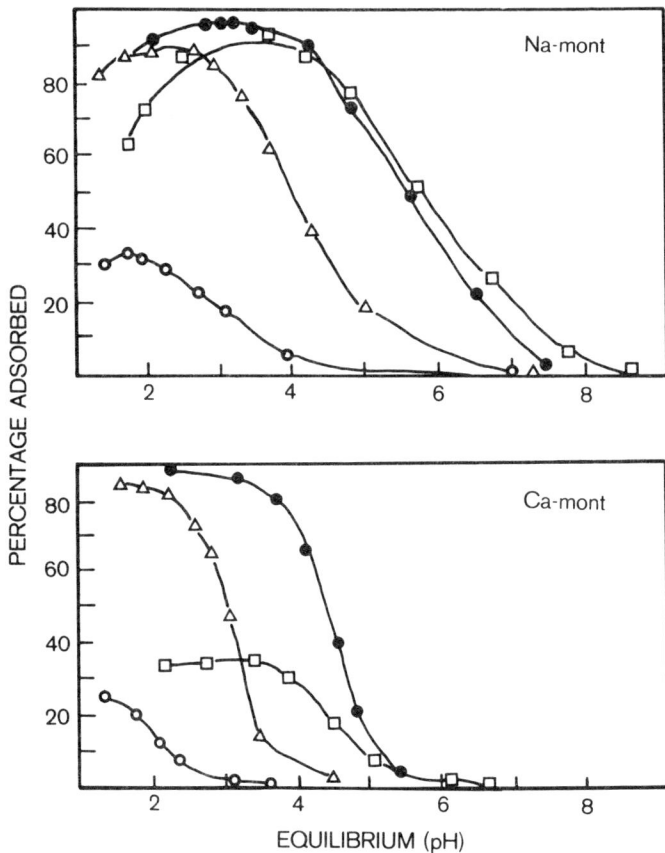

Fig. 69 Adsorption by sodium- and calcium-montmorillonite as a function of pH of adenosine ($\bullet$), inosine ($\bigcirc$), guanosine ($\triangle$) and cytidine ($\square$), from Lailach et al[111]

Fig. 70 Left: adenine adsorbed ($\bigcirc$) by sodium-montmorillonite and sodium ions liberated ($\bullet$) at different pH values; Right: adenine adsorbed ($\bullet$) by calcium-montmorillonite and calcium ions liberated ($\bullet$) at different pH values, from Lailach et al[111]

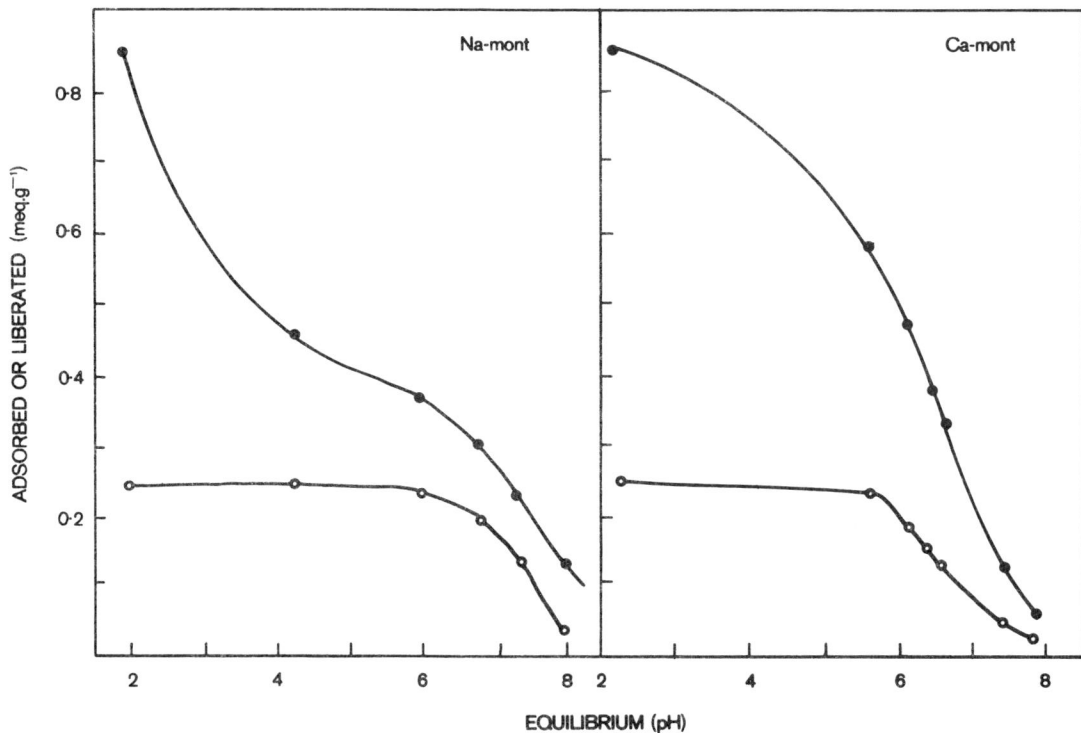

adenine adsorbed. This difference is ascribed to proton uptake. Similarly, at $pK_a < pH < 7$, Eqs. (7), (9), and (10) would describe the processes. However, the pH of the interlayer solution is likely to be higher than that of the bulk solution and hence, even when the latter is 1 to 2 units greater than pK_a, the interlayer pH may be comparable to pK_a. As expected, at $pH > 7$, little uptake occurs and the original cation-saturated montmorillonites, on the whole, preserve their identity.

The effect of molecular constitution is reflected in the observation that, all things being comparable, the steep decline in adsorption occurs at a higher equilibrium pH as basicity increases. This is illustrated in Figs. 67 and 68 and by the sequences: 6-chloropurine ($pK_a = 0.8$)-purine ($pK_a = 2.39$)-adenine ($pK_a = 4.22$), and inosine ($pK_a = 1.2$)-guanosine ($pK_a = 1.6$)-adenosine ($pK_a = 3.45$). Aromaticity is also an important molecular parameter influencing uptake. Thus, the sequence of hypoxanthine ($pK_a = 1.98$)-5-amino-6-methyluracil (AMU; $pK_a = 3.28$)-adenine is anticipated, whereas the observed order is AMU-hypoxanthine-adenine. This behaviour is attributed by Lailach *et al.* to the fact that AMU, both in the neutral and in the cationic form, is non-aromatic. On the other hand, hypoxanthine and adenine are aromatic. It is also interesting to note that hypoxanthine shows a rise in adsorption in the range of pH 7 to 9 for the divalent cation clay systems. This could be due to the formation of the following complex:

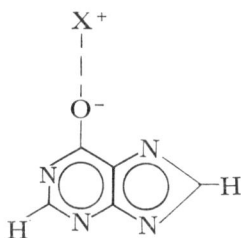

in the montmorillonite interlayers. The influence of molecular size shows itself in the markedly strong adsorption of caffeine, particularly at $pH > 7$, despite its weak basic character ($pK_a = 0.61$); and that of molecular structure by the observation that nucleosides, being non-planar, are usually less adsorbed than the purines and pyrimidines of lower molecular weight. The basal spacing of ~1.33 nm given by the adenosine complex after drying over P_2O_5 and when small amounts are intercalated requires a considerable amount of compression of the molecules in accord with the previously noted tendency for the adsorbed species to adopt a flat conformation.

Figure 70 shows that at low pH, adenine is taken up as the cation, the amount adsorbed being similar for sodium- and calcium-montmorillonite. In the range of pH 3 to 6.5, however, more calcium is desorbed than sodium and this must be due to Eq. (9) operating to a larger extent when $X = \frac{1}{2}Ca$ than when $X =$

193

Na. Lailach *et al.* suggested that the sodium ions tended to be more closely associated with the oxygen ions of the silicate layer, being partially keyed into the
ditrigonal depression, whereas the divalent cations tended to occupy positions
midway between two opposing layers. The complexes of sodium- and calcium-
montmorillonite with adenine both yield a basal spacing of $\sim$1·25 nm. This
would allow the residual Na^+ ions to occupy their normal positions with respect
to the oxygen surface, each cation probably being coordinated to 3 oxygens and
3 water molecules. However, the unexchanged Ca^{2+} ions, being constricted to
interlayer positions in a 1·25 nm complex, cannot maintain their normal coordination requirement of 6 water molecules and hence are squeezed out of the
interlayer space. To offset this process, hydronium ions which fit the geometry
more appropriately are thought to be taken in.

The nature of the interlayer cation also indirectly affects interlayer adsorption
in that it determines the state of dispersion of the clay. The uptake of the relatively large nucleoside compounds, for example, is greater for the monovalent
than for the divalent clay systems, presumably because the silicate layers in the
former instance are separated by larger distances than in either calcium- or
magnesium-montmorillonite where interlayer expansion in an aqueous medium
is restricted (cf. Tables 3 and 4).

Lailach *et al.*[112] have extended this study to montmorillonite containing
Co^{2+}, Ni^{2+}, Cu^{2+}, and Fe^{3+} ions. Comparison of the pH isotherms shows that
adsorption of purines and pyrimidines by the cobalt- and nickel-saturated
clays at pH < 6 and by copper-montmorillonite at pH < 3 is similar to that
previously noted for the calcium sample, in which the process is principally one
of cation exchange. Under weakly acidic and weakly basic conditions complex
formation with the transition metal cations occurs, the strength of the complexes
decreasing in the order $Cu \gg Ni > Co \gg Ca$. Complexation is indicated by
the presence of well-defined maxima in the isotherms in the range of pH 7 to 9
(Co and Ni) and at pH > 4 (Cu) and, for the cobalt and copper systems, by
pronounced colour changes.

The question arises as to which atoms or groups in, say, the purine derivatives
are involved in chelation to the exchangeable cations assuming that bidentate
complexes are formed. Some workers[115,116] have suggested that the substituent
in the 6 position and the nitrogen atom in the 7 position (N-7) are involved,
while others[117-119] favour N-3 and N-9 as being the atoms participating in
chelation.

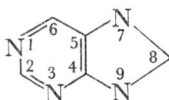

Lailach *et al.* have resolved this question by simply blocking either the 7- or

194

the 9-position by a methyl group and measuring the extent of uptake. They found that 7-methyladenine is more strongly adsorbed than adenine, and 9-methyladenine is not adsorbed under basic conditions. These observations support the view that the N-3 and N-9 positions are preferred in complex formation, in accord with the conclusions of Weiss and Venner[117,118] and Sletten[119]. The strong complexes with adenine, hypoxanthine, and purine are similarly interpreted. These compounds and 6-chloropurine differ only in the nature of the substituent in the 6-position which are NH_2, OH, H, and Cl, respectively. The presence of the electron-withdrawing chlorine atom results in weaker adsorption when compared with the other 3 derivatives, indicating that the basicity of the coordinating positions is an important factor determining the strength of the complex. As we shall see later, the dependence of adsorption on the nature of the substituent in this position is more clearly defined when the non-expanding layer silicate, illite, is used as the adsorbent. This behaviour is again reminiscent of that shown by the s-triazines where adsorption is controlled by the substituent group in the 2-position. Coordination of transition metal ions to groups in the 6-position and N-7 (or possibly N-1) is by no means excluded but the complexes so formed, e.g. with guanosine and inosine, are apparently weak. Little, if any, uptake of adenosine is observed, while cytosine and cytidine give rise to very weak complexes with copper-montmorillonite.

Iron (III)-montmorillonite shows quite a different behaviour from the other cation-exchanged samples. Complex formation may occur under weakly alkaline conditions but it is difficult to distinguish clearly between cation exchange and complexing processes from the nature of the isotherms. The NH_2 and/or the N-1 position of adenosine and the N-3 of cytidine seem to be involved in bonding.

In a subsequent study, Lailach and Brindley[113] have described the phenomenon of specific association or coadsorption of certain purines and pyrimidines taken up from aqueous solutions by sodium- and calcium-montmorillonites at pH 1 to 6. For example, thymine by itself fails to adsorb on montmorillonite but is taken up in the presence of adenine or hypoxanthine, both of which are intercalated. Although infra-red measurements were not carried out on these complexes, the adsorption data indicate that the mechanism underlying their formation is one of hydrogen-bonding interaction between the protonated and uncharged base (Fig. 71). Intermolecular hydrogen bonding of this type has been discussed previously (Section 3.4) and is of general occurrence in systems where organic bases are intercalated by montmorillonite containing interlayer organic cations (cf. Tables 28 and 34).

The basal spacing of the organic-association montmorillonite complexes is about 1·25 nm, indicating the presence in the interlayer space of a single layer of cation-base assemblages in a flat atomic arrangement. This observation contrasts with the conclusions of Ts'o and co-workers[120-122] who have presented evidence that purines, pyrimidines, and their nucleosides in aqueous solutions interact by vertical stacking of the molecules with partial ring overlap rather than by horizontal hydrogen bonding.

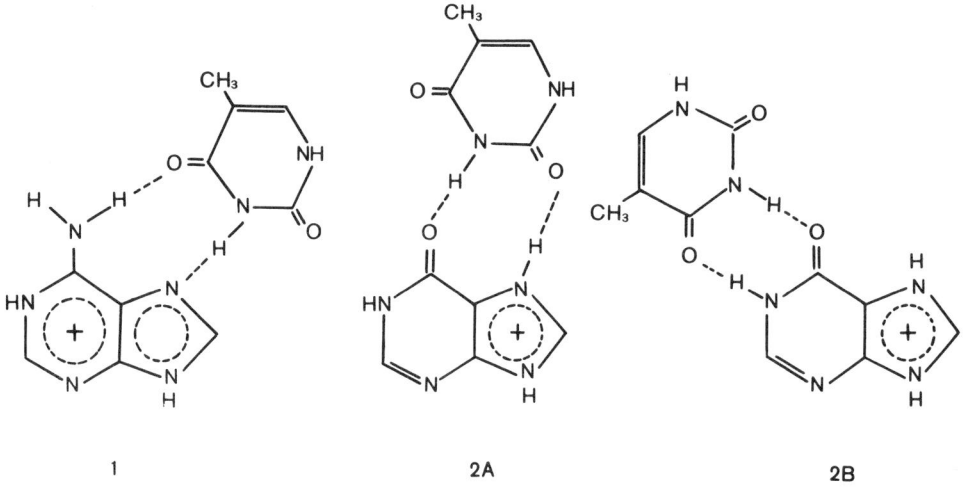

Fig. 71 Possible configurations of association (or coadsorption) complexes between (1) adenine and thymine; (2, A, B) between hypoxanthine and thymine, after Lailach and Brindley[113]

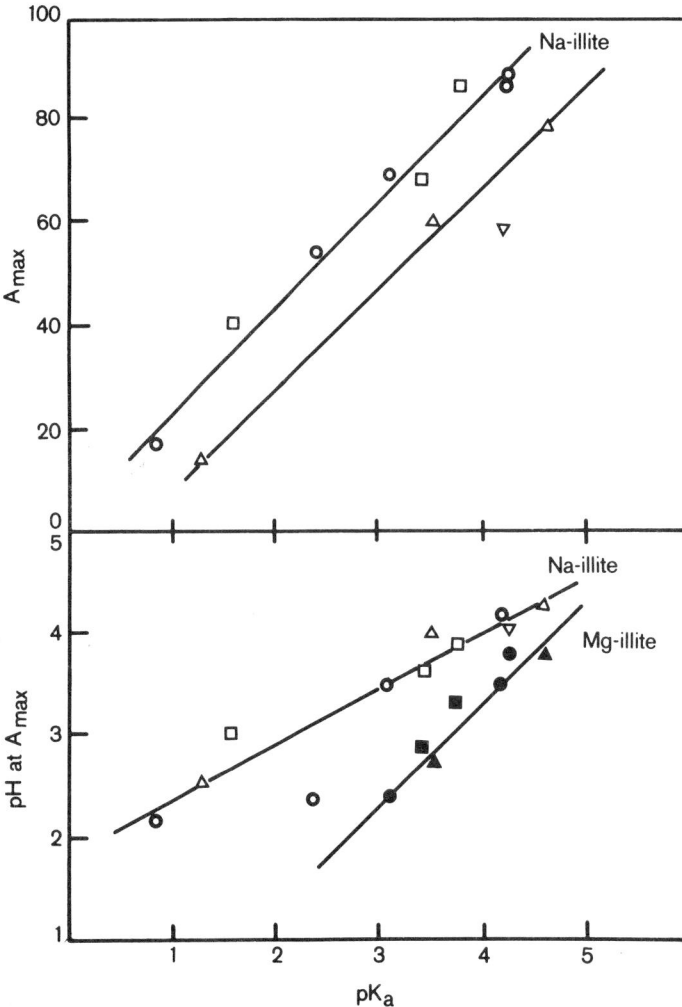

Fig. 72 Variation of pH at which adsorption is at a maximum, A_{max}, with pK_a of the organic compounds (bottom figure); and of A_{max} with pK_a (top figure). Open symbols refer to sodium-illite, solid symbols to magnesium-illite. The compounds examined are purines ($\bigcirc$), pyrimidines ($\triangle$), purine nucleosides ($\square$) and pyrimidine nucleoside (∇). Data from Thompson and Brindley[114]

As would be expected, the adsorption of purines, pyrimidines, and nucleosides from an aqueous medium by sodium-, magnesium-, and copper-illites[114] occurs by mechanisms similar to those controlling the uptake of these compounds by the corresponding montmorillonites. However, since illite is a non-expanding 2:1 layer silicate, adsorption is here confined to external crystal surfaces.

Under acidic conditions, cation exchange and proton transfer reactions predominate, the pH isotherms showing well-defined maxima. Both the pH values at which uptake is greatest and the height of the maxima are linearly related to the pK_a of the respective heterocyclic base or nucleoside (Fig. 72). The maximum amount adsorbed under comparable conditions also increases with molecular weight, reflecting the contribution of van der Waals interactions to the adsorption energy.

When the exchange positions are occupied by copper ions, complex formation between the organic molecule and the cation is indicated at near-neutral conditions (pH 6·5–8). For the purines, the maximum amount adsorbed can be correlated with the electron-releasing ability of the group attached at the 6-position of the purine ring. This is illustrated in Fig. 73 where maximum uptake is plotted against the respective Hammett σ_p coefficient, which is a measure of the electron-releasing (negative σ_p) or electron-withdrawing (positive σ_p) ability of the substituent group relative to hydrogen. The subscript p refers to the *para* relation between the substituent group and N-7 in the purine nucleus. For the derivatives examined the following σ_p values can be assigned: 0·66 (NH_2), 0·46 (OH), 0·17 (CH_3), 0·00 (H), and $-0·23$ (Cl).

The pyrimidines having a single ring structure do not appear to form a complex with copper ions. Adsorption of these compounds by copper-illite resembles that shown by the magnesium- and sodium-saturated materials in that it occurs principally by a cation-exchange process.

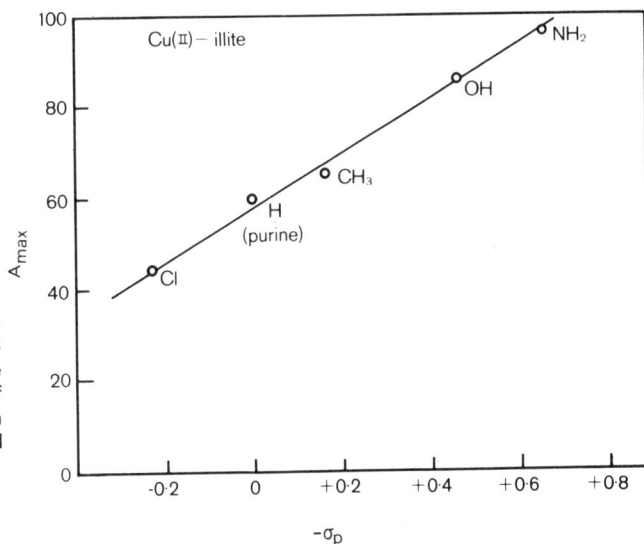

Fig. 73 Relation between maximum adsorption, A_{max}, of purines at pH 6·5 to 8 by copper (II)-illite and the electron-releasing coefficient,$-\sigma_p$, of the substituent groups in the 6-position of the purine ring. After Thompson and Brindley[114]

4.5. Complexes with Fatty Acids and Fats

There is much interest in the reactions of fatty acids with layer silicates because of the observation that clay minerals can catalyse the conversion of these compounds into petroleum hydrocarbons.[123,124] Details on the role played by clays in catalysing the reactions and transformations of adsorbed organic species are given in a later chapter.

By analogy with, say, acidic pesticides it might be expected that fatty acids would not be readily adsorbed by the negatively charged clay surface. However, with the possible exception of formic acid, alkane carboxylic acids show very limited dissociation in aqueous media and they exist, for the most part, in the uncharged state. Furthermore, complexes of these compounds with clays are usually formed in the absence of water. Even so, the preparation of clay complexes with fatty acids,[125] like those with the corresponding n-alcohols,[126] remains something of an art and the complexes produced are extremely sensitive to the mode of formation.

Brindley and Moll[125] have, nevertheless, been able to prepare two series of montmorillonite-fatty acid complexes, the short-spacing $(d(001) < 1\cdot7 \text{ nm})$ and long-spacing $(d(001) > 2\cdot0 \text{ nm})$ types. The former were obtained by reacting heat-treated (523°K, 45 mins) clay samples directly with the acid (up to and including C_9) and the latter resulted from first separating the silicate layers with n-hexanol or n-octanol before introducing the appropriate fatty acid (from C_{10} to C_{18}). The long-spacing complexes only exist at temperatures above the melting points of the free acids.

The X-ray diffraction data for the complexes with natural montmorillonite and synthetic fluormontmorillonite, both in the calcium-saturated form, are presented in Table 58. A plot of the long spacings, d, against the number of carbon atoms in the molecules, n, gives a straight line with $\Delta d/\Delta(2n) = 0\cdot2278$ nm. This value is less than the increase in chain length per two carbon atoms ($\sim0\cdot252$ nm). Following Brindley and Ray,[126] the angle ϕ at which the chain is inclined to the silicate surface is given by $\sin^{-1}(0\cdot2278/0\cdot252) = 1\cdot134$ rad (65°) if a single layer of fully extended chains is intercalated and assuming that the molecular configuration remains unaltered as n increases from 10 to 18. Similarly, the chain inclination is calculated as $\sin^{-1}(0\cdot2278/0\cdot504) = 0\cdot471$ rad (27°) if a double layer of organic molecules is present between two opposing silicate layers. The latter value, however, is too low to be compatible with any feasible model. We therefore infer that for the long-spacing complexes, the fatty acids are intercalated as single layers of molecules standing at $1\cdot134 \pm 0\cdot105$ rad to the montmorillonite surface. If this model is adopted, the length of the organic chain still falls short by 0·3 to 0·6 nm of the observed overall interlayer separation. This has led Brindley and Moll to propose that the fatty acid molecules are attached by their active carboxyl groups to the oxygens of the silicate surface, leaving the methyl terminations relatively free in a head-to-tail arrangement in pairs, with some lateral displacement due to the bulky carboxyl ends (Fig. 74). A more detailed analysis shows that an angle of tilt of about 1·222 rad (70°) besides giving rise to good intermolecular packing is

198

TABLE 58

Basal spacings (nm) of montmorillonite complexes with fatty acids at temperatures above the melting points of the free acids.[125]

Fatty acid		M.p. (°K)	Natural montmorillonite		Synthetic fluormontmorillonite		
C_2	Acetic acid	289·7	$1·65 \pm 0·01$	(5) f	$1·58_5 \pm 0·01$ $1·56_0 \pm 0·01$	(8) (11)	vg[1] vg
C_4	Butyric acid	267·5	$1·41 \pm 0·01$	(3) p	$1·47_0 \pm 0·01_5$ $1·46_0 \pm 0·02$	(7) (7)	g[1] g
C_6	Caproic acid	269·1	$1·52$	(1) p	$1·50_5 \pm 0·00_5$ $1·49_0 \pm 0·02$	(8) (7)	g[1] g
C_8	Caprylic (octoic) acid	289·5	$1·44 \pm 0·01$	(3) p	$1·44_0 \pm 0·09$	(7)	f
C_9	Pelargonic (nonoic) acid	285·5	$1·43 \pm 0·01$	(3) p	$1·31_5 \pm 0·05$	(7)	f
C_{10}	Capric (decoic) acid	304	$2·92 \pm 0·02_5$	(2) p	$2·99_5 \pm 0·01$ $3·06 \pm 0·02_5$ $3·06_5 \pm 0·02$	(11) (11) (11)	g[2] f[2] g
C_{11}	Undecoic acid	303	$3·12 \pm 0·14$	(2) p	$3·11_5 \pm 0·00_5$	(11)	f
C_{12}	Lauric acid	317	$3·22 \pm 0·07$	(7) p	$3·17_5 \pm 0·01_5$ $3·19_5 \pm 0·01$ $3·18 \pm 0·02_5$	(11) (11) (11)	f f g[3]
C_{13}	Tridecoic acid	318·5	$3·28 \pm 0·07$	(3) p	$3·26_5 \pm 0·01_5$	(12)	f
C_{14}	Myristic acid	326·6	$3·31 \pm 0·05$	(3) p	$3·46_5 \pm 0·01_5$ $3·34 \pm 0·03$ $3·42_5 \pm 0·02_5$	(11) (10) (12)	p p f[3]
C_{16}	Palmitic acid	335·6	$3·67 \pm 0·02$	(3) p	$3·68_5 \pm 0·01_5$ $3·66_5 \pm 0·03_5$	(13) (13)	f f
C_{18}	Stearic acid	342·6	$3·93 \pm 0·07$	(4) p	$4·05_5 \pm 0·02_5$ $3·93_5 \pm 0·03_5$ $3·86 \pm 0·02$	(18) (18) (13)	p p f[3]

1 Obtained after a few hours' contact only.
2 Obtained at about 301°K.
3 Prepared through the n-octanol-clay complex
   vg = very good;   g = good;   f = fair;   p = poor
   Number in parentheses refers to highest order observed.

also consistent with van der Waals contacts and O—H . . . O—Si bonding. Such an arrangement is also compatible with the a and b parameters of the silicate surfaces.

Using a series of 2:1 type layer silicates containing interlayer n-alkylam-

199

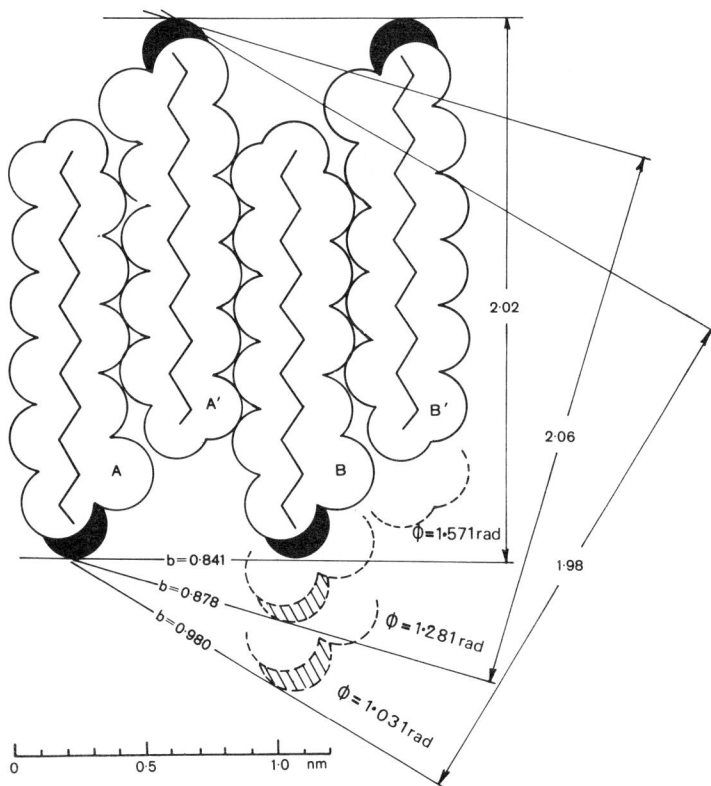

Fig. 74 Brindley and Moll's[125] proposal of close-packing of fatty acid molecules in pairs with head-to-tail arrangement in the interlayer space of montmorillonite. Dashed lines show various displacements of the pair B,B' with respect to A,A'. Spacings of terminal planes are shown and the corresponding angles of inclination, ϕ, of chain axes with respect to the terminal planes. Hydrogens in terminal hydroxyls are shaded

monium ions (RNH_3^+) and introducing into this system a variety of n-alkyl compounds (X) such as alcohols, amines, aldehydes, and carboxylic acids, Weiss[127] has observed that the sum of $(RNH_3^+ + X)$ in all the association complexes was remarkably similar, about 2·07 per $(Al, Si)_4O_{10}$ unit (formula unit, Table 1) or about 0·21 nm^2 per molecule of X. This value is comparable with that derived from Brindley and Moll's[125] model in which two fatty acid molecules are present per unit silicate cell of about 0·90 × 0·52 nm^2 or 0·23 nm^2 per molecule.

As for the short-spacing complexes, no regular trend is observed with an increase in chain length. Whether single or double layers of molecules, lying flat on the silicate surface, are intercalated cannot be deduced with any certainty from the observed basal spacings. In the absence of adsorption data, it is also difficult to remark on the mechanism controlling the intercalation of fatty acids in these complexes. However, by analogy with benzoic acid[75], it seems probable that short-chain fatty acids, at least, would interact through their carboxyl groups with the exchangeable calcium ions.

200

Brindley and Moll have also examined the stability of the long-spacing complexes against heat treatment and solvent extraction. Curiously, these complexes are stable above, but not below, the melting point of the respective free fatty acid. It may be recalled that the corresponding n-alcohols yield long-spacing complexes which can exist both above and below the melting point of the free compounds (cf. Fig. 19). Although the thermal behaviour of montmorillonite-fatty acid complexes contrasts with that shown by mono-layers of fatty acid molecules, which disorganize when the temperature of the system is raised above the melting point of the crystalline acids,[128] it is con-sistent with the view that the intercalated molecules are, on the whole, weakly held by the clay surface. Washing the complexes with ethanol, for example, causes the long spacings to collapse. Some high spacings (1·3–1·4 nm), however, persist even after heating up to 523°K for 70 hours, indicating that a portion of the intercalated acids or their decomposition products are strongly retained by the clay.

Using stearic and behenic acids and a predominantly calcium saturated montmorillonite, Sieskind and Ourisson[129] have recently shown that the part which is resistant to solvent (benzene) extraction is adsorbed at the edges of the clay crystals. The infra-red spectrum of the washed complex with stearic acid shows absorption bands characteristic of the carboxylate ion rather than of the un-ionized carboxyl group. This observation led these workers to suggest that octahedrally coordinated aluminium ions, exposed at crystal edges, provided the sites at which the (residual) fatty acid anions are held. The amount retained (0·2 meq g^{-1}), however, seems rather high to be accounted for by this mechanism alone. It seems probable that the carboxylate group of the anion is bonded to the clay through a calcium ion bridge while salt formation may also contribute to the overall uptake. This proposal is indirectly sup-ported by the observation that the sodium form of the clay sample retains considerably less stearic acid (0·04 meq g^{-1}) under similar conditions.

Sieskind and Siffert[130] have subsequently extended the above study to a synthetic hectorite in which the aluminium ions in the octahedral sheet of the silicate had been replaced by nickel ions. Using diffuse reflectance and infra-red spectroscopy, they confirmed that the nickel ions, exposed at crystal edges, were involved in binding the stearate anion to the clay. The precise mechanism is not clear, but it is inferred that an exchange reaction takes place between the organic anion and the hydroxyl groups initially attached to edge nickel ions. It is clear, however, that the crystal edges possess active sites at which organic anions may be firmly adsorbed.

Equally relevant in this context is the work of Kaufherr et al.[131] who studied the reaction of a commercial sample of K-Cu chlorophyllin with montmoril-lonite containing different exchangeable cations (Na, Cs, Mg, Co, Cu, and Al). The organic sample used is a mixture consisting of chlorophyllin in the tri-carboxylated form (Fig. 75(A)), chlorophyllin in the keto form (Fig. 75(B)), and an aliphatic carboxylate salt derived from the phytyl group. During uptake of this mixture, potassium replaces the exchangeable cations (except cesium)

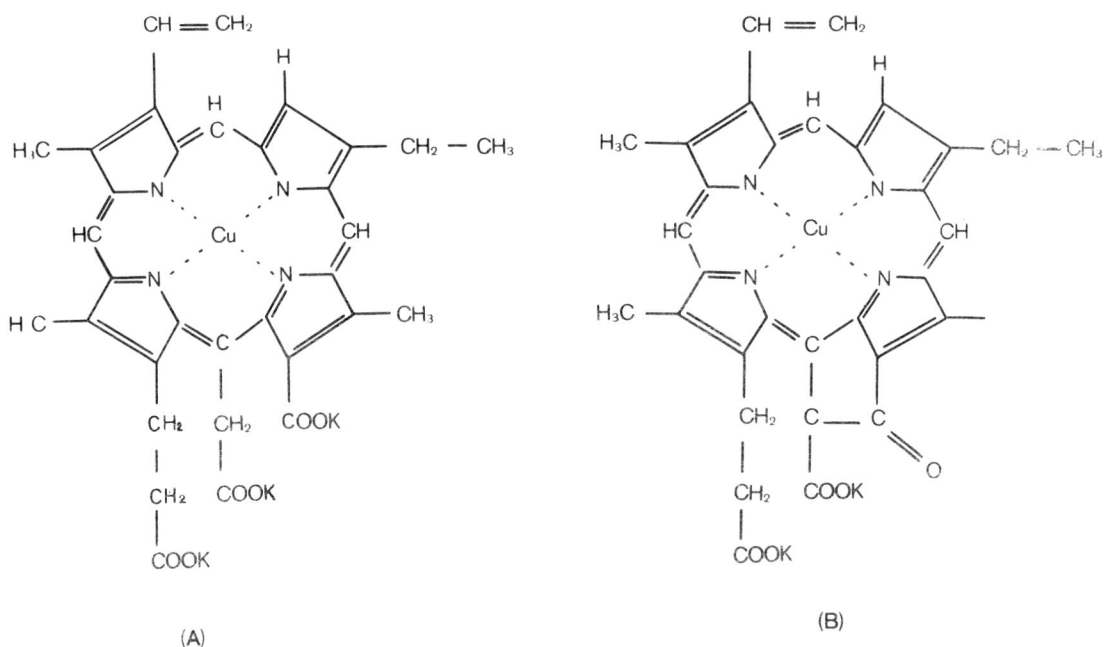

Fig. 75 Structure of chlorophyllin, (A) tricarboxylated form, and (B) α-*keto* form, after Kaufherr *et al*[131]

and the organic anion is adsorbed on external surfaces, presumably the crystal edges. Chemical and infra-red data indicate that the three components of the mixture are taken up selectively and to different extents by the various cationic forms of the montmorillonite. The cesium clay, for example, adsorbs chlorophyllin only, whereas the other samples of clay take up varying amounts of the carboxylic acid salts.

Weiss and Roloff[132] have earlier reported that porphyrins, presented in the cationic form, can penetrate the interlayer space of montmorillonite. Intercalation modifies the thermal stability of the hemin. The extent of this modification depends on such factors as the type of complex formed, layer charge, and the amount, size, and orientation of the organic species present. For montmorillonite and hectorite complexes with guanidine and its derivatives, Beck and Brunton[133] have observed that the decomposition temperature of the complexes can exceed that of the parent clays by as much as $548\,^{\circ}K$ as measured by the oscillating-heating X-ray technique.

The importance of clay minerals as a concentrating substrate or medium for fatty acids present in natural bodies of water has been investigated by Meyers and Quinn.[134] Using a saline solution comparable to sea water, these workers found that the isotherm for the adsorption of fatty acids such as heptadecanoic acid by montmorillonite was linear up to a concentration of $120\ \mu g\ 1^{-1}$. The amount of acid removed from solution by adsorption onto the clay is therefore directly related to the organic concentration. The amount of fatty acids removed from sea water by this means, however, is consistently less than from the artificial saline solution. This discrepancy is ascribed to interference and com-

202

TABLE 59

Basal spacings (nm) of various 2 : 1 type layer silicates containing n-dodecylammonium ions after intercalation of some triglycerides in relation to the charge per formula unit and the area available to the fat molecules.[135]

Triglyceride	Molecular area (nm²)*	n-Dodecylammonium-triglyceride complexes with					
		Hectorite	Montmorillonite (Wyoming)	Montmorillonite (Geisenheim)	Montmorillonite (Cyprus)	Beidellite (Unterrupsroth)	Vermiculite (S. Africa)
Triacetin	0·54	2·79	2·82	2·83	2·82	2·87	2·88
Tributyrin	0·98	2·84	2·84	2·84	2·86	2·86	3·20
Tricaproin	1·52	2·85	3·21	3·25	3·28	3·39	3·68
Tricaprylin	2·04	3·50	3·70	3·78	3·80	3·89	4·20
Charge per formula unit		0·25	0·32	0·34	0·40	0·43	0·68
Equivalent area or area per unit charge (nm²)		1·00	0·83	0·75	0·62	0·57	0·37
Area available for intercalated triglyceride molecule (nm²)**		1·80	1·46	1·30	1·04	0·94	0·54

* Estimated from minimum and maximum dimensions of model compound.
** Twice the equivalent area–area per alkylammonium ion, assuming close packing of steeply inclined chains. Broken line indicates the limit at which the area of the fat molecule exceeds the available area.

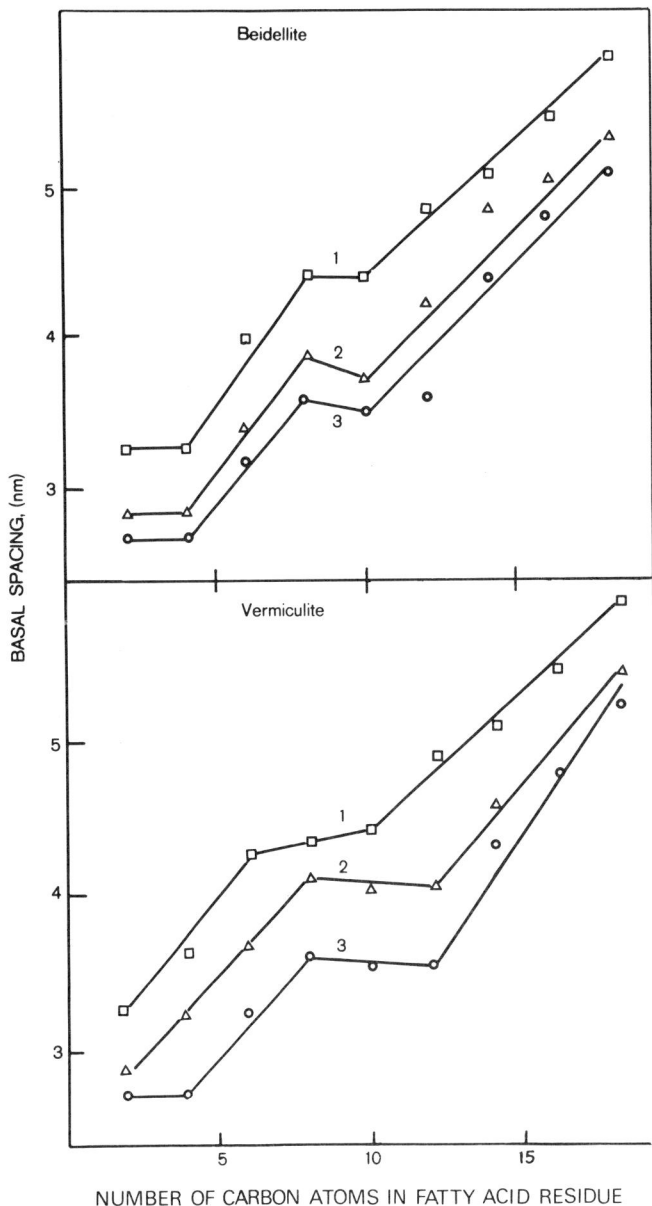

Fig. 76 Variation of basal spacing of two 2:1 layer silicates saturated with long-chain *n*-alkylammonium ions after intercalation of some symmetric triglyceride molecules, with number of carbon atoms in the fatty acid residue. *Top*: beidellite from Unterrupsroth saturated with *n*-octadecylammonium (1), *n*-dodecylammonium (2), and *n*-decylammonium (3) ions. *Bottom*: vermiculite from S. Africa saturated with *n*-octadecylammonium (1), *n*-dodecylammonium (2), and *n*-decyl-ammonium (3) ions. Data from Weiss and Roloff[135]

petition from other organic substances dissolved in sea water and to the presence of ionic species other than sodium and chloride in the natural environment of the sea.

The intercalation of symmetric triglycerides (fats) by montmorillonites and vermiculties has been examined by Weiss and Roloff[135] using X-ray diffraction techniques. As for the long-chain fatty acids, some difficulty is encountered in the preparation of complexes with montmorillonite containing interlayer sodium and calcium ions and the results are not always reproducible. This is probably due to the fact that part of the intercalated compounds may become hydro-lysed giving rise to a mixture of mono-, di-, and triglycerides, glycerol, and fatty acids. High-charge silicates such as the vermiculites fail to react with fats.

There is no such difficulty with the intercalation of fats by $2:1$ type layer silicates containing interlayer n-alkylammonium ions. Regular complexes are formed showing rational orders of basal reflections. We have already referred to the near-constancy of the sum of the pair of n-alkylammonium ion $+$ n-alkyl compound per $(Al, Si)_4O_{10}$ unit. Indeed, the basal spacing of such organic association clay complexes is related in a simple manner to the length of the fatty acid chain in the triglyceride molecule, the length of the alkylammonium ion, and the surface charge density of the mineral. These interrelationships are summarized in Table 59 and depicted in Fig. 76 (for two mineral species). Clearly the change in basal spacing on intercalation of fat molecules is deter-mined by both the total charge of the silicate and by the size of the triglyceride compound.

The break in the curves (Fig. 76) for C_{10} (in the beidellite sample) and C_{12} (in the vermiculite material), both of these compounds having melting points near the mean room temperature, suggests to Weiss and Roloff that interlayer expansion is temperature-dependent. It appears that, below the melting point of the fat, there is a small but regular increase in basal spacing with tempera-ture. Near the melting point of the intercalated compound the spacing abruptly decreases, this transition being more pronounced for the higher members of the series (Fig. 77). This behaviour is akin to that shown by the complexes with

Fig. 77 Dependence on temperature of basal spacing of n-dodecylammonium-beidellite after intercalation of some symmetric triglycerides, (1) triacetin; (2) tributyrin; (3) tricaproin; (4) tricaprylin; (5) tricaprin; (6) trilaurin; (7) trimyristin; (8) tripalmitin; and (9) tristearin. Data from Waiss and Roloff[135]

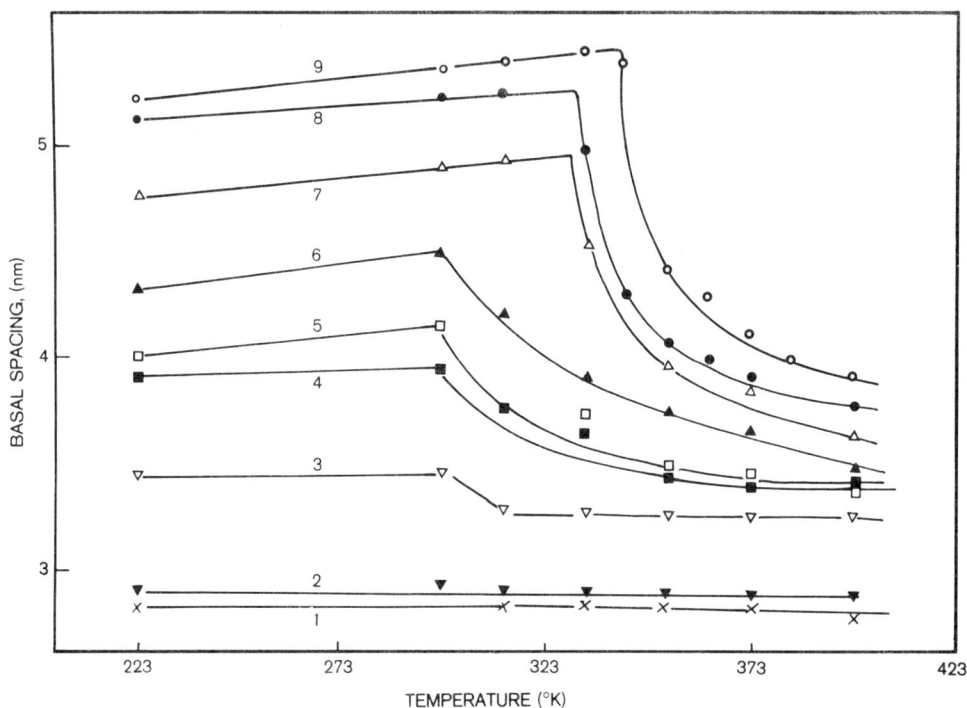

n-alcohols discussed in Section 3.1 (cf. Table 6). Unlike the *n*-alcohol complexes, however, this phase change is apparently reversible and hence is unlikely to be caused by a net loss of material. Rather, it is due to a change in configuration of the triglyceride molecule near or at its melting point, as proposed earlier by Brindley and Ray[126] for the *n*-alcohol-montmorillonite complexes.

4.6. Complexes with Saccharides

In studying the interaction between montmorillonite and a variety of polyfunctional organic compounds, Bradley[136] noted that sucrose was intercalated, yielding a complex with a basal spacing of about 1·83 nm. Greenland[137] later extended Bradley's work to a large number of mono- and oligo-saccharides and their derivatives using montmorillonite saturated with hydrogen, sodium, potassium, and calcium ions. He found that each sugar formed a complex having two and sometimes three different basal spacings, but the spacing for any one complex was characteristic of the saccharide used. Evidently, when more than one type of complex forms, two sets of basal spacings are observed indicating that within a single crystal each type is segregated rather than either randomly or regularly interstratified. The relative amount of the one and two layer types of complex can be estimated from the relative intensity of the basal reflections on the assumption that equal quantities of each type give rise to two sets of reflections of equal intensity, and also that the intensity of the higher orders of reflection relative to that of the first order varies in the same way as in a single-type system.

The X-ray data for complexes with some saccharides, together with the probable types of complex formed and their ratios, are listed in Table 60. The dependence of adsorption on sugar concentration is what would be expected for a process which is principally one of physical adsorption. However, the extent of adsorption does not appear to go beyond the formation of a double layer of sugar molecules in the interlayer space, even when the sugar concentration is increased to 20 per cent. The influence of the interlayer cation is also clearly shown. With glucose, for example, adsorption increases in the order K < Ca < Mg < H < Na under comparable conditions of sugar and clay concentrations. Omitting hydrogen and sodium, this order corresponds to that of the polarizing power of the cations and accords with the behaviour of nonionic compounds at clay mineral surfaces where ion-dipole interactions are the determining factor (Chapter 3). The position of hydrogen may be ascribed to the presence of substantial amounts of aluminium ions released from the silicate structure and migrating to exchange sites. The large adsorption with sodium-montmorillonite is associated with the high dispersion of this sample, giving rise to large interlayer separations and hence facilitating entry of the saccharide molecules into the clay interlayers.

Because adsorption depends on both the sugar concentration and the nature of the interlayer cation, these factors must be kept constant if the affinity of the

TABLE 60

Basal spacing (nm) of some montmorillonite-saccharide complexes as influenced by the exchangeable cation and the initial concentration of clay and saccharide.[137]

Exchangeable cation	Saccharide compound	Montmorillonite concentration %	Saccharide concentration %	Basal spacing		Type of complex	Ratio of types
H	glucose	5	2·5	1·85		double layer	—
Na	glucose	5	2·5	1·81		double layer	—
K	glucose	5	2·5	1·43		single layer	—
Ca	glucose	5	2·5	1·45		single layer	—
H	glucose	2·5	1	1·40;	1·22	single and O layer	2:1
Na	glucose	2·5	1	1·85		double layer	—
K	glucose	2·5	1	1·33(d)		single and O layer	2:1
Ca	glucose	2·5	1	1·42;	1·21	single and O layer	2:1
Mg	glucose	2·5	1	1·79;	1·45	double and single layer	1:2
Na	mannose	3	1·5	1·92;	1·59	double and single layer	1:2
Ca	mannose	5	2·5	1·60		single layer	—
Na	sucrose	3	1·5	1·95;	1·64	double and 1½ layer	3:2
Ca	sucrose	2·5	1	1·65		1½ layer	—
Na	galacturonolactone	3	1·5	1·80		double layer	—
Ca	galacturonolactone	4	2	1·75;	1·52	double and single layer	1:1

d = diffuse

organic compounds for the clay surface is to be compared. If we further assume that those saccharides which give a double-layer complex from the more dilute solutions are the more strongly adsorbed, while those sugars requiring high initial concentrations to form a complex are more weakly held, the organic compounds can be arranged in order of decreasing affinity (adsorption energy) as set out in Table 61.

We make the following general conclusions: methylated glucoses are very strongly adsorbed as compared with the unsubstituted sugars; carboxyl or

TABLE 61

Complexes* of montmorillonite with some sugars and related compounds together with Δ values and minimum van der Waals thicknesses of the molecules.[137]

| Compound[a] | Δ value (nm) | | Minimum molecular thickness (nm) |
	Single layer	Double layer	
2,4,6-Trimethyl glucose	0·51	0·92	0·53
4,6-Dimethyl glucose	0·56	1·09	0·53
Arabinose	—	0·90	—
Ascorbic acid	—	0·83	—
Fucose	—	0·94	—
N-acetylglucosamine	—	0·81 (and 1·0?)	—
Cellobiose	0·50	0·88	0·51
Lactose	0·48	0·87	0·53
Maltose	0·54	0·93	0·54
Melibiose	0·53	0·88	0·53
Galacturonolactone	0·58	0·85	—
Raffinose	0·58	0·91	—
Rhamnose	0·64	1·28	0·59
Sucrose	0·56	0·91	0·56
α-methyl glucoside	0·58	—	0·57
Sorbose	0·53	—	0·52
Erythritol	0·52	0·91	—
Galactose	0·58; 0·50	—	0·53
Heptulose	0·62	—	—
i-Inositol	0·49	—	—
Xylose	0·47; 0·51	0·96	0·48; 0·52
Glucose	0·48; 0·52	0·93	0·47; 0·53[b]
Mannose	0·67	1·04	0·59
Mannitol	0·50 (and 0·68?)	—	—
Gluconolactone	0·57	—	—
Glucuronolactone	0·57	—	—
Glucosamine	0·48	—	0·47; 0·53

* Prepared from a 2·5 per cent suspension of calcium-montmorillonite in a 1 per cent solution of the sugar.
[a] Listed in decreasing order of adsorption energy of affinity.
[b] α-glucopyranose (0·53).
  β-glucopyranose (0·47).

amino-substituted glucoses show stronger adsorption than the unsubstituted sugars; disaccharides have a greater affinity for montmorillonite than the monosaccharides.

Table 61 also compares Δ values with minimum molecular thicknesses of the intercalated saccharides. These are derived from the dimension of the pyranose and furanose rings found in crystals of some of the sugars, using Pauling's van der Waals radii (with hydrogen equals 0·12 nm) to estimate the size of the appropriate groups around the rings. There is clearly good agreement between Δ value and minimum molecular thickness for most of the compounds examined. This further supports the view that the smallest observed spacing for any complex is ascribable to the presence of a single layer of sugar molecules in the interlayer region, with the molecules lying as flat as possible between two opposing silicate layers. Only with rhamnose and 4,6-dimethylglucose does the second Δ value correspond exactly to twice the first; with the other sugars the former is slightly less than twice the first Δ value. This indicates that where a double layer of sugar molecules is intercalated the layers tend to interpenetrate. It is also of interest to note that the Δ values for single-layer complexes with β-glucose, 2,4,6-trimethyl-β-glucose, and cellobiose are not much different, being similar to the corresponding minimum molecular thicknesses. Since, in other respects, their dimensions are widely different, this observation provides strong evidence for the flat configuration of the intercalated molecules.

Chemical studies[138] on the sugar complexes confirm the X-ray data in that methylated derivatives are more strongly adsorbed than the unsubstituted sugars or sugars containing other substituents such as amino or carboxyl groups. Several factors may contribute to the strong adsorption of, for example, methylated glucose as compared with glucose. O—H . . . O—Si hydrogen bonding although unlikely to be strong, is absent in the methylated derivative. Further, methyl-substituted sugars tend to polymerize,[139] a tendency which would be promoted by close packing or favourable juxtaposition of the intercalated molecules. A more compelling reason for the weak adsorption of glucose is that the molecule is strongly hydrated in aqueous solutions[140] and hence would be less capable of displacing water molecules from the montmorillonite surface. Methylated glucose, being less solvated, is more readily adsorbed under comparable conditions.

Greenland has also determined the pH-uptake isotherms for gluconic acid, glucuronic acid, and glucosamine with calcium-montmorillonite. All three compounds show a more or less defined maximum in the range of pH 4 to 5. The isotherm for glucosamine closely resembles that of the purines and pyrimidines (Fig. 67) and can be similarly interpreted. The high adsorption at pH 3 to 6 can be ascribed to cation exchange and proton transfer processes. For the sugar acids, adsorption on crystal edges may be postulated as described in the preceding section. The rise in adsorption of the acids at pH > 7 may be due to the formation of insoluble salts or bridge linkages or by anion exchange involving edge aluminium ions.

Using a natural and calcium-saturated bentonite, Mitra et al.[141] have noted

209

that glucose adsorption was less with the latter material.. This again reflects the importance of the degree of disperson of the clay; the natural sample presumably being predominantly in the sodium form is capable of taking up a greater amount of the sugar. The adsorption data for the calcium system conformed to the Freundlich isotherm (Eq. (4)) and the formation of a complex between glucose and calcium ions was postulated.

Very recently, Jepson and Williams[142] have measured the adsorption of glucose on sodium kaolinite following an earlier report by Davis and Worrall[143] that this sugar showed negative adsorption from aqueous solution with kaolinite clays. The former workers found that adsorption was very small but was probably positive, attributing the discrepancy in observation to the fact that Davis and Worrall's experimental techniques were not sufficiently precise.

5

Interactions with Positively Charged Organic Species

5.1. Mechanisms of Formation

Due to isomorphous substitution, clay minerals carry a permanent negative charge in their structural framework (Chapter 1). In their natural state, this positive charge deficiency is balanced by sorption of an equivalent amount of extraneous inorganic cations, such as sodium, potassium, calcium and magnesium. Because of this property, it was predicted that positively charged organic species would likewise be capable of neutralizing the anionic framework of layer silicates. Not surprisingly, many of the early references to clay-organic systems were concerned with attempts at reacting clays with organic cations or at neutralizing acid-treated minerals with organic bases. However, the number and variety of organic compounds which can acquire a positive charge or act as a base is limited as compared with uncharged, polar species. Almost invariably, such compounds contain nitrogen in their molecule, of which the amines form perhaps the largest single class.

As early as 1916, Lloyd[1] observed that alkaloids presented as their aqueous salt solution were effectively removed by some samples of fuller's earth. But it was not until the 1930's that a systematic investigation began into the reactions of clays and (defined) organic cations and bases.

In 1934 Smith[2] reported that montmorillonite was capable of taking up hydrazine, ammonia, n-amylamine, piperidine, di-n-amylamine, nicotine and strychnine from their aqueous hydrochloride solution, adsorption increasing in this order. The adsorption process was essentially one of cation exchange, since washing the clay complexes with water alone did not materially displace the organic molecules, whereas a 1 per cent aqueous solution of sodium chloride released appreciable quantities of sorbed species.

Following up this work, Gieseking and co-workers[3-6] examined the interaction of montmorillonite clays with a number of organic bases, cations and proteins. These compounds were found either to replace the inorganic cations initially present or to neutralize the hydrogen ions at the clay surface. With proteins, such as gelatine, uptake under acid conditions gave rise to a marked reduction in the exchange capacity of the montmorillonite samples used. X-ray diffraction measurements showed that the basal spacing of the air-dry and oven-dry (378°K) complexes exceeded that of the parent calcium and hydrogen clays, indicating that intercalation had occurred. Besides being insensitive to the water content of the sample, the basal spacing was little correlated with the dimension of the intercalated organic cation.

Hendricks[7] extended Gieseking's[4] study to other aliphatic and aromatic amines, alkaloids, purines and nucleosides using hydrogen-montmorillonite as adsorbent. He was able to show that small organic bases neutralized the hydrogen ions up, or close to, the exchange capacity of the montmorillonite (determined by exchange with barium). On the other hand, large alkaloids, such as brucine and codeine, failed to neutralize all the hydrogen ions present. This led Hendricks to postulate that the difference between the total amount of hydrogen ions and that available for reacting with the alkaloid represents the quantity 'covered' by the organic base. This cover-up effect comes into operation when the size of the adsorbed organic molecules exceeds the area per exchange site (~ 0.8 nm^2 for montmorillonite). A similar effect has also been reported by Chakravarti[8] for a number of large quaternary ammonium and pyridinium salts in montmorillonite.

An important point to emerge from Hendricks' study is the recognition that the adsorption of organic cations by clays is influenced by both electrostatic (coulombic) and van der Waals attractive forces. The basal spacing of the interlayer complexes formed is, for this reason, frequently determined more by the thickness than by the overall dimension of the intercalated molecule. In montmorillonites, at least, adsorbed organic cations and bases tend to adopt the flattest possible conformation. Long chain alkylammonium ions adsorbed in the interlayer space of high-charge vermiculites usually adopt an extended conformation in which the alkyl chain is inclined at a high angle to the silicate surface. This arrangement is imposed by the layer charge on the mineral and by the requirement of close van der Waals contact between adjacent hydrocarbon chains.

The influence of van der Waals forces on the adsorption of large organic cations by clays was subsequently confirmed by Grim *et al.*[9] who determined the uptake of *n*-butyl-, *n*-dodecyl-, and ethyldimethyloctadecenylammonium ions by potassium-saturated samples of a kaolinite, an illite, and a montmorillonite. Uptake of these organic ions was shown to be accompanied by a stoichiometric replacement of potassium from the exchange sites. For the relatively small *n*-butylammonium the maximum amount adsorbed did not go beyond the cation exchange capacity (c.e.c.) of the clays even when a large excess of the organic ion was present in the external solution. Under similar conditions, the two larger organic ions were taken up in excess of the c.e.c., this excess being adsorbed by van der Waals forces only. For the quaternary ammonium ion used, the quantity so held was present principally as its corresponding salt, since bromide was concomitantly adsorbed and there was little change in suspension pH. For *n*-dodecylammonium, however, the equilibrium pH

212

($\sim$5·6) was appreciably less than that of the original solution (pH $\sim$7) so that the excess was partly present in the form of the corresponding free amine.

Adsorption exceeding the c.e.c. of the mineral has also been noted by Kurilenko and Mikhalyuk[10] for methyloctadecyl-, n-dodecyl-, tri-n-decyl-, and trimethyloctadecyl-ammonium ions in sodium- and calcium-montmorillonites.

Similarly Morel and Hénin[11] have reported 'excess' uptake of hexamethylenediamine, dodecylpropyldiamine, and polyvinylamine cations in kaolinite, montmorillonite, and sepiolite, the amount adsorbed beyond the c.e.c. increasing with molecular weight.

Despite the inherent complexity of the systems, several workers have attempted to characterize the exchange process using the thermodynamic approach. As Barrer and Kelsey[12] have pointed out, at least qualitatively correct interpretation of prominent physical aspects could be obtained by application of thermodynamic functions to adsorption processes in clay-organic systems.

Cowan and White[13] have determined the isotherms at 293°K for the adsorption of an homologous series of primary n-alkylammonium ions from ethyl-(C_2) to tetradecyl-ammonium (C_{14}) by sodium-montmorillonite. They showed that the isotherms belonged to the L or Langmuir type[14] of which both the initial slope and the plateau adsorption increased with molecular weight. Above C_8, adsorption beyond the c.e.c. of the clay occurred. Similar conclusions were reported earlier by Sieskind and Wey.[15]

The exchange process can be represented by the following relationship where A and AH^+ refer to the free amine and the amine cation (alkylammonium), respectively:

$$Na^+{}_{clay} + AH^+{}_{soln} \rightleftharpoons AH^+{}_{clay} + Na^+{}_{soln} \tag{1}$$

and

$$AH^+{}_{clay} + AHCl_{soln} \rightleftharpoons (AH^+{}_{clay})(AHCl_{clay}) \tag{2}$$

$$(AH^+{}_{clay})(AHCl_{clay}) \overset{H_2O}{\rightleftharpoons} AH^+{}_{clay} + A_{clay} + HCl_{soln} \tag{3}$$

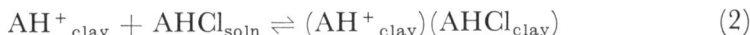

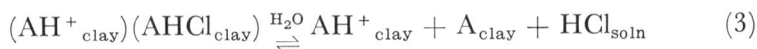

Reaction (1) applies to all the amine cations and proceeds until the exchangeable sodiums are completely, or nearly completely, replaced. Reactions (2) and (3) represent situations where adsorption in excess of the c.e.c. occurs, the excess adsorbate being present either as the amine salt (2), or as the free amine (3). If reaction (2) predominates, the corresponding anion (in this case chloride) would also be taken up, whereas reaction (3) would give a decrease in pH. The latter seems to be the dominant mechanism in Cowan and White's system.

By applying the mass-action relationship to reaction (1), the non-standard free energy change of the exchange process, $-\Delta G^m$, was derived (cf. Section 4.2). There is a regular increase of $-\Delta G^m$ values with molecular weight, amounting to $\sim$0·63 kJ mol^{-1} per CH_2 group from C_2 to C_5 and $\sim$1·67 kJ mol^{-1} from C_5 to C_{10}. This relationship between free energy change and molecular weight (or chain length) is of general applicability to the adsorption of organic compounds by montmorillonite, being attributed to the increased contribution of van der Waals forces to the adsorption energy.

Theng et al.[16] have extended Cowan and White's study to the di-, tri-, and tetra-alkylammonium ions (with an alkyl chain ranging from C_1 to C_4) in an attempt to clarify the effect of cation size and shape on relative adsorption affinity. The exchange isotherms for the ethyl (Et) derivatives with sodium-montmorillonite are shown in Fig. 78 where the equivalent fraction of the organic cation in solution, S_{AH^+}, is plotted against the corresponding equivalent

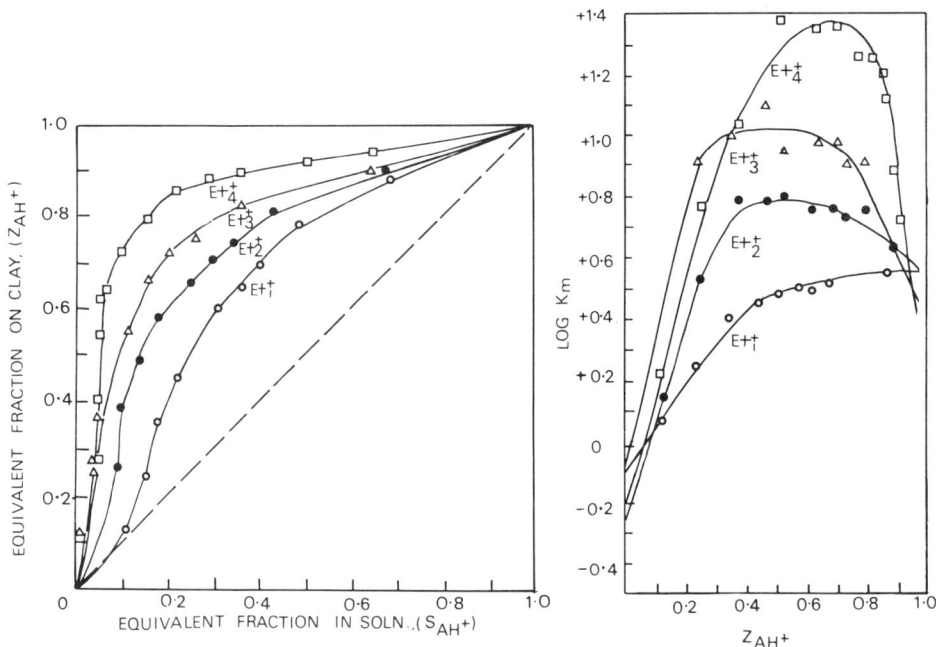

Fig. 78 *Left:* Isotherms for the exchange of monoethyl-(Et_1), diethyl-(Et_2), triethyl(Et_3), and tetraethyl-ammonium (Et_4) ions for sodium ions in montmorillonite at 298°K. *Right* Variation of the logarithm of the corrected molar selectivity coefficients with equivalent fraction of adsorbed ethylammonium ions (after Theng *et al*[16])

fraction on the clay, Z_{AH^+}. The diagonal of the square plot thus represents the hypothetical situation where the clay shows no preference for either one of the exchanging cations. The isotherms, all of which are of the L type, indicate that the alkylammonium ion is preferred to sodium at all values of S_{AH^+}. This preference increases with molecular weight, confirming previous findings[13,15] for the monoalkylammonium series.

The equilibrium constant, K, of the reaction is estimated using the simplified form of the equation developed by Gaines and Thomas:[17]

$$\ln K = \int_0^1 \ln K' dZ_{AH^+} \qquad (4)$$

214

where K' is the 'corrected' rational selectivity coefficient,[18] defined as

$$K' = K_m \gamma_{Na^+}/\gamma_{AH^+} \qquad (5)$$

γ_{a^+} and γ_{AH^+} are the activity coefficients of the respective cation in solution and may be replaced, to a good approximation, by the mean activity coefficients for the corresponding chloride salt. Since for the dilute solutions (<0.1 M) used, $\gamma_{NaCl}/\gamma_{AHCl}$ does not deviate significantly from unity,[19,20] $K' \simeq K_m$, and hence K is obtained from a plot of $\log K_m$ against Z_{AH^+} (Fig. 78) by graphical integration[21] between the limits $Z_{AH^+} = 0$ and $Z_{AH^+} = 1$.

It is implicitly assumed that reaction (1) is reversible, that is, the same equilibrium conditions would be obtained by approach from either side. As Fripiat et al.[22] have demonstrated, this type of macroscopic reversibility is seldom realized in montmorillonite systems primarily because the inter-layer separation in a crystal varies during the course of the reaction. Whether the system will be more or less dispersed compared with its original state depends on the nature of the interacting cations. Basal spacing measurements[16] of the moist alkylammonium-sodium-montmorillonite complexes indicate that the replacement of sodium by alkylammonium leads to interlayer con-traction, that is, to a less dispersed state. However, in a recent paper Vansant and Uytterhoeven[23] have indicated that reaction (1) involving short-chain primary n-alkylammonium ions (C_1 to C_4) was essentially reversible.

In the absence of confirmation, K is best regarded as an affinity coefficient rather than as a true thermodynamic equilibrium constant. The (non-standard) free energy change of the exchange process, $-\Delta G^m$, is estimated from the relationship

$$-\Delta G^m = RT \ln K \qquad (6)$$

where R is the gas constant and T the absolute temperature. As might be pre-dicted from the isotherms, $-\Delta G^m$ increases regularly with molecular weight of the alkylammonium ions (Fig. 79) although the absolute values of $-\Delta G^m$ differ from those reported by Cowan and White for the same ions.

Using mono-n-butyl-, di-n-butyl-, and n-octylammonium ions in calcium-montmorillonite, Slabaugh and Kupka[24] have compared K_m values at a single point of the isotherm, corresponding to $Z_{AH^+} = 0.7$. They noted that K_m for n-octyl was greater than that for its isomer di-n-butyl. They attributed this to the presence of fewer hydrogen atoms attached to nitrogen in the secondary amine cation, capable of hydrogen-bonding to the oxygens of the silicate surface. However, since K_m may vary considerably with Z_{AH^+} (Fig. 78), as Slabaugh and Kupka's data indicate, such comparisons would not be valid for the whole isotherm. Previously, Slabaugh[25] had reported thermodynamic data for the exchange adsorption of some primary n-alkylammonium ions (C_1, C_4, C_8, C_{10}, C_{14} and C_{18}) by sodium-montmorillonite. In this instance, $-\Delta G^m$ was derived from the mean K_m values determined using different Na^+/AH^+ ratios at constant Na^+ concentration. Although the ionic strength was variable, this approach would give a better estimate of relative affinities.

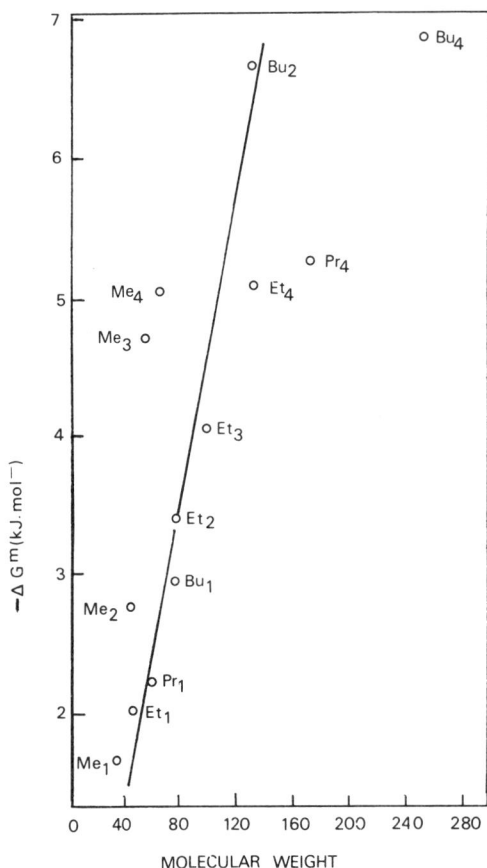

Fig. 79 Variation of $-\Delta G^m$ with molecular weight of some alkylammonium ions replacing sodium ions in montmorillonite (after Theng et al[16])

Weiss[26,27] has stated that the affinity of alkylammonium ions for montmoril-lonite decreased in the order $R_1NH_3^+ > R_2NH_2^+ > R_3NH^+$ but no experimental data or reference was given in support of the statement.

The data of Fig. 79 refer to $-\Delta G^m$ values derived from integration of the whole isotherm so that internal comparisons can be made. We find an order of affinity which is the reverse of that claimed by Weiss. Thus for the ethyl series, $-\Delta G^m$ increases in the order $Et_1 < Et_2 < Et_3$. A similar order obtains when alkylammonium ions of identical molecular weight are compared, for example $Et_1 < Me_2$, $Bu_1 < Et_2$ and $Pr_1 < Me_3$ (the symbols Me, Et, Pr, and Bu refer to methyl-, ethyl- -n-propyl-, and n-butyl-ammonium, respec-tively; the subscripts 1, 2, 3 denote the mono-, di-, and trialkyl derivative, in that order). Assuming that the increase in affinity with cation size is due to the increased contribution of van der Waals forces to the adsorption energy, the above observation suggests that a contribution from this quarter more than offsets the reduction in affinity due to a decrease in the number or strength of cation-to-surface hydrogen bonds in the series primary to secondary to tertiary amine cations. This accords with the data of Laby and Walker[28] who have

216

shown that N—H . . . O—Si hydrogen bonding in primary n-alkylammonium-vermiculites is only weak.

The shape of the adsorbed amine cation also influences affinity. Thus, the position of the points with respect to the line drawn in Fig. 79 indicates that the methyl derivatives are more strongly adsorbed than would be expected from their molecular weight, whereas the reverse is true for the Pr_4^+ and Bu_4^+ ions. This is because the shape of the cation determines the relative extent to which van der Waals contact can be made between cation and clay surface or between the cations themselves. The importance of the shape factor in controlling the magnitude of physical adsorption forces has been noted previously for the uptake of some amino acid and peptide cations by montmorillonite (Section 4.2). Comparison between observed Δ values and minimum molecular thickness indicates that, with the possible exception of Pr_4^+ and Bu_4^+, the alkylammonium ions examined by Theng et al. can be keyed to different extents into the oxygen surface of the montmorillonite layer. The compact methylammonium ions can presumably be more effectively keyed in this way. Rowland and Weiss[29] have shown that the ammonium group in Me_1^+ and Me_2^+ ions is partially embedded into the ditrigonal depression of the silicate surface. Close contact with the surface cannot be effected by the bulky and irregularly shaped Pr_4^+ and Bu_4^+ ions and, hence, their affinity for the clay is much less than for monoalkylammonium ions of comparable molecular weight.[13,24]

A more direct demonstration of the influence of van der Waals forces and of keying on the interaction of some short-chain alkylammonium ions with montmorillonite has recently been given by Fripiat et al.[30] using infra-red spectroscopy. These workers have examined the infra-red spectra of such cations as Me_1^+, Me_2^+, Et_1, and Et_2^+ present in the montmorillonite interlayers after outgassing the respective complexes at 298°K and again after introducing water vapour into the evacuated system.

In addition to the usual shifts in the stretching and deformation vibrations of the $—NH_3^+$ and $—NH_2^+$ groups when they hydrogen-bond to water, the intensities of the CH_2 and CH_3 deformation bands are markedly altered on exposing the clay film to water vapour (Table 62). These intensity changes were particularly pronounced for the dialkyl derivative, presumably because it is relatively weakly keyed so that the aliphatic chain can readily disengage from the oxygen surface. Since the frequency of the CH_2 and CH_3 bands remains sensibly constant whether or not water is present, hydrogen bonding between these groups and water and between these and surface oxygens is not important. Hydration also causes the basal spacing of the Et_2^+ complex to increase from about 1·287 nm (for the outgassed state) to 1·304 nm.

From these observations Fripiat et al. proposed that, in the absence of water, the aliphatic chains are tightly held between the oxygen atoms of the silicate surface with the chain axes lying parallel to the (001) planes. The keying effect is reflected in the absence of the CH_3 symmetric deformation band in the Me_1^+ complex and of the wagging mode in the Et_1^+ and Et_2^+ complexes (Table 62). Hydration of the complexes with dialkylammonium causes the

TABLE 62

Relative intensities of the CH_2 and CH_3 deformation bands in the infra-red spectra of some short-chain alkylammonium ions adsorbed by outgassed and hydrated montmorillonites.[30]

Deformation mode	Alkylammonium ion on clay							
	Mono-methyl		Di-methyl		Mono-ethyl		Di-ethyl	
	outgassed[a]	hydrated[b]	outgassed[a]	hydrated[b]	outgassed[a]	hydrated[b]	outgassed[a]	hydrated[b]
(CH_2) wagging	—	—	—	—	n.v.	n.v.	n.v.	n.v.
(CH_2) scissoring	—	—	—	—	weak (1470 cm^{-1})	weak (1470 cm^{-1})	weak (1481 cm^{-1})	weak (1477 cm^{-1})
(CH_3) symmetric	n.v.	n.v.	weak	n.v.	strong (1401 cm^{-1})	strong (1401 cm^{-1})	strong (1391 cm^{-1})	medium (1390 cm^{-1})
(CH_3) asymmetric	strong (1466 cm^{-1})	strong (1466 cm^{-1})	medium (1470 cm^{-1})	strong (1470 cm^{-1})	strong (1459 cm^{-1})	strong (1459 cm^{-1})	weak (1453 cm^{-1})	strong (1460 cm^{-1})

(a) clay film outgassed at 298°K.
(b) outgassed film exposed to water vapour at a pressure of 3200 Nm^{-2}.
n.v. = not visible.

alkyl chain to rotate, but this treatment has no effect on the orientation of the monoalkylammonium ions.

Persuasive as these arguments are as regards the operation and importance of van der Waals interactions in alkylammonium-montmorillonites, recent thermodynamic data on the adsorption of Me_1^+, Et_1^+, Pr_1^+, and Bu_1^+ ions by sodium-montmorillonite indicate that ion hydration effects are equally, if not more, important than van der Waals forces in these systems. By determining the exchange isotherms for these cations at three different temperatures (277°, 298°, and 328°K), Vansant and Uytterhoeven[23] confirmed that ΔG increased with molecular weight. However, intra- and inter-comparisons of values for the increment of ΔG per CH_2 group showed appreciable variations. This would not be expected if such increments were only due to the increase in van der Waals interactions with chain length, although the absolute values of ΔG might vary with experimental conditions and with the montmorillonite sample used. They also showed that the isotherms were insensitive to temperature, the curves being coincident in every instance indicating zero enthalpy change. The effect of an increase in chain length is therefore to increase the entropy change (ΔS) from 2·4 e.u. (for Me_1^+) to 3·2 e.u. (for Bu_1^+). These observations led them to conclude that the combination of variation in hydration status of the cations and electrostatic interactions between cation and clay surface played an important part in causing the observed increase in ΔG with molecular weight.

5.2. Interlayer Organization of Adsorbed Species

In measuring basal spacings of outgassed samples of sodium-montmorillonite containing different proportions of Me_1^+ and Me_4^+ ions, Barrer and Brummer[31] have observed that the spacings increased with the amount of organic cations occupying exchange positions. This behaviour, together with the water sorption characteristics of the partially exchanged clays, led them to propose that out of five possible ways in which the alkylammonium and sodium ions might be distributed within a given crystal, that of random interstratification of 'organic-rich' and 'sodium-rich' layers was the dominant one. Using a wider range of alkylammonium ions and sodium-montmorillonite, Theng et al.[16] have observed a similar relationship between basal spacing and degree of saturation by the organic cations, confirming Barrer and Brummer's conclusions. But Vansant and Uytterhoeven[23] have recently suggested that, on thermodynamic grounds, this type of layer segregation would be less stable than a system in which there is homogeneous mixing of the two exchanging cations in every layer. Demixing may, however, occur when the appropriate complexes (containing varying ratios of sodium to alkylammonium ion) are dried.

Basal spacing data for the fully exchanged alkylammonium montmorillonites examined by Theng et al.[16] are listed in Table 63. These values refer to samples dried at 343°K and are greater by 0·01–0·06 nm than those reported previously for similar complexes.[32-34] There are a number of reasons for these discre-

TABLE 63

Basal spacings and Δ values for montmorillonite complexes with some alkylammonium ions together with minimum van der Waals thicknesses of the cations estimated from 'Catalin' molecular models.[16]

Cation[a]	Basal spacing (nm)		Δ value[b] (nm)	Min. thickness[c] (nm)	Difference (nm)
	(1)[d]	(2)	(3)	(4)	(5) = (4) − (3)
Me_1^+	1·216	1·25	0·30	0·37	0·07
Et_1^+	1·266	1·27	0·32	0·39	0·07
Pr_1^+	—	1·27	0·32	0·40	0·08
Bu_1^+	—	1·28	0·33	0·40	0·07
Me_2^+	1·252	1·28	0·33	0·39	0·06
Et_2^+	1·281	1·30	0·35	0·40	0·05
Pr_2^+	1·315	—	—	—	—
Bu_2^+	—	1·32	0·37	0·42	0·05
Me_3^+	1·284	1·33	0·38	0·43	0·05
Et_3^+	1·326	1·32	0·37	0·47	0·10
Pr_3^+	1·336	—	—	—	—
Bu_3^+	1·351	—	—	—	—
Me_4^+	1·385	1·38	0·43	0·53	0·10
Et_4^+	1·417	1·40	0·45	0·55	0·10
Pr_4^+	—	1·45	0·50	—	—
Bu_4^+	—	1·62	0·67	—	—

[a] Me, Et, Pr, and Bu refer to methyl-, ethyl-, n-propyl-, and n-butyl-ammonium; 1, 2, 3 and 4 represent mono-, di-, tri-, and tetra-derivative.
[b] obtained by subtracting 0·95 (the assumed thickness of a montmorillonite layer) from the observed basal spacing.
[c] Corrected (see text). No value can be assigned to Pr_4^+ and Bu_4^+ because these cations can assume a number of different configurations.
[d] Diamond and Kinter's[34] values.

pancies. The higher values observed by Theng *et al.* can, at least in part, be ascribed to interlayer adsorption of atmospheric water during X-ray diffraction analysis, since no precaution was taken to exclude moisture from the sample chamber. On the other hand, the spacings reported by Diamond and Kinter[34] and by Barrer and co-workers[32,33] could also be too low due to a greater loss of interlayer water or to heat decomposition of the adsorbed organic cations, since their specimens were dried by heating at 383°K (in air) or by evacuation at 323°K. Using infra-red spectroscopy, Chaussidon and Calvet,[35] for example, have shown that alkylammonium ions (e.g. Me_1^+, Pr_1^+, Bu_1^+, Et_2^+, and Et_3^+) intercalated in montmorillonite were degraded into ammonium and a mixture of hydrocarbons when their respective clay complexes were heated at temperatures below 523°K. Degradation reactions of this type catalysed by clays are discussed more fully in Chapter 7. Examination of the infra-red spectra of the oven-dry (343°K) complexes in Table 63 showed no evidence for such decomposition. Moreover, comparisons between the spectra of the cations in aqueous solution and those in the adsorbed phase indicated that the conformation of the cations was not significantly altered on intercalation.[36]

If the thickness of a methyl or methylene group is taken as 0·4 nm,[37] the Δ values indicate that only a single layer of organic cations is present in the interlayer region. Except perhaps for Pr_4^+ and Bu_4^+, none of the cations in Table 63 takes up an area larger than the area per exchange site. There is therefore no difficulty in accommodating the full complement of these ions within a single layer. The minimum thickness of the cations derived from molecular ('Catalin') models (corrected by taking 0·12 nm for the van der Waals radius of hydrogen),[37] is always larger than the corresponding Δ value. In Chapter 2 we discussed the possible causes for this apparent contraction in the thickness of adsorbed organic species. Since C—H . . . O—Si hydrogen bonding is not important in these systems and there is no significant compression of the cation at the surface, the shortening of apparent contact distances observed here and elsewhere[32-34] must be due to keying of the cation into the oxygen surface of the montmorillonite layer. It can readily be demonstrated by the aid of molecular models of the montmorillonite surface and of the appropriate alkylammonium ions that the cations can be keyed to give contractions equal to or greater than those observed (Table 63, column 5).

From a survey of the literature, Brindley and Hoffmann[38] have concluded that uncharged organics having strongly polar groups preferred an orientation in which the plane of the carbon zig-zag is parallel to the silicate surface. This arrangement (α_{II}) gives rise to basal spacings of 1·30–1·31 nm. The d(001) spacings of the complexes with the alkylammonium ions (excepting Me_4^+, Et_4^+, Pr_4^+ and Bu_4^+) indicate that this arrangement is adopted. Because of their tetrahedral configuration, the orientation of the tetra-alkylammonium ions cannot be deduced with any certainty from the basal spacing of their respective complexes but the 'flattest possible' arrangement is also indicated by the X-ray data.

Previously, the basal spacing of montmorillonite complexes with primary n-alkylammonium ions of varying chain length (from C_3 to C_{18}) had been measured by Jordan.[39] He observed that from C_1 to C_{10} the basal spacing remained sensibly constant at $\sim$1·36 nm, corresponding to an interlayer separation of $\sim$0·4 nm (the van der Waals thickness of a methyl or methylene group). Beyond C_{10} the spacing increased to $\sim$1·76 nm, maintaining this value to C_{18}. This represented a step of 0·4 nm also (Fig. 80). These observations were explained in terms of the formation of single (d(001) $\sim$ 1·36 nm) and double layer (d(001) $\sim$ 1·76 nm) complexes with the alkyl chain lying parallel to the silicate surface. From the theoretical surface area and the exchange capacity of the montmorillonite sample, the area per exchange site in the interlayer space calculates as $\sim$1·65 nm². Assuming an α_{II} arrangement, the area covered by each amine cation can be deduced. These values expressed as a percentage of the planar internal area of the clay are set out along the upper abscissa of Fig. 80. If the area of the cation does not exceed half the area per exchange position, the amine cation adsorbed on one surface can be fitted into

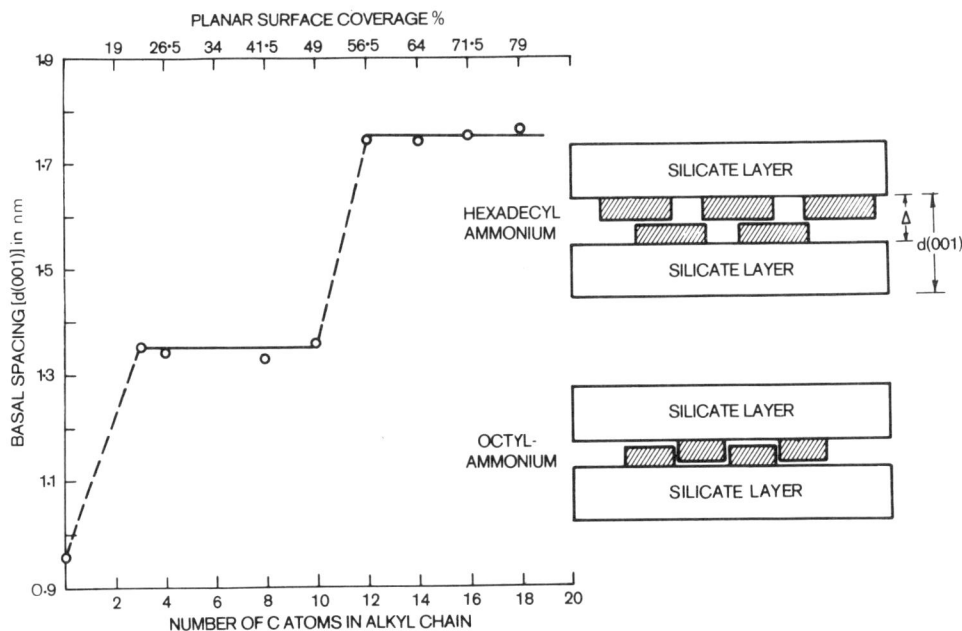

Fig. 80 Variation in basal spacing with number of carbon atoms in alkyl chain and with percentage of internal basal plane area occupied by the cation, in complexes of montmorillonite with primary *n*-alkylammonium ions (after Jordan[39])

the gaps between the cations lying on the opposing surface, as depicted in Fig. 80 for *n*-octylammonium. If, however, the cation area is greater than half of the area per exchange site, interpenetration of this kind cannot be realized and double-layer complexes form (e.g. for *n*-hexadecylammonium). These considerations apply to complexes fully saturated with the amine cations, that is, the amount adsorbed is equal or close to the exchange capacity of the clay.

Step-wise changes in basal spacing also obtain when a long-chain cation, such as *n*-octadecylammonium, is added in increasing amounts to montmorillonite, up to twice the exchange capacity[40] (Fig. 81). The first and second plateaux are comparable with those of the previous figure. When the exchange capacity ($\sim$1 mmol g^{-1}) is exceeded, the chains in the double-layer complex stand at an angle to the silicate surface. At still higher amounts adsorbed, the alkyl chains are 'crowded' into a near-vertical position with respect to the surface, giving rise to a close-packed arrangement which allows extensive van der Waals interactions between adjacent alkyl chains to be established. Greenland and Quirk[41] have reported similarly for cetylpyridinium ions in montmorillonite.

The arrangement and packing of primary *n*-alkylammonium ions in the interlayer region of vermiculite and mica minerals have been examined by a number of authors.[42-45] Weiss and his collaborators,[46-49] in particular, have paid considerable attention to these systems and their extensive data have been summarized by Weiss in a series of reviews.[26,27,50]

Besides being dependent on the alkyl chain length of the cation and the

222

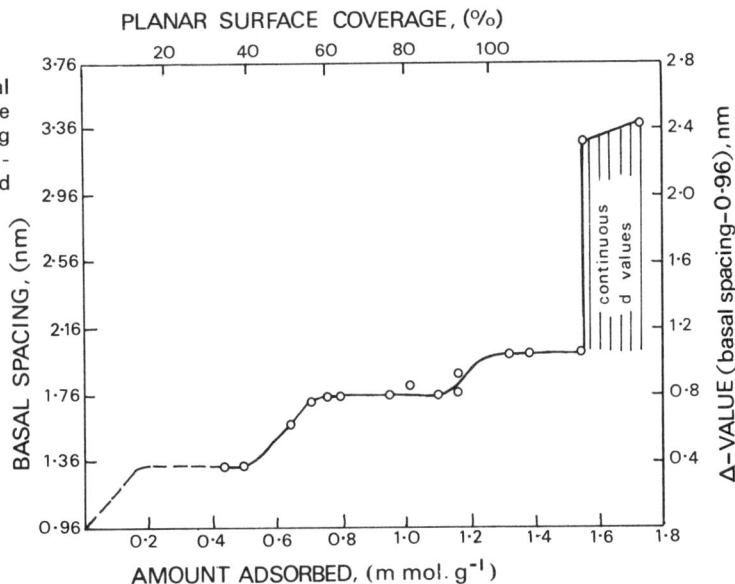

PLANAR SURFACE COVERAGE, (%)

Fig. 81 Changes in basal spacing of montmorillonite complexes with increasing amount of n-octadecyl-ammonium ions adsorbed (after Jordan et al[40])

charge on the mineral layer, the arrangement of intercalated n-alkylammonium ions is also influenced by the method of preparation of the complex. On the reasonable assumption that the charge on the silicate layer is discrete rather than 'smeared out' over the surface, an increase in layer charge must lead to a decrease in the distance separating the negatively charged sites on the mineral layer. This intercharge distance imposes an upper limit on the length of the cation beyond which interlayer expansion will occur. In high charge minerals with $x \sim 1$ (x is the charge per formula unit; cf. Table 1), for example, inter-layer swelling is observed from ethyl onwards. With $x \sim 0.66$ this expansion does not occur until an alkyl chain length of four carbon atoms is exceeded, while in low charge layer silicates (e.g. montmorillonite) with $x \sim 0.33$, the basal spacing remains sensibly constant at 1·3–1·4 nm from ethyl- to n-decyl-ammonium. This dependence of basal spacing on layer charge in mica-type minerals containing n-alkylammonium ions forms the basis of a method, pro-posed by Weiss and Kantner,[49] of estimating surface charge densities from basal spacing measurements.

Weiss[26] has reported that the basal spacing of vermiculites, such as the 'batavite' sample from Kropfmühl with $x = 0.67$, containing n-alkylammonium ions of increasing chain length (C_5 to C_{12}) increases in a step-wise fashion. (Fig. 82, dashed line). This behaviour was thought to arise from the fact that on passing from an odd to the next even number of C atoms (odd → even transition) the interlayer distance increases by 0·20–0·21 nm, but that on passing from an even to the next higher odd number of C's (even → odd transition) there is virtually no increase in this distance.

The significance of this stepped increase in spacing in this sequence is dis-cussed below. Here it is instructive to summarize the data and conclusions of

223

Walker[45] on the adsorption of *n*-alkylammonium ions by some vermiculite samples. Walker points out that, during the formation of the complexes, alkyl chains from opposing silicate surfaces interpenetrate to varying degrees. When the concentration of the treating solution is sufficiently high, complete double layers of cations (with free amine and probably also water and amine salt in the gaps between) are formed in the interlayer region, the average increment per C atom being about 0·204 nm (Fig. 82, line I.) In a *trans-trans* conformation of the alkyl chain and with the N—C bond normal to the silicate layer, the chain makes an angle of 0·973 rad (54° 44′) with the (001) plane. Taking the radius of the terminal methyl group as 0·2 nm and assuming no chain overlap, the theoretical basal spacings may be calculated (Table 64, column 3). These values are smaller by about 0·03 nm for even-C complexes

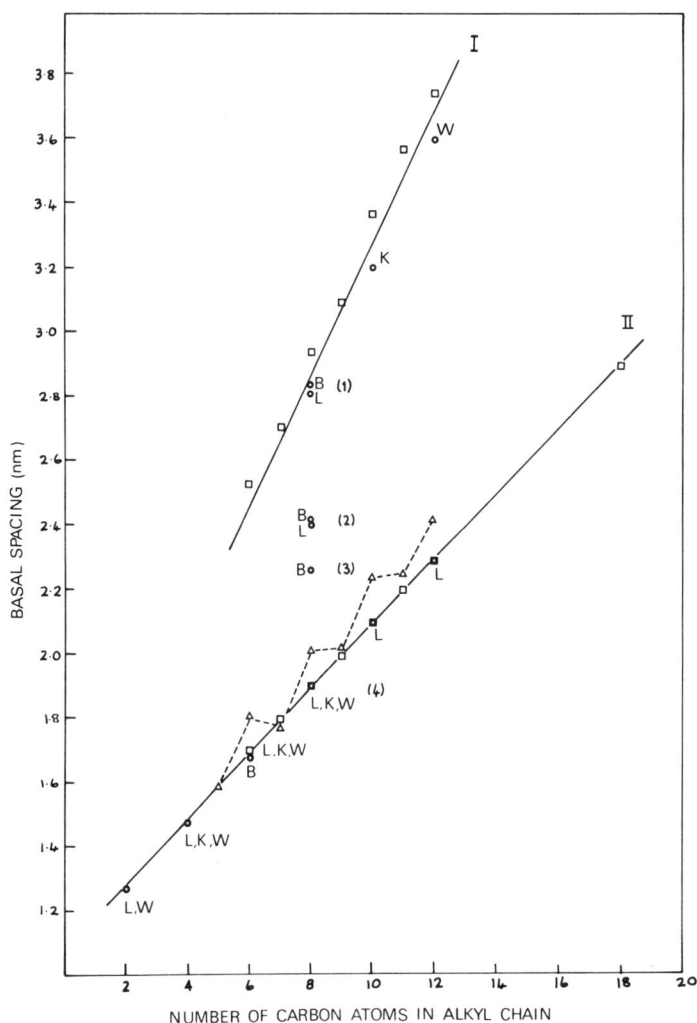

Fig. 82 Basal spacing of complexes between some vermiculites and primary *n*-alkylammonium ions as a function of alkyl chain length: line I, double-layer complexes; line II, single-layer complexes; dashed line, 'Weiss complexes', □ data from Walker[45] for Llano vermiculite (x = 0·98) ○ data from Johns and Sen Gupta[44] for Llano (L), batavite (B; x ~ 1), West Chester (W; x = 0·7), and Kenya (K; x = 0·65) samples. △ data from Weiss[26] for batavite (x = 0·67)

TABLE 64

Observed and calculated basal spacings (nm) of vermiculite (Llano) complexes with primary *n*-alkylammonium ions of varying chain length.[45]

Number of carbon atoms in alkyl chain	Double Layer		Single Layer	
	Observed	Calculated*	Observed	Calculated
6	2·53	2·56 (0·3)	1·70	1·71
7	2·71	2·87 (1·6)	1·80	1·81
8	2·94	2·97 (0·3)	1·90	1·90
9	3·10	3·28 (1·8)	2·00	2·00
10	3·37	3·38 (0·1)	2·10	2·10
11	3·57	3·69 (1·2)	2·20	2·20
12	3·75	3·79 (0·4)	2·29	2·29
18	—	—	2·90	2·88

* No methyl overlap has been allowed for. For details of calculation, see reference (45). Figures in parentheses refer to difference between calculated and observed values.

and by 0·15 nm for odd-C complexes. It is therefore reasonable to suppose that terminal methyl groups overlap at the centre of the interlayers.

On washing and drying of the double-layer complex, the interlayer space contracts (as excess amine, salt, and water is removed) and alkyl chains adsorbed on opposite surfaces move over one another and interpenetrate to form a single layer of ions. Again, assuming *trans-trans* conformation the spacings may be calculated (Table 64, column 5). In even-C complexes the terminal CH_3 lies at the same level as the β-methylene group of the opposing chain; in odd-C complexes, it lies at the level of the opposing α-methylene, (Fig. 83). The average depth of keying of the terminal CH_3 for even- and odd-C complexes is calculated at about 0·01 and 0·05 nm, respectively.

Of the three possible ways in which two *trans-trans* alkyl chains in van der Waals contact can be juxtaposed, only in the one where contact occurs in the plane parallel to the zig-zag of the C—C bonds do the terminal methyls lie at the levels depicted in Fig. 83. The staggered arrangement of vertical and non-vertical C—C bonds from opposing chains explains why the spacings of the single-layer complexes fall on a straight line (Fig. 82, line II). Only if the contact between opposing chains occurs along the plane perpendicular to the carbon zig-zag chain, giving rise to a coincidence of vertical with vertical and non-vertical with non-vertical bonds in opposing chains, will there be a stepped curve with steps of ~0·2 nm for *even → odd* transition, that is, the reverse of Weiss's[26] observation. Nevertheless, like Walker, Weiss assumed that the N—C bond is normal to the silicate layer and that the alkyl chains adopt the extended conformation. To account for this step 'reversal' Weiss suggests that only the terminal CH_3 group of even-C atoms can key into the ditrigonal hole of the silicate surface and, hence, only CH_3 groups of odd-C chains give rise to an increase in basal spacing by about one van der Waals radius of a CH_3 group. As Walker has pointed out, this argument is not valid even if the suggested

SILICATE LAYER

A

TERMINAL

SILICATE LAYER

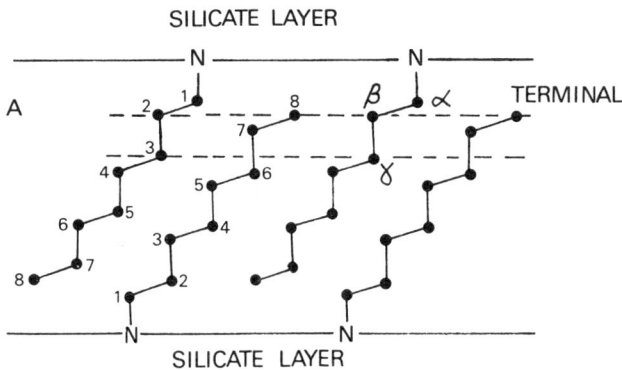

Fig. 83 Idealized diagram of the final
arrangement of the vermiculite complexes with an interpenetrating monolayer of n-alkylammonium ions attached
to opposing silicate surfaces. $-NH_3^+$
groups electrostatically bonded to the
silicate surfaces are indicated by N:
(A) even-C complexes; (B) odd-C
complexes (after Walker[45])

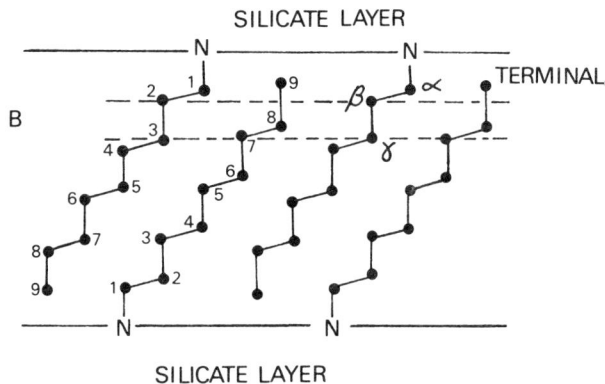

SILICATE LAYER

B

TERMINAL

SILICATE LAYER

keying mechanism were to operate, because it would account only for steps
from even to odd and not the reverse as observed. Incidentally, the arrangement
depicted in Fig. 83 shows that, in fact, a terminal CH_3 at the end of a vertical
C—C bond is more deeply keyed (odd-C complexes) than one at the 'blunt'
end of a near-horizontal C—C bond (even-C complexes).

It is, of course, possible that the N—C bond is at a low angle instead of being
perpendicular to the (001) plane and that the bonds between the α and β,
γ and δ methylene groups and so on, are normal to this plane. This arrangement, however, is highly improbable if, as indicated by Fourier analysis, the
$-NH_3^+$ group is appreciably keyed into the ditrigonal hole and the alkyl
chain is extended (to account for the average increment of 0·104 nm per C
atom in basal spacings). Alternatively, reversed steps can be obtained by
keeping the N—C bond normal to the silicate surface and rotating the C—C
bonds joining the α–β and β–γ methylene groups such that the γ-methylene is
close to the surface instead of standing away from it. Such an arrangement is
shown in Fig. 84. The α–β methylene bond is perpendicular to the plane of the
drawing, all others being parallel to it. The coincidence of vertical with
vertical and non-vertical with non-vertical C—C bonds in opposing chains
provides the stepping required.

Clearly, the structure of Weiss complexes is more the exception rather than

226

SILICATE LAYER

Fig. 84 Walker's[45] proposal for the arrangement of interlayer *n*-alkylammonium ions in
'Weiss complexes' (cf. Fig. 82) The bond join-
ing C_1 (α) and C_2 (β) is at right angles to the
plan of the drawing, the remainder are parallel
to this plane. —NH_3^+ groups electrostatically
bonded to the silicate surfaces are indicated
by N: (A) even-C complexes; (B) odd-C
complexes

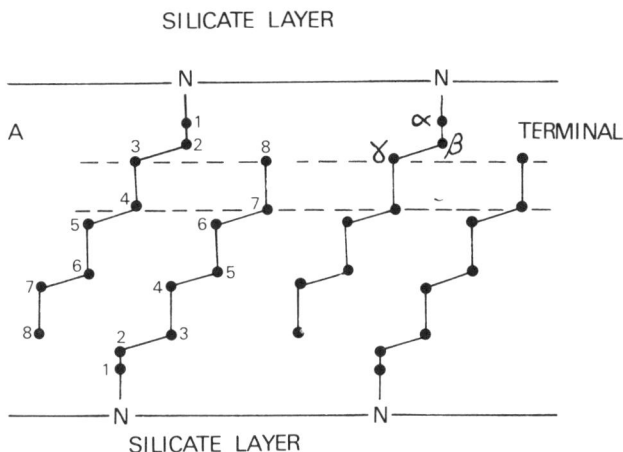

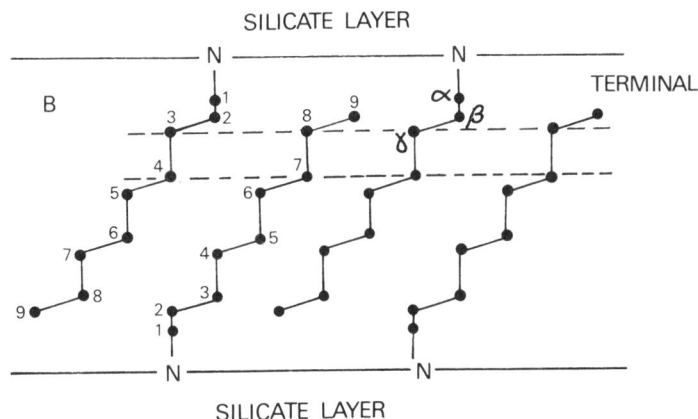

the rule. The reason why Weiss obtained the observed step reversal may be related
to the mild grinding of the samples during their preparation, although Walker's
attempts to reproduce Weiss's results by this means were only partially successful.

Walker's general conclusions regarding the arrangement of the intercalated *n*-alkylammonium ions in vermiculite crystals have been confirmed
by Johns and Sen Gupta[44] who independently have examined similar complexes
with various vermiculite samples using one-dimensional Fourier syntheses.
These workers also observed different degrees of interpenetration during complex formation, each of which corresponds to a distinct crystalline phase. In
each instance the alkyl chain is extended in the *trans-trans* conformation making
an angle of 0.96 ± 0.087 rad with the silicate surface. Further, the N—C
bond is normal to the silicate surface with the N atom located at 0.465 nm from
the centre of the silicate layer. Only after extensive washing and drying do the
basal spacings of the alkylammonium complexes reach their respective minimum
values, all of which fall on a straight line coincident with that obtained by
Walker for his single-layer complexes. (Fig. 82, line II). The different crystal-

227

line phases corresponding to the various degrees of interpenetration of alkyl chains may be represented by the idealized formulae

1. $(Mg_6\, Al_2\, Si_6\, O_{20}\, (OH)_4)^{-2\cdot00}\ ^+(H_3N-R)_2 . 2\,R-NH_2$

2. $(Mg_6\, Al_2\, Si_6\, O_{20}\, (OH)_4)^{-2\cdot00}\ ^+(H_3N-R)_2 . 1\,R-NH_3\,Cl$

3. $(Mg_6\, Al_2\, Si_6\, O_{20}\, (OH)_4)^{-2\cdot00}\ ^+(H_3N-R)_2 . 2\,H_2O$

4. $(Mg_6\, Al_2\, Si_6\, O_{20}\, (OH)_4)^{-2\cdot00}\ ^+(H_3N-R)_2$

The sequence of events represented by these phases, (1) to (4), is illustrated in Fig. 82 where, for R = octyl, the basal spacing corresponding to each of these phases is indicated by the points 1, 2, 3, and 4, respectively. The last of these phases (4) corresponds to the arrangement depicted in Fig. 83 showing that two cations are associated with one unit cell, that is, each second ditrigonal hole on a given silicate surface is occupied by the charged ($-NH_3^+$) end of the cation.

The likelihood of hydrogen-bond formation between $-NH_3^+$ and surface oxygens has been discussed by Johns and Sen Gupta on the basis of the shortening of N—O distances. By assuming that the $-NH_3^+$ group is symmetrically positioned at the centre of a ditrigonal hole, they have calculated a value of 0·30 nm for this distance. Weiss et al.[48] reported a minimum value of 0·285 nm in n-hexylammonium-batavite and Haase et al.[51] of 0·282 nm in hexamethylene-diammonium-vermiculite.

From a comparison with N—O distances in a number of ammonium salts, Laby and Walker[28] have concluded that the reported distances in vermiculite complexes are not consistent with strong hydrogen bonding. They also showed that for the shorter chain cations (C_6 and less), the assumption that the N—C bond is normal to the silicate surface is only very approximately correct. Indeed, the infra-red data for complexes of a vermiculite sample from Kenya ($x = 0\cdot65$) with a series of primary n-alkylammonium ions from C_1 to C_{18}, reveal that there are two distinct environments for the terminal $-NH_3^+$ group in these systems. Environment (I) is present in C_1 to C_6 complexes and is characterized by a mean N—H stretching frequency, $\nu_{(N-H)}$, of 3,164 $\pm$ 4 cm^{-1}, and a mean N—H symmetric deformation frequency, $\delta_{s(N-H)}$, of 1,505 $\pm$ 5 cm^{-1}. For this series of complexes, the N—C bond is significantly less than 1·57 rad (90°) to the silicate surface and the *trans-trans* conformation is not preserved. Environment (II) is present in complexes C_8 to C_{18} and with C_4 which has been expanded with n-butylamine. Here the mean value for $\nu_{(N-H)}$ is 3,148 $\pm$ 3 cm^{-1} and that for $\delta_{s(N-H)}$ is 1,545 $\pm$ 5 cm^{-1}. The N—C bond is perpendicular to the silicate surface and the alkyl chains are extended in the *trans-trans* conformation making an angle of about 0·97 rad to the silicate surface. Both environments may coexist in the C_8 to C_{18} series of complexes as indicated by the presence of a shoulder at 1,500 cm^{-1} in their infra-red spectra.

In environment (I) the plane of three hydrogens in the terminal $-NH_3^+$ group is not parallel to the silicate layer whereas in that of (II) it is parallel, the —NH_3^+ group being keyed more symmetrically into the ditrigonal depression of the oxygen surface. It is therefore concluded that although hydrogen bonds are weak in both series of complexes, they are weaker in those with C_1 to C_6 where the N—C bond is not normal to the (001) plane.

Weiss[50] has presented X-ray data to show that the difference in interlayer arrangement between even-C and odd-C cations extends to the alkyl-(α, ω)-diammonium ions. For the same batavite sample (from Kropfmühl), even-C cations adopt the 'normal' extended *trans-trans* conformation, the alkyl chains making an angle of 0·97 rad with the silicate surface. On the other hand, the odd-C ions are apparently capable of standing vertically on the (001) plane and consequently, the basal spacing for these odd-C complexes is greater than for those with the corresponding even-C ions.

The basal spacings of montmorillonite complexes with 1-*n*-alkyl pyridinium ions have been reported by Greenland and Quirk[41]. The complexes with 0·6–1·0 mmol g^{-1} of the butyl, octyl and dodecyl compounds give spacings in the range of 1·40–1·43 nm, corresponding to Δ values of 0·45–0·48 nm. This indicates that the alkyl chains lie parallel to the silicate surface adopting an α_1 arrangement, probably as a result of some degree of rotation around the N—C

TABLE 65

Basal spacings of montmorillonite complexes containing different amounts of cetylpyridinium bromide. Data refer to samples dried in air at 378 °K for 16 hours.[41]

Amount present (m mol g^{-1})	Cation initially present	Basal spacing (nm)	Higher orders observed
0·6	Na	4·2 and 2·1	8 and 3
0·7	Na	1·80	nil
0·98	Na	(4·2) and 2·1	(1) and 1
1·07	Na	2·1	1
1·14	Na	1·70	nil
1·16	Na	4·2 and 2·1	3 and 1
1·25	Na	(4·2) and 2·1	3 and 1
1·36	Na	(4·2) and 2·1	3 and 1
1·48	Na	2·1	1
2·10	Na	4·2 and 2·1	9 and 4
2·64	Na	4·2 and 2·1	11 and 4
2·99	Na	4·2 and 2·1	8 and 4
3·05	Na	4·2 and (2·1)	13 and (6)
0·7	Ca	1·78	nil
1·54	Ca	4·2 and 2·1	4 and 2
1·91	Ca	4·2 and 2·1	7 and 3
2·06	Ca	4·2 and (2·1)	9 and (3)

* Cation exchange capacity of the montmorillonite used was 0·9 m mol g^{-1} (determined by the conventional ammonium acetate leaching method). Figures in parentheses refer to very weak reflections.

bond. On the other hand, the spacings for the methyl and ethyl compounds (1·26–1·28 nm) give Δ values which are not only less than the minimum thickness of the aliphatic chain (0·4 nm) but are also smaller than the thickness of the pyridinium ring (0·37 nm). This observation recalls Greene-Kelly's[52] earlier findings for pyridine and its derivatives and may be explained in terms of keying of the small methyl and ethyl groups into the ditrigonal holes of the silicate surface and to the effect of attractive forces between the aromatic ring and the clay surface.

The dodecyl and cetyl compounds are adsorbed beyond the exchange capacity of the clay and the corresponding X-ray diffraction data are considerably more complicated than for the lower members of the series. Table 65 summarizes the basal spacing data for the cetyl complexes with different amounts adsorbed. Spacings of 2·1 nm and less are probably due to interstratification of 1-, 2-, and 3- layer complexes. The 4·2 nm spacing is consistent with an arrangement in which the alkyl chains adsorbed on opposing surfaces interpenetrate (cf. Fig. 83) but with the chain and the pyridine ring perpendicular to the (001) plane.

5.3. Some Properties of Cationic Complexes

Unlike the sodium system (Fig. 79), the values of $-\Delta G^m$ for the replacement of calcium ions by short-chain alkylammonium ions in montmorillonite are negative.[16] That exchangeable calcium (or magnesium) ions are much more difficult to replace than the monovalent cations had earlier been demonstrated by several workers.[10,41,53] Using n-octadecylammonium and large quaternary ammoniums, such as dimethyldioctadecyl-(DMDO) and dimethylbenzyl-lauryl-ammonium (DMBL) ions, McAtee[53] has observed an almost stoichiometric replacement of sodium ions in montmorillonite. Under the same conditions, considerably less calcium ions were exchanged by the organic cations. By the same token, once intercalated these large ions are strongly retained.

McAtee[54,55] has examined the desorption of DMBL ions from montmorillonite and hectorite using a number of primary, secondary, and tertiary amine and DMDO cations in an organic solvent medium (isooctane-isopropanol mixture). The amount of DMBL replaced appears to be influenced by both the basicity and the size of the replacing cation, desorption being favoured by an increase and decrease in these parameters, respectively. However, as Jordan[56] has pointed out the solvents themselves, particularly those of high polarity, are capable of extracting an appreciable proportion of the adsorbed quaternary ammonium ions. The observation[57] that, under certain conditions, considerable replacement of one organic ion for another may occur, has an important bearing on the use of montmorillonite complexes of this type as gelling agents in grease compositions. Since the rheological properties of the grease depends on the kind of organic compound associated with the clay, the above process accounts for and provides a means of obtaining either thicker or thinner gels.

The desorption of cobaltic hexammine cations in montmorillonite by some inorganic and quaternary ammonium salts in an aqueous environment has been reported by Das Kanungo and co-workers.[58,59] They found that the replacing ability of the various ions was related to the unhydrated size of the ion, the lyotropic series being followed in the order $Li^+ < Na^+ < K^+ < NH_4^+ < Rb^+ < Cs^+$ and $Mg^{2+} < Ca^{2+} < Ba^{2+}$. For the organic ions the observed order, at least at low concentrations, is $Me_4^+ < Et_4^+ <$ cetyltri-methyl-ammonium < cetylpyridinium. Under similar conditions, the organic ions are usually more effective than the inorganic species in desorbing $Co(NH_3)_6^{3+}$.

With a few notable exceptions (which we shall presently discuss in full), the replacement of the inorganic cations initially present at the clay surface by large organic ions brings about a marked reduction in the water uptake capacity of the mineral. This was first pointed out by Gieseking[4] who noted that his montmorillonite complexes with large aliphatic and aromatic amine cations failed to show a significant increase in basal spacing when exposed to water vapour or liquid water. Similar observations have since been reported by many authors[7,9,60,61,62] for montmorillonite complexes with a wide range of organic cations and bases.

Hendricks[7] has suggested that, steric considerations aside, a large proportion of the clay surface, being occupied by the organic species, would no longer be available to water. In addition, the alkyl groups of the cation would tend to repel rather than attract water molecules. In accord with Hendricks' postulate, recent measurements by Theng et al.[62] on the swelling in water of sodium-montmorillonite containing increasing proportions of exchangeable alkylam-monium ions showed that there was a regular decrease in interlayer swelling with surface coverage.

On the other hand, for montmorillonite saturated with small organic ions (e.g. Me_1^+, Me_4^+) which take up a relatively small fraction of the total surface of the mineral, adsorption of some polar, but especially of non-polar organic species, from the vapour or gaseous phase may actually increase as compared with that occurring on the untreated sample. This is ascribed to the fact that the intercalated organic ions are capable of propping the silicate layers apart and so confer to the crystals a 'permanent' interlayer porosity.[32-34]

The transformation of clays from what is an essentially hydrophilic material to one with hydrophobic (organophilic) properties when complexed with alkylammonium ions has led to the wide use of such complexes as thickeners or gelling agents of organic systems. Jordan, Granquist, Slabaugh, and their co-workers, inter alia,[39,40,63-67] have paid considerable attention to the colloidal and surface properties of montmorillonite complexes with long-chain n-alkylammonium and large quaternary ammonium ions.

As we have noted, these cations are generally adsorbed beyond the exchange capacity of the clay. The hydrophilic nature of the mineral would be expected to decrease as the inorganic cations initially present are progressively replaced by the organic ions, reaching a minimum point near the exchange capacity beyond which the system behaves as a positively charged, hydrophobic colloid. Many properties of clay-organic-water systems therefore show an inversion or reversal at this *equivalence point*. For example, the viscosity of an aqueous dispersion of sodium-montmorillonite increases to a maximum at this point and then declines when increasing amounts of n-octadecylammonium acetate are added to the system.[39] The initial increase in viscosity can be ascribed to flocculation of the montmorillonite particles brought about by a diminution of (osmotic) repulsive interactions between silicate layers, leading to the collapse of diffuse double layers. Beyond the exchange capacity, the clay particles, now coated with a layer of close-packed positively charged ions, will again repel each other and deflocculation occurs.

Chakravarti[68] has reported similarly for the sedimentation volumes of complexes with large quaternary ammonium and pyridinium ions. The reversal in charge (from negative to positive) at the equivalence point is reflected by a minimum in the zeta-potential curve. The heat of wetting also follows a similar trend but in the opposite sense to that observed for viscosity, decreasing with the amount of organic ions present.[10,69]

The swelling in organic liquids of montmorillonite complexes with a series of primary n-alkylammonium ions has been examined by Jordan et al.[40] who noted that an alkyl chain of at least 10 carbon atoms was required before appreciable gelation occurred. Reference to Fig. 80 shows that this onset of large swelling coincides with the point of 50 per cent surface coverage at which there is a monolayer of close-packed interpenetrating cations adsorbed on opposing silicate surfaces.

Besides being dependent on chain length, gelation is also influenced by the polarity of the solvent. Polar (e.g. alcohols, ketones) and non-polar (e.g. benzene) liquids are relatively ineffective as swelling agents. On the other hand, gelation is excellent in nitrobenzene and binary mixtures of polar and non-polar liquids (e.g. methanol/toluene), that is, in compounds or media which combine highly polar with non-polar (organic) characteristics. This observation is generally attributed to the solvation of the 'uncoated' or exposed portion of the mineral surface by the polar component of the swelling molecule (or the mixture) and to the van der Waals type interactions between the alkyl groups of the cation and the non-polar component of the liquid, both effects contributing towards a further expansion of the interlayer space and hence towards a large gel volume.

Using montmorillonite and hectorite complexes with dimethylaryloctadecylammonium (DMAO) and dimethyldioctadecylammonium (DMDO) ions in acetone-heptane and n-alcohol-heptane mixtures, Slabaugh and St. Clair[67] have observed that both the gel volume and the heat of immersion increased with the weight per cent of acetone in the mixture and decreased with the carbon chain length of the alcohol. Besides supporting the hypothesis proposed by Jordan et al. regarding the mechanism of gelation, these results also confirm that solvation and adsorption are exothermic processes.

The data of Granquist and co-workers[63,64] on the gelation of hydrocarbons by DMDO-montmorillonite in the presence of varying amounts of methanol acting as a dispersant have indicated that electrostatic effects are also important in determining the final strength and volume of the gels produced. These workers have proposed a mechanism based on the electric double-layer theory, which can qualitatively account for the variation in gel strength and flow properties as a function of methanol concentration. In essence, the mechanism involves the adoption by the dispersant molecules of a preferred orientation across the disperse phase/hydrocarbon interface, promoting the dissociation of the quaternary ammonium-clay complex and the development of a diffuse double layer. This results in a gelation/redispersion situation as the dispersant concentration in the system approaches and then exceeds the optimum.

Following up the work of Barrer and co-workers,[31–33,70] Slabaugh and Hiltner[65] have determined the vapour phase adsorption isotherms of some alcohols, n-heptane, nitromethane and ethyl acetate on montmorillonite containing n-alkylammonium ions with alkyl chain length from 4 to 18 carbon atoms. Concurrently, the basal spacing of the intercalation complexes was measured. After adsorption of methanol and ethanol below $P/P_0 = 0.8$, the spacings of complexes with C_{12}, C_{14}, C_{16}, and C_{18} remain virtually constant at ~1·8 nm corresponding to a double layer of n-alkylammonium ions in the interlayer space in an arrangement as depicted in Fig. 80 for C_{18}. Beyond $P/P_0 = 0.8$ there is an abrupt increase in basal spacings, the values of which indicate an arrangement of interpenetrating alkyl chains in a single-layer complex with the chain standing perpendicular to the silicate surface. It is therefore deduced that swelling is a two-step process. The first step involves the adsorption of the molecules onto the external crystal and the interlayer surfaces not already occupied by the amine cations; little, if any, interlayer expansion is observed. In the second step, occurring at high adsorbate pressures, the alkyl chains of the cations are displaced from the silicate surface. This creates new surfaces on which the polar molecules can adsorb and, at the same time, 'liberates' alkyl chains which are then solvated by the alcohol. Besides the alcohols, such molecules as nitromethane and ethyl acetate can produce this type of swelling. Water vapour would only influence and take part in the first step, as indicated by the infra-red data of Fripiat et al.[30] referred to earlier. Paraffins, on the other hand, would only interact with the alkyl chains of the cation. This explains why 'strictly' polar and non-polar molecules do not give rise to appreciable swelling, while compounds possessing both polar and non-polar groups in the molecule can form large gel volumes.

Previously, Barrer and Millington[70] had observed increases in basal spacing of montmorillonite and hectorite complexes with a series of n-alkylammonium and α,ω-alkyldiammonium ions on intercalation of benzene, n-heptane, acetonitrile, and α,ω-diaminopropane gases and vapours. With all the forms of organo-clays, α,ω-diaminopropane and benzene gave complexes having d(001) spacings of ~1·43 and 1·50 nm, respectively. For acetonitrile, two series of complexes were obtained with spacings of ~1·63 and ~1·99 nm, respectively. These observations indicate that the final interlayer separation is controlled by the penetrant rather than by the valency or size of the organic cation. If, as suggested above, the adsorbate molecules can displace the hydrocarbon chains from the silicate surface, the arrangement of the organic cations must also be determined by the intercalated species.

In examining the relative dispersing power of a series of polar organic liquids for alkylammonium-montmorillonites, Vold and Phansalkar[71] have observed a limiting degree of dispersion in most liquids. However, no simple correlation seems to exist between chemical constitution of the liquids and their dispersing power. They pointed out that in measuring gel volumes a distinction should be made between sedimentation and swelling volumes. In some liquids the clay forms a coherent solvent-retaining structure, but not in others. Thus, the complexes may be present in a flocculated form in one liquid and be deflocculated in another although their floc size, deduced from sedimentation rate data, may be comparable in both instances.

When the alkyl chains of interlayer n-alkylammonium ions are already 'detached' from the silicate surface as they are in vermiculite complexes (Fig. 83), a wide variety of polar and non-polar organic liquids (e.g. phenols, esters, ethers, ketones, halogen and nitro compounds, hydrocarbons) are readily intercalated, giving rise to appreciable additional interlayer expansions.[26,50] Intercalation of these compounds, particularly those possessing n-alkyl groups in their molecules, evidently causes the interpenetrating n-alkylammonium chains in the original single-layer complex to separate. This process may thus be regarded as the reverse of that observed during the contraction of a double-layer complex with n-alkylammonium ions after washing and drying treatments, discussed in the preceding section. In this way, gaps are created between the extended chains attached to opposing silicate surfaces into which the molecules of the swelling liquid can be accommodated. Remarkably, the sum of $(RNH_3^+ + X)$ per formula unit (where X represents an n-alkyl-alcohol, amine, aldehyde, or carboxylic acid) is about 2·07 irrespective of the layer charge on the mineral. This indicates the presence of a close-packed arrangement of double layers of cation-neutral molecule assemblages in the interlayer space with the alkylammonium chains standing upright, resembling Johns and Sen Gupta's[44] composition as represented by formula (1) on page 228.

Water may also intercalate into mica-type complexes containing n-alkylammonium ions[26]. Whether additional interlayer expansion occurs depends on the layer charge of the mineral and the length of the cation. Minerals with a layer charge, x, of 0·42 swell in water only when the alkyl chain length exceeds eight carbon atoms. In higher-charge minerals ($x \sim 0.66$) swelling occurs with C_2–C_4 onward while the low-charge montmorillonites may not swell appreciably until there are chain lengths of 12 or more carbon atoms.

Vermiculites saturated with certain short-chain n-alkylammonium ions, such as Bu_1^+, behave exceptionally in that they show extensive interlayer swelling ('macroscopic swelling') when the crystals are immersed in water.[43,72] Table 66 lists the basal spacings observed for complexes of various vermiculites

234

TABLE 66

Basal spacings (nm) for some vermiculites saturated with n-alkylammonium
ions on immersion in water.[10]

Alkyl group in cation	Vermiculite source			
	Young River ($x = 0.6$)	Kenya ($x = 0.65$)	West Chester ($x = 0.7$)	Macon County ($x = 0.8$)
Methyl	1·22	1·22	1·23	1·26
Ethyl	1·26	1·26	1·27	1·26
n-Propyl	M.S.	M.S.	M.S.	M.S.
n-Butyl	M.S.	M.S.	M.S.	M.S.
Isoamyl	M.S.	M.S.	M.S.	M.S.
n-Amyl	1·98	2·02	2·00	1·93
n-Hexyl	2·12	—	—	2·14
n-Heptyl	2·22	2·26	2·23	2·20

M.S. = macroscopic swelling.

with a series of n-alkylammoniums after treatment with liquid water. Macroscopic swelling, which is most marked in the n-butyl series, appears to be restricted to cations with alkyl chains of 3 to 4 carbon atoms. Crystals which show this type of swelling are coherent and have a gel-like consistency, allowing the mean interlayer distances to be estimated by directly measuring the thickness of the swollen complex.

Macroscopic swelling also occurs in dilute (<0.06 M) aqueous solutions of n-butylammonium chloride, a linear relationship being observed between interlayer distance and the inverse square root of concentration (Fig. 85). At high concentrations, basal spacings of 1·92 nm are found remaining constant until the concentration is reduced to about 0·06 M when there is an abrupt 'jump' in swelling (d(001) ∼ 17 nm).

It is suggested that ordered water structures ('icebergs')[73] build around the

Fig. 85 Variation of interlayer spacing of n-butylammonium-vermiculite complexes with concentration of n-butylammonium chloride in external solution; C = mol 1^{-1} (after Garrett and Walker[43])

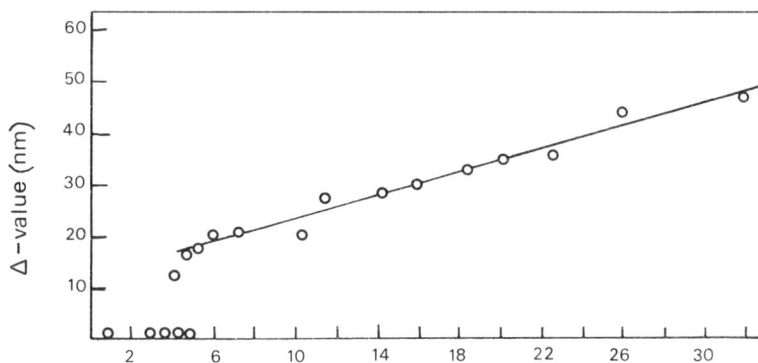

Bu_1^+ ion, that is, water molecules act as hosts of the cations that act as supports for the water. The formation of such structures, referred to as *clathrate compounds*, in the vermiculite interlayers evidently promotes the dissociation of the interlayer cations from the silicate surfaces, giving rise to the formation of diffuse double layers.

Following up this work, Rausell-Colom[74] has made a low-angle X-ray diffraction study of Bu_1^+-vermiculite single crystals swollen by immersion in dilute *n*-butylammonium chloride solutions. He confirmed that macroscopic swelling arose from the interaction of diffuse double layers formed on the vermiculite interlayer surfaces. By measuring the basal spacing of the swollen gel as a function of an externally applied load, the swelling pressure, and hence the repulsive forces involved in expanding the complex, can be estimated (cf. Section 4.2). The difference between the calculated repulsion and the experimentally derived swelling pressure, at any given equilibrium interlayer separation, was found to be of the same order of magnitude as the van der Waals attraction between two opposing silicate layers.

The behaviour of Bu_1^+-vermiculite shows close similarities to that of the lithium-saturated mineral.[75] In both systems, there is a range of electrolyte concentrations in which only limited interlayer expansion takes place. When a given critical level of dilution is reached there is a sudden transition from limited to macroscopic swelling giving rise to gel formation. Unlike the lithium clay, however, the swelling of the butylammonium material is very temperature sensitive.[43] Thus a collapsed crystal of Bu_1^+-vermiculite fails to swell extensively when immersed in water kept at 323°K and above, presumably because the cation-water clathrate is thermally less stable than the lithium-water association.

Another point of interest is the formation of aqueous colloidal dispersions showing strong film-forming characteristics, when a shearing force is applied to the swollen crystals.[76] Such dispersions are stable over long periods when stored in the absence of electrolytes. Addition of even small amounts of inorganic cations such as Ca^{2+} ($\sim 10^{-4}$ M) causes flocculation.

Unlike its high-charge counterpart, montmorillonite saturated with *n*-butylammonium ions does not show macroscopic swelling in water. Indeed, none of the alkylammonium ions listed in Table 63 gives the extensive interlayer expansion of the type encountered with Bu_1^+-vermiculite, that is, these cations show a 'normal' behaviour. However, when increasing amounts of short-chain primary *n*-alkylammonium ions are introduced into calcium-montmorillonite and the complexes allowed to wet at a hydrostatic suction of 1,500 N m^{-2}, macroscopic swelling can occur.[62,77] Figure 86 shows that as calcium was replaced by Me_1^+, Et_1^+, Pr_1^+, and Bu_1^+ ions, the swelling of the complexes rose to a maximum and then decreased. The water content of the complex containing approximately equal proportions of calcium and mono-alkylammonium ions (9 kg kg^{-1} at 1,500 N m^{-2}) was comparable with that for sodium montmorillonite (12 kg kg^{-1}) whereas the parent calcium clay took up only 2 kg kg^{-1} at this suction.

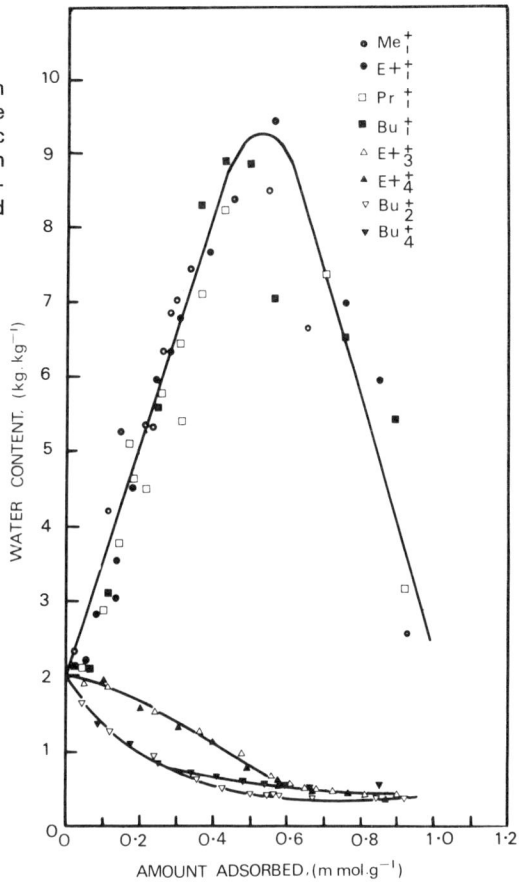

Fig. 86 Effect of replacement of calcium ions by some alkylammonium ions on the water uptake by cores wet to a hydrostatic suction of 1500 Nm^{-2}, expressed as a function of degree of saturation of the exchange complex by the organic ions. (after Greenland et al[77])

Much of the initial increase in swelling can again be attributed to inter-layer expansion due to the formation of diffuse double layers. As the proportion of alkylammonium ions in the complex increased, the 1·9 nm line of the parent calcium clay became progressively less sharp. When about 45 mmol g^{-1} had been adsorbed, this line was absent in the X-ray diffraction pattern of the swollen complex, the pattern in this region becoming diffuse.

On the basis of hydration energy, calcium-montmorillonite would be expected to show macroscopic swelling in water similar to that obtained for the lithium- and sodium-saturated samples. That the interlayer expansion in water of calcium-montmorillonite is limited to about ~1·9 nm (Table 3) is ascribed to the larger radius of dielectric saturation of the Ca^{2+} ion compared with the Na^+ ion.[78] Introduction of the small n-alkylammonium ions may cause the disruption of the ordered water structures around the divalent cation (cf. Figs. 10 and 11). This would increase the entropy of the system and so decrease the electrostatic attraction between cation and clay surface from which it can then dissociate.

Macroscopic swelling of this kind appears to be restricted to alkylammonium ions containing less than 5 carbon atoms. With the larger ions there is a pro-

237

gressive decrease in swelling as the amount present increases (Fig. 86). However, even for the mono-alkylammonium systems extensive interlayer swelling is not always observed. The reason for this discordant behaviour is not clearly apparent although Garrett and Walker's[43] work has indicated that optimum swelling is influenced by the conditions under which the appropriate complexes are prepared.

In the hypothesis advanced to explain the extensive swelling of mixed ion systems, it is implicit that the calcium and alkylammonium ions are more or less randomly distributed within each interlayer region where diffuse double layers form. If the interlayers tend to be filled very largely by either Ca^{2+} or RNH_3^+ ions, the swelling will be the mean of those of the calcium and the alkylammonium clays, neither of which shows macroscopic swelling. Vansant and Uytterhoeven[23] have pointed out that a mixed system is thermodynamically more stable than one in which there is layer segregation. Their data also indicate that the former situation tends to obtain in an aqueous environment but that on drying the complex layer segregation may occur. If this view is accepted, the ability of a mixed ion system to swell extensively would clearly be dependent on whether or not the complex has been subjected to a drying regime during its formation.

We have already referred to the ability of alkylammonium clays to adsorb various organic species because these complexes are inherently porous. When such complexes are used as gas-chromatographic separation media clear-cut separations of organic mixtures are achieved. Barrer and co-workers[32,33,70,79] have published extensively in this area, using principally montmorillonite containing relatively small alkylammonium ions such as the methyl and ethyl derivatives. Complexes with long chain quaternary ammonium ions, commonly referred to by their trade names of 'Bentones', have also been widely investigated for their suitability in acting as stationary phases in gas-liquid chromatographic columns.[80-82] The retention values appear to be related not only to the available pore space and steric factors but also to the magnitude of the layer charge and the location of this charge in the silicate structure.

6

Complexes with the Kaolinite Group of Minerals

6.1. Halloysite

In the preceding chapters we referred to the work of MacEwan[1,2] on interlayer complexes of halloysites (and montmorillonite) with a wide variety of non-ionic, polar organic compounds. In the following discussion, the term 'halloysite' is used to denote the fully hydrated member while the dehydrated form is referred to as metahalloysite.

Apart from a brief note by Caillère et al.[3] in which they described the formation of expanded halloysite using propanol-water (d(001) = 1·4 nm) and saturated barium hydroxide-ethanol (d(001 = 1·7 nm) mixtures, MacEwan's data published in the late 1940s stood alone for a long time. In 1963 Weiss and Russow[4] reported that formamide, hydrazine, and urea were capable of forming interlayer complexes with halloysite. Subsequently, Camazano and Garcia[5] and Olejnik et al.[6] noted that dimethyl sulphoxide (DMSO) formed an interlayer complex with halloysite, giving a basal spacing of 1·118 nm. Similar complexes with some ammonium and potassium salts have been prepared by Wada[7,8] and by Garrett and Walker[9]. It was not until 1971, however, that substantial additional data on the halloysite-organic interaction appeared when Carr and Chih[10] reported the results of their systematic investigations on the formation and properties of halloysite complexes with a large number and variety of organic molecules.

We noted in Chapter 1 that, whereas montmorillonite may intercalate up to three layers of organic molecules, never more than a single layer of molecules is taken up by halloysite (Fig. 5). MacEwan has attempted to correlate this behaviour of montmorillonite with the ratio of the dipole moment to the volume (parachor) of the organic molecule. He suggested that the reason halloysite intercalates, at most, only one layer of organic molecules is that the halloysite layer is amphoteric, one side being made up of a network of oxygen ions carrying negative charges while the other side may be likened to an α-hydroxide surface and is positively charged.[11] In repeating MacEwan's experiments and in extending his data to many more organic substances, Carr and Chih have been able to relate interlayer complex formation to molecular and structural properties.

The usual criterion for intercalation and one which is implicitly adopted by MacEwan, is that the Δ value of the complex should exceed that of the original, fully hydrated mineral (that is, the water complex). In other words,

$$\Delta_{complex} > \Delta_{H_2O} \sim 0\cdot28 \text{ nm}$$

However, as pointed out in Chapter 1 and by Churchman et al.[12] halloysite (d(001) $\sim 1\cdot0$ nm) and metahalloysite (d(001) $\sim 0\cdot72$ nm) represent but the end members of a continuous series of hydration states. Only the end members show basal reflections which are reasonably sharp and symmetrical in profile. Because the partially hydrated forms are essentially randomly interstratified structures, their X-ray diffraction peaks tend to be broad and asymmetric, extending across the region between the 1·0 and 0·72 nm (001) peaks without showing higher orders of reflections. Organic complexes with halloysite might be expected to show a parallel behaviour. For this reason, Carr and Chih have introduced a new criterion for complex formation. According to their criterion a sample is fully complexed when, besides showing a sharp, symmetrical (001) reflection, higher (rational) orders of basal reflections are present. Mere conformity to the rule of $\Delta_{complex} > \Delta_{H_2O}$ only points to a tendency for complex formation. Similarly, intercalation cannot be unequivocally demonstrated when $\Delta_{complex} \leqslant \Delta_{H_2O}$.

Table 67 summarizes Carr and Chih's results with Te Puke (New Zealand) halloysite having a tabular particle morphology and using the range of organic compounds examined by MacEwan. There is reasonably good agreement between the two sets of data except that Carr and Chih failed to observe complex formation with ethanol, acetone, acetaldehyde, and acetonitrile, and obtained generally lower Δ values. These variations were ascribed to a difference in the method of preparing the respective complexes, MacEwan exposing his halloysite sample for 1 hour to the organic liquid whereas Carr and Chih allowed a time of contact of at least 3 days. We have already noted that kinetic factors play an important role in the formation of montmorillonite complexes with n-alkanes (Section 3.6). They assume greater importance in the intercalation of organic molecules by kaolinite as we shall discuss in the following section.

Besides kinetics, structural factors also control complex formation. This is illustrated by the behaviour of Kauri halloysite which has a smaller b dimension than the Te Puke specimen and shows the tubular morphology normally associated with halloysite. In general, the Kauri clay forms a complex more slowly and reaches equilibrium at a smaller Δ value, indicating that it holds interlayer water more tightly compared with the tabular Te Puke material. Since MacEwan obtained complexes with ethanol, acetaldehyde, acetone and acetonitrile, his clay specimen (from Hungary) must have been more reactive. A similar explanation can account for the fact that Camazano and Garcia[5] and Olejnik et al.[6] were able to form an interlayer halloysite complex with dimethyl

240

TABLE 67

Interlayer complexes of halloysite (Te Puke) with organic compounds used by MacEwan.[10]

Compound	X-ray reflections	$\Delta_{observed}$ (nm)	$\Delta_{min(calc)}$ (nm)	$\Delta_{MacEwan}$ (nm)
Methanol	001 (asym)	0·2	0·37	0·34
Ethanol	001 (d,asym)	<0·03	0·40	0·28 (d)
1-Propanol	001 (d,asym)	<0·03	0·40	0·03
1-Butanol	001 (d,asym)	<0·03	0·40	0·03
Ethylene glycol	001, 002, 003	0·36	0·40	0·37
1,3-Propanediol	001, 003	0·38	0·40	0·44
1,3-Butanediol	001 (d,asym)	0·35 (d)	0·40	0·40 (d)
Glycerol	001, 003	0·40	0·44	0·38
1,4-Dioxane	001 (asym)	n.c.	0·42	(d)
Methyl cellosolve	001, 003	0·38	0·44	0·35
Ethyl cellosolve	001, 003	0·37	0·44	0·36 (d)
Butyl cellosolve	001 (d,asym)	(d)	0·44	0·31 (d)
Diethylene glycol	001, 003	0·39	0·39	0·36
Triethylene glycol	001, 003	0·38	0·40	0·35
2-Chloroethanol	001, 003	0·36	0·40	0·36
1,2-Ethanediamine	001, 003	0·45	0·40	0·45
1,2-Propanediamine	001, 002	0·42	0·45	(d)
Diethylamine	001, (d,asym)	0·03	0·40	0·03
Acetone	001 (d,asym)	0·12 (d)	0·40	0·39
Acetaldehyde	001 (asym)	n.c.	0·40	0.36
Acetonitrile	001 (d,asym)	0·05	0·37	0·34
Nitromethane	001, 003	0·30	0·37	0·33
Nitrobenzene	001 (asym)	n.c.	0.37	(d)

d = diffuse (Carr and Chih[10]; MacEwan[1]); or double peak MacEwan.
asym = asymmetric peak shape.
n.c. = no complex

sulphoxide (DMSO) whereas Carr and Chih failed to observe intercalation of this compound.

In an attempt to correlate complex formation with molecular properties, Carr and Chih tested 127 different compounds. Only a relatively small proportion of these could be intercalated (Table 68). Since halloysite samples obtained by displacement with water of intercalated inorganic salts, such as potassium acetate, appear to be less stable than the naturally occurring minerals[13,14] Carr and Chih also examined complex formation with such rehydrated specimens, using those organic compounds which either failed or only showed a tendency to react. The results, set out in Table 69, further support the postulate that structural factors are important in determining reactivity (compare, for example, the data for acetaldehyde and acetonitrile).

Although the polarity of the organic compounds is important in controlling complex formation, no simple relationship is noted between complexing ability and either dielectric constant or dipole moment of the organic species examined. However, intercalation must involve the formation of hydrogen bonds between the organic molecule and the clay surface. Infra-red measurements indicate that of the two prominent hydroxyl stretching vibrations[14,15] at about 3,620

TABLE 68

New interlayer complexes of halloysite (Te Puke) with organic compounds.[10]

Compound	X-ray reflections	$\Delta_{observed}$ (nm)	$\Delta_{min(calc)}$ (nm)
Ethanol	001 (dd)	0·35	0·40
Crotyl alcohol	001, 003	0·43	0·40
1,2-Propanediol	001, 002, 003	0·44	0·42
2-Butyne-1, 4-diol (in diethyl ether)	001, 003	0·35	0·40
Formaldehyde	001, 003	0·31	0·30
Acetyl acetone	001, 003	0·40	0·40
1,2-Cyclohexanedione (in acetone)	001, 002 (d), 003 (d)	0·54	0·50
Acetic acid	001, 003	0·44	0·40
Propionic acid	001, 003	0·49	0·40
Lactic acid	001, 003	0·46	0·42
Malonic acid (in water)	001, 003	0·35	0·39
Formamide	001, 003	0·30	0·30
Acetamide	001, 003	0·39	0·40
Urea	001, 003	0·38	0·30
Semicarbazide HCl (in water)	001, 003	0·32	0·36
Methylamine HCl (in water)	001, 003	0·45	0·36
Ethylamine HCl (in water)	001, 003	0·55	0·36
Diethylene triamine	001, 003	0·38	0·45
Hydrazine hydrate	001, 003	0·33	0·39
Hydroxylamine HCl	001, 003	0·30	0·36
Aniline HCl	001 (dd)	0·78	0·36
Ethanolamine	001, 003	0·35	0·40
Isopropanolamine	001, 003	0·38	0·45
N-acetylethanolamine	001, 003	0·40	0·40
Triethanolamine	001, 003	0·40	0·49
Glycine (in water)	001, 003	0·32	0·40
β-alanine (in water)	001, 003	0·45	0·40
Betaine HCl (in water)	001 (dd)	0·59	0·57
Sarcosine (in water)	001 (dd)	0·48	0·40

d = diffuse.
dd = very diffuse.

TABLE 69

Complexes formed with rehydrated halloysite (Te Puke) obtained by washing the potassium acetate complex with water.[10]

Compound	Exposure time (days)	X-ray reflections	$\Delta_{observed}$ (nm)	$\Delta_{min(calc)}$ (nm)	Behaviour at room temperature
2,4-dinitrophenyl-hydrazine	8	001 (dd,asym)	0·28	0·45	—
Resorcinol	8	001 (dd,asym)	0·28	0·37	—
Picric acid	8	001 (dd,asym)	0·08	0·37	—
Pyridine	4	001,002,003	0·50	0·37	Extensively dissociated after 20 hr
1,3-Butanediol	8	001 (dd)	0·34	0·40	—
Aniline HCl	8	001 003	0·29	0·36	Dissociated after 2 hr
Methanol	4	001 003	0·35	0·37	Dissociated rapidly
Acetaldehyde	8	001	0·28	0·40	Dissociated rapidly
Acetone	8	001	0·28	0·40	Dissociated after 2 hr
1,4-Dioxane	8	001 (dd)	0·05	0·42	—
Acetonitrile	10	001 003	0·33	0·37	Dissociated after 2 hr
Oxalic acid	4	001 (d,asym)	0·27	0·25	—
Pyruvic acid	4	001 (d)	0·51	0·37	—
Dicyandiamide	8	001 (dd,asym)	0·28	0·33	—

d=diffuse
dd=very diffuse
asym=asymmetric peak shape

and 3,690 cm^{-1} in the spectrum of the Te Puke material, the 3,690 cm^{-1} band weakens on organic intercalation. This accords with the finding of Olejnik *et al.*[6] who reported the presence of a broad band between 3,498 and 3,530 cm^{-1} together with two shoulders at 3,690 and 3,648 cm^{-1} in the spectrum of a halloysite-DMSO complex. The first two bands are common to DMSO complexes with kaolin-type minerals, increasing in intensity while the 3,690 cm^{-1} peak weakens as intercalation progresses. They suggested that the DMSO molecule hydrogen bonds to some of the 3,690 cm^{-1} hydroxyls (of the mineral) via its oxygen atom. The intercalation of DMSO into kaolin minerals is discussed more fully in the following section.

In the absence of pure, fully hydrated samples, however, we can make no unambiguous statements regarding specific organic-to-clay interaction, other than to say that changes in hydrogen bonding take place during interlayer complex formation. The fact that in some instances the observed Δ values are less than the corresponding minimum calculated thicknesses of the intercalated species may not necessarily indicate bond shortening because of the uncertainties concerning the assignment of van der Waals radii and molecular orientation. Further, as we stressed in Chapter 2, there is considerable doubt about the importance of C—H . . . O type bonding in these systems.

Nevertheless, Carr and Chih's data permit the following general conclusions to be drawn, the main ones being: (i) particle morphology (that is, whether tabular or tubular) influences the kinetics of complex formation; (ii) the compounds which intercalate are polar and are usually either acids or bases. As a rule, they contain two functional groups, are relatively small, and have *at least* one functional group for every two carbon atoms but do not usually include cyclic and aromatic compounds. Intercalation is favoured by the presence of —OH and/or —NH$_2$ substituent groups; (iii) complex formation involves the formation of multiple hydrogen bonds between the organic molecule and the halloysite layer.

6.2. Kaolinite

Although there is evidence to indicate that interlayer complexes of kaolin minerals with organic substances were known to the ancient Chinese[16] it was not until very recently that the formation and nature of such complexes were systematically investigated. The lack of information in this segment of clay-organic systems is perhaps hardly surprising since, unlike the smectites and vermiculites, the minerals of the kaolinite group do not contain interlayer cations capable of being solvated by water and polar organic liquids.

The observation made by Wada[13] in 1961 that certain inorganic salts, such as potassium acetate, could penetrate the interlayer space of kaolinite and so expand the crystal from a basal spacing of $\sim$0·72 nm to about 1·42 nm, has triggered considerable research into the formation and properties of interlayer complexes of kaolinites. It is now generally accepted that under certain conditions kaolin minerals can intercalate a variety of organic compounds. Weiss and his collaborators[17-20], in particular, have been instrumental in elucidating many of the mechanisms underlying the intercalation of polar organic molecules into kaolinite.

On the basis of their mode of intercalation, Weiss et al.[18] and more recently Olejnik et al.[21] have been able to distinguish between three main groups of polar organic compounds: (A) those which intercalate directly from either the liquid, the melt, or the concentrated aqueous solution; (B) those which can enter the interlayer space by means of an 'entraining agent'; and (C) those which can only be intercalated by displacement of a previously intercalated compound.

Kaolin minerals are on the whole well crystallized, the crystals showing a high degree of order. For this reason, the intercalation process can be examined by X-ray diffraction. The ratio of the intensity, I, of the (001) peak of the complex to the sum of the intensity of the (001) peaks due to the complex and the unexpanded mineral, that is,

$$\frac{I_{(001)complex}}{I_{(001)complex} + I_{(001)kaolinite}}$$

at any given time of contact with the organic liquid, gives the proportion of expanded layers within a single crystal. The intercalation ratio therefore serves as a useful, and readily measurable, index of the reaction or intercalation rate. Using dimethyl sulphoxide as an intercalating agent in kaolinite, Mata-Arjona et al.[22] have derived an expression for the intercalation ratio which takes into account the structural, Lorentzian, and polarization factors.

In general, the intercalation ratio approaches unity, although with some organic compounds the reaction proceeds slowly and remains incomplete even after prolonged exposure of the mineral to the organic liquid; that is, the basal reflection at about 0·72 nm for the unexpanded parent material never completely disappears from the diffractogram of the complex. However, for a given period of exposure, this ratio increases with the temperature and the addition of water to the system, passing through a maximum with water content (Fig. 87).[21]

As we pointed out in Section 1.2, pairing of oxygen and hydroxyl between contiguous kaolinite layers within a crystal gives rise to strong interlayer hydrogen bonding. Not surprisingly, early attempts at intercalating organic species into kaolinite have involved compounds such as urea, hydrazine, and formamide. Besides being capable of breaking interlayer hydrogen bonds, these substances can themselves become hydrogen-bonded to the kaolinite surface.

Using a concentrated (>10 M) aqueous solution, Weiss[17] and Weiss et al.[18] have noted rapid and substantial intercalation of urea into kaolinite yielding a complex with a basal spacing of about 1·06 nm. The interlayer species, however, is relatively weakly held, being removed by washing with water or by heat treatment at ~420°K which decomposes it into ammonia and carbon dioxide. Hydrazine[19] and formamide[18] react similarly, causing the kaolinite crystals to expand to d(001) spacings of 1·04 and 1·0 nm, respectively. Hydrazine appears

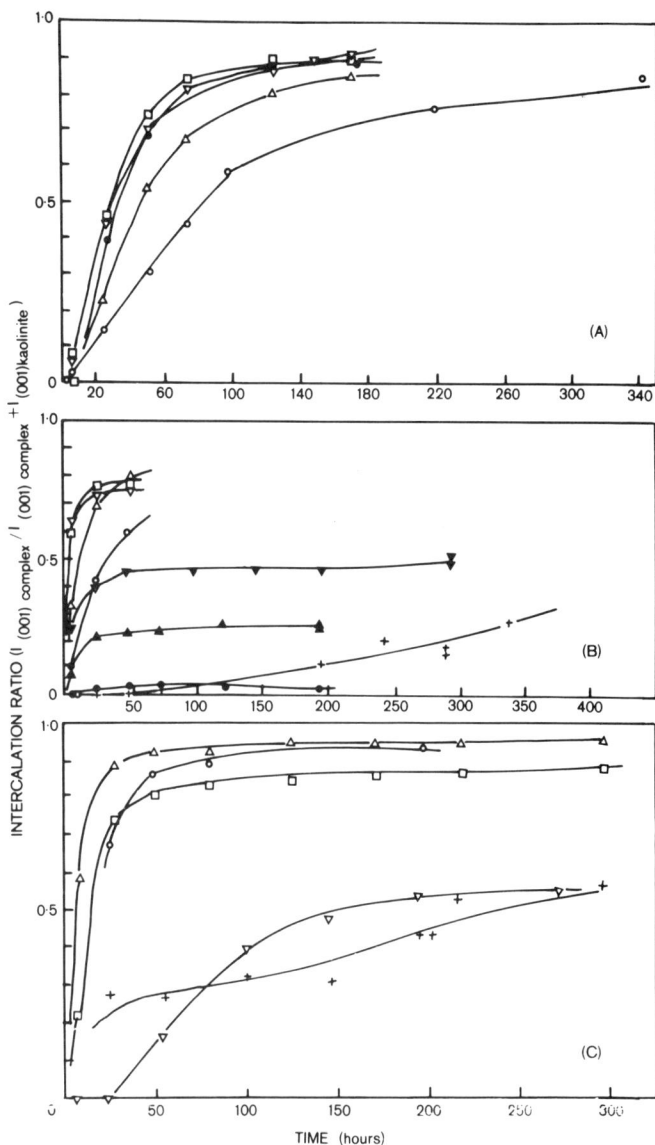

Fig. 87 (A) Rate of intercalation of dimethylsulphoxide (DMSO) into kaolinite (API-9) in DMSO and water-DMSO solutions. $\triangle$, 0·99% H_2O; ∇ 4·77% H_2O; $\square$, 9·09% H_2O; $\bullet$, 20% H_2O; $\bigcirc$, pure DMSO (after Olejnik et al[6])

(B) Rate of intercalation of formamide from the liquid and from aqueous formamide solutions into API-9 kaolinite. Except for the pure formamide-kaolinite system, the temperature was 293°K. $\bigcirc$, 1·96% H_2O; $\triangle$, 4·76%, H_2O; $\square$, 9·1% H_2O; ∇, 20·3% H_2O; $\blacktriangledown$, 30·3% H_2O; $\blacktriangle$, 40·1% H_2O; $\bullet$, 50% H_2O; +, pure formamide (after Olejnik et al[21])

(C) Rate of intercalation into API-9 kaolinite of some amides under the specified conditions. $\bigcirc$, pure N-methylformamide (NMF); $\triangle$, NMF + 9·1% H_2O; $\square$, NMF + 40·1% H_2O; ∇, N-methylacetamide + 9·1% H_2O; +, pure acetamide at 378°K (after Olejnik et al[21])

to intercalate at a faster rate than either urea or formamide but the resultant complex is unstable in air.

Weiss et al.[18] have also published one-dimensional Fourier syntheses of these complexes showing that the N—H . . . O distances are of the order of 0·281–0·284 nm and, further, that the carbonyl group of the molecules is directed towards the hydroxyl surfaces of the kaolinite layer. This suggests that N—H . . . O—Si and C=O . . . H—O hydrogen bonding is the mechanism controlling intercalation.

Urea, formamide, and particularly hydrazine, can act as entraining agents. The data of Weiss et al.[18] indicate that except for ammonium salts, hydrazine

245

can entrain any neutral molecule or salt, the only requirement being that the entrained species is soluble in aqueous hydrazine solutions. The appropriate complex with the entrained compound may then be obtained by selectively removing the hydrazine, for example, by evaporation over concentrated sulphuric acid, exposure to air, or heat treatment. Table 70 lists some compounds

TABLE 70

Basal spacing of intercalation complexes of kaolinite with some compounds prepared by means of entrainment in 24 per cent aqueous hydrazine solution.[18]

Compound	Basal spacing (nm)	
	in presence of entraining agent	after removal of entraining agent
Potassium acetate	1·04	1·40
Sodium acetate	1·04	1·00$_6$
Potassium oxalate	1·04	1·02$_8$
Potassium salt of glycine	1·07	1·24$_4$; 0·93$_1$[b]
Potassium salt of alanine	1·07	1·24$_5$
Potassium salt of lysine	1·06	1·48$_8$
Potassium lactate	1·06	1·11$_8$
Glycerol	1·04	1·05
n-Octylamine	3·17$_1$	3·17$_1$
Benzidine[a]	1·06	2·07$_8$[c]

[a] entrained in 24 per cent solution of hydrazine in water-ethanol.
[b] after prolonged heating at 383°K.
[c] after treatment with the melt.

(B group) which can intercalate indirectly through entrainment in 24 per cent aqueous hydrazine.

Other compounds (C group) such as nitrobenzene, acetonitrile, glycol, and long-chain alkylamines, fail to intercalate directly into kaolinite but will penetrate the mineral interlayers when they have already been expanded, for example by ammonium acetate[18,20] or dimethyl sulphoxide (DMSO)[5] which is displaced during the process.

Using a series of amides, pyridine-N-oxide (PNO), DMSO, and acetone, Olejnik et al.[6,21] have been able to correlate the intercalation process with molecular properties which, in turn, are influenced by the temperature and the water content of the system. Some of their results are shown in Fig. 87.

They noted that the complexes with compounds which intercalated directly from their corresponding liquid (or melt) were indefinitely stable as long as contact with the pure liquid was maintained. For those complexes obtained by the displacement method (e.g. dimethylacetamide) the intercalation ratio is independent of the organic compound in the interlayer space. This suggested to

246

them that when equilibrium is slowly attained the limitation is probably kinetic rather than thermodynamic. Limitation ascribable to the latter cause is shown by water, which removes the intercalated organic species without itself becoming intercalated. This behaviour of water in kaolinite may be compared with that in halloysite systems which, as discussed in the preceding section, can be rehydrated on eluting the intercalated species with water.

The possession by the organic molecule of a large dipole moment and a small size favours its intercalation. High polarity, however, is usually associated with a large molecular size and hence these properties tend to act in opposition. The compounds examined are all highly polar, having large dipole moments (3·71 to 5·37 Debye). For the amides, these moments do not vary greatly with increasing methyl substitution. However, because of the electron-donating inductive effect of the methyl groups, the basicity of the carbonyl oxygen increases with substitution and so causes marked variations in the intercalation of amides. The formation and stability of the complexes, once formed, are enhanced by methyl substitution on the carbon atom.

While a large dipole moment tends to enhance complex formation, it may also cause extensive association in the liquid state and this would decrease the intercalation rate. Below a limiting concentration (e.g. ∼10 to 11 M for urea and hydrazine[18]) the solute is highly solvated and relatively few unsolvated molecules are present or available for intercalation. Increasing the solute concentration will initially bring about an increase in the intercalation rate until a point is reached beyond which the intercalation ratio falls. This observation is attributed to the self-association of the organic molecules at very high concentrations, leaving relatively few 'free' molecules to effect intercalation. Addition of water at this stage disrupts this association and the reaction rate again increases. Further addition of water beyond a certain level, however, will cause the rate to decline for reasons mentioned above. Structural changes of this nature would account for the observation that the rate passes through a maximum with concentration and/or water content (Fig. 87). The liquid structure can also be disrupted by raising the temperature and, hence, an increase in temperature gives rise to a faster rate of intercalation (Fig. 87).

The size and shape of the molecules also influences the intercalation rate. As might be expected, the rate usually increases with a decrease in size. The effect of shape is illustrated by comparing the rate for dimethylformamide (DMF) with that for its isomer, N-methylacetamide (NMA). The former is more slowly intercalated than NMA, presumably because of the difficulty in accommodating the two methyl groups when they are attached to a single atom.

Comparison of Δ values with minimum molecular thicknesses of the amide (Table 71) indicates that only a single layer of organic molecules is intercalated with a considerable amount of keying. In the case of PNO, Weiss et al.[20] have suggested two possible arrangements: (i) a double layer of molecules with the plane of the aromatic ring lying parallel to the kaolinite surface; and (ii) a single layer with the ring and the z axis (cf. Fig. 15) perpendicular to the silicate

TABLE 71
Δ values of kaolinite complexes with some organic compounds.[21]

Compound	$\Delta_{observed}$[a] (nm)	$\Delta_{min(calc)}$[b] (nm)
Pyridine-N-oxide (PNO)	0·53	0·76
Formamide	0·29	0·47
N-methylformamide (NMF)	0·35	0·51
Dimethylformamide (DMF)	0·49	0·59
Acetamide	0·37	0·53
N-methylacetamide (NMA)	0·41	0·53
Dimethylacetamide (DMA)	0·51	0·61

[a] The thickness of the kaolinite layer was assumed to be 0·72 nm.
[b] The structures and dimensions of these compounds were taken from the literature referred to by Olejnik et al.[21] The values for the amides were derived assuming a perpendicular orientation of the plane of the molecule to the (001) plane of kaolinite. For PNO, $\Delta_{min(calc)}$ is for the molecule oriented with its principal (z) axis perpendicular to the kaolinite surface (cf. Fig. 15).
The following van der Waals radii were used: C (=0·17 nm); aromatic ring H (= 0·10 nm);   CH$_3$ (= 0·35 nm).

layer and the oxygen atom of the molecule in contact with the inner-surface hydroxyl. Neither orientation, however, is compatible with the observed Δ value and the pleochroic behaviour of the PNO bands in the infra-red spectrum of its complex.[23] The amount of shortening of contact distances (0·23 nm) can partly be accounted for by keying of the hydrogen atoms into the oxygen surface and of the oxygen atom of the molecule into the hydroxyl surface, and partly by tilting of the ring to give an inclination of about 0·89 to 0·99 rad to the hydroxyl surface. Similarly, the extent to which amides are keyed is at least of the same order and often larger than the usually accepted range obtained with montmorillonite complexes for an α_I orientation (Chapter 2). This discrepancy is less if the plane of the molecule is allowed to tilt with respect to the silicate surface. Indeed, the Fourier analyses of Weiss et al.[18,20] and the infra-red data of Ledoux and White[24,25] indicate that the tilted arrangement is more the rule rather than the exception. Comparison of the Δ values observed for primary, secondary, and tertiary formamides and acetamides suggests that methyl substitution at nitrogen rather than at carbon determines the effectiveness of keying of the NH groups into the tetrahedral oxygen surface of the kaolinite layer.

Although X-ray diffraction data have indicated that the intercalated organic molecules interact by hydrogen bonding to the kaolinite surface, there is no unequivocal answer to whether a particular functional group in the organic molecule will preferentially hydrogen bond to one or the other of the exposed kaolinite surfaces. It is therefore not surprising that classical X-ray diffractometry has, in recent years, been increasingly supplemented and even supplanted by infra-red spectroscopy for examining kaolinite intercalation complexes. Following Ledoux and White,[15,26] we may distinguish between four types of OH groups in the kaolinite structure (see Fig. 1). These are the outer hydroxyls which include hydroxyls situated at the crystallite edges (A) and those belonging to the 'upper' surface of the crystallite (B), both being located externally; the inner-surface hydroxyls (C); and the inner hydroxyls situated in a plane common to both the tetrahedral and octahedral sheets (D). The infra-red spectrum of kaolinite samples usually shows four bands in the OH stretching region, near 3,695, 3,670, 3,650 and 3,620 cm^{-1}, the 3,695 and 3,620 cm^{-1} bands being of similar intensity and much stronger than those occurring near 3,670 and 3,650 cm^{-1} (Fig. 88).

Although conflicting views have been expressed as to their respective assign-

Fig. 88   Infra-red spectra of (A) kaolinite (Hydrite No. 10, Georgia Kaolin Company), (B) kaolinite-hydrazine complex at 298°K, (C) hydrazine complex heated at 383°K for 2 minutes, and (D) hydrazine complex heated at 383°K for 15 minutes (after Ledoux and White[24])

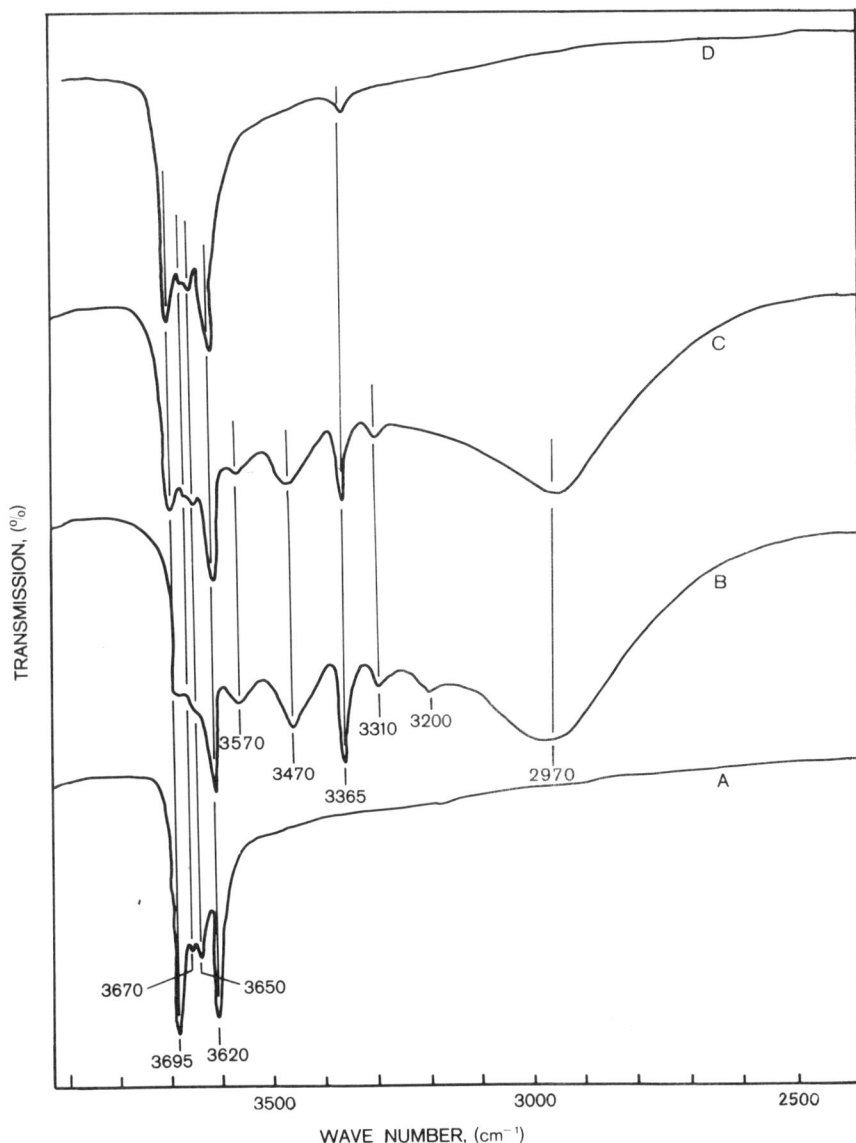

ments,[15] we shall assume that absorption at 3,695, 3,670, and 3,650 cm^{-1} principally arise from vibrations of the inner-surface hydroxyls,[15] whereas hydroxyls of type D give rise to the 3,620 cm^{-1} band.[14,15,26] The first three bands are pleochroic, that is, their intensity is influenced by the way the clay film is placed with respect to the direction of the incident infra-red radiation (Section 2.4.1), but the 3,620 cm^{-1} vibration is not. Further, the 3,695 and 3,670 cm^{-1} hydroxyls appear to orient almost perpendicularly to the basal (001) plane. The direction of the dipole moment change for the 3,650 cm^{-1} hydroxyl makes a high, but less than 1·57 rad (90°), angle to this plane.[14,15,26] The 3,620 cm^{-1} hydroxyls have their dipole moment change inclined at about 0·26 rad to the *ab* plane, the proton being directed towards the vacant octahedral site. Also, the 3,670 and 3,650 cm^{-1} hydroxyls are thought to be involved in interlayer hydrogen bonding within a kaolinite crystal, whereas the 3,695 cm^{-1} hydroxyls are relatively free.

Ledoux and White[24] reported that intercalation of hydrazine reduces the intensity of the 3,695, 3,670, and 3,650 cm^{-1} bands and additional bands of lesser intensity appear at 3,570, 3,470, 3,365, 3,310, 3,200, and 2,970 cm^{-1} (Fig. 88). With minor variations, this observation appears to be of general applicability to the intercalation of polar compounds into kaolinite. In the case of hydrazine, for example, the new band occurring at 3,365 cm^{-1} is attributed to the ν_{NH_2} stretching mode. In liquid hydrazine this mode absorbs at 3,338 cm^{-1}. The broad band at 2,970 cm^{-1} in the spectrum of the complex is probably due to the formation of a O—H . . . NH$_2$ type bond. Intercalation of urea principally affects the 3,695 cm^{-1} band, causing it to weaken considerably; additional bands appearing at 3,520, 3,500, 3,415, and 3,380 cm^{-1} are ascribed to ν_{NH_2} stretching vibrations, the 3,520 and 3,415 cm^{-1} modes being due to the asymmetric and symmetric stretching modes of NH$_2$ groups weakly hydrogen bonded to inner-surface hydroxyls. The 3,500 and 3,380 cm^{-1} bands may correspond to NH$_2$ groups interacting with the oxygens located on the tetrahedral surface.

The infra-red spectra of kaolinite complexes with formamide, N-methylformamide (NMF), and dimethylformamide (DMF) have been reported by White and co-workers.[24,27] Their results and conclusions are in general agreement with those of Olejnik *et al.*[28] who subsequently examined the same complexes in some detail. Only the data for the NMF system are shown here (Fig. 89 and Table 72) since, although the reactions of NMF show features common to those displayed by formamide and DMF, the behaviour of NMF towards the inner-surface hydroxyls of kaolinite differs in one important detail.

Olejnik *et al.* observed a decrease in intensity of the 3,690 cm^{-1} band for all three amides. Although the 3,664 and 3,648 cm^{-1} bands also weaken on intercalation, the extent of this intensity decrease varies from one complex to another. This indicates that each amide interacts differently with the inner-surface hydroxyls of kaolinite. Formamide, for example, hydrogen bonds to the 3,690 and 3,664 cm^{-1} hydroxyls whereas NMF tends to bond to both the

250

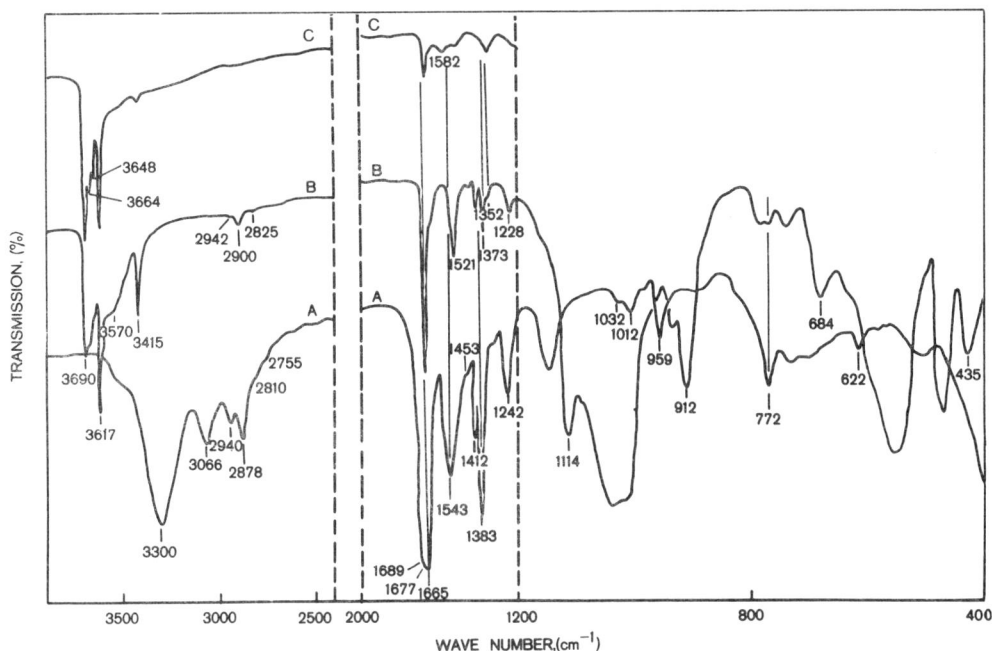

Fig. 89 Infra-red spectra of (A) a thin film of liquid N-methylformamide (NMF), (B) kaolinite (API-9)—NMF complex, and (C) NMF complex after standing in air for 16 hours (after Olejnik et al[28])

3,690 and 3,648 cm^{-1} hydroxyls. Assuming that the intensity of the 3,617 cm^{-1} hydroxyl stretching band is unaffected on intercalation, the ratio of intensities of other bands to the 3,617 cm^{-1} band in the complex can be used as an index of intensity changes occurring on complex formation. For the NMF complex the intensity ratio of the 3,617 cm^{-1} band to the 3,664 cm^{-1} band is 2·0. These values may be compared with those given by the unexpanded kaolinite sample which are 1·99 and 2·0, respectively. It may therefore be inferred that the hydroxyl groups giving rise to the 3,664 cm^{-1} band are not involved in hydrogen bonding to NMF. By similar reasoning, the 3,648 hydroxyls do not appear to participate in bonding to formamide (3,617/3,648 $\sim$ 2). On the other hand, there is no appreciable weakening of the hydroxyl stretching bands for the DMF complex although hydrogen-bonding interactions must also occur here since a new band appears at 3,625 cm^{-1} on intercalation.

Correlation of ν_{OH} bands in the untreated kaolinite with the hydrogen-bonded ν_{OH} in the amide complexes is difficult although the data suggest that bonding to the 3,690 and 3,664 cm^{-1} hydroxyls of kaolinite gives rise to hydrogen-bonded hydroxyl bands at about 3,626 and 3,596 cm^{-1}, respectively, in the spectrum of the complexes. Further support for hydrogen-bond formation is provided by the weakening of the Al—O—H bands ($\sim$941 and $\sim$917 cm^{-1}) on complex formation.

251

TABLE 72

Band frequencies (cm^{-1}) and assignments for N-methylformamide and its intercalation complex with kaolinite.[28]

| Complex | N-methylformamide | | | | Assignment[c] |
	Liquid	Liquid[a]	Dil. soln.	Vapour[b]	
3690					
3664 (sh)					
3648					Kaolinite ν_{OH}
3617					
3570					
				3504	
				3494	ν_{NH}
			3466[d]	3471	
			3429[d]		
3415					ν_{NH}
	3300	3300			Associated ν_{NH}
	3066	3070			Fermi resonance band
2942 (vw)	2940	2952		2939	$\nu_{as}(CH_3)N$
2900				2858	
				2847	ν_{CH}
				2832	
2825 (vw)	2810 (sh)			2763	
	2755 (sh)			2750	$\nu_s(CH_3)N$
				2736	
	1689 (sh)			1749	
1675	1677			1733	Amide I, mostly $\nu_{C=O}$
	1665	1675	1698[e]	1726	
1521	1543	1546	1490[e]	1494	Amide II $(\delta_{NH} + \nu_{CN})$
1452 (vw)	1453	1455		1460	$\delta_{as}(CH_3)$
1412	1412	1414		1409	$\delta_s(CH_3)$
1373	1383	1384		1376	δ_{CH}
	1321 (sh)				
1228	1242	1248		1213	
			1200[e]	1201	Amide III $(\nu_{CN} + \delta_{NH})$
				1190	
	1148	1150		1152	CH_3 rocking
		1040		1040	CH_3 out-of-plane rocking
	1012	1011		1015	π_{CH}
	959	960		961	
				949	ν_{CH_3N}
				935	
773	772	771		616	
					Amide IV (δ_{OCN})
				587	
		720		500	Amide V

a,b,c,d,e These frequencies and assignments are taken from the literature referred to by Olejnik *et al.*[28]

sh = shoulder;   vw = very weak.

From a one-dimensional Fourier synthesis of the NMF complex, Weiss et al.[20] have deduced that the C=O bond in the organic molecule is directed and slightly inclined to the hydroxyl surface of the kaolinite layer to form C=O . . . H—O bonds. If this is so, the amide I band in the complex may be expected to shift to a lower frequency as compared with that in the liquid. However, little, if any, shift of this band is observed. This is interpreted in terms of the depolymerization of associated structure in the liquid[29,30] and the breaking of N—H . . . O=C intermolecular bonds. Both effects would tend to increase the amide I band frequency and so offset the expected lowering of this band on formation of C=O . . . H—O hydrogen bonds. Besides this type of interaction N—H . . . O hydrogen bonding is clearly possible, at least for formamide and NMF, as is indeed indicated by the Fourier analyses of Weiss et al.[18,20] remarked on earlier. The ν_{NH} frequencies in the formamide and NMF complexes (Table 72) are intermediate between those in the liquid and in dilute non-polar solution. This observation is consistent with N—H . . . O—Si bonding while formation of intermolecular N—H . . . O=C hydrogen bonds in the kaolinite interlayers may also be a contributory factor. Which alternative is the most important is difficult to evaluate from the infra-red data. It seems clear from the lowering of the amide I and III band frequencies in the complexes, however, that the intercalated amide molecules exist in a state intermediate between that in the corresponding liquid and a dilute solution of the compound in a non-polar solvent.

As Olejnik et al. have pointed out, their results differ in some respects from those of Cruz et al.[27] First, the intensity decrease of the 3,695, 3,670, and 3,650 cm^{-1} hydroxyl stretching bands on intercalation in general follows the order formamide > NMF > DMF for the samples used by Cruz et al. whereas Olejnik et al. observed a large intensity decrease in the 3,690 and 3,664 cm^{-1} bands for formamide and a comparable weakening of the 3,690 and 3,638 cm^{-1} bands for the NMF complex. The results for DMF are less clear owing to peak overlap. Secondly, the frequency shift of the perturbed hydroxyls assigned by Cruz et al. suggests that the stability of the complexes decreases in the order DMF > NMF > formamide. However, from the changes observed in the infra-red spectra on exposure to air or on heating of the complexes (Fig. 89) Olejnik et al. deduce a reverse order of stability, that is, DMF < NMF < formamide. The order of DMF > NMF > formamide might be expected on the ground that it reflects the order of decreasing electron-donating ability. Steric hindrance, due to molecular size and configuration, however, may possibly offset this effect. Thirdly, Cruz et al. state that DMF is hydrogen-bonded to the hydroxyls of the kaolinite layer through both C=O and N. In view of the fact that the oxygen rather than the nitrogen is the active site for protonation and coordination (Section 3.5) and that two methyl groups are attached to the nitrogen atom in DMF, hydrogen-bonding through N would seem unlikely.

Olejnik et al.[29] have extended their investigation to acetamide, N-methyl-acetamide (NMA), and dimethylacetamide (DMA). The infra-red spectra

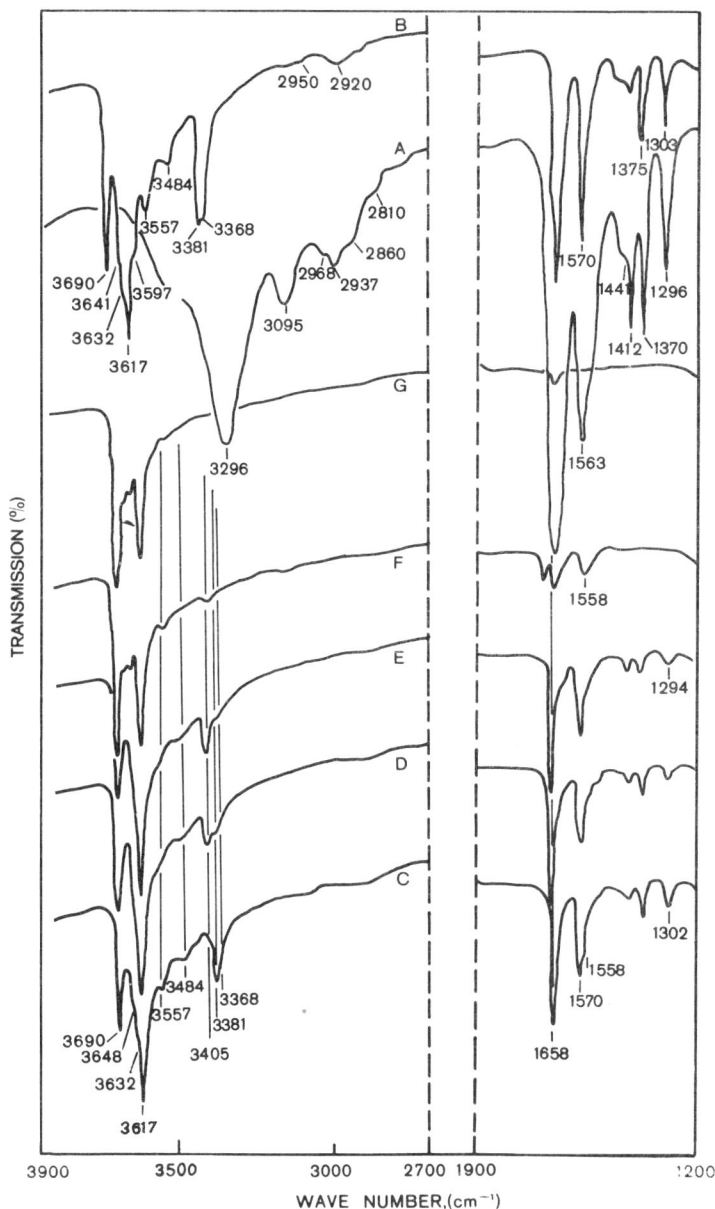

Fig. 90 Infra-red spectra of (A) a thin film of liquid N-methylacet-amide (NMA), (B) kaolinite—NMA complex, (C) NMA complex evacuated at 308°K for 1 hour, (D) NMA complex exposed to air for 12 minutes, (E) NMA complex exposed to air for ∼ 30 minutes, (F) NMA complex heated at 335°K for 30 minutes, and (G) sample (F) further heated at 377°K for 45 minutes (after Olejnik et al[29])

of the NMA complex recorded under the specified conditions are shown in Fig. 90 while Table 73 summarizes the data and gives the assignments.

The acetamides interact chiefly with the 3,690 and 3,664 cm⁻¹ hydroxyls of the kaolinite layer but do not affect the 3,648 cm⁻¹ hydroxyls. It therefore seems that this observation is in general applicable to the intercalation of amides by kaolinite, with the possible exception of NMF. This may be compared with the earlier data of Ledoux and White[24] who reported substantial weaken-

TABLE 73

Band frequencies (cm^{-1}) and assignments for N-methylacetamide and its intercalation complex with kaolinite.[29]

| Complex | N-methylacetamide | | | | Assignment[e] |
	Liquid[a]		Dil. soln.	Vapour[d]	
3690	—	—	—	—	} Kaolinite ν_{OH}
3644 (sh)	—	—	—	—	
3632 (sh)	—	—	—	—	H-bonded kaolinite ν_{OH}
3617	—	—	—	—	Kaolinite ν_{OH}
3597 (sh)	—	—	—	—	H-bonded Kaolinite ν_{OH}
3557	—	—	—	—	H-bonded Kaolinite ν_{OH}
—	—	—	—	3501	} ν_{NH}
—	—	—	—	3489	
3484	—	—	—	—	
—	—	—	3472[b]		} ν_{NH}
—	—	—	3440[b]	—	
3381	—	—	—	—	} bonded ν_{NH}
3368	—	—	—	—	
—	3296	3300	—	—	associated ν_{NH}
—	3095	3110	—	—	Fermi resonance band
—	—	—	—	3001	} $\nu_{as}(CH_3)N$
2950	2968 (sh)	—	—		
2920	2937	—	—	2940	
—	2880 (sh)	—	—	—	$\nu_{as}(CH_3)C$
—	—	—	—	2830	$\nu_s(CH_3)C$
—	2810 (sh)	—	—	2815	$\nu_s(CH_3)N$
				1731	} Amide I mainly $\nu_{C=O}$
1658	1658	—	1691[c]	1713	
1636 (sh)	—	—	—	—	
1570	1563	—	1486[c]	1497	Amide II ($\delta_{NH} + \nu_{CN}$)
1558 (sh)	—	—	—	—	
1456 (w)	—	—	—	—	$\delta_{as}(CH_3)C$
1441 (w)	1441 (sh)	—	—	1426	$\delta_{as}(CH_3)N$; $\delta_{as}(CH_3)C$
1427 (w)	—	—	—	—	
1414	1412	—	—	—	$\delta_s(CH_3)N$
1375	1370	—	—	1377	$\delta_s(CH_3)C$
1303	1296	—	1250	1257	Amide III ($\nu_{CN} + \delta_{NH}$)
1158	1158	—	—	1181	$(CH_3)N$ rocking

[a,b,c,d,e] These frequencies and assignments are taken from the literature referred to by Olejnik et al.[29]

sh = shoulder;   w = weak.

ing of the 3,695, 3,670, and 3,650 cm^{-1} bands on complex formation with a number of organic molecules including formamide. That the 3,690 cm^{-1} band invariably weakens on intercalation is perhaps not surprising since it is supposed that the 3,664 and 3,648 cm^{-1} hydroxyls are involved in interlayer hydrogen bonding within a kaolinite crystal, the 3,690 cm^{-1} hydroxyls being relatively free. Nevertheless, as the available data suggest, all inner-surface hydroxyls can react with, and hence are frequently equally accessible to, organic molecules capable of expanding the crystal.

Another general feature is that there are no large frequency shifts of the hydrogen-bonded hydroxyls in the complexes from their frequencies in kaolinite, presumably because the amides and the compounds used by Ledoux and White, such as urea and hydrazine, hydrogen bond to the kaolinite hydroxyls through a similar functional group. There is also an apparent relationship between the particular inner-surface hydroxyl involved in hydrogen bonding and the stability of the resulting complex. The complexes with formamide, acetamide, NMA, and DMA are the most stable of the amide complexes examined. These compounds also react with the 3,690 and 3,664 cm^{-1} hydroxyls. The NMF complex by comparison is less stable and, remarkably, NMF hydrogen bonds to the 3,690 and 3,648 cm^{-1} hydroxyls.

Complex formation also distorts the Si-O (1,117 cm^{-1}) and other vibrations of the silicate structure and rapid removal of the intercalated species does not usually lead to a complete restoration of the infra-red spectrum of the original, unexpanded mineral.

That the acetamides usually show larger lowering in the frequency of the amide I band than the corresponding formamides may be attributed to the greater basicity (electron-donating ability) of the carbonyl group in the acetamides because of the presence of additional methyl groups which exert an electron-releasing inductive effect. The stronger bonding of the acetamides to kaolinite is reflected, for example, in the greater stability of their complexes compared with those of the formamides.

The interaction of kaolinite with dimethyl sulphoxide has received much attention because DMSO is readily intercalated and the complex so formed can be used as a starting material to prepare complexes with other compounds by the displacement method. The intercalation of DMSO into kaolin minerals has been examined by Olejnik et al.[6] using infra-red spectroscopy. The spectra of DMSO and its mineral complexes taken under the specified conditions are shown in Fig. 91.

Intercalation of DMSO into kaolinite leads to an appreciable decrease of the 3,690 cm^{-1} hydroxyl band and the appearance of three addition sharp bands at 3,658, 3,535 and 3,499 cm^{-1}. This observation may be compared and contrasted with the behaviour of formamide, hydrazine (Fig. 88) and urea which appear to react with the 3,695, 3,670 and 3,650 cm^{-1} hydroxyls, or the amides which decrease the intensity of both the 3,690 and either the 3,670 or the 3,650 cm^{-1} bands of kaolinite. As the intercalation ratio for DMSO increases, the intensity of the 3,658, 3,535 and 3,499 cm^{-1} also rises while the 3,690 cm^{-1} band further weakens. Removal of the intercalated species by heating the complex, for example, gives rise to a collapse of the crystal from d(001) $\sim$ 1·116 to 0·714 nm restoring the original hydroxyl peaks of the unexpanded mineral.

It appears that DMSO interacts with the relatively free inner-surface hydroxyls (3,690 cm^{-1}) of kaolinite and, additionally, new hydrogen bonds are formed. Some of the oxygen atoms of the molecule may take up similar positions in relation to the surface hydroxyl sheet as the oxygen atoms of the opposing tetrahedral sheet in the parent kaolinite. If this leads to an increase in

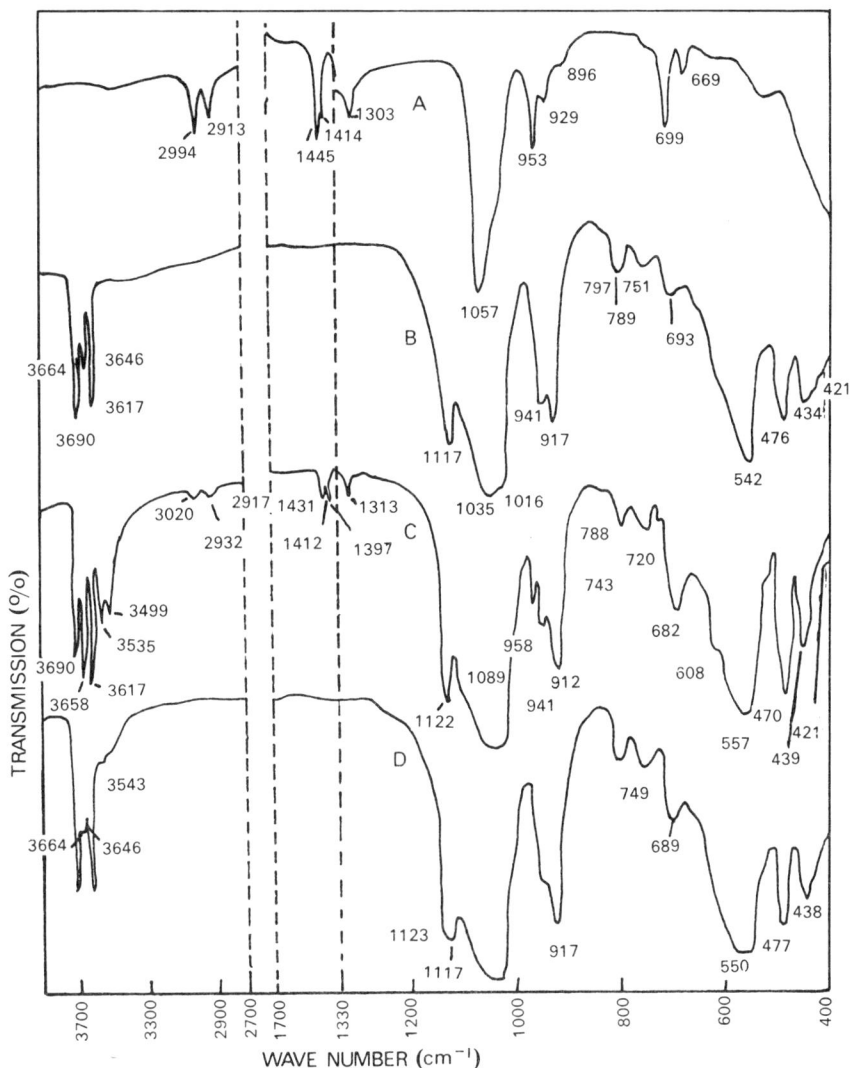

Fig. 91 Infra-red spectra of (A) liquid dimethylsulphoxide (DMSO), (B) API-9 kaolinite, (C) kaolinite—DMSO complex, and (D) kaolinite—DMSO complex heated at 373°K for 15 minutes (after Olejnik *et al*[6])

the number of weak hydrogen bonds, the intensity of the 3,658 cm^{-1} band would increase. The sharpness of the new peaks at 3,535 and 3,499 cm^{-1} might be due to distortion in the kaolinite structure, symmetry factors, or degeneracy of the vibrations. DMSO is more mobile than, say, formamide which bonds to both the oxygen and hydroxyl surface of kaolinite. This property may also permit different arrangements of the DMSO oxygens with respect to the hydroxyl sheet and so give rise to band splitting (two sharp peaks) rather than to one broad band. Such a splitting, for example, is not observed for the halloysite complex.

257

Instead, two closely overlapping bands at 3,530 and 3,498 cm^{-1} and a shoulder at ~3,648 cm^{-1} are observed, probably because of the reduced crystallinity of halloysite and the more random manner in which the individual layers are stacked within a crystal. With the more highly crystalline polymorph dickite, two bands at 3,528 and 3,483 cm^{-1} are observed. Their large intensity, overlap, and frequency shift from the corresponding positions at 3,535 and 3,499 cm^{-1} in the kaolinite complex indicate that hydrogen-bonding interactions are stronger in dickite.

DMSO can bond to the kaolinite hydroxyls either through the sulphur or the oxygen atoms. The infra-red spectrum indicates that the latter mechanism is the more likely alternative, assuming that the 1,030 cm^{-1} band is due to SO stretching and that the peak near 720 cm^{-1} can be assigned to the asymmetric CS stretching. The corresponding peaks in the spectrum of liquid DMSO occur at 1,057 and 699 cm^{-1} (Fig. 91). These observations, together with the absence of the symmetric CS vibration in the spectrum of the DMSO complex, favour H-bonding through oxygen rather than sulphur.

There is also an upward shift in CH stretching vibrations in all the complexes examined, an observation incompatible with the concept of C—H . . . O—Si bonding proposed by MacEwan[1] (Section 2.4.2).

The Δ value of the DMSO complex indicates a shortening of apparent contact distances of the intercalated species so that appreciable keying must occur. Assuming a trigonal pyramidal configuration with the sulphur atom at the apex[30], the DMSO molecule appears to be intercalated with its apex directed towards the tetrahedral oxygen sheet and its base lying nearly parallel to the hydroxyl surface of the opposite layer.

Differences in interlayer hydrogen bonding and crystallinity of minerals in the kaolinite group are also reflected in intercalation rates. Olejnik et al.[6] observed that for DMSO these rates increase in the order of dickite < kaolinite < halloysite. The infra-red spectra in the hydroxyl stretching region of these minerals are consistent with the view that the strength and number of interlayer hydrogen bonds are greatest for dickite and least for halloysite. The intensity of the band near 3,650 cm^{-1}, for example, markedly increases in the order halloysite < kaolinite < dickite. Indeed, the dickite sample used by Olejnik et al. failed to intercalate DMSO unless previously ground.

Like the halloysites which were discussed previously, there are also differences in reactivity between the kaolinites themselves. Reduction in particle size alone cannot account for the observed enhancement in reactivity, and structural factors seem to be equally important in causing such variations. However, as Lloyd and Conley[31] have shown, crystallinity of the kaolinites within a geographic deposit exhibits an inconsistent pattern with respect to particle size, although there is a tendency for crystallinity to increase towards the fine particle size range. Small variations in unit cell dimensions, for example, can cause marked differences in the reactivity of halloysites. Thus, a smaller b dimension here is associated with a tendency to hold the interlayer water more firmly and to assume a tubular form more easily.[32] Likewise, Weiss et al.[18,19]

have earlier reported that b axis disordered kaolinites, sometimes referred to as 'fire-clay' kaolins,[33,34] show less rather than greater tendency to intercalate such compounds as urea, hydrazine, and formamide. The breaking of inter-layer hydrogen bonds in kaolinites by the intercalated species and the reforming of such links with the interlayer molecules would be expected to alter the relationship between contiguous layers within a crystal. In general, complex formation makes it easier to displace the kaolinite layers with respect to each other. Using electron microscopy Weiss and Thielepape[35], for example, have shown that on grinding the complex with ammonium acetate the crystallite becomes progressively thinner and tends to fold. Exfoliation may eventually lead to the formation of either tubules or extremely thin montmorillonite-like layers. Such morphological changes are accompanied by marked increases in the plasticity and dry bending strength of the material. Parenthetically, examination of historical notes and records has led Weiss[16] to postulate that the 'secret' of the art of porcelain manufacture by the ancient Chinese lies in their recognition and application of the ability of kaolin minerals to intercalate organic compounds. The term kaolin derives from the Chinese *kao-ling* or high ridge from which the clay was quarried and mixed with human urine in large pits. The system was left to age so that intercalation could take place. The attendant increase in plasticity and dry bending strength of such modified kaolin enabled the material to be worked and shaped into the fine, egg-shell porcelain articles for which imperial China of medieval times was well known. Clay-organic complex formation has thus left a mark on human civilization.

Using DMSO as the intercalated agent, Camazano and Garcia[36] have noted that complex formation with kaolinite similarly induces the hexagonal plates of the mineral to curl, giving rise to halloysite-like morphology. On the other hand, Jacobs and Sterckx[37] in confirming the general conclusions reached by Olejnik *et al.* (regarding the intercalation and mode of bonding of DMSO to the kaolinite surface) have observed that after removal of the interlayer DMSO by evacuation, the infra-red spectrum of the hydroxyl stretching region resembles that of dickite.

Aromatics, such as pyridine-N-oxide (PNO), can also intercalate into kaolinite[20,23] increasing its basal spacing from 0·71 to 1·253 nm. PNO bonds through the NO group to the 3,690 and 3,664 cm^{-1} hydroxyls of kaolinite.[23] This results in an intensity decrease of these bands and also causes the appearance of a single hydrogen-bonded hydroxyl band at about 3,522 cm^{-1}.

The orientation of the molecule in the kaolinite interlayers can be deduced from the intensity changes of certain bands on rotating the film of the clay complex. Like pyridine itself, PNO is assumed to belong to the C_{2v} symmetry class (Section 2.4.1). Inclining the film at 0·785 rad (45°) to the infra-red beam increases the intensity of the 3,522 cm^{-1} band by about 23 to 25 per cent and the A_1 (in-plane) vibrations at 1,466 and 1,248 cm^{-1}, arising from a dipole moment change parallel to the z axis, by about 48 per cent. This indicates that the NC_4 axis is perpendicular to the plane of the kaolinite surface, similar to the situation described by Serratosa[38] for the vermiculite-pyridinium ion complex (Fig. 18). In this orientation the N—O bond is pointing to the hydroxyl surface and N—O . . . H—O hydrogen-bonds are formed.

The appearance of three additional sharp bands in the complex with DMSO and the large frequency lowering of the hydrogen-bonded hydroxyls in the PNO complex can therefore be regarded as being characteristic of and specific to these compounds, such features not being observable with polar organic molecules in general. The difference in intercalation behaviour between DMSO and PNO might be due to the fact that the former has a trigonal pyramid configuration as opposed to the planar shape of the PNO molecule.

7

Organic Reactions Catalysed by Clay Minerals

7.1. Colour Reactions

It has long been known that many aromatic amines convert to their coloured derivatives when they are brought into contact with clays,[1-5] but only relatively recently have the mechanisms underlying this process come to be understood.

By examining and testing a large number of amines for their ability to produce coloured complexes with layer silicate minerals, Hauser and Leggett[1] have established some general rules, the most important being: (a) only certain amines of the aniline type give the reaction; (b) the benzidines give rise to blue, the anilines to green, and the toluidines to pink or yellow colours—that is, the colour appears to be specific to the amines; (c) the colour is produced with both 1:1 and 2:1 type layer minerals, although the intensity of the colour depends on the mineral species; (d) in some instances, drying the complex causes a lightening of colour; a similar reduction in colour intensity can be brought about by adding acid to the coloured material.

Subsequently, Weil-Malherbe and Weiss[5] showed that both acid-base interactions and oxidation-reduction reactions were involved in the formation of coloured clay complexes with certain aromatic amines. The former mechanism predominates with acid-washed clays which can act as proton donors (Brønsted acids) since the activity of the mineral can largely be suppressed by treatment with a base, such as ammonia. Oxidation-reduction reactions, on the other hand, are relatively insensitive to either acid or base pretreatment of the clay but are markedly influenced by the presence of reducing agents, such as stannous chloride. The operative mechanism here is apparently one of electron transfer from the adsorbed species to the clay acting as an electron acceptor (Lewis acid). Following Briegleb[6] such association between clays and organic compounds may thus be termed *electron-donor-acceptor* complexes.

We have already cited a number of examples of the capacity of cation-exchanged montmorillonites to act as strong Brønsted acids, the reactive protons being derived from the dissociation of water molecules because of the

261

TABLE 74

Amounts of ammonium ions (meq. g^{-1}) formed in montmorillonite and nontronite, over NH_3—H_2O (0·7 per cent NH_4OH) solutions as influenced by the nature of the exchangeable cation and relative humidity.[7]

Exchangeable cation	Montmorillonite[a]		Nontronite[b]	
	Low water content[c]	High water content[d]	Low water content[c]	High water content[d]
Li	0·23	0·17	0·68	0·34
Na	0·16	0·10	0·49	0·16
K	0·10	0·11	0·18	0·16
Ca	0·80	0·16	0·62	0·14
Mg	1·01	0·74	0·75	0·64
Al	1·01	1·00	0·74	0·72

(a) The layer charge originates principally from isomorphous replacement (Mg^{2+} for Al^{3+}) in the octahedral sheet.
(b) Source of layer charge resides chiefly in the tetrahedral sheet (Al^{3+} replacing Si^{4+}).
(c) relative humidity $(P/P_0) = 0·20$.
(d) relative humidity $(P/P_0) = 0·98$.
The greatest change in proton-donating ability due to hydration effects is observed for the calcium-saturated sample, indicated by dashed line (cf. Table 75).

polarization by the exchangeable cations. Besides being dependent on the nature of the cation, the proton-donating ability of clays is, as Mortland and Raman[7] have shown, also influenced by the hydration status of the clay and the source of the layer charge (Table 74).

In the presence of much water, for example at 98 per cent relative humidity, the polarization effect exerted by the cation on the water is presumably dissipated among a large number of water molecules and the pK then approaches that of the ion in aqueous solution. As the water content of the system decreases, the forces of polarization act on the few, residual water molecules causing a marked increase in the extent of their dissociation and, hence, in their proton-donating property. These situations may be expressed in terms of the following equilibria,[7]

$$[M(H_2O)_x]^{n+} \overset{K_1}{\rightleftharpoons} [M(H_2O)_{x-1}(OH)]^{(n-1)+} + H^+ \tag{1}$$

and

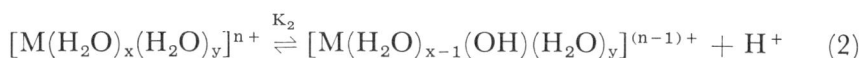

$$[M(H_2O)_x(H_2O)_y]^{n+} \overset{K_2}{\rightleftharpoons} [M(H_2O)_{x-1}(OH)(H_2O)_y]^{(n-1)+} + H^+ \tag{2}$$

where M is the exchangeable cation of valency, n; x and y represent the amount of water in the inner and outer coordination spheres of the cation, respectively. K_1 and K_2 are the ionization (dissociation) constants of the two systems. Equilibrium (1) thus represents the 'dry' system corresponding to, say, an environment of about 20 per cent relative humidity where $K_1 > K_2$. This type of Brønsted acidity may be estimated by sorption of ammonia and organic bases and measuring the amount of the corresponding protonated species formed using infra-red spectroscopy. The data of Mortland and Raman derived in this way are listed in Table 75. Alternatively, the proton-donating ability of the 'dry' mineral may be estimated by titration, in a liquid hydro-

TABLE 75

Proportion of water, ammonia, and ammonium ions present in calcium-montmorillonite films at equilibrium, on exposure to varying pressures of water and ammonia.[7]

Water content of film* (%)	mmol H_2O per g clay	mmol NH_3 per g clay	mmol NH_4^+ per g clay	Equilibrium constant, K_e**
40·7	22·6	0·26	0·16	4.4×10^{-3}
39·0	21·7	0·33	0·26	9.4×10^{-3}
31·1	17·3	1·08	0·51	13.9×10^{-3}
5·9	3·27	1·63	0·80	120×10^{-3}

* Expressed on an oven-dry basis (378°K).

** $K_e = \dfrac{[NH_4^+]^2}{[H_2O][NH_3]}$

carbon medium, of the clay with an organic base in the presence of Hammett indicators[8,9] (cf. Tables 47 and 48).

In the colour reactions of clays with aromatic amines both Lewis and Brønsted acidities are involved. Adsorption of the organic molecules is controlled by electron transfer and by cation exchange processes occurring at both the edge and the basal surfaces of the clay crystals. These observations have led Solomon et al.[10,11] to propose a model based on charge transfer between the mineral and the adsorbed organic species. Besides being able to account for the influence of clays on colour formation with certain organic compounds, this model is also successful in rationalizing the activity of clay minerals in catalysing the polymerization of adsorbed organic monomers.[12-14] These systems will therefore be discussed consecutively.

The colour reaction of benzidine with layer silicates is perhaps the best known and most widely studied system and serves as an outstanding example of the manner by which clays activate reactions of this type (Fig. 92).

263

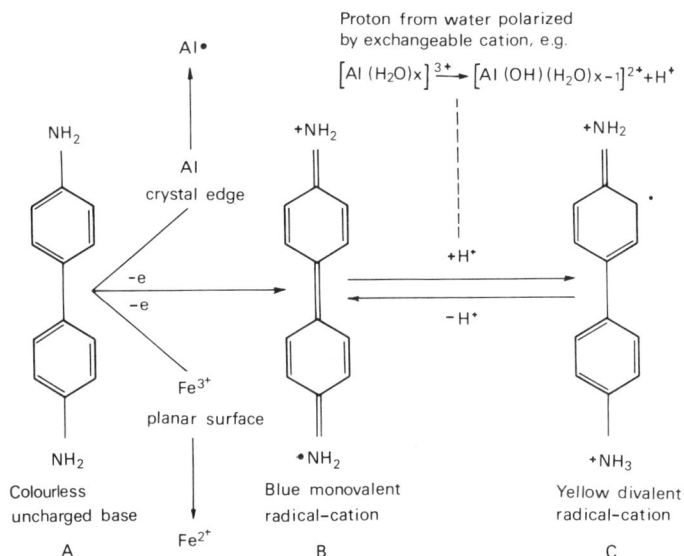

Fig. 92 Diagram illustrating the influence of clays and pH on the transformation of benzidine into its blue and yellow radical-cation forms. (after Theng[13])

The conversion of the colourless neutral diamine (A) to its blue derivative (B) when an aqueous solution of benzidine is added to certain clays has been reported by various workers[1,5,15-19]. It is generally agreed that this transformation involves the transfer of an electron (that is, oxidation) from the diamine to the mineral yielding the blue monovalent radical-cation (*semiquinone*). The stability and blue colour of this species are thought to arise from conjugation of the unpaired $2p_z$ electron from nitrogen with the π electron system of the aromatic ring, reinforced by resonance involving both rings. The blue species, however, only exists between pH 2·5 and 6. When the pH of the system falls below about 2, it can accept a proton on the lone pair of electrons of the nitrogen atom, giving the yellow divalent radical-cation. The yellow colour indicates a limited number of possible resonance structures.

In an attempt to find a ready and convenient method of distinguishing between the various groups of clay minerals (Table 1), Hendricks and Alexander[15] have examined the reaction of some aromatic diamines with clays. They noted that the minerals of the kaolinite group, as opposed to the montmorillonites (smectites), failed to give the blue colour with benzidine. Subsequently, Bosazza[18] and Vedeneeva[19] have reported similarly, counter to the earlier findings of Hauser and Leggett[1] and of Weil-Malherbe and Weiss[5] who observed that their kaolin specimens reacted positively, although the blue colour produced was much less intense than that given by montmorillonite-type minerals.

In re-examining the colour reaction of a number of soils and clays with benzidine, Page[20] has pointed out that simple inorganic oxidants, such as $FeCl_3$, H_2O_2, and bromine-water, also give a blue colour when benzidine is added. The presence of such impurities, together with that of organic matter in a given soil or clay sample, would thus mask the mineral's own activity. Page therefore concluded that the benzidine blue test for montmorillonite-type minerals was unreliable or, at best, only of limited use.

Using a range of representative clay minerals, Solomon et al.[10] have demonstrated that, with the possible exception of talc, each clay specimen produced a blue colour of varying intensity when brought into contact with a saturated aqueous solution of benzidine hydrochloride. Montmorillonites, previously treated with sodium hexametaphosphate (*calgon*) which specifically adsorbs on the edge surface of the clay crystals,[21] showed an appreciable decrease in activity whereas kaolinite and pyrophyllite, similarly treated, failed to react.

These observations indicated to Solomon et al. that in some minerals, such as kaolinite, the crystal edges have oxidizing (electron-accepting) sites while in others (e.g. montmorillonite) these sites are located both at the edge and at the planar surfaces of the crystals. They suggested that the active sites at the crystal edges consist of exposed, octahedrally coordinated aluminiums capable of accepting electrons from the benzidine molecules, that is, of acting as Lewis acids.

That ferric ions forming part of the montmorillonite layer structure can accept electrons has earlier been hinted at by other workers.[5,15,20,22] Weil-Malherbe and Weiss[5], for example, noted that reduction of their reactive clay specimens using stannous chloride inhibited colour formation whereas the mere removal of loosely bound iron using conventional extractants had little effect on reactivity. Similarly, Krüger and Oberlies[23] observed marked enhancement of the blue colour of their montmorillonite-benzidine complexes when the clay was pre-heated in air at 373°K.

By using a number of montmorillonites containing varying amounts of iron, subjecting them to oxidation (e.g. heating in air) or reduction (e.g. treatment with hydrazine) regimes, and following the change in the valency state of the iron by electron spin resonance (ESR) spectroscopy, Solomon et al.[10] unequivocally demonstrated that ferric ions occupying octahadral sites within the silicate layer profoundly influenced colour formation with benzidine (Table 76). Since these ions are located in the interior of the clay structure—apart from the few exposed at the edges—the activity of the clay may, in this case, be ascribed to the planar surfaces as distinct from that due to the crystal edges.

Table 77 summarizes the data of Solomon et al. on the benzidine blue reaction of different clays. Minerals of the montmorillonite group clearly show both crystal edge and planar surface activity, since treatments designed to deactivate either the crystal edge (e.g. by adsorption of polyphosphate) or the planar surface (e.g. by reduction of the mineral) fail to inhibit colour formation entirely. The activity of montmorillonites can only be completely suppressed by first reducing and then treating the minerals with polyphosphate. Interestingly, the hectorite sample used by Solomon et al. gave a pale blue colour with benzidine, a behaviour not expected from the 'ideal' composition of hectorite given by Grim[24]. The slight activity observed might be due to either the presence of surface-adsorbed contaminants that can accept electrons or to a limited

TABLE 76

Colour reaction with benzidine in aqueous medium of montmorillonites[a] as influenced by iron content and different pretreatments.[10]

	Montmorillonite sample	
Treatment of mineral	Wyoming bentonite (Volclay)	California bentonite (Otay)
	Fe_2O_3 content $= 3\cdot54\%$	Fe_2O_3 content $= 0\cdot97\%$
As received (untreated)	Dark blue	Light blue
After oxidation[b]	Very dark blue	Dark blue
After reduction[b]	Very pale blue	Very pale blue
After oxidation and polyphosphate treatment	Medium blue	Very pale blue
After reduction and polyphosphate treatment[c]	No colour	No colour

(a)   15 montmorillonites were examined giving results similar to those set out above.
(b)   the oxidized mineral showed an increase in the signal at $g = 4\cdot3$ (due to ferric ion) in the ESR spectrum as compared with the reduced sample.
(c)   with some samples a trace of blue colour was detected after a few minutes.

amount of isomorphous replacement of magnesium (or lithium) by ferric ions in the octahedral sheet.

The intensity of the blue colour of the clay-benzidine complex is clearly influenced by both the concentration and the source (location) of electron-accepting sites on the mineral. Recognition of this fact can adequately account for and reconcile the seemingly conflicting experimental data on the benzidine blue reaction of clay minerals. Since the transfer of an electron from the neutral base to the clay gives rise to the corresponding monovalent radical-cation (Fig.

TABLE 77

Colour reactions of some representative minerals with aqueous benzidine solution. The location of active sites is indicated.[10]

Mineral	Location of principal active sites*	Colour after one minute exposure
Talc	No active sites	No colour
Pyrophyllite (Robins, N. Carolina, USA)	Crystal edge	Pale blue
Hectorite	Planar surface	Very pale blue
Muscovite	Crystal edge	Very pale blue
Nontronite	Planar surface	Very pale blue
Illite (Fithian, Illinois, USA)	Crystal edge	Medium blue
Palygorskite (attapulgite)	Edge and planar	Medium blue
Kaolinite (Mt Egerton, Australia)	Crystal edge	Very light blue

* determined on the basis of the reaction shown by the mineral after oxidation, reduction, and polyphosphate treatment of the oxidized and reduced samples.

92) this species would be expected to adsorb on the mineral surface by replacement of the exchangeable (inorganic) cations initially present. Using a series of buffer solutions of pH 0·5 to 6·5, Dodd and Ray[25] have reported that adsorption (expressed on the basis of moles per unit weight of clay) increased with a rise in pH. This accords with ion exchange being the principal mechanism controlling adsorption because the (positive) charge on the molecule is pH-dependent.

More recently, Lahav and Raziel[26] have examined the influence of pH on the uptake of benzidine by montmorillonite. They noted that the adsorption data at pH $\sim$ 7 could be described by the Langmuir equation (see Chapter 4, Eq. 27). Measurements of the amount of inorganic cations (calcium) replaced indicates an approximate stoichiometric relationship between uptake and release so that the process involved is essentially one of cation exchange, confirming Dodd and Ray's conclusions. The exchange reaction, however, is apparently irreversible in agreement with Hauser and Leggett's finding that the blue colour of the benzidine-clay complex shows great stability. Evidently, factors other than simple ion exchange are involved in the adsorption process. One such factor may be molecular geometry of both substrate and adsorbate, since interaction between the orbitals of donor and acceptor atoms or groups is required for charge transfer to occur. The importance of steric factors is illustrated by the work of Conley and Lloyd[27] on the adsorption of some amines by hydrogen-kaolinite where the crystal edge surface is principally involved in adsorption.

That benzidine interacts strongly with clay mineral surfaces is also reflected in the observation that addition of concentrated NaCl or CaCl$_2$ solutions to the montmorillonite suspension does not sensibly prevent colour formation.[26] On the other hand, addition of polyphosphate before mixing the clay suspension with the benzidine solution markedly reduces the rate of colour development, probably as a result of decreased adsorption and inhibition of electron transfer from the benzidine molecule to edge aluminium.[26]

Besides influencing the amount adsorbed, the pH of the system also determines the colour of the resultant complex since the divalent-radical-cation formed below pH $\sim$ 2 is yellow in colour (Fig. 92). As reaction (1) indicates, the acidity of the system may be greatly enhanced by decreasing the hydration status of the clay. This increase in Brønsted acidity is responsible for the yellow colour observed when an anhydrous benzene, ethanol, or acetone solution of benzidine is mixed with dry montmorillonite[10,28,29]. Addition of water to the system or exposure of the yellow complex to water vapour restores the blue colour[10,28]. Similarly, the blue complex with benzidine turns yellow when strong acid is added[10] or water is removed.[19,28] Following Dodd and Ray[25] the role of water in colour formation may be represented by the equilibrium

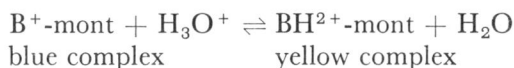

$$\text{B}^+\text{-mont} + \text{H}_3\text{O}^+ \rightleftharpoons \text{BH}^{2+}\text{-mont} + \text{H}_2\text{O}$$
$$\text{blue complex} \qquad\qquad \text{yellow complex}$$

where B refers to benzidine.

Following on the work of Hasegawa[29], Hakusui et al.[30] have examined the diffuse reflection spectra of benzidine and some related diamines adsorbed on acid clays (of unspecified mineralogical composition, probably of a montmorillonite type). The results indicate that both the monovalent and the divalent-radical-cations are present at the clay surface, the equilibrium between the two ionic forms being influenced by pH. What is perhaps significant is that the monovalent species tends to aggregate and this would contribute to the apparent irreversibility of the adsorption process, the strong retention of benzidine by the clay, and the stability of the blue colour (within certain pH limits).

In contrast to benzidine, the colour reactions of the stable reddish-purple free radical 2,2'-diphenylpicrylhydrazyl (DPPH·) and some leuco dyes (e.g. malachite green) with dry montmorillonite are influenced by the type of solvent used.[11] These compounds react positively in a benzene medium as shown by the loss of colour of DPPH· and the development of colours characteristic of the oxidized form of the respective dyes. When ethanol is substituted for benzene, however, no such colour changes are observed in the presence of the dry mineral, presumably because ethanol, being a stronger electron donor (Lewis base), is preferentially taken up over DPPH· and the leuco dyes which are thereby prevented from approaching the active (Lewis acid) sites at the clay surface. Note that DPPH· is a weak electron acceptor and is used as such in organic analysis.[31] That it donates rather than accepts electrons to and from the clay must be because the (oxidized) clay is a stronger electron acceptor. Indeed, strong electron-accepting organic species such as tetracyanoethylene (TCNE), can abstract electrons from (reduced) montmorillonite acting as an electron donor. Reduction of the uncharged TCNE molecule by planar surface ferrous ions gives rise to the red TCNE radical-anion.[11]

Since the mechanism controlling the uptake of benzidine and related compounds by montmorillonite is essentially one of ion exchange, the nature of the exchangeable cation at the clay surface clearly influences the adsorption process. The saturating cation also affects the rate of colour formation in that it determines the state of dispersion of the clay, that is, the extent of interlayer expansion (cf. Table 4), and hence the rate at which the organic molecule is intercalated.[10] Transition metal cations exert a more indirect effect on colour reactions. For example, the site of reaction for DPPH· is normally the edge of the clay crystal since calcium-montmorillonite pretreated with polyphosphate fails to decolourize a benzene solution of DPPH·. On the other hand, the cobalt-saturated clay shows some activity under similar conditions. This leads Solomon et al.[11] to propose that the cobalt ion can act as a 'bridge' across which an electron from DPPH· is transferred to a ferric ion at the planar surface.

As we have mentioned, Hendricks and Alexander[15] carried out their investigation on coloured clay-organic complexes with a view to using the colour reactions of clays to detect the presence or absence of certain groups of minerals. Despite its limitation,[20] the benzidine blue test may still be a useful diagnostic criterion for the montmorillonite-type clays if a few simple modifications (e.g. preheating the material in air followed by washing with polyphosphate) are introduced. However, the general availability of modern analytical instruments, such as X-ray diffractometers and infra-red spectrophotometers, has largely superseded those methods of clay mineral analysis based on colour formation involving organic compounds.

But the colour reactions of clays still enjoy substantial application in the paper industry, notably in the manufacture of pressure-sensitive, carbon-free paper[32-35]. Basically, the process involves bringing together a suitable electron-donating organic compound and a clay capable of accepting electrons (present as a coating over the paper) when pressure is applied. The manufacture of light-fast printing paper is a variant of this process in which the organic substance (e.g. triphenylmethane) is reacted with the clay coating. This leads to a marked increase in the stability of the dye and a slowing down of its decomposition by light.[36]

Another field of application of clay minerals as electron-donating or electron-accepting agents is in organic analysis. Complex formation with such compounds as DPPH· and TCNE, for example, gives rise to a change in free radical concentration which may be followed by electron spin resonance techniques or by simply measuring the change in colour intensity.

7.2. Polymerization Reactions

The initiation of polymerization of organic monomers by clays involves the conversion of the appropriate monomer to a reactive intermediate. In this respect, the behaviour of clays resembles that of chemical initiators. However, clays differ from most conventional initiators in that they can also inhibit the formation of a reactive species.

Although the ability of clays to influence and interfere with polymerization reactions has been recognized for many years,[37-39] it is only relatively recently that the underlying mechanisms have been clarified. As we pointed out in the preceding section, the charge transfer theory involving crystal edge and planar surface activation, developed by Solomon and co-workers,[12,40-42] can adequately account for many of the experimental observations on the polymerization of organic monomers catalysed by layer silicates.

The polymerization of styrene monomers in the presence of montmorillonite and other minerals has received much attention. Using montmorillonite which had been acid-washed and dried under vacuum at 553°K, Bittles et al.[43,44] obtained a substantial yield of polystyrene. They suggested that the reaction was initiated by a carbonium ion formed by proton transfer from sites associated with tetrahedrally coordinated aluminium to the monomer. There seems to be little need, however, to invoke any special active sites at the mineral surface in this system since a supply of readily accessible protons is available.

Acid-treated montmorillonite is known to be essentially unstable, the protons readily displacing aluminiums (and other octahedrally coordinated ions) from the interior of the silicate layers, causing them to migrate to the exchange sites.[45] [46]

In an attempt to separate the acidity arising from protons from that due to aluminium ions, Kusnitsyna and Ostrovskaya[47] have prepared a series of montmorillonites containing different proportions of exchangeable hydrogen and aluminium ions, storing the samples under benzene to preserve stability. As might be expected (cf. Eq. 1), both the hydrogen and aluminium forms were found to initiate styrene polymerization although the former was more active. This was thought to be due to the presence of a larger number of initiating points per unit weight of clay in the hydrogen system. These workers further noted that the rate of polymerization was greatly increased when clay was present compared with that using conventional initiators in a homogeneous medium. They ascribed this to the orienting effect of the clay surface on the adsorbed monomer molecules.

Solomon and Rosser[40] have previously demonstrated that cation-exchanged clays after drying can catalyse styrene polymerization. Using a range of clay minerals, they found that dry kaolinite and palygorskite showed greater activity towards styrene than either montmorillonite or pyrophyllite, while talc was completely inert. Pretreatment of the minerals with polyphosphate brought about a marked reduction in activity. These findings are therefore entirely analogous to those observed for the colour reaction of such minerals with benzidine (Table 77) and may be explained in terms of the presence at the crystal edges of electron-accepting sites (Lewis acids) consisting of aluminiums in octahedral coordination. Since the reaction shows features common to both a free radical and an ionic mechanism, Solomon and Rosser suggested that an electron transfers from the adsorbed styrene monomer to the aluminium, producing a radical-cation (cf. benzidine to benzidine blue transformation; Fig. 92) which rapidly forms a dimer. Both the radical-cation and its dimer are involved in the initiation process but propagation is probably cationic. This mechanism explains why alcohols and amines, for example, inhibit polymer formation[9,44] since these substances are strong Lewis bases and would be preferentially adsorbed over styrene.[27]

Matsumoto et al.[48] have recently proposed that Brønsted rather than Lewis acidity is responsible for initiating the polymerization of styrene by montmorillonite. Their data, however, are not self-consistent. For example, replacement of the exchangeable hydrogen ions by sodium or ammonium ions almost completely inhibits polystyrene formation while rewashing the clay with acid restores activity, all of which accords with a Brønsted acid initiation concept. On the other hand, treatment of the clay with trityl chloride, which appears in be selectively adsorbed on Lewis sites, brings about a marked decrease to polymer formation. Since the degree of polymerization increases as the dielectric constant of the medium rises, and is insensitive to the initial concentration of styrene, Matsumoto et al. propose a cationic mechanism for the initiation process. However, these observations are equally applicable to the propagation step and do not rule out the possibility that initiation occurs by a radical-cation mechanism.

As we have noted, both Brønsted and Lewis acid sites are present at the clay surface but the ratio of Brønsted to Lewis sites varies with the moisture content of the system[9] since the number of protons, derived from the dissociation of residual water molecules under the polarizing influence of the exchangeable cations, (Eq. 1) increases as the clay dries out (Table 75). To ascribe the activity of the clay to either Lewis or Brønsted acidity alone is oversimplifying the situation.

The available evidence[40,49] indicates that styrene either cannot penetrate the interlayer space of the montmorillonite crystal or intercalates only with diffi-culty. The fact that styrene is a non-polar molecule does not, of course, preclude its intercalation since both aliphatic and aromatic hydrocarbons can enter the montmorillonite interlayers (Section 3.6). However, under conditions favour-able to intercalation, such as drying or evacuating the mineral, styrene under-goes a vigorous catalysed polymerization. Friedlander and Frink[50] have stated that partial intercalation of styrene occurred by leaving the dry mineral in contact with excess monomer for two weeks but their evidence is not entirely convincing. Similarly, the data of Pezerat and Mantin[51] do not unequivocally demonstrate that styrene can enter the interlayer space of montmorillonite. Although most of the work dealing with the initiation of polymerization by clay minerals has been carried out using styrene as the monomer, olefins, dienes, and 4-vinylpyridine can also be polymerized in this way.[50,52−54]

Inhibition of polymer formation by clays is also explicable in terms of charge transfer between the monomer and the mineral surface with the clay acting as an electron acceptor. The reaction of methyl methacrylate serves as an example of the inhibitory effect of clays[41]. Methyl methacrylate is normally polymerized by free radical initiation, the reactive species being formed either by heat treatment or by decomposition of conventional chemical initiators. In the presence of montmorillonite, however, this monomer fails to polymerize even after heating the mixture at $373°K$. Other types of clay, such as kaolinites, reduce the yield and increase the molecular weight of the polymer produced. This indicates that the number of free radicals available for chain initiation or propagation is reduced when clay is present. The data can be rationalized by postulating that the reactive intermediate is adsorbed by the mineral edge and there deactivated. Electron transfer from the methyl methacrylate radical to edge aluminium would give rise to the corresponding cation which, in this case, is a non-propagating species.

Planar surface activation occurs with certain hydroxymethacrylate mono-mers which form interlayer complexes with montmorillonite.[42] These com-pounds polymerize spontaneously if the montmorillonite contains a transition metal ion (usually iron) in the lower valency state in the octahedral sheet of the silicate layer. Preheating in air or treatment with oxidizing agents deactivates the mineral. Conversely, reduction of the oxidized clay, for example by treat-

271

ment with aqueous hydrazine or by electrochemical means, restores its ability to initiate polymerization. The polymerization of interlayer hydroxyethyl methacrylate, for example, occurs by electron transfer from the clay to the monomer which transforms into the corresponding radical-anion. Addition of a proton from a suitable proton donor such as water converts this radical-anion to a free radical which propagates between the montmorillonite layers and outwards to the external monomer liquid. Evidently the active sites, in this instance, are identifiable with transition metal ions in the octahedral sheet of the silicate layer acting as electron donors. By analogy with the colour reactions catalysed by montmorillonite-type minerals, the initiation of polymerization may be said to occur by 'planar surface' activation.

The various mechanisms by which clays, such as montmorillonite, activate or inhibit polymerization of adsorbed monomers, are illustrated in Fig. 93.

Some monomers which fail to polymerize spontaneously in the presence of clay minerals may be induced to do so by chemical initiators or by ionizing radiation. For example, 2,2'-azobisisobutyramidine (ABIB) in the cationic form can be introduced into the exchange sites of montmorillonite. The ABIB-clay complex, when gently heated, decomposes to yield free radicals.[55] Using this complex, Dekking[56] has been able to polymerize such monomers as methyl methacrylate, acrylamide, vinyl acetate, and 4-vinylpyridine, the rate of polymer formation being greater than that for ABIB alone. This observation was ascribed by Dekking to the higher rate of decomposition of the initiator-clay complex as compared with that of ABIB itself, and to a decrease in the rate of chain termination. The greater accessibility together with the availability to the monomer molecules of the initiator 'extended' on the silicate surface must also contribute significantly to the increased efficiency.

In a series of papers,[49,57-61] Blumstein and co-workers have reported on the polymerization of interlayer methyl methacrylate (and methyl acrylate) in montmorillonite using γ-rays, attention being directed to the properties of the polymer formed rather than to the mechanism of polymerization. The inter-layer polymers were resistant to the usual solvent extraction procedures, necessitating the use of hydrofluoric acid to dissolve out the montmorillonite layers and so liberate the polymer.[49] Such a harsh treatment, however, is liable to damage most polymers so that the technique is only of limited scope. The heat stability[57] and dilute solution properties[58] of the interlayer polymers markedly differ from those of the corresponding bulk polymers. This observation leads the Blumstein group to suggest that the interlayer poly(methyl methacrylate) has a two-dimensional sheet structure when a cross-linking agent is incorporated in the monomer system. Individual polymer sheets tend to aggregate into packets of 8 to 10 molecular thicknesses, reflecting the amount of layer stacking in the crystal of the montmorillonite sample.[59] Treatment with 30 per cent chloroform at 333°K gave rise to disaggregation of the polymer[59,60] which appeared to consist of short, predominantly isotactic stereosequences.[61] That the poly(methyl methacrylate) so formed could show small scale stereoregularity of this kind might perhaps be expected since ion-

272

$$CH=CH_2 \qquad \overset{+}{CH}-\overset{\bullet}{CH_2}$$

Styrene

Radical Cation
(Propagating Species)

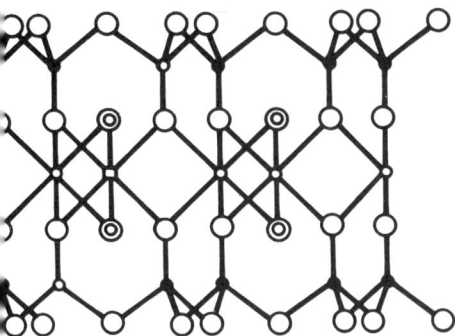

TERLAYER REGION—Activation by
ctron transfer to monomer from Fe^{2+} in
ahedral coordination in silicate layer.

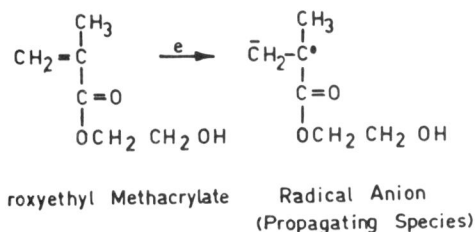

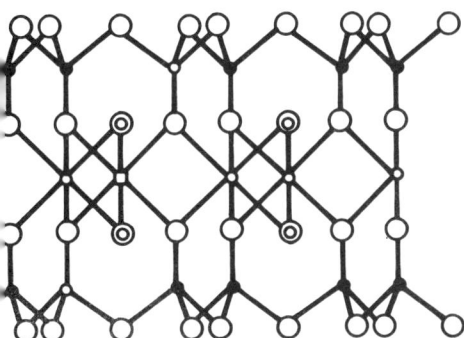

$$CH_2=\overset{\overset{\displaystyle CH_3}{|}}{\underset{\underset{\displaystyle OCH_2\ CH_2\ OH}{|}}{\underset{|}{C}}}\xrightarrow{\quad e \quad}\bar{C}H_2-\overset{\overset{\displaystyle CH_3}{|}}{\underset{\underset{\displaystyle OCH_2\ CH_2\ OH}{|}}{\underset{|}{C^\bullet}}}$$

roxyethyl Methacrylate   Radical Anion
(Propagating Species)

CRYSTAL EDGE—Deactivation by electron
transfer to edge Al from propagating radical.

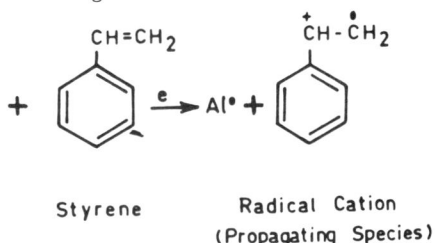

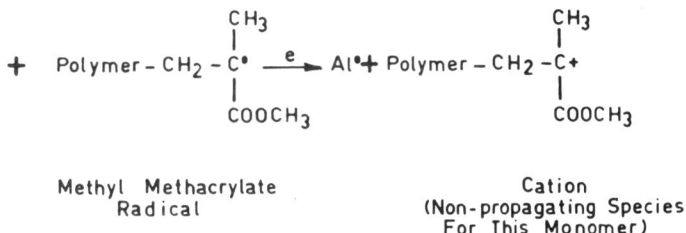

$$+ \ \ Polymer-CH_2-\overset{\overset{\displaystyle CH_3}{|}}{\underset{\underset{\displaystyle COOCH_3}{|}}{C^\bullet}} \xrightarrow{\quad e \quad} Al^\bullet+ Polymer-CH_2-\overset{\overset{\displaystyle CH_3}{|}}{\underset{\underset{\displaystyle COOCH_3}{|}}{C+}}$$

Methyl Methacrylate   Cation
Radical   (Non-propagating Species
For This Monomer)

XTERNAL PLANAR SURFACE—Activa-
on by proton transfer to monomer from
xchange site or from H_2O polarized by ex-
hange cation e.g. $[Al(H_2O)_x]^{3+} \rightarrow [Al(OH)$
$H_2O)_{x-1}]^{2+} + H^+$

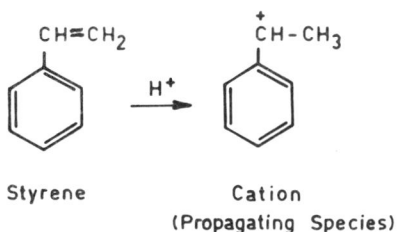

KEY: ◯ Oxygen

◉ Hydroxyl

● Silicon

◯ Aluminium

▢ Ferrous Iron

$$CH=CH_2 \qquad \overset{+}{CH}-CH_3$$

H$^+$

Styrene   Cation
(Propagating Species)

Fig. 93 Diagram illustrating the mechanisms of
monomer activation/deactivation by clays (e.g.
montmorillonite) and the location of active sites
on the silicate layer (after Theng and Walker[14])

dipole interactions would tend to direct the ester carbonyl group of the mono-mer towards the exchangeable (sodium) ion.

The formation of stereoregular poly(acrylonitrile) and poly(acrylic acid) in the interlayer space of montmorillonite using similar techniques had earlier been claimed by Glavati et al.[62] and Glavati and Polak.[63] On the basis of X-ray diffraction, viscosity and light scattering measurements on the clay-polymer complexes and on the interlayer polymers after dissolution of the clay, these workers inferred that the polymers were syndiotactic but their evidence is not unambiguous.

The formation of stereoregular macromolecules by polymerizing adsorbed monomers in the montmorillonite interlayers *in situ*, however, is by no means unlikely because organic molecules are oriented and packed in some particular arrangement when they enter the interlayer space of the mineral (Chapter 2). The conformation of the intercalated monomer is clearly an important factor in determining whether or not polymerization occurs. Thus, unlike the hydroxy methacrylates, methyl methacrylate itself and the amino methacrylates inter-calated in montmorillonite fail to polymerize spontaneously, presumably be-cause the orientation adopted is unfavourable to the formation of intermole-cular bonds. Inhibition of polymer formation by steric factors has also been observed in urea clathrates[64] and in nickel cyanide[65] complexes. Orientation effects may likewise modify the products formed from polymerizing adsorbed monomers. For example, exposure of isoprene to γ-rays normally leads to the formation of cyclic products. When absorbed on vermiculite, however, acyclic polymers containing multiples of the basic five-carbon skeleton are produced by this treatment.[66]

The practical difficulty of separating the interlayer polymer from the clay support has led Theng[67] to use calcium aluminate hydrate as a substrate for producing two-dimensional organic macromolecules. This mineral, having the composition $4CaO . Al_2O_3 . H_2O$, is particularly suited for this purpose because it can be made synthetically in pure form, is capable of intercalating a wide variety of organic compounds and monomers, and is readily soluble in dilute hydrochloric acid.[68] Polymerization of interlayer monomers, for example 1,3-butyleneglycol dimethacrylate, by means of γ-rays produces thin plates of polymer about 20 nm thick. Since individual polymer sheets are of the order of 1 nm, considerable face-to-face aggregation, as reported by Blumstein *et al.*[59,60] for interlayer poly (methyl methacrylate) in montmorillonite, must also occur here.

The hypothesis that complex organic molecules can arise from such simple precursors as methane, nitrogen, hydrogen, and water under conditions which may have existed in the primeval Earth's atmosphere and oceans by the action of various forms of energy, has been amply substantiated by simulation experiments (see, for example, Calvin[69] for a summary of the literature on chemical evolution.) It is generally accepted that the origin of life involves three cumulative processes: (a) the formation of monomers; (b) the synthesis of polymeric compounds of biological importance; and (c) the organization of such polymers into cellular systems. The origin of life on Earth has been fully discussed by Oparin[70] and Cairns-Smith.[71] We are only concerned here with the role of clay minerals as substrates and catalysts in the production of bio-polymers.

As Bernal[72] has pointed out, the prime difficulty in going from step (a) to (b), such as may have occurred in the free oceans, lies in the extreme dilution of the system, estimated on thermodynamic grounds to be about 10^{-15} to 10^{-30} M[73]. At such levels of monomer concentration, polymerization in sea water systems would be extremely unlikely even if the rate of monomer accumu-lation could be raised by a few orders of magnitude. Bernal therefore suggested that adsorption on clay minerals provided the favourable conditions for concen-tration and protection of the primordial photochemical products, necessary for chemical evolution to proceed. This proposal has stimulated a great deal of research into the ability of clay minerals to act as a concentrating surface and catalyst in the synthesis of biologically important monomers and polymers.

The synthesis of amino acids by means of ultra-violet irradiation of solutions of formaldehyde and ammonium salts in the presence of clays has been reported by Pavlovskaya et al.[74] Similarly, Yoshino et al.[75] have synthesized a range of amino acids (glycine, α- and β-alanine, sarcosine, aspartic acid, glutamic acid, arginine, histidine, lysine and ornithine) by reacting CO, H_2 and NH_3 at 473 to 973°K in the presence of montmorillonite. Purines, pyrimidines, and hydrocarbons also appear to be formed together with the amino acids.

The use of clays as a support for polypeptide synthesis has received much attention. Degens and Mathéja,[76,77] for example, have obtained a number of polypeptides of high molecular weight ($>10,000$) by condensing L-amino acids on silicate surfaces, notably those of kaolinites, at temperatures below the boiling point of water. Elevated reaction temperatures are not essential for condensation, as had been shown earlier by Fripiat et al.[78] using infra-red spectroscopy (see Section 4.2).

More recently, Paecht-Horowitz et al.[79] have observed that montmorillonite could catalyse the formation of polypeptides from amino acid adenylates, such as alanyl adenylate. These derivatives, rather than the amino acids as such, were chosen on the ground that amino acid adenylates were produced in the first step of protein synthesis by cellular systems. Interestingly, kaolinites showed little, if any, activity in this instance, presumably because a certain level of surface concentration of the active monomers was required before appreciable polycondensation could occur. Although basal spacing data were not reported it seemed probable that, like the corresponding amino acids, the adenylate derivatives were intercalated by montmorillonite. On this assump-tion it can be shown that at maximum adsorption ($\sim 10^{-3}$ mol per g clay) some 30 per cent of the clay surface was covered by adenyl adenylate ions.

Unlike polycondensation in a homogeneous solution, which is preceded by hydrolysis yielding the corresponding free amino acids, heterogeneous poly-merization does not usually give rise to the liberation of the amino acids. The first important effect of surface adsorption is therefore to inhibit hydrolysis, as Bernal[72] had previously predicted on theoretical grounds. Furthermore, an appreciable proportion of the polypeptides formed contains terminal adenylic acid.

The polycondensation process taking place in the presence of montmorillonite, as envisaged by Paecht-Horowitz *et al.*, is given by the following scheme, where A and P represent amino acid and adenylate, respectively.

(i) *Propagation step*

$$A \sim P + A \sim P \rightarrow A_2 \sim P + P$$
$$A_2 \sim P + A \sim P \rightarrow A_3 \sim P + P$$
$$A_2 \sim P + A_3 \sim P \rightarrow A_5 \sim P + P$$

and so on, or in general

$$A_i \sim P + A_j \sim P \rightarrow A_{i+j} \sim P + P$$

where A_i and A_j denote that i- and j-meric peptide.

(ii) *Termination step*

$$A_i \sim P \rightarrow A_i - P$$

where the active phosphoanhydride bond ($\sim$P) is converted to the inactive ester bond ($-$P).

Sephadex gel filtration of the reaction medium yields fractions with an irregular but discrete distribution of molecular weights of peptide adenylates, ranging from 2,000 to 4,000 and corresponding to a degree of polymerization between 30 and 56. In agreement with the proposed scheme, all the peptides carry terminal adenylic acid residues, and little or no monomer hydrolysis is observed. This provides experimental verification of the protective effect of adsorption, thus allowing the monomers to polymerize. The free adenylic acid produced during polymerization tends to inhibit polycondensation by suppressing the dissociation of the non-reactive $-NH_3^+$ group into a proton and the highly reactive terminal $-NH_2$. Thus, rapid polycondensation gives rise to a high local concentration of adenylic acid which arrests the process until this acid diffuses out. When this diffusion is complete, a second pulse of polypeptide formation occurs and this will, in time, cease for the stated reasons.

Besides acting as a supporting and protective medium for polypeptide formation there is evidence to show that clay minerals, at least the kaolinites, preferentially polymerize the L optical isomers of amino acids over those of the D enantiomers.[80,81] Although the underlying mechanism is as yet not fully understood, this selective capacity of clays may well be the reason why most, if not all, proteins in living systems consist of L-amino acid sequences. It has been pointed out[82-84] that both the stability and the maximum length attainable by the primary protein structure would be limited in an atactic chain (composed of randomly assorted L and D optical monomers) owing to steric hindrance involving the amino acid side chains and intramolecular repulsion. For the same reason, the orderly arrangement of secondary structures (e.g. the α helix) could not be realized if the polypeptide chain were to be atactic. In an

276

isotactic polymer (consisting of either D or L optical isomers), on the other hand, steric and repulsive effects are absent or minimal, allowing the protein molecule to adopt the highly organized arrangement and complex structure characteristic of and, perhaps even essential to, living systems.[85]

By reacting the L, D, and DL forms of such amino acids as aspartic acid, serine and phenylalanine with a kaolinite clay at 333 to 363°K for varying periods of time, Jackson[81] has demonstrated that the L optical isomers were polymerized at an appreciably higher rate and to a larger extent than the corresponding D forms, while the racemic mixture polymerized at an intermediate rate. Figure 94 shows the results for aspartic acid where

$$\text{per cent polymerized} = \frac{\text{total amino acid} - \text{free amino acid}}{\text{total amino acid}} \times 100$$

the 'total' and 'free' amino acid representing the concentration before and after hydrolysis, respectively. Although on this basis and under similar conditions

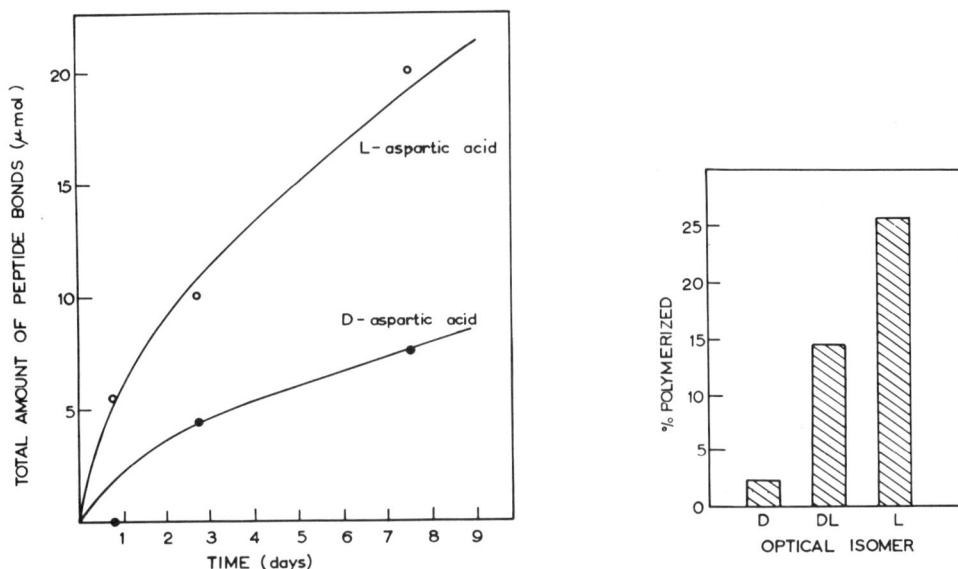

Fig. 94 *Left:* Rate of peptide bond formation in solutions of L- and D-aspartic acid at 363°K and in the presence of kaolinite *Right:* Relative amounts of aspartic acid optical isomers polymerized on kaolinite after 32 days, estimated from analysis of the supernatant solutions (after Jackson[81])

polypeptides also form in the absence of kaolinite, no consistent or significant relationship is observed between chirality and the extent of polymerization. The formation of peptide bonds is further indicated by the biuret test and by infra-red examination of the material extracted from the kaolinite complex using ammonia. Comparison of the amount of phenylalanine adsorbed at pH

2 and 5·8 suggests that the preferential uptake of the L isomer over the D form occurs at the crystal edges. This might perhaps be anticipated from the fact that the edges of kaolinite crystal are enantiomorphous because of the arrangement of oxygens, hydrogens, aluminiums, and octahedral vacancies so that kaolinite crystals are either left or right handed.[86]

Jackson[81] suggests that coordination to edge aluminium of the carboxyl and amino groups, and electrostatic and hydrogen-bonding interactions all play a part in the adsorption of the amino acids to the kaolinite edge surface. The presence of alternating positively and negatively charged sites along this surface could provide the required alignment of the amino acid molecules with their carboxyl and amino groups in juxtaposition, facilitating polycondensation.

Why should the L optical isomer rather than the D enantiomer be selected, assuming that the amino acids formed under conditions of the primitive Earth were racemic? Wald[82] has suggested that both L and D organisms were initially present but that the former gained ascendancy during the evolutionary process. Alternatively, the mineral templates on which polymerization occurred were dominantly of the L type so that polypeptides having the L configuration were preferentially formed. The latter alternative is more easily verified, because we can see whether most naturally occurring kaolinite minerals are indeed of the left-handed type.

Kaolinite can also catalyse the synthesis of carbohydrates and lipids. Harvey et al.[87] for example, have presented evidence for the formation of polysaccharides by incubating kaolinite crystals with paraformaldehyde in an aqueous medium at $353°K$ for 160 days. The sugars released on hydrolysis were separated by column chromatography and characterized spectrophotometrically after reaction with an alkaline solution of tetrazolium blue. The catalytic activity of kaolinite is more clearly in evidence here than in the amino acid system studied by Jackson, since in the absence of clay no sugar is detected in the hydrolysate. Octahedrally coordinated aluminiums at the crystal edges are thought to provide the active sites at which an aldol-type condensation reaction occurs. The sequence of possible reactions is shown schematically in Fig. 95.

These workers further postulate that a formaldehyde unit may condense with a sugar precursor, such as glyceraldehyde, one at a time until a five- or six-membered cyclic hemiacetal is yielded. On this basis, steric repulsion involving hydroxyl groups would give rise to the all-*trans* sugar idose (or its rearrangement isomer, sorbose) as the principal product. The pathway of threose → xylose → idose → sorbose may thus be anticipated. This and other possible sequences are illustrated in Fig. 96.

The validity of the proposed reaction scheme is supported by the observation that on mixing arabinose with an aqueous suspension of kaolinite at $353°K$ for 2 weeks followed by extraction of the clay residue with sulphuric acid, 6-deoxymannose (rhamnose), glucose and arabinose are found in the hydrolysate.

Kaolinite can also act as a catalyst for the esterification of fatty acids in an

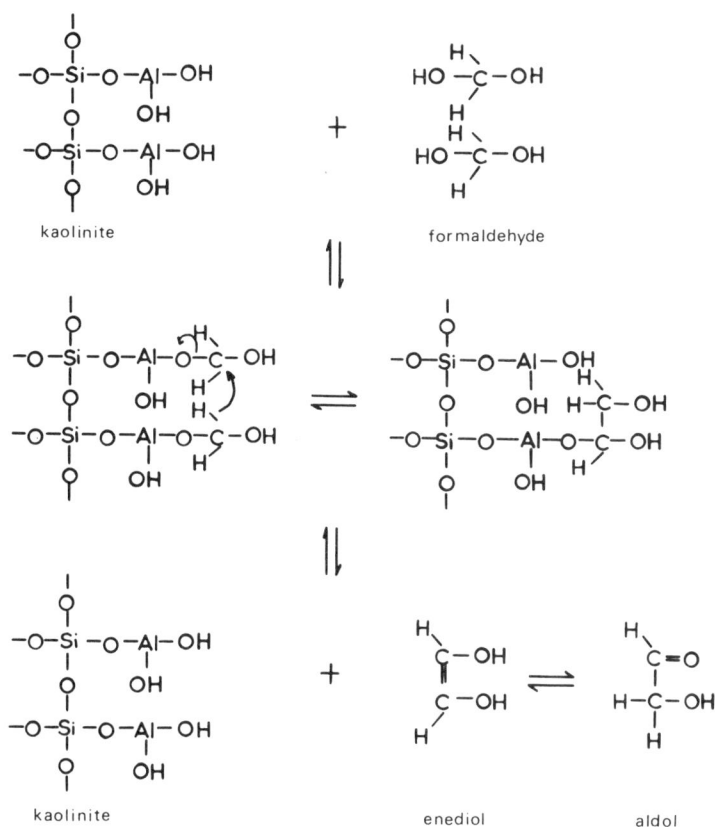

Fig. 95 Possible reaction sequence in the formation of some sugar precursors from formaldehyde, activated by kaolinite (after Harvey *et al*[87])

aqueous environment yielding glycerides.[87] For example, various quantities of tripalmitin, 1,2-dipalmitin and monopalmitin were recovered from a system of kaolinite, glycerin, palmitic acid and water (after incubation at $358°K$ for 2 weeks in an evacuated flask). The glycerides formed were identified by their melting point, and infra-red and mass spectra. Similarly, Ponnamperuma[88] has reported to have formed a number of biologically important sugars on refluxing formaldehyde over illite samples.

Harvey *et al.*[87] have further claimed to have observed the formation of phospholipid monolayers on the basal (hexagonal) surfaces of kaolinite crystals when the complex obtained by adding calcium phosphate to an aqueous mixture of kaolinite, glycerol and palmitic acid was examined under the electron microscope. If substantiated, this finding may have important implications since phospholipids are stabilized by association with hexagonal protein crystals in biological membrane. The protein could substitute for the silicate template from which the phospholipid monolayer might be stripped off intact.

Previously, Degens and co-workers[76,77] had reported that kaolinite catalysed

279

CHO

CH₂OH

glyceraldehyde

CHO

CH₂OH

erythrose

CHO

CH₂OH

threose

CHO

CH₂O

ribose

CHO

CH₂O

arabinose

CHO

CH₂OH

xylose

CHO

CH₂OH

lyxose

CHO   CHO   CHO   CHO   CHO   CHO   CHO

CH₂OH   CH₂OH   CH₂OH   CH₂OH   CH₂OH   CH₂OH   CH₂OH   CH₂OH

allose   altrose   glucose   mannose   gulose   idose   galactose   tallose

sorbose

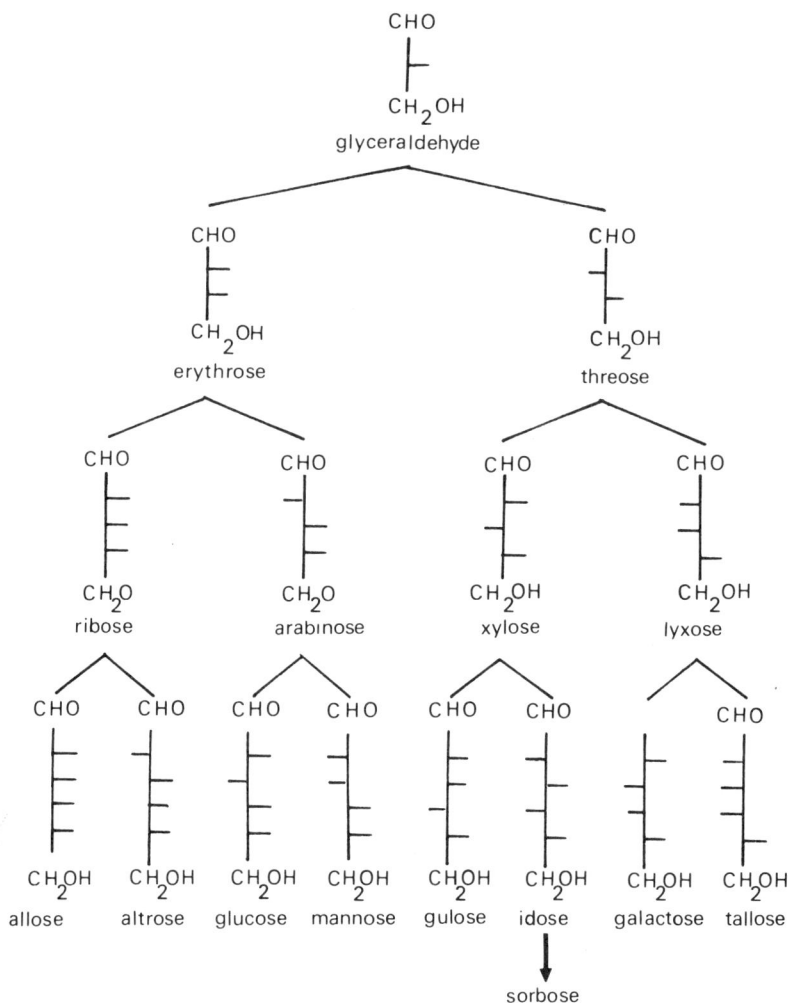

Fig. 96   Distribution of sugars by addition of one formaldehyde unit, one at a time, to D-glyceraldehyde in the presence of kaolinite. Steric repulsion between hydroxyl groups would influence the distribution (after Harvey et al[87])

the formation of polypeptides starting from carbon monoxide and ammonia, according to the general scheme

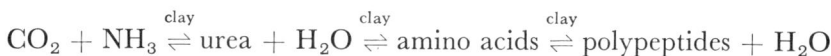

$$CO_2 + NH_3 \underset{\text{clay}}{\rightleftharpoons} \text{urea} + H_2O \underset{\text{clay}}{\rightleftharpoons} \text{amino acids} \underset{\text{clay}}{\rightleftharpoons} \text{polypeptides} + H_2O$$

Subsequently, cytosine, uracil, and cyanuric acid were also found in detectable amounts in this system for which plausible reaction pathways have been suggested.[89]

More recently, Ibanez et al.[90] have noted that in the presence of montmorillonite, condensation of mononucleotides by cyanamide gave rise to oligodeoxyribonucleotides of longer chain dimensions than those produced in homo-

280

geneous aqueous solutions. Although no explanation was offered for this observation, it seems probable that surface adsorption, as remarked on earlier, may protect the active species from being hydrolysed. In addition, adsorption may impose on the reactants a favourable steric configuration so that condensation becomes more extensive than would otherwise be possible.

7.3. Transformation and Decomposition Reactions

The ability of clay minerals to activate a wide variety of organic reactions leading to the transformation and/or decomposition of the adsorbed species has long been recognized.[38,91,92] The cracking of petroleum is perhaps the best known example of a degradation reaction catalysed by clays. The chemistry of the process is beyond the scope of this book; it has been dealt with by a number of authors.[93-95]

A normal paraffin can be cracked by either a *thermal* or a *catalytic* mechanism.[95] The former reaction is initiated by the loss of a hydrogen atom from the hydrocarbon molecule. The hydrocarbon radical so produced may immediately crack or may undergo radical isomerization which involves the change in position of a hydrogen atom to yield an energetically more favourable radical. Cracking occurs at the C—C bond in the β position to the carbon atom lacking the hydrogen atom, forming a primary radical and an α-olefin. Thus, the major products of thermal cracking of an *n*-paraffin consists of normal hydrocarbons, α-olefins and ethylene.

Catalytic cracking invariably requires the presence of acidic catalysts, such as clays, capable of supplying available protons which initiate the reaction. Initiation involves the abstraction of a hydride ion from the *n*-paraffin molecule resulting in the formation of a carbonium ion. The carbonium ion produced may then undergo a number of rearrangements and reactions, such as β splitting, methyl group shift, hydride ion abstraction, and hydrogen shift.

Both types of cracking reaction are possible mechanisms in the generation of petroleum. Catalytic cracking is favoured by Brooks,[96] for example, whereas Jurg and Eisma[97] believe that thermal cracking is the principal mechanism. However, both processes may operate together in a given system and it is often difficult to distinguish between them clearly.

The thermal alteration of long chain *n*-alkanes, such as *n*-octacosane ($C_{28}H_{58}$), in a homogeneous system and in the presence of montmorillonite has been studied by Henderson *et al.*[98] who heated the *n*-paraffin to temperatures of or above 473°K and examined the decomposition products by gas-liquid chromatography (GLC). Heating at 648°K for varying periods of time yielded a series of *n*-alkanes and *n*-alkenes with a smooth carbon-number distribution centring around nC_{21} (alkanes) and at C_{22} and C_{18} (alkenes). With increasing time of heating the yield of *n*-alkanes increased, whereas both the proportion of branched and cyclic alkanes and the yield of alkenes and aromatics tended to fall. In addition, the carbon-number maxima for the alkanes and alkenes became lower. Heating at 473°K for 1,000 hours *in vacuo* gave rise to a marked decrease in the extent of alteration (<0·1 per cent), presumably because the activation energy of fission of C—C bonds was high and, consequently, the rate of reaction decreased. However, under the same conditions but with bentonite present about 1 per cent of the *n*-octacosane was converted to other alkanes, alkenes and aromatics, that is, there was more than a ten-fold increase in the extent of alteration. Raising the temperature to 548°K resulted in more than 90 per cent conversion of the starting hydrocarbon to an insoluble, black carbonaceous material. Only a small proportion of the *n*-octacosane remained in the solvent-soluble products which were more abundant in alkanes and aromatics compared with those produced in a homogeneous medium.

Of greater interest is the high concentration of branched or cyclic alkanes generated in the presence of bentonite, the repeating pattern of GLC peaks resembling that given by the alkane fractions of crude petroleum.

These results clearly demonstrate that even 'brief'—in terms of geological time—heating at $\geqslant 473°$K leads to extensive alteration, indicating that the thermal degradation of an *n*-alkane (C_nH_{2n+2}) gives rise to the complete homologous series of *n*-alkanes (C_mH_{2m+2}) where $m < n$, and to the corresponding *n*-alkenes $(C_{n-m}H_{2(n-m)})$.

Greensfelder *et al.*[99] had previously shown that olefins and alkanes could undergo thermal cracking at the same rate, but that in the presence of bentonite the former yielded carbonium ions more readily and the rate of cracking was higher. Olefins formed by thermal cracking may, of course, isomerize, undergo rearrangements, cyclization, and hydrogen transfer reactions when clays are present. Thermal cracking therefore tends to generate a high concentration of *n*-alkanes whereas catalytic cracking would produce mainly branched and cyclic alkanes.

The hypothesis that fatty acids are precursors of *n*-paraffins in petroleum has stimulated a considerable amount of research into the influence of clays on the catalytic degradation of adsorbed fatty acid molecules. Jurg and Eisma[97] heated behenic acid $(C_{21}H_{43}COOH)$ with bentonite, in the presence and absence of water, at 473°K for 89 and 760 hours and obtained hydrocarbons containing 3, 4, or 5 carbon units. The total amount of saturated hydrocarbons increased with time of heating, whereas that of the olefinic type decreased. The presence of water gave rise to a different distribution of the degradation products and also altered the ratio of the various components. For example, the ratio of normal to branched chain C_4 and C_5 alkanes was about 0·1 when water was present and about 4 in the absence of water. Similar values were observed for the ratio of iso- to normal pentane. When behenic acid was heated at 473°K without the bentonite clay, no hydrocarbons were produced.

Tables 78 and 79, taken from a later study by Eisma and Jurg,[95] show the amounts of various low molecular weight hydrocarbons produced on heating behenic acid with kaolinite at different temperatures and for varying periods of time. Besides the compounds listed, C_{15} to C_{34} *n*-alkanes and C_{15} to C_{35} fatty acids were detected in the GLC pattern. It is of interest that the relative

TABLE 78
Amounts (μmol) of low molecular weight hydrocarbons generated when
1 g behenic acid is heated in the presence of 2·5 g kaolinite at 473 °K for
varying periods of time.[95]

Products	Time of heating (hours)				
	94	283	330	976	1848
Ethane + ethene	0·03	0·07	—	0·14	0·14
Propane	0·32	0·84	1·00	1·97	2·94
Propene	0·58	0·75	0·64	0·41	0·34
Isobutane	3·04	4·28	3·94	5·94	10·14
n-Butene	0·08	0·19	0·23	0·45	0·93
Isobutane + 1-butene	0·13	0·09	0·06	0·04	0·05
2-Butene-trans	0·08	0·12	0·11	0·09	0·10
2-Butene-cis	0·05	0·06	0·07	0·05	0·07
Isopentane	1·45	2·62	4·13	4·62	8·27
n-Pentane	0·17	0·14	0·28	0·36	0·51
1-Pentene	0·01	0·01	0·01	0·01	0·01
2-Me-1-butene	0·03	0·06	0·03	0·03	0·01
2-Pentene-trans	0·06	0·02	0·09	0·04	0·04
2-Pentene-cis	0·02	0·11	0·03	0·02	0·01
2-Me-2-butene	0·15	0·10	0·14	0·07	0·05
2-Me-pentane	0·66	1·23	2·41	3·69	4·16
3-Me-pentane	0·21	0·43	0·88	1·31	1·47
n-Hexane	0·08	0·12	0·22	0·31	0·39
Total	7·1	11·2	14·2	19·5	29·6

TABLE 79

Amounts (μmol) of low molecular weight hydrocarbons generated when
1 g behenic acid is heated in the presence of 2·5 g kaolinite and 7·5 g water
at different temperatures (°K) and for varying periods of time (hours).[95]

Products	Temperature	473	523	538	538	548
	Time	75	275	625	1300	330
Ethane + ethene		0·06	0·04	0·21	0·26	0·24
Propane		0·01	0·03	0·08	0·12	0·14
Propene		0·07	0·08	—	0·25	0·16
Isobutane		—	0·02	0·03	0·04	0·03
n-Butane		0·01	0·02	0·07	0·09	0·13
Isobutane + 1-butene		0·01	0·03	0·06	0·08	0·06
2-Butene-trans		—	0·02	0·07	0·10	0·08
2-Butene-cis		—	0·01	0·05	0·06	0·06
n-Pentane		0·02	0·05	0·06	0·09	0·13
1-Pentene		—	0·01	0·01	0·09	—
Isopentane		—	0·01	0·01	0·03	0·02

amount of C_2 alkanes together with the percentage of saturated and unsaturated hydrocarbons increased markedly with heating temperature.

To account for these observations, Eisma and Jurg[95] have proposed the following reaction scheme:

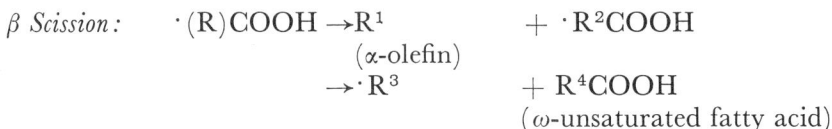

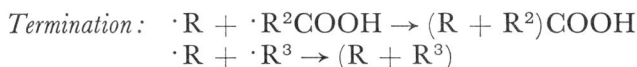

Initiation: $\quad$ RCOOH → ·R

Propagation: $\quad$ ·R + RCOOH → ·(R)COOH + RH

β Scission: $\quad$ ·(R)COOH → R¹ $\qquad\qquad$ + ·R²COOH
$\qquad\qquad\qquad$ (α-olefin)
$\qquad\qquad\qquad$ → ·R³ $\qquad\qquad\qquad$ + R⁴COOH
$\qquad\qquad\qquad\qquad\qquad$ (ω-unsaturated fatty acid)

Termination: $\quad$ ·R + ·R²COOH → (R + R²)COOH
$\qquad\qquad$ ·R + ·R³ → (R + R³)

The radicals ·R³ and ·R²COOH may abstract a hydrogen atom from another molecule to form the short-chain n-alkanes and fatty acids.

According to this scheme, the initiation step consists of the decarboxylation of the fatty acid to yield the corresponding alkyl radical, which then reacts with the original fatty acid giving an n-alkane and a secondary radical of the fatty acid. The latter then splits up by β scission into four components.

Support for the proposed scheme comes from the observation that for the behenic acid-clay system, the C_{21} n-alkane is always the major hydrocarbon formed and that the amount of low molecular weight hydrocarbons decreases with an increase in heating time, indicating the mediation of unsaturated species in the process. Moreover, when an initiator, such as 2,2'-azopropane, that can produce ·C_3H_7 radicals is added to the system, an $n·C_{24}$ radical is found to be the dominant species formed by combination of $n·C_{21}$ and $n·C_3$.

The data are therefore consistent with the view that (free) radicals are intermediates in the formation of long chain n-alkanes and fatty acids in the clay-behenic acid-water system. The generation of low molecular weight hydrocarbons may be similarly explained since skeletal isomerization does not occur to an appreciable extent in this system (Table 79). On the other hand, considerable isomerism of this kind is observed when water is absent (Table 78) indicating that, in this situation, carbonium ions play an important part in the formation of low molecular weight species.

Similar experiments have recently been performed by Shimoyama and Johns[100] using behenic acid and calcium-montmorillonite. They found that the principal reaction product was an organic-insoluble, kerogen-like material deposited on the clay. In agreement with Eisma and Jurg's findings, the major hydrocarbon produced was $C_{21}H_{44}$ formed by decarboxylation of behenic acid. Smaller amounts of other long-chain paraffins down to C_{16} and in the C_{21}–C_{28} range were also detected in the gas-liquid chromatogram (Fig. 97). In contrast to Eisma and Jurg, however, no fatty acids were found among the products of degradation; at least, if formed, they escaped detection. Failure

284

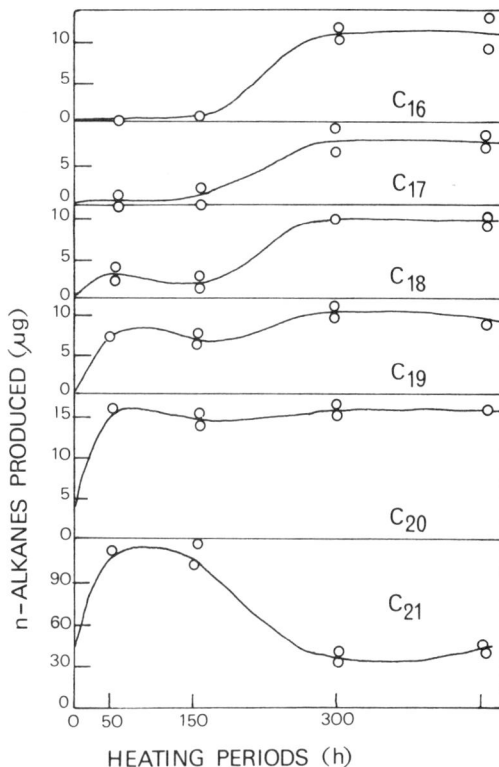

Fig. 97 Amounts of normal hydrocarbons produced from behenic acid in the presence of calcium-montmorillonite at 532°K, as a function of heating time (after Shimoyama and Johns[100])

to detect new fatty acids may, of course, be due to their adsorption by the clay (cf. Section 4.4) or by the kerogen-like substance.

The data of Fig. 97 may be explained by postulating that the generation of $C_{21}H_{44}$ is inhibited by 'poisoning' of the montmorillonite catalyst through deposition of the kerogen-like material and that the C_{16} and C_{17} n-paraffins are formed by decomposition of $n \cdot C_{21}$. In the early stages, C_{18}, C_{19}, and C_{20} paraffins are likely to arise from the fatty acid and later from $C_{21}H_{44}$ degradation, that is, these paraffins may have a dual origin.

Modern sediments usually contain less hydrocarbons and more fatty acids than ancient sediments.[101,102] The ratio of even to odd carbon number n-fatty acids is also higher in modern sediments than in ancient ones.

On the other hand, the hydrocarbons in petroleum and in ancient sediments do not show marked odd or even carbon-number preponderance.[101] The extent of diagenesis and maturation of n-paraffins in sediments can therefore be deduced from their so-called *carbon-preference index* (c.p.i.) which increases with age and maturity.

For the system under study, Shimoyama and Johns have defined the c.p.i. as

$$\text{c.p.i.} = \frac{C_{17} + C_{19} + C_{21}}{C_{16} + C_{18} + C_{20}}$$

For the C_{16} to C_{20} range the c.p.i. shows a comparable decrease with an increase in time of heating and temperature, presumably because of the increased cracking of the C_{21} component. The resultant values of 1·5–1·7 (at 523°K and >300 hours) fall within the range reported for ancient shales. This implies that C_{21} n-fatty acids can, in time, be converted to n-paraffins with c.p.i. values similar to those of paraffins in ancient sediments (since the hydrocarbons initially formed have c.p.i. values comparable with those found for material in modern sediments). The time required for the formation of petroleum-like n-paraffins from n-fatty acids is thus a function of the cracking reaction rate, this rate being much slower than that for the initial conversion of the fatty acid. This conclusion is further supported by calculations based on the data which suggest that mature n-paraffins distributions can be attained under conditions approaching the geological requirements of time and temperature.

Waples[103] has suggested a slightly different interpretation of Shimoyama and Johns' data in that the early reactions might proceed by a step-wise degradation mechanism since the suite of hydrocarbons obtained in the early stages (50 hours) shows an uneven distribution, not expected if random cracking is the most important paraffin-producing reaction. Waples also suggested that since cracking did not become significant until 150 hours, the catalyst itself might require activation.

In their reply, Shimoyama and Johns[104] stated that the decarboxylation reaction is catalysed by Lewis acids at the edges of the montmorillonite crystallites (discussed in the preceding sections) but that cracking was activated by Brønsted acids arising from dissociation of residual water molecules due to the polarization by the exchangeable cations (Eq. (1); Section 1).

This type of Brønsted acidity has, indeed, been held responsible for many decomposition reactions catalysed by clays. We have already referred to the work of Chaussidon and Calvet[105] (Section 3.6) who showed that alkylammonium ions, such as methyl-, ethyl-, n-propyl-, n-butyl-, diethyl-, and triethyl-ammonium, intercalated in montmorillonite, decomposed into a mixture of alkanes and alkenes on heating the clay complexes at temperatures up to 523°K. This conclusion was based on the GLC analyses of the desorption products and on the changes observed in the infra-red spectra of the respective alkylammonium-montmorillonite complexes. The spectra of the complex with ethylammonium ions, for example, showed a progressive weakening of the bands due to the organic cation, notably the NH_3^+ deformation vibration at about 1,515 cm^{-1}, with an increase in heating temperature. Concomitantly, the bands at ~1,430 and ~3,280 cm^{-1} ascribed to NH_4^+ intensified until at 523°K the infra-red spectrum was identical with that given by the montmorillonite saturated with NH_4^+ ions. Similar measurements on the complexes with aromatic amine cations (e.g. anilinium, o-, m-, and p-toluidinium) indicated that these cations degraded to ammonium ions and phenols.

The overall reaction may thus be written as

$$ANH_3^+ - clay + H^+ + OH^- \rightarrow NH_4^+ - clay + AOH \qquad (3)$$

where A is an alkyl or a phenyl group, so that AOH represents either an alcohol or a phenol. The fact that no primary n-alcohols were detectable in the gas-liquid chromatogram or the infra-red spectra was ascribed to the inherent instability of these compounds, which would have decomposed into hydrocarbons by dehydration. The phenols so formed, however, were apparently resistant to dehydration and only the C—N bond of aromatic amines was split under these conditions.

The general conclusions of Chaussidon and Calvet have been substantiated and confirmed by recent studies of Galwey[106-108] and of Durand et al.[109] The decomposition of primary n-alcohols (such as ethanol, n-propanol, n-octanol, n-dodecanol, and n-octadecanol) and of n-heptadecene and n-octacosane activated by montmorillonite has been reported by Galwey in a series of papers.

Using mild heat treatment ($<523°K$) with continuous product condensation at $78°K$, n-hexanol yielded an equilibrium mixture of n-hexene isomers with no appreciable cracking below $473°K$. Both n-dodecanol and n-octadecanol cracked to form nC_5-nC_8 alkenes as well as a mixture of n-hexenes; no significant yield of any alkanes was found. The products of degradation for adsorbed n-heptadecene-1 and n-octacosane were very similar to those for the long-chain n-alcohols.

Following Chaussidon and Calvet[105], dehydration and double-bond migration were postulated for n-hexanol during thermal desorption. For the long-chain reactants an additional factor was operative since appreciable cracking also occurred. Galway suggested that these molecules might be bonded to the clay surface at more than a single point and that the positions of such bonds on the hydrocarbon chain are mobile if the location of the double bond in the alkene is to be determined by the hydrocarbon-clay bond in the adsorbed radical. When two such substrate-surface bonds approach each other there is considerable local strain and chain rupture occurs. Adsorption on Lewis acid sites was thought to be the mechanism involved here since, under the conditions of continuous thermal desorption, the level of available protons was kept low and alkane formation was inhibited.

If, on the other hand, the decomposition products and water vapour were allowed to be readsorbed (as would have occurred when these compounds were not rapidly removed or not continuously condensed at $78°K$), the same alcohols yielded mixtures consisting predominantly of iso-alkanes (C_4 to at least C_8 from reaction $<473°K$) while yields of alkenes were small. In addition, complex surface processes occurred, such as polymerization, cracking, skeletal isomerization, and hydrogenation. However, all the reactants examined gave rise to similar product distributions although the quantities produced increased with chain length of the alcohol up to n-dodecanol. These observations were interpreted in terms of a carbonium ion mechanism involving Brønsted acid sites.

The thermal transformation of some alkylammonium ions in the interlayer space of montmorillonite under an inert atmosphere at temperatures below $523°K$, and for varying periods of time up to 270 days, has been studied by Durand et al.[109] From infra-red measurements of the clay complexes and the

GLC analyses of the reaction products they proposed a plausible reaction scheme based principally on transalkylation catalysed at Brønsted acid sites. The proton (in reality the hydronium ion) at such sites interacts with an alkyl chain (R) to give a reactive species, *viz.* an oxonium ion. The scheme they envisaged may be represented by the following equilibria, where M stands for the montmorillonite surface:

monoalkylamines
$$M—RNH_3^+ + H_3O^+ \rightleftharpoons M—NH_4^+ + RH_2O^+ \tag{4}$$

dialkylamines
$$M—R_2NH_2^+ + H_3O^+ \rightleftharpoons M—RNH_3^+ + RH_2O^+ \tag{5}$$

trialkylamines
$$M—R_3NH^+ + H_3O^+ \rightleftharpoons M—R_2NH_2^+ + RH_2O^+ \tag{6}$$

quaternary
    ammonium ions
$$M—R_4N^+ + H_3O^+ \rightleftharpoons M—R_3NH^+ + RH_2O^+ \tag{7}$$

alcohols
$$ROH + H_3O^+ \rightleftharpoons RH_2O^+ + H_2O \tag{8}$$

ethers
$$R_2O + H_3O^+ \rightleftharpoons ROH + RH_2O^+ \tag{9}$$

unsaturated
    hydrocarbons
$$(R—H) + H_3O^+ \rightleftharpoons RH_2O^+ \tag{10}$$

Equations (4) to (10) represent transalkylation with the various alkylamines, quaternary ammoniums, alcohols, ethers, and unsaturated hydrocarbons. The production of alcohols (reaction 8) becomes important at relatively high humidity while that of unsaturated hydrocarbons would require a high concentration of RH_2O^+ (reaction (10) from right to left) and may follow on (4) to (7). With quaternary ammonium-montmorillonite a higher rate of (R—H) production was observed since no further transalkylation was possible, and hence RH_2O^+ ions became available for reaction (10) to proceed from right to left. Besides transalkylation, polymerization may occur, for example:

$$C_2H_4 + C_2H_5 \cdot H_2O^+ \rightleftharpoons C_4H_9 \cdot H_2O^+ \tag{11}$$

followed by

$$C_4H_9 \cdot H_2O^+ \rightleftharpoons C_4H_8 + H_3O^+ \tag{12}$$

Such reactions may give rise to high molecular weight species which would be strongly retained by the clay.

It is also clear that the sequence of reactions depicted by equilibria (4) to (10) will be inhibited if water is absent or if it is present in excess, since in the former situation protons are no longer generated while in the latter instance the concentration of protons will be too low.

Earlier, Chou and McAtee[110] examined the thermal decomposition of some alkylammonium ions occupying the exchange positions in montmorillonite and hectorite using differential thermal analysis and heating-oscillating X-ray diffraction methods.

The principal reaction between 453° and 623°K was thought to be one of dehydrogenation followed by hydrolysis leaving a layer of 'carbon' on the

clay surface. This 'carbon' was oxidized at about 823°K and the montmorillonite interlayers collapsed to a basal spacing of ~0·96 nm. Since no attempt
was made to characterize the products of decomposition or to examine the infra-
red spectra of the complexes at each stage of heat treatment, the extent of
dehydrogenation could not be deduced from their data. That some dehydro-
generation occurred had earlier been indicated by the infra-red studies of
Chaussidon and Calvet[105] who observed a progressive weakening of the C—H
vibrational bands as the heating temperature was raised. However, these bands
were still present in the spectra of the complexes which had been heated to
523°K. Of greater interest is the observation made by Chou and McAtee
that the activation energy of the process is correlated more with geometrical
factors (e.g. C—C bond distance) than with the chemical and electronic
properties of the organic cations.

Some idea of the variety of hydrocarbons produced on heating various
alkylammonium-montmorillonites at 473°K in air has been given by Weiss and
Roloff.[111] Using complexes with n-hexylammonium, n-octylammonium,
hexamethylenediamonium and cyclohexylammonium ions, these workers
reported observing a series of alkanes up to C_8, cyclic and aromatic hydrocarbons as well as olefins in the GLC traces. They suggested that the primary reaction in thermal decomposition was the formation of ammonium-montmorillonite and an olefin. The olefin may then undergo such reactions as isomerization, alkylation, cracking, condensation, and polymerization to yield the various
hydrocarbons.

Clearly, the degradation of even simple organics at clay surfaces frequently
gives rise to a variety of compounds which may enter into other reactions.
Only in relatively few instances has the reaction been followed and the products
of decomposition unequivocally characterized. One such reaction is the
decomposition of cobaltic hexammine and cobaltic chloropentammine cations
adsorbed in montmorillonite.[112] The infra-red spectra of these complex ions
on the clay closely resemble those of the ions in the corresponding chloride
salts. However, when the montmorillonite sample containing these cations was
(partially) dehydrated either by evacuation at room temperature or by mild
heat treatment (<373°K), profound changes occurred in the spectra of the
adsorbed species. Notably, the band at $1,432$ cm^{-1} due to ammonium ions
progressively intensified, while the intensity of the NH_3^+ symmetric deformation vibration (at $1,358$ cm^{-1} for $Co(NH_3)_6^{3+}$ and at $1,330$ cm^{-1} for $[CoCl
(NH_3)_5]^{2+}$) decreased. These observations parallel those reported by Chaussidon
and Calvet for alkylammonium-montmorillonites treated similarly and may be
ascribed to the creation of Brønsted acid sites by the dissociation of residual
water molecules at the silicate surface.

From a combination of infra-red and chemical data, Fripiat and Helsen[112]
were able to write the following equations for the decomposition process,

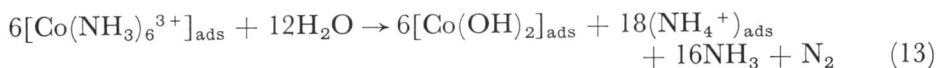

$$6[Co(NH_3)_6^{3+}]_{ads} + 12H_2O \rightarrow 6[Co(OH)_2]_{ads} + 18(NH_4^+)_{ads}$$
$$+ 16NH_3 + N_2 \qquad (13)$$

and

$$6[CoCl(NH_3)_5{}^{2+}]_{ads} + 6H_2O \rightarrow 3[Co(OH)_2]_{ads} + 12(NH_4{}^+)_{ads}$$
$$+ 3CoCl_2 + 16NH_3 + N_2 \qquad (14)$$

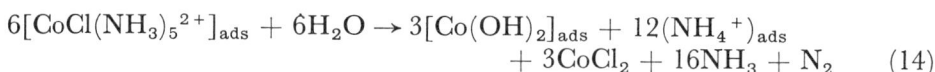

so that at the completion of the process the negative charges on the silicate surface are balanced by ammonium ions. The apparent activation energies of reactions (13) and (14) derived from kinetic measurements by application of the Arrhenius equation (Chapter 4; Eq. 26) are 180 and 85·8 kJ mol^{-1}, respectively. The latter figure may be compared with the value of 117·2 kJ mol^{-1} obtained by Wendlandt and Smith[113] for the decomposition of [CoCl-(NH$_3$)$_5$]Cl$_2$.

Brønsted acid catalysis is also responsible for the formation of the triphenyl-carbonium ion when montmorillonite containing triphenylcarbionol is evacuated at 373°K. The reaction is reversible, triphenylcarbinol being restored by rehydration of the system:[114]

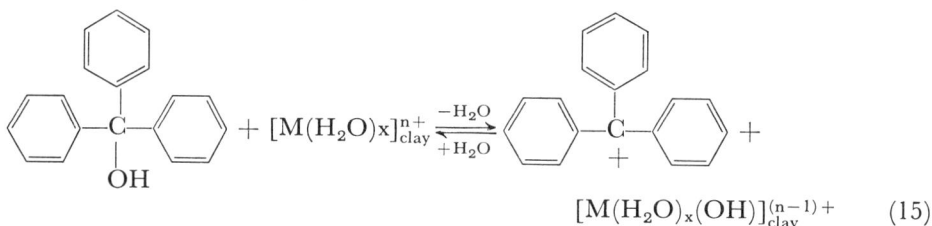

$$[M(H_2O)_x(OH)]_{clay}^{(n-1)+} \qquad (15)$$

The catalytic decomposition of glycerol adsorbed by vermiculite and montmorillonite has been reported by Walker.[115] Decomposition is visible by the dark colours produced when the mineral is immersed in boiling glycerol (vermiculite) or the complex is heated until the specimen dries out (montmorillonite). The reaction apparently occurs only when the glycerol molecule is in contact with two silicate surfaces simultaneously, since a double-layer glycerol complex fails to darken when treated similarly.

On the basis of the depth of colour produced, Walker found the following order of relative efficiency of various interlayer cations in promoting glycerol decomposition:

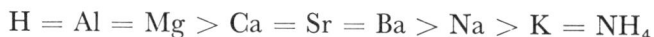

$$H = Al = Mg > Ca = Sr = Ba > Na > K = NH_4$$

Note that this is roughly the order of the polarizing power of the cations, that is, the smaller and more highly charged ions show the greater efficiency.

Related to this is the observation made by Kumada and Kato[116] that various clay minerals as well as soil clays can promote the 'browning' reaction of pyrogallol. The implication here is that clays may influence the humification process under field conditions.

Only limited information is available on the catalytic activity of clays other than montmorillonite and kaolinite. Where a large surface concentration of reactants is required, the montmorillonite group of minerals is clearly better suited than the kaolinites. More particularly, the montmorillonites with their high exchangeable cation content can act as strong Brønsted acids at low hydration status.

Palygorskite with its open structure and large surface area would be expected to show considerable catalytic activity. The available evidence indicates that this is so. Jupp and Rau,[117] for example, have reported that palygorskite (attapulgite) is highly effective in catalysing the thermal decomposition of terphenyls. Both the amount of terphenyls decomposed and the yield of products are substantially increased by adding the mineral to the system. Whereas in the absence of clay the process appears to follow a radical mechanism, the clay-catalysed reaction involves the participation of carbonium ions as intermediates.

Kaolinite, halloysite, and montmorillonite also find application in organosilicon chemistry where they act as efficient catalysts in the inter-change of hydrogen and siloxy ligands on silicon, promoting the redistribution and polymerization of organosiloxanes.[118,119]

Other examples of organic reactions activated by clays are the formation of aldehydes from alcohols[120] and of butadiene from ethanol.[121] Transesterification and esterification of waxes and carboxylic acids in the presences of methanol or ethanol[122] and alkylation of aromatic hydrocarbons[123] have also been reported to be catalysed by clays.

References

Introduction

1. GRIM, R. E., 1953. *Clay Mineralogy.* McGraw-Hill, New York, N.Y.
2. MACKENZIE, R. C., and MITCHELL, B. D., 1966. Clay mineralogy. *Earth Sci. Revs.,* **2**: 47–91.
3. GRIM, R. E., 1962. *Applied Clay Mineralogy.* McGraw-Hill, New York, N.Y.
4. SIDDIQUI, M. K. H., 1968. *Bleaching Earths.* Pergamon Press, Oxford.
5. DEMOLON, A., and BARBIER, G., 1929. Conditions de formation et constitution du complexe argilo-humique des sols. *C.R. Acad. Sci., Paris,* **188**: 654–656.
6. DEMOLON, A., and BRIGANDO, J., 1932. Sur la fixation des protéides par le sol. *C.R. Acad. Sci., Paris,* **194**: 311–313.
7. MATTSON, S., 1932. The laws of soil colloidal behaviour. VII. Proteins and proteinated complexes. *Soil Sci.,* **23**: 41–72.
8. WAKSMAN, S. A., and IYER, K. R. N., 1933. Contribution to our knowledge of the chemical nature and origin of humus. IV. Fixation of proteins by lignins and formation of complexes resistant to microbial decomposition. *Soil Sci.,* **36**: 69–82.
9. GREENLAND, D. J., 1965. Interaction between clays and organic compounds in soils. I. Mechanisms of interaction between clays and defined organic compounds. *Soils Fertilizers,* **28**: 415–425.
10. GREENLAND, D. J., 1965. Interaction between clays and organic compounds in soils. II. Adsorption of organic compounds and its effect on soil properties. *Soils Fertilizers,* **28**: 521–532.
11. JORDAN, J. W., 1963. Organophilic clay-base thickeners. *Clays Clay Minerals,* **10**: 299–308.
12. NAHIN, P. G., 1963. Perspectives in applied organo-clay chemistry. *Clays Clay Minerals,* **10**: 257–271.
13. SOLOMON, D. H., 1968. Clay minerals as electron acceptors and/or electron donors in organic reactions. *Clays Clay Minerals,* **16**: 31–39.
14. THENG, B. K. G., 1970. Interactions of clay minerals with organic polymers. Some practical applications. *Clays Clay Minerals,* **18**: 357–362.
15. THENG, B. K. G., 1971. Mechanisms of formation of colored clay-organic complexes. A review. *Clays Clay Minerals,* **19**: 383–390.
16. THENG, B. K. G., and WALKER, G. F., 1970. Interactions of clay minerals with organic monomers. *Israel J. Chem.,* **8**: 417–424.
17. WEISS, A., 1963. Mica-type layer silicates with alkylammonium ions. *Clays Clay Minerals,* **10**: 191–224.
18. BRINDLEY, G. W., 1970. Organic complexes of silicates. *Reunion Hispano-Belga de Minerales de la Arcilla, Madrid,* 55–66.
19. FARMER, V. C., 1971. The characterization of adsorption bonds in clays by infrared spectroscopy. *Soil Sci.,* **112**: 62–68.
20. MORTLAND, M. M., 1970. Clay-organic complexes and interactions. *Advan. Agron.,* **22**: 75–117.

Chapter 1

1. GRIM, R. E., 1953. *Clay Mineralogy.* McGraw-Hill, New York, N.Y.
2. BROWN, G. (Editor), 1961. *The X-ray Identification and Crystal Structures of Clay Minerals,* 2nd ed. Mineral. Soc., London.
3. HADDING, A., 1923. Eine röntgenographische Methode kristalline und kryptokristalline Substanzen zu identifizieren. *Z. Krist.,* **58**: 108–112.
4. RINNE, F., 1924. Röntgenographische Untersuchungen an einigen feinzerteilten Mineralien Kunstprodukten und dichten Gesteinen. *Z. Krist.,* **60**: 55–69.
5. ROSS, C. S., 1927. The mineralogy of clays. *Trans. 1st Int. Congr. Soil Sci., Washington,* **4**: 555–561.
6. HENDRICKS, S. B., 1929. Diffraction of X-ray radiation from crystalline aggregates. *Z. Krist.,* **71**: 269–273.
7. HENDRICKS, S. B., and FRY, W. H., 1930. The results of X-ray and microscopal examinations of soil colloids. *Soil Sci.,* **29**: 457–478.
8. PAULING, L., 1930. Structure of the micas and related minerals. *Proc. Natl. Acad. Sci. U.S.,* **16**: 123–129.
9. PAULING, L., 1930. Structure of the chlorities. *Proc. Natl. Acad. Sci. U.S.,* **16**: 578–582.
10. MARSHALL, C. E., 1949. *The Colloid Chemistry of the Silicate Minerals.* Academic Press, New York, N.Y.
11. MACKENZIE, R. C., and MITCHELL, B. D., 1966. Clay Mineralology. *Earth Sci. Rev.,* **2**: 47–91.
12. BAILEY, S. W., BRINDLEY, G. W., JOHNS, W. D., MARTIN, R. T., and ROSS, M., 1971. Summary of national and international recommendations on clay mineral nomenclature. 1969–70 CMS Nomenclature Committee. *Clays Clay Minerals,* **19**: 129–132.
13. BRINDLEY, G. W., 1961. Kaolin, serpentine and kindred minerals. In: G. Brown (Editor), *The X-ray Identification and Crystal Structures of Clay Minerals.* Mineral. Soc., London, pp. 51–131.
14. BAILEY, S. W., 1963. Polymorphism of the kaolin minerals. *Am. Mineralogist,* **48**: 1196–1209.
15. SMITH, J. V., and YODER, H. S., 1956. Experimental and theoretical studies of the mica polymorphs. *Mineral. Mag.,* **31**: 209–235.
16. MACEWAN, D. M. C., RUIZ-AMIL, A., and BROWN, G., 1961. Interstratified clay minerals. In: G. Brown (Editor), *The X-ray Identification and Crystal Structures of Clay minerals.* Mineral. Soc., London, pp. 393–445.
17. RADOSLOVICH, E. W., 1963. Cell dimension studies on layer—lattice silicates: A summary. *Clays Clay Minerals,* **11**: 225–228.
18. CAILLÈRE, S., and HÉNIN, S., 1961a. Sepiolite. In: G. Brown (Editor), *The X-ray Identification and Crystal Structures of Clay Minerals.* Mineral. Soc., London, pp. 325–342.
19. CAILLÈRE, S., and HÉNIN, S., 1961b. Palygorskite. In: G. Brown (Editor), *The X-ray Identification and Crystal Structures of Clay Minerals.* Mineral. Soc., London, pp. 343–353.
20. PEDRO, G., 1970. Report of the AIPEA Nomenclature Committee. *AIPEA Newsletter No. 4*: 3–4.
21. BIRRELL, K. S., and FIELDES, M., 1952. Allophane in volcanic ash soils. *J. Soil Sci.,* **3**: 156–166.
22. FIELDES, M., 1955. Clay mineralogy of New Zealand soils. Part II. Allophane and related mineral colloids. *N.Z. J. Sci. Tech.,* **B37**: 336–350.
23. MITCHELL, B. D., FARMER, V. C., and McHARDY, W. J., 1964. Amorphous inorganic materials in soils. *Advan. Agron.,* **16**: 327–383.
24. WADA, K., 1967. A structural scheme of soil allophane. *Am. Mineralogist,* **52**: 690–708.
25. UDAGAWA, S., NAKADA, T., and NAKAHIRA, M., 1969. Molecular structure of allophane as revealed by its thermal transformation. *Proc. Int. Clay Conf., Tokyo,* **1**: 151–159.
26. MILESTONE, N. B., 1971. Allophane, its structure and possible uses. *J. N.Z. Inst. Chem.,* **35**: 191–197.

27. AOMINE, S., and KOBAYASHI, Y., 1964. Effects of allophane on the enzymatic activity of a protease. *Soil Sci. Plant Nutr. (Tokyo)*, **10**: 28–32.

28. WADA, K., and INOUE, T., 1967. Retention of humic substances derived from rotten clover leaves in soils containing montmorillonite and allophane. *Soil Sci. Plant Nutr. (Tokyo)*, **13**: 9–16.

29. THENG, B. K. G., 1972. Adsorption of ammonium and some primary *n*-alkylammonium cations by soil allophane. *Nature*, **238**: 150–151.

30. SCHOFIELD, R. K., and SAMSON, H. R., 1953. The deflocculation of kaolinite suspensions and the accompanying changeover from positive to negative chloride adsorption. *Clay Minerals Bull.*, **2**: 45–51.

31. ROBERTSON, R. H. S., BRINDLEY, G. W., and MACKENZIE, R. C., 1954. Kaolin clays from Pugu, Tanganyika. *Am. Mineralogist*, **39**: 118–139.

32. QUIRK, J. P., 1960. Negative and positive adsorption of chloride by kaolinite. *Nature*, **188**: 253–254.

33. SCHOFIELD, R. K., 1949. Effect of pH on electric charges carried by clay particles. *J. Soil Sci.*, **1**: 1–8.

34. HOFMANN, U., BOEHM, H.-P., and GROMES, W., 1961. Die Abmessungen der Kristalle der Tomminerale. *Z. anorg. allgem. Chem.*, **308**: 143–154.

35. WEISS, A., and RUSSOW, J., 1963. Über die Lage der austauschbaren Kationen bei Kaolinit. *Proc. Int. Clay Conf., Stockholm*, **1**: 203–213.

36. FOLLETT, E. A. C., 1965. Retention of amorphous colloidal "ferric hydroxide" by kaolinites. *J. Soil Sci.*, **16**: 542–548.

37. CASHEN, G. H., 1959. Electric charges of kaolin. *Trans. Faraday Soc.*, **55**: 477–486.

38. GREENE-KELLY, R., 1964. The specific surface areas of montmorillonites. *Clay Minerals Bull.*, **5**: 392–400.

39. EDWARDS, D. G., and QUIRK, J. P., 1962. Repulsion of chloride by montmorillonite. *J. Colloid Sci.*, **17**: 872–882.

40. SOLOMON, D. H., 1968. Clay minerals as electron acceptors and/or electron donors in organic reactions. *Clays Clay Minerals*, **16**: 31–39.

41. THENG, B. K. G., 1971. Mechanisms of formation of colored clay—organic complexes. A review. *Clays Clay Minerals*, **19**: 383–390.

42. SOLOMON, D. H., and ROSSER, M. J., 1965. Reactions catalyzed by minerals. I. Polymerization of styrene. *J. Appl. Polymer Sci.*, **9**: 1261–1271.

43. THENG, B. K. G., and WALKER, G. F., 1970. Interactions of clay minerals with organic monomers. *Israel J. Chem.*, **8**: 417–424.

44. WADA, K., 1964. Ammonium chloride—kaolin complexes. *Clay Sci.*, **2**: 42–56.

45. WEISS, A., THIELEPAPE, W., GÖRING, G., RITTER, W., and SCHÄFER, H., 1963. Kaolinit-Einlagerungs-Verbindungen. *Proc. Int. Clay Conf., Stockholm*, **1**: 287–305.

46. OLEJNIK, S., POSNER, A. M., and QUIRK, J. P., 1970. The intercalation of polar organic compounds into kaolinite. *Clay Minerals*, **8**: 421–434.

47. HENDRICKS, S. B., and JEFFERSON, M. E., 1938. Structures of kaolin and talc—pyrophyllite hydrates and their bearing on water sorption of the clays. *Am. Mineralogist*, **24**: 729–771.

48. BATES, T. F., HILDEBRAND, F. A., and SWINEFORD, A., 1950. Morphology and structure of endellite and halloysite. *Am. Mineralogist*, **35**: 463–484.

49. RADOSLOVICH, E. W., 1963. The cell dimensions and symmetry of layer—lattice silicates. VI. Serpentine and kaolin morphology. *Am. Mineralogist*, **48**: 368–378.

50. MACEWAN, D. M. C., 1948. Complexes of clays with organic compounds. I. Complex formation between montmorillonite and halloysite and certain organic liquids. *Trans. Faraday Soc.*, **44**: 349–367.

51. MACEWAN, D. M. C., 1961. Montmorillonite minerals. In: G. Brown (Editor), *The X-ray Identification and Crystal Structures of Clay Minerals*. Mineral. Soc., London, pp. 143–207.

52. WEISS, A., 1963. Mica-type layer silicates with alkylammonium ions. *Clays Clay Minerals*, **10**: 191–224.

53. GREENLAND, D. J., 1965. Interaction between clays and organic compounds in soils. I. Mechanism of interaction between clays and defined organic compounds. *Soils Fertilizers*, **28**: 415–425.

54. WALKER, G. F., 1961. Vermiculite minerals. In: G. Brown (Editor), *The X-ray Identification and Crystal Structures of Clay Minerals*. Mineral. Soc., London, pp. 297–324.

55. HAASE, D. J., WEISS, E. J., and STEINFINK, H., 1963. The crystal structure of a hexamethylene—diamine—vermiculite complex. *Am. Mineralogist*, **48**: 261–270.

56. JOHNS, W. D., and SEN GUPTA, P. K., 1967. Vermiculite—alkylammonium complexes. *Am. Mineralogist*, **52**: 1706–1724.

57. WEISS, A., 1958. Der Kationenaustausch bei den Mineralen der Glimmer-, Vermikulit- und Montmorillonitgruppe. *Z. anorg. allgem. Chem.*, **297**: 257–286.

58. SCOTT, A. D., 1968. Effect of particle size on interlayer potassium exchange in micas. *Trans. 9th Int. Congr. Soil Sci., Adelaide*, **2**: 649–660.

59. WELLS, C. B., and NORRISH, K. 1968. Accelerated rates of release of interlayer potassium from micas. *Trans. 9th Int. Congr. Soil Sci., Adelaide*, **2**: 683–694.

60. MACINTOSH, E. E., LEWIS, D. G., and GREENLAND, D. J., 1971. Dodecylammonium—mica complexes—I. Factors affecting the exchange reaction. *Clays Clay Minerals*, **19**: 209–218.

61. WEISS, A., MEHLER, A., and HOFMANN, U., 1956. Kationenaustausch und innerkristallines Quellungsvermögen bei den Mineralen der Glimmergruppe. *Z. Naturforschung*, **116**: 435–438.

62. HOFMANN, U., ENDELL, K., and WILM, D., 1933. Kristalstruktur und Quellung von Montmorillonit. *Z. Krist.*, **86**: 340–348.

63. MARSHALL, C. E., 1935. Layer lattices and base—exchange clays. *Z. Krist.*, **91**: 433–449.

64. MAEGDEFRAU, E., and HOFMANN, U., 1937. Die Kristalstruktur des Montmorillonits. *Z. Krist*, **98**: 299–323.

65. HENDRICKS, S. B., 1942. Lattice structure of clay minerals and some properties of clays. *J. Geol.* **50**: 276–290.

66. MOONEY, R. W., KEENAN, A. G., and WOOD, L. A., 1952. Adsorption of water vapor by montmorillonite. *J. Am. Chem. Soc.*, **74**: 1367–1374.

67. NORRISH, K., 1954. Swelling of montmorillonite. *Disc. Faraday Soc.*, **18**: 120–134.

68. MATHIESON, A. McL., and WALKER, G. F., 1954. Structure of magnesium—vermiculite. *Am. Mineralogist*, **39**: 231–255.

69. GRAHAM, J., 1964. Adsorbed water on clays. *Rev. Pure Appl. Chem.*, **14**: 81–90.

70. MORTLAND, M. M., 1970. Clay—organic complexes and interactions. *Advan. Agron.*, **22**: 75–117.

71. EDELMAN, C. H., and FAVEJEE, J. C. L., 1940. On the crystal structure of montmorillonite and halloysite. *Z. Krist.*, **A102**: 417–431.

72. McCONNELL, D., 1950. The crystal chemistry of montmorillonite. *Am. Mineralogist*, **35**: 166–172.

73. GRIM, R. E., and KULBICKI, G., 1961. Montmorillonite: high temperature reaction and classification. *Am. Mineralogist*, **46**: 1329–1369.

74. SCHULTZ, L. G., 1969. Lithium and potassium absorption, dehydroxylation temperature, and structural water content of aluminous smectites. *Clays Clay Minerals*, **17**: 115–149.

75. BERGER, G., 1941. Struktuur van montmorilloniet. *Chem. Weekbl.*, **38**: 42–43.

76. DEUEL, H., 1952. Organic derivatives of clay minerals. *Clay Minerals Bull.*, **1**: 205–212.

77. DEUEL, H., 1959. Reactions of silicates with organic compounds. *Makromol. Chem.*, **34**: 206–215.

78. BROWN, G., GREENE-KELLY, R., and NORRISH, K., 1952. Organic derivatives of montmorillonite. *Clay Minerals Bull.*, **1**: 214–220.

79. GREENLAND, D. J., and RUSSELL, E. W., 1955. Organo-clay derivatives. *Trans. Faraday Soc.*, **51**: 1300–1307.

80. UYTTERHOEVEN, J., 1960. Organic derivatives of silicates and aluminosilicates. *Silicates inds.*, **25**: 403–409.

296

81. KUKHARSKAYA, E. V., and FEDOSEEV, A. D., 1963. Organic derivatives of laminar silicates. *Russian Chem. Revs.*, **32**: 490–495.
82. UYTTERHOEVEN, J., 1962. Determination of the surface hydroxyl groups of kaolinite by organometallic compounds (CH_3MgI and CH_3Li). *Bull. Groupe Fr. Argiles*, **13**: 69–76.
83. WALKER, G. F., and COLE, W. F., 1957. The vermiculite minerals. In: R. C. Mackenzie (Editor), *The Differential Thermal Investigation of Clays*. Mineral. Soc., London, pp. 191–206.
84. BARSHAD, I., 1952. Factors affecting the interlayer expansion of vermiculite and mont-morillonite with organic substances. *Proc. Soil Sci. Soc. Am.*, **16**: 176–182.
85. THENG, B. K. G., GREENLAND, D. J., and QUIRK, J. P., 1967. Adsorption of alkylammonium cations by montmorillonite. *Clay Minerals*, **7**: 1–17.
86. WALKER, G. F., 1967. Interactions of n-alkylammonium ions with mica-type layer lattices. *Clay Minerals*, **7**: 129–143.
87. BRADLEY, W. F., and GRIM, R. E., 1961. Mica clay minerals. In: G. Brown (Editor), *The X-ray Identification and Crystal Structures of Clay Minerals*. Mineral. Soc., London, pp. 208–241.
88. GAINES, G. L., Jr., and TABOR, D., (1956). Surface adhesion and elastic properties of mica. *Nature*, **178**: 1304–1305.

Chapter 2

1. MacEWAN, D. M. C., 1961. Montmorillonite minerals. In: G. Brown (Editor), *The X-ray Identification and Crystal Structures of Clay Minerals*. Mineral. Soc., London, pp. 143–207.
2. WALKER, G. F., 1961. Vermiculite minerals. In: G. Brown (Editor), *The X-ray Identification and Crystal Structures of Clay Minerals*. Mineral. Soc., London, pp. 297–324.
3. GREENLAND, D. J., 1965. Interaction between clays and organic compounds in soils. I. Mechanisms of interaction between clays and defined organic compounds. *Soils Fertilizers*, **28**: 415–425.
4. LITTLE, L. H., 1966. *Infrared Spectra of Adsorbed Species*. Academic Press, London.
5. MORTLAND, M. M., 1970. Clay-organic complexes and interactions. *Advan. Agron.*, **22**: 75–117.
6. FARMER, V. C., 1971. The characterization of adsorption bonds in clays by infrared spectroscopy. *Soil Sci.*, **112**: 62–68.
7. FRIPIAT, J. J., 1968. Surface fields, and transformation of adsorbed molecules in soil colloids. *Trans. 9th Int. Congr. Soil Sci., Adelaide*, **1**: 679–689.
8. MORTLAND, M. M., 1968. Protonation of compounds at clay mineral surfaces. *Trans. 9th Int. Congr. Soil Sci., Adelaide*, **1**: 691–699.
9. RUSSELL, J. D., and FARMER, V. C., 1964. Infra-red spectroscopic studies of the dehydration of montmorillonite and saponite. *Clay Minerals Bull.*, **5**: 443–464.
10. CHAUSSIDON, J., and PROST, R., 1967. Infrared valence vibration spectrum of water adsorbed on montmorillonite. *Bull. Groupe Fr. Argiles*, **19**: 25–38.
11. FARMER, V. C., and RUSSELL, J. D., 1967. Infrared adsorption spectrometry in clay studies. *Clays Clay Minerals*, **15**: 121–142.
12. PROST, R., and CHAUSSIDON, J., 1969. The infrared spectrum of water adsorbed on hectorite. *Clay Minerals*, **8**: 143–149.
13. LEONARD, R. A., 1970. Infrared analysis of partially deuterated water adsorbed on clay. *Soil Sci. Soc. Am. Proc.*, **34**: 339–343.
14. FARMER, V. C., and RUSSELL, J. D., 1971. Interlayer complexes in layer silicates. *Trans. Faraday Soc.*, **67**: 2737–2749.
15. DRYANSKI, P., and KECKI, Z., 1969. Effect of perchlorates on the infrared bands of water. *Rocz. Chem.*, **43**: 1053–1061.
16. BRINK, G., and FALK, M., 1970. Infrared studies of water in crystalline hydrates: $NaClO_4$.-H_2O, $LiClO_4.3H_2O$, and $Ba(ClO_4)_2.3H_2O$. *Can. J. Chem.*, **48**: 2096–2103.

17. TARASEVICH, YU. I., and OVCHARENKO, F. D., 1969. Infrared spectroscopic study of the thermal dehydration of cation-substituted vermiculite. *Kolloidn. Zh.*, **31**: 451–458.

18. TARASEVICH, YU. I., and OVCHARENKO, F. D., 1969. Infrared spectral study of the adsorption of heavy water by cation—substituted vermiculite. *Dokl. Akad. Nauk.* S.S.S.R., **184**: 142–143.

19. MORTLAND, M. M., FRIPIAT, J. J., CHAUSSIDON, J., and UYTTERHOEVEN, J. B., 1963. Interaction between ammonia and the expanding lattices of montmorillonite and vermiculite. *J. Phys. Chem.*, **67**: 248–258.

20. DONER, H. E., and MORTLAND, M. M., 1969. Intermolecular interaction in montmorillonites: NH-CO systems. *Clays Clay Minerals*, **17**: 265–270.

21. FARMER, V. C., and MORTLAND, M. M., 1965. Infrared study of complexes of ethylamine with ethylammonium and copper ions in montmorillonite. *J. Phys. Chem.*, **69**: 683–686.

22. LABY, R. H., and WALKER, G. F., 1970. Hydrogen bonding in primary alkylammonium-vermiculite complexes. *J. Phys. Chem.*, **74**: 2369–2373.

23. WADDINGTON, T. C., 1958. Infrared spectra, structure, and hydrogen bonding in ammonium salts. *J. Chem. Soc.*, 4340–4344.

24. CABANA, A., and SANDORFY, C., 1962. The infrared spectra of solid methylammonium halides. *Spectrochim. Acta.*, **18**: 843–861.

25. LABY, R. H., and THENG, B. K. G., unpublished results.

26. FARMER, V. C., and MORTLAND, M. M., 1966. An infrared study of the coordination of pyridine and water to exchangeable cations in montmorillonite and saponite. *J. Chem., Soc.*, 344–351.

27. YARIV, S., RUSSELL, J. D., and FARMER, V. C., 1966. Infrared study of the adsorption of benzoic acid and nitrobenzene in montmorillonite. *Israel J. Chem.*, **4**: 201–213.

28. PARFITT, R. L., and MORTLAND, M. M., 1968. Ketone adsorption on montmorillonite. *Soil Sci. Soc. Am. Proc.*, **32**: 355–363.

29. THENG, B. K. G., 1971. Mechanisms of formation of colored clay-organic complexes. A review. *Clays Clay Minerals*, **19**: 383–390.

30. THENG, B. K. G., and WALKER, G. F., 1970. Interactions of clay minerals with organic monomers. *Israel J. Chem.*, **8**: 417–424.

31. DURAND, B., PELET, R., and FRIPIAT, J. J., 1972. Alkylammonium decomposition on montmorillonite surfaces in an inert atmosphere. *Clays Clay Minerals*, **20**: 21–35.

32. HOFFMANN, R. W., and BRINDLEY, G. W., 1960. Adsorption of non-ionic aliphatic molecules from aqueous solutions on montmorillonite. Clay-organic studies II. *Geochim. Cosmochim. Acta*, **20**: 15–29.

33. GERMAN, W. L., and HARDING, D. A., 1969. The adsorption of aliphatic alcohols by montmorillonite and kaolinite. *Clay Minerals*, **8**: 213–227.

34. BRADLEY, W. F., 1945. Molecular associations between montmorillonite and some polyfunctional organic liquids. *J. Am. Chem. Soc.*, **67**: 975–981.

35. MACEWAN, D. M. C., 1944. Complexes of clays with organic compounds. I. Complex formation between montmorillonite and halloysite and certain organic liquids. *Trans. Faraday Soc.*, **44**: 349–367.

36. HOFFMANN, R. W., and BRINDLEY, G. W., 1961. Adsorption of ethylene glycol and glycerol by montmorillonite. *Am. Mineralogist*, **46**: 450–452.

37. BRINDLEY, G. W., and RUSTOM, M., 1958. Adsorption and retention of an organic material by montmorillonite in the presence of water. *Am. Mineralogist.* **43**: 627–640.

38. DOEHLER, R. W., and YOUNG, W. A., 1962. Some conditions affecting the adsorption of quinoline by clay minerals in aqueous suspensions. *Clays Clay Minerals*, **9**: 468–483.

39. BOWER, C. A., 1963. Adsorption of *o*-phenanthroline by clay minerals and soils. *Soil Sci.*, **95**: 192–195.

40. DE BOER, J. H., and HELLER, G., 1937. The anisotropy of van der Waals forces. *Physica*, **4**: 1045–1057.

41. HAMAKER, H. C., 1937. London-van der Waals attraction between spherical particles. *Physica*, **4**: 1058–1072.

42. GREENLAND, D. J., 1963. The adsorption of polyvinyl alcohols by montmorillonite. *J. Colloid Sci.*, **18**: 647–664.

43. PARFITT, R. L., and GREENLAND, D. J., 1970. The adsorption of poly(ethylene glycols) on clay minerals. *Clay Minerals*, **8**: 305–315.

44. GREENE-KELLY, R., 1955. Sorption of aromatic organic compounds by montmorillonite. Part 1. Orientation studies *Trans. Faraday Soc.*, **51**: 412–424.

45. BRINDLEY, G. W., and HOFFMANN, R. W., 1962. Orientation and packing of aliphatic chain molecules on montmorillonite. Clay-organic studies VI. *Clays Clay Minerals*, **9**: 546–556.

46. WALKER, G. F., 1958. Reactions of expanding-lattice minerals with glycerol and ethylene glycol. *Clay Minerals Bull.*, **3**: 302–313.

47. BRADLEY, W. F., ROWLAND, R. A., WEISS, E. J., and WEAVER, C. E., 1958. Temperature stabilities of montmorillonite—and vermiculite—glycol complexes. *Clays Clay Minerals*, **5**: 348–355.

48. BRADLEY, W. F., WEISS, E. J., and ROWLAND, R. A., 1963. A glycol-sodium vermiculite complex. *Clays Clay Minerals*, **10**: 117–122.

49. REYNOLDS, R. C., Jr., 1965. An X-ray study of an ethylene glycol-montmorillonite complex. *Am. Mineralogist*, **50**: 990–1001.

50. HAASE, D. J., WEISS, E. J., and STEINFINK, H., 1963. The crystal structure of a hexamethylenediamine-vermiculite complex. *Am. Mineralogist*, **48**: 261–270.

51. STEINFINK, H., WEISS, E. J., HAASE, D. J., and ROWLAND, R. A., 1963. An X-ray diffraction study of a hexamethylenediamine-vermiculite complex. *Proc. Int. Clay Conf., Stockholm*, **1**: 343–348.

52. SUSA, K., STEINFINK, H., and BRADLEY, W. F., 1967. The crystal structure of a pyridine-vermiculite complex. *Clay Minerals*, **7**: 145–153.

53. SERRATOSA, J. M., 1966. Infrared analysis of the orientation of pyridine molecules in clay complexes. *Clays Clay Minerals*, **14**: 385–391.

54. SERRATOSA, J. M., 1968. Infrared study of benzonitrile (C_6H_5CN)-montmorillonite complexes. *Am. Mineralogist*, **53**: 1244–1251.

55. GREENE-KELLY, R., 1956. The sorption of saturated organic compounds by montmorillonite. *Trans. Faraday Soc.*, **52**: 1281–1286.

56. HERZBERG, G., 1945. *Molecular Spectra and Molecular Structure. II. Infrared and Raman Spectra of Polyatomic Molecules.* D. Van Nostrand, New York, N.Y.

57. WILMHURST, J. K., and BERNSTEIN, H. J., 1957. The vibrational spectra of pyridine, pyridine-4-d, pyridine-2,6-d_2, and pyridine-3,5-d_2. *Can. J. Chem.*, **35**: 1183–1194.

58. GREENE-KELLY, R., 1955. Sorption of aromatic organic compounds by montmorillonite. Part 2. Packing studies with pyridine. *Trans. Faraday Soc.*, **51**: 425–430.

59. PAULING, L., 1960. *The Nature of the Chemical Bond.* 3rd ed. Cornell University Press, Ithaca, N.Y.

60. GATINEAU, L., and MÉRING, J., 1958. Prècision sur la structure de la muscovite. *Clay Minerals Bull.*, **3**: 238–243.

61. SUTOR, D. J., 1962. The C—H . . . O hydrogen bond in crystals. *Nature*, **195**: 68–69.

62. DONOHUE, J., 1968. Selected topics in hydrogen bonding. In: A. Rich and N. Davidson (Editors), *Structural Chemistry and Molecular Biology.* Freeman, San Francisco, Calif., pp. 443–465.

63. GOEL, A., and RAO, C. N. R., 1971. Hydrogen bonds formed by C—H groups. *Trans. Faraday Soc.*, **67**: 2828–2832.

64. LABY, R. H., 1962. *Adsorption of Amino Acids and Peptides by Montmorillonite.* Ph.D. Thesis, University of Adelaide, South Australia.

65. TENSMEYER, L. G., HOFFMANN, R. W., and BRINDLEY, G. W., 1960. Infrared studies of some complexes between ketones and calcium montmorillonite. Clay-organic studies. Part III. *J. Phys. Chem.*, **64**: 1655–1662.

66. MORTLAND, M. M., 1966. Urea complexes with montmorillonite: An infrared absorption study. *Clay Minerals*, **6**: 143–156.

67. FARMER, W. J., and AHLRICHS, J. L., 1969. Infrared studies of the mechanism of adsorption of urea-d$_4$, methylurea-d$_3$, and 1,1-dimethylurea-d$_2$ by montmorillonite. *Soil Sci. Soc. Am. Proc.*, **33**: 254–258.
68. RADOSLOVICH, E. W., 1963. Cell dimension studies on layer-lattice silicates: A summary. *Clays Clay Minerals*, **11**: 225–228.
69. GREENLAND, D. J., LABY, R. H., and QUIRK, J. P., 1962. Adsorption of glycine and its di-, tri-, and tetra-peptides by montmorillonite. *Trans. Faraday Soc.*, **58**: 829–841.
70. THENG, B. K. G., GREENLAND, D. J., and QUIRK, J. P., 1967. Adsorption of alkylammonium cations by montmorillonite. *Clay Minerals*, **7**: 1–17.

Chapter 3

1. GERMAN, W. L., and HARDING, D. A., 1969. The adsorption of aliphatic alcohols by montmorillonite and kaolinite. *Clay Minerals*, **8**: 213–227.
2. HOFFMANN, R. W., and BRINDLEY, G. W., 1960. Adsorption of non-ionic aliphatic molecules from aqueous solutions on montmorillonite. Clay-organic studies II. *Geochim. Cosmochim. Acta*, **20**: 15–29.
3. DOWDY, R. H., and MORTLAND, M. M., 1967. Alcohol-water interactions on montmorillonite surfaces. I. Ethanol. *Clays Clay Minerals*, **15**: 259–271.
4. RUSSELL, J. D., and FARMER, V. C., 1964. Infra-red spectroscopic studies of the dehydration of montmorillonite and saponite. *Clay Minerals Bull.*, **5**: 443–464.
5. AYLMORE, L. A. G., and QUIRK, J. P., 1962. The structural status of clay systems. *Clays Clay Minerals*, **9**: 104–130.
6. GILES, C. H., MACEWAN, T. H., NAKHWA, S. N., and SMITH, D., 1960. Studies in adsorption. XI. A system of classification of solution adsorption isotherms and its use in diagnosis of adsorption mechanisms and in measurement of specific surface areas of solids. *J. Chem. Soc.*, 3973–3993.
7. GREENLAND, D. J., LABY, R. H., and QUIRK, J. P., 1962. Adsorption of glycine and its di-, tri-, and tetra-peptides by montmorillonite. *Trans. Faraday Soc.*, **58**: 829–841.
8. THENG, B. K. G., 1972. Adsorption of ammonium and some primary *n*-alkylammonium cations by soil allophane. *Nature*, **238**: 150–151.
9. BISSADA, K. K., JOHNS, W. D., and CHENG, F. S., 1967. Cation-dipole interactions in clay organic complexes. *Clay Minerals*, **7**: 155–166.
10. MORTLAND, M. M., 1970. Clay-organic complexes and interactions. *Advan. Agron.*, **22**: 75–117.
11. FRANK, H. S., and EVANS, M. W., 1945. Free volume and entropy in condensed systems. III. Entropy in binary liquid mixtures; partial molal entropy in dilute solutions; structure and thermodynamics in aqueous electrolytes. *J. Chem. Phys.*, **13**: 507–532.
12. THENG, B. K. G., GREENLAND, D. J., and QUIRK, J. P., 1968. The effect of exchangeable alkylammonium ions on the swelling of montmorillonite in water. *Clay Minerals*, **7**: 271–293.
13. BRINDLEY, G. W., and RAY, S., 1964. Complexes of Ca-montmorillonite with primary monohydric alcohols (Clay-organic studies VIII). *Am. Mineralogist*, **49**: 106–115.
14. GERMAN, W. L., and HARDING, D. A., 1971. Primary aliphatic alcohol-homoionic montmorillonite interactions. *Clay Minerals*, **9**: 167–175.
15. MACEWAN, D. M. C., 1948. Complexes of clays with organic compounds. I. Complex formation between montmorillonite and halloysite and certain organic liquids. *Trans. Faraday Soc.*, **44**: 349–367.
16. BARSHAD, I., 1952. Factors affecting the interlayer expansion of vermiculite and montmorillonite with organic substances. *Soil Sci. Soc. Am. Proc.*, **16**: 176–182.
17. GLAESER, R., 1954. *Complexes Organo-argileux et Rôle des Cations Échangeables*. Thèse, Paris.

18. WEISS, A., 1969. Organic derivatives of clay minerals, zeolites, and related minerals. In: G. Eglinton and M. T. J. Murphy (Editors), *Organic Geochemistry*. Springer Verlag, Berlin, pp. 737–781.

19. EMERSON, W. W., 1957. Organo-clay complexes. *Nature*, 180: 48–49.

20. DONOHUE, J., 1968. Selected topics in hydrogen bonding. In: A. Rich and N. Davidson (Editors), *Structural Chemistry and Molecular Biology*. Freeman, San Francisco, Calif., pp. 443–465.

21. LABY, R. H., and WALKER, G. F., 1970. Hydrogen bonding in primary alkylammonium-vermiculite complexes. *J. Phys. Chem.*, 74: 2369–2373.

22. GREENE-KELLY, R., 1955. Sorption of aromatic organic compounds by montmorillonite. Part 1. Orientation studies. *Trans. Faraday Soc.*, 51: 412–424.

23. GREENLAND, D. J., and QUIRK, J. P., 1962. Adsorption of 1-*n*-alkylpyridinium bromides by montmorillonite. *Clays Clay Minerals*, 9: 484–499.

24. THENG, B. K. G., 1963. Unpublished results.

25. FARMER, V. C., 1971. The characterization of adsorption bonds in clays by infrared spectroscopy. *Soil Sci.*, 112: 62–68.

26. RUNDLE, R. E., NAKAMOTO, K., and RICHARDSON, J. W., 1955. Concerning hydrogen positions in aquo-complexes—$CuCl_2 \cdot 2H_2O$. *J. Chem. Phys.*, 23: 2450–2451.

27. DRUSHEL, H. V., SENN, W. L., Jr., and ELLERBE, J. S., 1963. Effect of structure on the methyl and methylene group absorptivities in aliphatic alcohols. *Spectrochim. Acta*, 19: 1915–1930.

28. GLAESER, R., and MÉRING, J., 1952. Les propriétes des associations organo-argileuses. *Comp. rend. Congr. Geol. Int., Alger*, 117–121.

29. BRINDLEY, G. W., 1971. Organic complexes of silicates. Mechanisms of formation. *Reunion Hispano-Belga de Minerales de la Arcilla, Madrid*, 1970, pp. 55–66.

30. BENSON, S. W., and KING, J. W., Jr., 1965. Electrostatic aspects of physical adsorption: implications for molecular sieves and gaseous anesthesia. *Science*, 150: 1710–1713.

31. BERNAL, J. D., and FOWLER, R. H., 1933. A theory of water and ionic solution with particular reference to hydrogen and hydroxyl ions. *J. Chem. Phys.*, 1: 515–548.

32. MACKENZIE, R. C., 1964. Hydratationseigenschaften von Montmorillonit. *Ber. Deut. Keram. Ges.*, 41: 696–708.

33. BRINDLEY, G. W., and HOFFMANN, R. W., 1962. Orientation and packing of aliphatic chain molecules on montmorillonite. *Clays Clay Minerals*, 9: 546–556.

34. RADUL, N. M., and OVCHARENKO, F. D., 1971. Method for determining the orientation of adsorbed alcohol molecules in the interlayer space of montmorillonite. *Ukr. Khim. Zh.*, 37: 775–777.

35. BRADLEY, W. F., 1945. Molecular associations between montmorillonite and some polyfunctional organic liquids. *J. Am. Chem. Soc.*, 67: 975–981.

36. BRINDLEY, G. W., 1966. Ethylene glycol and glycerol complexes of smectites and vermiculites. *Clay Minerals*, 6: 237–259.

37. MacEWAN, D. M. C., 1946. Identification of montmorillonite. *J. Soc. Chem. Ind.*, 65: 298–305.

38. BRADLEY, W. F., 1945. Diagnostic criteria. *Am. Mineralogist*, 30: 704–713.

39. CARROLL, D., 1970. *Clay Minerals: A Guide to Their X-ray Identification*. Geol. Soc. Am. Special Paper 126.

40. WALKER, G. F., 1957. On the differentiation of vermiculties and smectites in clays. *Clay Minerals Bull.*, 3: 154–163.

41. WALKER, G. F., 1958. Reactions of expanding-lattice clay minerals with glycerol and ethylene glycol. *Clay Minerals Bull.*, 3: 302–313.

42. MEHRA, O. P., and JACKSON, M. L., 1959. Constancy of the sum of mica unit cell potassium surface and interlayer sorption surface in vermiculite-illite clays. *Soil Sci. Soc. Am. Proc.*, 23: 101–105.

43. MEHRA, O. P., and JACKSON, M. L., 1959. Specific surface determination by duo-inter-layer and mono-interlayer glycerol sorption for vermiculite and montmorillonite analysis. *Soil Sci. Soc. Am. Proc.*, **23**: 351–354.

44. KUNZE, G. W., 1955. Anomalies in the ethylene glycol solvation technique used in X-ray diffraction. *Clays Clay Minerals*, **3**: 88–93.

45. WEAVER, C. E., 1958. The effects and geologic significance of potassium 'fixation' by expandable clay minerals derived from muscovite, biotite, chlorite and volcanic material. *Am. Mineralogist*, **43**: 839–861.

46. JONAS, E. C., and THOMAS, G. L., 1960. Hydration properties of potassium deficient clay micas. *Clays Clay Minerals*, **8**: 183–192.

47. GRIM, R. E., and KULBICKI, G., 1961. Montmorillonite: high temperature reactions and classification. *Am. Mineralogist*, **46**: 1329–1369.

48. GREENE-KELLY, R., 1953. Irreversible dehydration in montmorillonite. II. *Clay Minerals Bull.*, **2**: 52–56.

49. GONZALEZ GARCIA, F., and GONZALEZ GARCIA, S., 1953. Modificaciones producidas por tratamiento termico en las propiedades fisicoquimicas de los silicatos de la serie iso-morfa montmorillonita-beidellita. *An. Edafol. Fisiol. veg.*, **12**: 925–992.

50. HOFMANN, U., and KLEMEN, R., 1950. Verlust der Austauschfähigkeit von Lithiumionen an Bentonit durch Erhitzung. *Z. anorg. allgem. Chem.*, **262**: 95–99.

51. JOHNS, W. D., and TETTENHORST, R. T., 1959. Differences in the montmorillonite solvating ability of polar liquids. *Am. Mineralogist*, **44**: 894–896.

52. QUIRK, J. P., and THENG, B. K. G., 1960. Effect of surface density of charge on the physical swelling of lithium montmorillonite. *Nature*, **187**: 967–968.

53. BRINDLEY, G. W., and ERTEM, G., 1971. Preparation and solvation properties of some variable charge montmorillonites. *Clays Clay Minerals*, **19**: 399–404.

54. BRINDLEY, G. W., WIEWIORA, K., and WIEWIORA, A., 1969. Intracrystalline swelling of montmorillonite in some water-organic mixtures. (Clay-organic studies XVII). *Am. Mineralogist*, **54**: 1635—1644.

55. RUIZ-AMIL, A., and MacEWAN, D. M. C., 1957. Interlamellar sorption of mixed liquids by montmorillonite: The system montmorillonite-water-acetone-NaCl. *Kolloid Z.*, **155**: 134–135.

56. HARWARD, M. E., and BRINDLEY, G. W., 1965. Swelling properties of synthetic smectites in relation to lattice substitutions. *Clays Clay Minerals*, **13**: 209–222.

57. HARWARD, M. E., CARSTEA, D. D., and SAYEGH, A. H., 1969. Properties of vermiculites and smectites: expansion and collapse. *Clays Clay Minerals*, **16**: 437–447.

58. TETTENHORST, R. T., BECK, C. W., and BRUNTON, G., 1962. Montmorillonite-polyalcohol complexes. *Clays Clay Minerals*, **9**: 500–519.

59. WEAR, J. L., and WHITE, J. L., 1951. Potassium fixation in clay minerals as related to crystal structure. *Soil Sci.*, **71**: 1–14.

60. BARSHAD, I., 1950. The effect of interlayer cations on the expansion of the mica type of crystal lattice. *Am. Mineralogist*, **35**: 225–238.

61. SAYEGH, A. H., 1964. *Changes in Lattice Spacing of Expanding Clay Minerals*. Ph.D. thesis, Oregon State University, Corvallis, Oregon.

62. DYAL, R. S., and HENDRICKS, S. B., 1950. Total surface of clays in polar liquids as a characteristic index. *Soil Sci.*, **69**: 421–432.

63. DYAL, R. S., and HENDRICKS, S. B., 1952. Formation of mixed-layer minerals by potassium fixation in montmorillonite. *Soil Sci. Soc. Am. Proc.*, **16**: 45–51.

64. BOWER, C. A., and GSCHWEND, F. B., 1952. Ethylene glycol retention by soils as a measure of surface area and interlayer swelling. *Soil. Sci. Soc. Am. Proc.*, **16**: 342–345.

65. MARTIN, R. T., 1955. Ethylene glycol retention by clays. *Soil Sci. Soc. Am. Proc.*, **19**: 160–164.

66. MORTLAND, M. M., and ERICKSON, A. E., 1956. Surface reactions of clay minerals. *Soil Sci. Soc. Am. Proc.*, **20**: 476–479.

302

67. BOWER, C. A., and GOERTZEN, J. O., 1959. Surface area of soils and clays by an equilibrium ethylene glycol method. *Soil Sci.*, **87**: 289–292.

68. McNEAL, B. L., 1964. Effect of exchangeable cations on glycol retention by clay minerals. *Soil Sci.*, **97**: 96–102.

69. KINTER, E. B., and DIAMOND, S., 1958. Gravimetric determination of monolayer glycerol complexes by clay minerals. *Clays Clay Minerals*, **5**: 318–333.

70. DIAMOND, S., and KINTER, E. B., 1958. Surface areas of clay minerals as derived from measurements of glycerol retention. *Clays Clay Minerals*, **5**: 334–347.

71. HACH-ALI, P. F., and MARTIN VIVALDI, J. L., 1968. Estudio de complejos organicos de silicatos mediante la technica del analysis termico differencial. II. Vermiculata-H₂O y vermiculata-etilenglicol. *Rend. Soc. Ital. Mineral. Petrologia*, **25**: 35–44.

72. BRUNTON, G., TETTENHORST, R. T., and BECK, C. W., 1964. Montmorillonite-polyalcohol complexes. II. *Clays Clay Minerals*, **11**: 105–116.

73. BODENHEIMER, W., HELLER, L., and YARIV, S., 1966. Organo-metallic clay complexes. VII. Thermal analysis of montmorillonite-diamine and glycol complexes. *Clay Minerals*, **6**: 167–177.

74. DOWDY, R. H., and MORTLAND, M. M., 1968. Alcohol-water interactions on montmorillonite surfaces: II. Ethylene glycol. *Soil Sci.*, **105**: 36–43.

75. DAVIDSON, W. H. T., 1955. Infrared spectra and crystallinity. III. *J. Chem. Soc.*, 3270–3274.

76. KUHN, M., LUTTKE, W., and MECKE, R., 1959. Infrarotspektroskopische Untersuchungen über die Rotationsisomerie bei 2-substitutierten Äthanolen. *Z. anal. Chem.*, **170**: 106–114.

77. MIYAZAWA, T., FUKUSHIMA, K., and IDEGUCHI, Y., 1962. Molecular vibrations and structure of high polymers. III. *J. Chem. Phys.*, **37**: 2764–2776.

78. KRIMM, S., LIANG, C. Y., and SUTHERLAND, G. B. B. M., 1956. Infrared spectra of high polymers. V. *J. Polymer Sci.*, **22**: 227–247.

79. KUHN, L. P., 1952. The hydrogen bond. I. Intra- and intermolecular hydrogen bonds in alcohols. *J. Am. Chem. Soc.*, **74**: 2492–2499.

80. BELLAMY, L. J., 1958. *The Infra-red Spectra of Complex Molecules*. 2nd ed. Methuen, London.

81. RAO, C. N. R., 1963. *Chemical Applications of Infra-red Spectroscopy*. Academic Press, New York and London.

82. MACKENZIE, R. C., 1948. Complexes of clays with organic compounds. II. Investigation of the ethylene glycol-water-montmorillonite system using the Karl Fischer reagent. *Trans. Faraday Soc.*, **44**: 368–375.

83. HOFFMANN, R. W., and BRINDLEY, G. W., 1961. Adsorption of ethylene glycol and glycerol by montmorillonite. *Am. Mineralogist*, **46**: 450–452.

84. MORIN, R. E., and JACOBS, H. S., 1964. Surface area determination of soils by adsorption of ethylene glycol vapor. *Soil Sci. Soc. Am. Proc.*, **28**: 190–194.

85. HAJEK, B. F., and DIXON, J. B., 1966. Desorption of glycerol from clays as a function of glycerol vapor pressure. *Soil Sci. Soc. Am. Proc.*, **29**: 30–34.

86. MOORE, D. E., and DIXON, J. B., 1970. Glycerol vapor adsorption on clay minerals and montmorillonite soil clays. *Soil Sci. Soc. Am. Proc.*, **34**: 816–822.

87. BRUNAUER, S., EMMETT, P. H., and TELLER, E., 1938. Adsorption of gases in multimolecular layers. *J. Am. Chem. Soc.*, **60**: 309–319.

88. SOR, K., and KEMPER, W. D., 1959. Estimation of hydrateable surface area of soils and clays from the amount of absorption and retention of ethylene glycol. *Soil Sci. Soc. Am. Proc.*, **23**: 105–110.

89. MILFORD, M. H., and JACKSON, M. L., 1962. Specific surface determination of expansible layer silicates. *Science*, **135**: 929–930.

90. VAN DEN HEUVEL, R. C., and JACKSON, M. L., 1953. Surface determination of mineral colloids by glycerol sorption and its application to interstratified layer silicates. *Agron. Abst. Annual Mtg. Am. Soc. Agron.*, Texas.

91. MORTLAND, M. M., and KEMPER, W. D., 1965. Specific surface. In: *Methods of Soil Analysis*, Part 1. Am. Soc. Agron., Madison, Wisconsin, pp. 532–544.

92. GUYOT, J., 1969. Mésure des surfaces spécifiques des argiles par adsorption. *Ann. agron.*, **20**: 333–359.

93. VAN OLPHEN, H., 1969. Determination of surface areas of clays—evaluation of methods. In: D. H. Everett (Editor) *Surface Area Determination, Proc. Int. Symp.*, Butterworth, London, pp. 255–271.

94. QUIRK, J. P., 1955. Significance of surface areas calculated from water vapor sorption isotherms by use of the B.E.T. equation. *Soil Sci.*, **80**: 423–430.

95. CARTER, D. L., HEILMAN, M. D., and GONZALEZ, C. L., 1965. Ethylene glycol mono-ethyl ether for determining surface area of silicate minerals. *Soil Sci.*, **100**: 356–360.

96. LAWRIE, D. C., 1961. A rapid method for the determination of approximate surface areas of clays. *Soil Sci.*, **92**: 188–191.

97. BOWER, C. A., 1963. Adsorption of *o*-phenanthroline by clay minerals and soils. *Soil Sci.*, **95**: 192–195.

98. THENG, B. K. G., 1961. *Adsorption of Organic Cations and the Physical Structure of Clay Systems.* Honours thesis, University of Adelaide, South Australia.

99. GREENLAND, D. J., and QUIRK, J. P., 1962. Surface areas of soil colloids. *Proc. Int. Soc. Soil Sci.* Comm. 4–5, New Zealand, 79–87.

100. GREENLAND, D. J., and QUIRK, J. P., 1964. Determination of the total specific surface areas of soils by adsorption of cetylpyridinium bromide. *J. Soil Sci.*, **15**: 178–191.

101. PHAM, T. H., and BRINDLEY, G. W., 1970. Methylene blue absorption by clay minerals. Determination of surface areas and cation exchange capacities. (Clay-organic studies XVIII). *Clays Clay Minerals*, **18**: 203–212.

102. BROWN, G., 1950. A Fourier investigation of montmorillonite. *Clay Minerals Bull.*, **4**: 109–111.

103. BRADLEY, W. F., ROWLAND, R. A., WEISS, E. J., and WEAVER, C. E., 1958. Temperature stabilities of montmorillonite- and vermiculite-glycol complexes. *Clays Clay Minerals*, **5**: 348–355.

104. BRADLEY, W. F., WEISS, E. J., and ROWLAND, R. A., 1963. A glycol-sodium vermiculite complex. *Clays Clay Minerals*, **10**: 117–122.

105. BRINDLEY, G. W., 1956. Allevardite, a swelling double-layer mica mineral. *Am. Mineralogist*, **41**: 91–103.

106. REYNOLDS, R. C., 1965. An X-ray study of an ethylene glycol-montmorillonite complex. *Am. Mineralogist*, **50**: 990–1001.

107. REYNOLDS, R. C., 1969. Orientation of ethylene glycol monoethyl ether molecules on montmorillonite. *Am. Mineralogist*, **54**: 562–567.

108. GLAESER, R., 1948. On the mechanism of formation of montmorillonite-acetone complexes. *Clay Minerals Bull.*, **1**: 88–90.

109. PARFITT, R. L., and MORTLAND, M. M., 1968. Ketone adsorption on montmorillonite. *Soil Sci. Soc. Am. Proc.*, **32**: 355–363.

110. PULLIN, A. D. E., and POLLOCK, J. McC., 1958. Spectra of solutions of silver and lithium perchlorates in acetone. *Trans. Faraday Soc.*, **54**: 11–18.

111. FARMER, V. C., and MORTLAND, M. M., 1966. An infrared study of the co-ordination of pyridine and water to exchangeable cations in montmorillonite and saponite. *J. Chem. Soc. A.*, 344–351.

112. YARIV, S., RUSSELL, J. D., and FARMER, V. C., 1966. Infrared study of the adsorption of benzoic acid and nitrobenzene in montmorillonite. *Israel J. Chem.*, **4**: 201–213.

113. SERRATOSA, J. M., 1968. Infrared study of benzonitrile-montmorillonite complexes *Am. Mineralogist*, **53**: 1244–1251.

114. LARSON, G. O., and SHERMAN, L. R., 1964. Infrared spectrophotometric analysis of some carbonyl compounds adsorbed on bentonite clay. *Soil Sci.*, **98**: 328–331.

115. KOHL, R. A., and TAYLOR, S. R., 1961. Hydrogen bonding between the carbonyl group and Wyoming bentonite. *Soil Sci.*, **91**: 223–227.

116. VON MECKE, R., and FUNCK, E., 1956. Tautomerie und Infrarot-absorptionspektrum von Acetylaceton. *Z. Elektrochem.*, **60**: 1124–1130.

117. LAWSON, K. E., 1961. The infrared absorption spectra of metal acetyl acetonates. *Spectrochim. Acta*, **17**: 248–258.

118. NAKAMOTO, K., 1963. *Infrared Spectra of Inorganic and Coordination Compounds.* John Wiley & Sons, Inc., New York, N.Y.

119. TENSMEYER, L. G., HOFFMANN, R. W., and BRINDLEY, G. W., 1960. Infrared studies of some complexes between ketones and calcium montmorillonite. Clay-organic studies III. *J. Phys. Chem.*, **64**: 1655–1662.

120. THENG, B. K. G., 1971. Mechanisms of formation of colored clay-organic complexes. A review. *Clays Clay Minerals*, **19**: 383–390.

121. BRINDLEY, G. W., BENDER, R., and RAY, S., 1963. Sorption of non-ionic aliphatic molecules from aqueous solutions on clay minerals. Clay-organic studies VII. *Geochim. Cosmochim. Acta*, **27**: 1129–1137.

122. FARMER, V. C., and RUSSELL, J. D., 1967. Infrared absorption spectrometry in clay studies. *Clays Clay Minerals*, **15**: 121–142.

123. GREEN, J. H. S., 1961. Vibrational spectra of benzene derivatives. II. Assignment and calculated thermodynamic functions for benzonitrile. *Spectrochim. Acta*, **17**: 607–613.

124. TARASEVICH, YU. I., TELICHKUN, V. P., and OVCHARENKO, F. D., 1968. Infrared spectra of acetonitrile adsorbed on montmorillonite. *Dokl. Akad. Nauk., S.S.S.R.*, **182**: 141–143.

125. PINCHAS, S., SAMUEL, D., and SILVER, B. L., 1964. The infrared absorption spectrum of ^{18}O-labelled nitrobenzene. *Spectrochim. Acta*, **20**: 179–185.

126. HAXAIRE, A., and BLOCH, J. M., 1956. Sorption de molécules organiques azotées par la montmorillonite. Étude du mécanisme. *Bull. Soc. France Mineral. Crist.*, **79**: 464–475.

127. ROWLAND, R. A., and WEISS, E. J., 1963. Bentonite-methylamine complexes. *Clays Clay Minerals*, **10**: 460–468.

128. SERVAIS, A., FRIPIAT, J. J., and LÉONARD, A., 1962. Étude de l'adsorption des amines par les montmorillonites. I. Les processus chimiques. *Bull. Soc. Chim. France*, 617–625.

129. LÉONARD, A., SERVAIS, A., and FRIPIAT, J. J., 1962. Étude de l'adsorption des amines par les montmorillonites. II. La structure des complexes. *Bull. Soc. Chim. France*, 625–635.

130. FRIPIAT, J. J., SERVAIS, A., and LÉONARD, A., 1962. Étude de l'adsorption des amines par les montmorillonites. III. La nature de la liaison amine-montmorillonite. *Bull. Soc. Chim. France*, 635–644.

131. SIESKIND, O., and WEY, R., 1958. Influence du pH sur l'adsorption d'amines aliphatiques normales par la montmorillonite—H. *C.R. Acad. Sci., Paris*, **247**: 74–76.

132. SIESKIND, O., and WEY, R., 1958. Sur l'adsorption d'amines aliphatiques par la montmorillonite. *Bull. Groupe Fr. Argiles*, **10**: 9–14.

133. COLEMAN, N. T., and CRAIG, D., 1961. The spontaneous alteration of hydrogen clay. *Soil Sci.*, **91**: 14–18.

134. COWAN, C. T., and WHITE, D., 1958. The mechanism of exchange reactions occurring between sodium montmorillonite and various *n*-primary aliphatic amine salts. *Trans. Faraday Soc.*, **54**: 691–697.

135. GREENLAND, D. J., 1965. Interaction between clays and organic compounds in soils. 1. Mechanisms of interaction between clays and defined organic compounds. *Soils Fertilizers*, **28**: 415–425.

136. THENG, B. K. G., GREENLAND, D. J., and QUIRK, J. P., 1967. Adsorption of alkylammonium cations by montmorillonite. *Clay Minerals*, **7**: 1–17.

137. FRIPIAT, J. J., PENNEQUIN, M., PONCELET, G., and CLOOS, P., 1969. Influence of the van der Waals force on the infrared spectra of short aliphatic alkylammonium cations held on montmorillonite. *Clay Minerals*, **8**: 119–134.

138. FARMER, V. C., and MORTLAND, M. M., 1965. An infrared study of complexes of ethylamine with ethylammonium and copper ions in montmorillonite. *J. Phys. Chem.*, **69**: 683–686.

305

139. VANSANT, E. F., and UYTTERHOEVEN, J. B., 1972. Thermodynamics of the exchange of *n*-alkylammonium ions on Na-montmorillonite. *Clays Clay Minerals*, **20**: 47–54.

140. PALMER, J., and BAUER, N., 1961. Sorption of amines by montmorillonite. *J. Phys. Chem.*, **65**: 894–895.

141. BODENHEIMER, W., HELLER, L., and YARIV, S., 1966. Organometallic clay complexes. VI. Copper-montmorillonite-alkylamines. *Proc. Int. Clay Conf., Jerusalem*, **1**: 251–261.

142. BODENHEIMER, W., KIRSON, B., and YARIV, S., 1963. Organometallic clay complexes. I. *Israel J. Chem.*, **1**: 69–78.

143. BODENHEIMER, W., HELLER, L., KIRSON, B., and YARIV, S., 1962. Organo-metallic clay complexes. II. *Clay Minerals Bull.*, **5**: 145–154.

144. BODENHEIMER, W., HELLER, L., KIRSON, B., and YARIV, S., 1963. Organo-metallic clay complexes. III. Copper-polyamine-clay complexes. *Proc. Int. Clay Conf., Stockholm*, **1**: 351–363.

145. BODENHEIMER, W., HELLER, L., KIRSON, B., and YARIV, S., 1963. Organo-metallic clay complexes. IV. Nickel and mercury aliphatic polyamines. *Israel J. Chem.*, **1**: 391–403.

146. LAURA, R. D., and CLOOS, P., 1970. Adsorption of ethylenediamine (EDA) on montmorillonite saturated with different cations. I. Copper-montmorillonite: coordination. *Reunion Hispano-Belga de Minerals de la Arcilla, Madrid*, 76–86.

147. SANTOS, A., GONZALEZ GARMENDIA, J., and RODRIGUEZ, A., 1969. Complejos de vermiculita con aminas. *Anal. Quim.*, **65**: 433–442.

148. SANTOS, A., RODRIGUEZ, A., GONZALEZ GARMENDIA, J., and BARRIOS, J., 1970. Espectros infarrojos de muestras homoionicas de montmorillonita y vermiculita y sus complejos interlaminares con amines. *Reunion Hispano-Belga de Minerales de la Arcilla, Madrid*, 87–97.

149. FARMER, V. C., and MORTLAND, M. M., 1966. An infrared study of the co-ordination of pyridine and water to exchangeable cations in montmorillonite and saponite. *J. Chem. Soc.*, 344–351.

150. MORTLAND, M. M., 1966. Urea complexes with montmorillonite: an infrared absorption study. *Clay Minerals*, **6**: 143–156.

151. TAHOUN, S., and MORTLAND, M. M., 1966. Complexes of montmorillonite with primary, secondary, and tertiary amides: I. Protonation of amides on the surface of montmorillonite. *Soil Sci.*, **102**: 248–254.

152. RUSSELL, J. D., CRUZ, M. I., and WHITE, J. L., 1968. The adsorption of 3-aminotriazole by montmorillonite. *J. Agr. Food Chem.*, **16**: 21–24.

153. RAMAN, K. V., and MORTLAND, M. M., 1969. Proton transfer reactions at clay mineral surfaces. *Soil Sci. Soc. Am. Proc.*, **33**: 313–317.

154. CANO RUIZ, J., and MACEWAN, D. M. C., 1956. Graphitic acid. *Proc. 3rd Int. Congr. Reactions in Solid State*, **1**: 227–243.

155. ARAGON, F., CANO RUIZ, J., and MACEWAN, D. M. C., 1959. β-Type interlamellar complexes. *Nature*, **183**: 740–741.

156. SUTHERLAND, H. H., and MACEWAN, D. M. C., 1961. Organic complexes of vermiculite. *Clay Minerals Bull.*, **4**: 229–233.

157. MACEWAN, D. M. C., 1967. Complejos interlaminares de sorcion. La configuracion de las cadenas moleculares on complejos tipo β. *An. Edafol. Agrobiol.*, **26**: 1115–1127.

158. BRINDLEY, G. W., 1965. Clay-organic studies. X. Complexes of primary amines with montmorillonite and vermiculite. *Clay Minerals*, **6**: 91–96.

159. HACH-ALI, P. F., and MARTIN VIVALDI, J. L., 1969. Estructura de complejos organicos de silicatos. I. Complejos de montmorillonita con aminas alifaticas. *An. Real Soc. Esp. Fis. Quim.*, **65**: 355–362.

160. POWELL, D. B., and SHEPPARD, N., 1961. The assignment of infra-red absorption bands to fundamental vibrations in some metal-ethylenediamine complexes. *Spectrochim. Acta*, **17**: 68–76.

161. HADJILIADIS, N., DIOT, A., and THEOPHANIDES, T., 1972. A quantum chemical conformational analysis of the ethylenediamine molecule. *Can. J. Chem.*, **50**: 1005–1007.

162. YARIV, S., and HELLER, L., 1970. Sorption of cyclohexylamine by montmorillonite. *Israel J. Chem.*, **8**: 935–945.

163. SWOBODA, A. R., and KUNZE, G. W., 1968. Reactivity of montmorillonite surfaces with weak organic bases. *Soil Sci. Soc. Am. Proc.*, **32**: 806–811.

164. HENDRICKS, S. B., 1941. Base exchange of the clay mineral montmorillonite for organic cations and its dependence upon adsorption due to van der Waals forces. *J. Phys. Chem.*, **45**: 65–81.

165. BAILEY, G. W., and WHITE, J. L., 1964. Pesticide-soil colloid interactions. *Progress Report EF*-00055. *U.S. Public Health Service, Div. Env. Food Prot.* (quoted by R. D. Harter and J. L. Ahlrichs, ref. 166).

166. HARTER, R. D., and AHLRICHS, J. L., 1969. Effect of acidity on reactions of organic acids and amines with montmorillonite clay surfaces. *Soil Sci. Soc. Am. Proc.*, **33**: 859–863.

167. YARIV, S., HELLER, L., SOFER, Z., and BODENHEIMER, W., 1968. Sorption of aniline by montmorillonite. *Israel J. Chem.*, **6**: 741–756.

168. MORTLAND, M. M., FARMER, V. C., and RUSSELL, J. D., 1969. Reactivity of montmorillonite surfaces with weak organic bases. *Soil Sci. Soc. Am. Proc.*, **33**: 818.

169. HELLER, L., and YARIV, S., 1969. Sorption of some anilines by Mn-, Co-, Ni-, Cu-, Zn-, and Cd- montmorillonite. *Proc. Int. Clay Conf., Tokyo*, **1**: 741–755.

170. YARIV, S., HELLER, L., and KAUFHERR, N., 1969. Effect of acidity in montmorillonite interlayer on the sorption of aniline derivatives. *Clays Clay Minerals*, **17**: 301–308.

171. FRIPIAT, J. J., 1964. Surface properties of alumino-silicates. *Clays Clay Minerals*, **12**: 327–357.

172. MORTLAND, M. M., 1968. Protonation of compounds at clay mineral surfaces. *Trans. 9th Int. Congr. Soil Sci., Adelaide*, **1**: 691–699.

173. FARMER, V. C., 1971. The characterization of adsorption bonds in clays by infrared spectroscopy. *Soil Sci.*, **112**: 62–68.

174. MORTLAND, M. M., and RAMAN, K. V., 1968. Surface acidity of smectites in relation to hydration, exchangeable cation, and structure. *Clays Clay Minerals*, **16**: 393–398.

175. ALBERT, N. and BADGER, R. M., 1958. Infrared absorption associated with strong hydrogen bonds. *J. Chem. Phys.*, **29**: 1193–1194.

176. FRAENKEL, A., and FRANCONI, C., 1960. Protonation of amides. *J. Am. Chem. Soc.*, **82**: 4478–4483.

177. JANSSEN, M. J., 1961. The structure of protonated amides and ureas and their thio analogues. *Spectrochim. Acta*, **17**: 475–485.

178. KUTZELNIGG, W., and MECKE, R., 1962. Spektroskopische Untersuchungen an organischen Ionen. V. *Spectrochim. Acta*, **18**: 549–560.

179. COOK, D., 1964. Protonated carbonyl groups. IV. *Can. J. Chem.*, **42**: 2721–2727.

180. SMITH, C. R., and YATES, K., 1972. Kinetic evidence for predominant oxygen protonation of amides. *Can. J. Chem.*, **50**: 771–773.

181. TAHOUN, S. A., and MORTLAND, M. M., 1966. Complexes of montmorillonite with primary, secondary, and tertiary amides: II. Coordination of amides on the surface of montmorillonite. *Soil Sci.*, **102**: 314–321.

182. DONER, H. E., and MORTLAND, M. M., 1969. Intermolecular interaction in montmorillonites: NH—CO systems. *Clays Clay Minerals*, **17**: 265–270.

183. KAGARISE, R. E., 1955. Relation between the electronegativities of adjacent substituents and the stretching frequency of the carbonyl group. *J. Am. Chem. Soc.*, **77**: 1377–1379.

184. TAFT, R. W., Jr., 1956. Separation of polar, steric and resonance effects in reactivity. In: M. S. Newman (editor), *Steric Effects in Organic Chemistry*. Wiley, New York, pp. 586–675.

185. ADELMAN, R. L., 1964. Studies on the base strengths of N,N-disubstituted amides. *J. Org. Chem.*, **29**: 1837–1844.

186. MORTLAND, M. M. 1968. Pyridinium-montmorillonite complexes with ethyl N,N-di-*n*-propylthiolcarbamate (EPTC). *J. Agr. Food Chem.*, **16**: 706–707.

187. WEISMILLER, R. A., 1970. Effect of N-ethylacetamide (NEA) upon some physico-chemical properties of montmorillonite and vermiculite. *Diss. Abst.*, **30**: 4874-B.

188. GARRETT, W. G., and WALKER, G. F., 1962. Swelling of some vermiculite-organic complexes in water. *Clays Clay Minerals*, **9**: 557–567.

189. WALKER, G. F., and GARRETT, W. G., 1961. Complexes of vermiculite with amino acids. *Nature*, **191**: 1389.

190. PENLAND, R. B., MIZUSHIMA, S., CURRAN, C., and QUAGLIANO, J. V., 1957. Infrared absorption spectra of inorganic coordination complexes. X. Studies of some metal-urea complexes. *J. Am. Chem. Soc.*, **79**: 1575–1578.

191. MITSUI, S., and TAKATOH, H., 1963. An infra-red spectrophotometric analysis for a mechanism of adsorption of urea. *Soil Sci. Plant Nutr. (Tokyo)*, **9**: 103–110.

192. SPINNER, E., 1959. The vibration spectra and structure of the hydrochlorides of urea, thiourea, and acetamide. The basic properties of amides and thioamides. *Spectrochim. Acta*, **15**: 95–109.

193. FARMER, W. J., and AHLRICHS, J. L., 1969. Infra-red studies of the mechanism of adsorption of urea-d$_4$, methylurea-d$_3$, and 1,1-dimethylurea-d$_2$ by montmorillonite. *Soil Sci. Soc. Am. Proc.*, **33**: 254–258.

194. SHIGA, Y., 1961. Studies on the complexes of montmorines with urea and its derivatives. I. The influence of exchangeable cations on the interlayer adsorption of urea by montmorines. *Soil Sci. Plant Nutr. (Tokyo)*, **7**: 119–124.

195. LIBOR, O., and GRABER, L., 1969. Investigation of montmorillonites treated by urea solutions. *Ann. Univ. Sci. Budapest Rolando Eotvos Nominatae, Sect. Geol.*, **13**: 91–100.

196. LIBOR, O., GRABER, L., and DONATH, E., 1970. Investigation on montmorillonites contacting urea. II. Thermic examination of Na- and H-montmorillonites containing urea. *Agrokem. Talajtan*, **19**: 293–310.

197. LIBOR, O., GRABER, L., and DONATH, E., 1971. Montmorillonites treated with urea. I. Montmorillonites treated with urea solutions. *Acta Mineral. Petrogr.*, **20**: 97–111.

198. FARMER, W. J., 1967. Infrared studies of the mechanisms of adsorption of urea and its herbicidal derivatives by montmorillonite. *Diss. Abst.*, **27**: 3749-B.

199. KIM, J. T., 1970. The absorption of substituted urea compounds on montmorillonite. *Diss. Abst.*, **30**: 5326-B.

200. BARRER, R. M., and MacLEOD, D. M., 1954. Intercalation and sorption by montmorillonite. *Trans. Faraday Soc.*, **50**: 980–989.

201. BARRER, R. M., and MacLEOD, D. M., 1955. Activation of montmorillonite by ion exchange and sorption complexes of tetra-alkylammonium montmorillonites. *Trans. Faraday Soc.*, **51**: 1290–1300.

202. BARRER, R. M., and REAY, J. S. S., 1957. Sorption and intercalation by methylammonium montmorillonites. *Trans. Faraday Soc.*, **53**: 1253–1261.

203. BARRER, R. M., and REAY, J. S. S., 1957. Interlamellar sorption by montmorillonite. *Proc. 2nd Int. Congr. Surface Activity, London*, **2**: 79–89.

204. BARRER, R. M., and PERRY, G. S., 1961. Sorption of mixtures, and selectivity in alkylammonium montmorillonites. I. Monomethylammonium bentonite. *J. Chem. Soc.*, 842–849.

205. BARRER, R. M., and HAMPTON, M. G., 1957. Gas chromatography and mixture isotherms in alkylammonium bentonites. *Trans. Faraday Soc.*, **53**: 1462–1475.

206. TARAMASSO, M., and VENIALE, F., 1969. Gas-chromatographic investigations on dimethyldioctadecylammonium derivatives of different clay minerals. *Contr. Mineral. Petrol.*, **21**: 53–62.

207. TARAMASSO, M., and VENIALE, F., 1969. Gas chromatographic performance of long chain alkylammonium complexes with 'beidellite-type' clay minerals. *Chromatographia*, **2**: 239–241.

208. TARAMASSO, M., and VENIALE, F., 1970. Comportamiento de los complejos alquilamonio-biotita vermiculitizada como fase estacionaria en cromatografia de gases. *Reunion Hispano-Belga de Minerales de la Arcilla, Madrid*, 129–132.

308

209. TARAMASSO, M., LAGALY, G., and WEISS, A., 1971. Gas chromatographic use and surface properties of dimethyldioctadecylammonium derivatives of mica-type layer silicates capable of swelling. *Kolloid—Z.Z. Polym.*, **245**: 580–588.
210. CHAUSSIDON, J., and CALVET, R., 1965. Evolution of amine cations on montmorillonite with dehydration of the mineral. *J. Phys. Chem.*, **69**: 2265–2268.
211. GALWEY, A. K., 1969. Reactions of alcohols adsorbed on montmorillonite and the role of minerals in petroleum genesis. *J. Chem. Soc. D.*, 577–578.
212. GALWEY, A. K., 1970. Reactions of alcohols and of hydrocarbons on montmorillonite surfaces. *J. Catalysis*, **19**: 330–342.
213. EISMA, E., and JURG, J. W., 1969. Fundamental aspects of the generation of petroleum, In: G. Eglinton and M. T. J. Murphy (Editors), *Organic Geochemistry*. Springer Verlag, Berlin, pp. 676–698.
214. SHIMOYAMA, A., and JOHNS, W. D., 1971. Catalytic conversion of fatty acids to petroleum-like paraffins and their maturation. *Nature*, **232**: 140–144.
215. DURAND, B., PELET, R., and FRIPIAT, J. J., 1972. Alkylammonium decomposition on montmorillonite surfaces in an inert atmosphere. *Clays Clay Minerals*, **20**: 21–35.
216. ELTANTAWY, I. M., and ARNOLD, P. W., 1972. Adsorption of *n*-alkanes by Wyoming montmorillonite. *Nature Phys. Sci.*, **237**: 123–125.
217. DE BOER, J. H., and ZWIKKER, C., 1929. Adsorption as a result of polarization. The adsorption isotherm. *Z. physik. Chem.*, **B.3**: 407–418.
218. FRIPIAT, J. J., 1968. Surface fields and transformation of adsorbed molecules in soil colloids. *Trans. 9th Int. Congr. Soil Sci., Adelaide*, **1**: 679–689.
219. DONER, H. E., and MORTLAND, M. M., 1969. Benzene complexes with Cu (II) montmorillonite. *Science*, **166**: 1406–1407.
220. MORTLAND, M. M., and PINNAVAIA, T. J., 1971. Formation of copper (II) arene complexes on the interlamellar surfaces of montmorillonite. *Nature Phys. Sci.*, **229**: 75–77.
221. PINNAVAIA, T. J., and MORTLAND, M. M., 1971. Interlamellar metal complexes of layer silicates. I. Copper (II)-arene complexes on montmorillonite. *J. Phys. Chem.*, **75**: 3957–3962.
222. FUSON, N., GARRIGOU-LAGRANGE, C., and JOSIEN, M. L., 1960. Infrared spectra and vibrational assignments for the toluenes $C_6H_5CH_3$, $C_6H_5CD_3$ and $C_6D_5CD_3$. *Spectrochim. Acta*, **16**: 106–127.

Chapter 4

1. HIRT, R. C., and SCHMITT, R. G., 1958. Ultraviolet absorption spectra of derivatives of *s*-triazine. II. Oxo-triazines and their acyclic analogs. *Spectrochim. Acta*, **12**: 127–138.
2. BAILEY, G. W., and WHITE, J. L., 1970. Factors influencing the adsorption, desorption, and movement of pesticides in soil. *Residue Revs.*, **32**: 29–92.
3. WEBER, J. B., 1970. Mechanisms of adsorption of *s*-triazines by clay colloids and factors affecting plant availability. *Residue Revs.*, **32**: 93–130.
4. FRISSEL, M. J., 1961. *The Adsorption of Some Organic Compounds, especially Herbicides, on Clay Minerals*. Ph.D. Thesis, Wageningen. Published as Versl. Landbouwk. Onderz. N.R. 67.3.
5. FRISSEL, M. J., and BOLT, G. H., 1962. Interaction between certain ionizable organic compounds (herbicides) and clay minerals. *Soil Sci.*, **94**: 284–291.
6. HARRIS, C. I., and WARREN, G. F., 1964. Adsorption and desorption of herbicides by soil. *Weeds*, **12**: 120–126.
7. WEBER, J. B., PERRY, P. W., and UPCHURCH, R. P., 1965. The influence of temperature and time on the adsorption of paraquat, diquat, 2,4-D, and prometone by clays, charcoal, and an anion-exchange resin. *Soil Sci. Soc. Am. Proc.*, **29**: 678–687.

8. WEBER, J. B., 1966. Molecular structure and pH effects on the adsorption of 13 s-triazine compounds on montmorillonite clay. *Am. Mineralogist*, **51**: 1657–1670.

9. YAMANE, V. K., and GREEN, R. E., 1972. Adsorption of ametryne and atrazine on an oxysol, montmorillonite, and charcoal in relation to pH and solubility effects. *Soil Sci. Soc. Am. Proc.*, **36**: 58–63.

10. TALBERT, R. E., and FLETCHALL, O. H., 1965. The adsorption of some s-triazines in soils. *Weeds*, **13**: 46–52.

11. WEBER, J. B., and WEED, S. B., 1968. Adsorption and desorption of diquat, paraquat, and prometone by montmorillonitic and kaolinitic clay materials. *Soil Sci. Soc. Am. Proc.*, **32**: 485–487.

12. CRUZ, M., WHITE, J. L., and RUSSELL, J. D., 1968. Montmorillonite-s-triazine interactions. *Israel J. Chem.*, **6**: 315–323.

13. LAMBERT, S. M., 1967. Functional relationship between sorption in soil and chemical structure. *J. Agr. Food Chem.*, **15**: 572–576.

14. BRIGGS, G. G., 1969. Molecular structure of herbicides and their sorption by soils. *Nature*, **223**: 1288.

15. BAILEY, G. W., WHITE, J. L., and ROTHBERG, T., 1968. Adsorption of organic herbicides by montmorillonite: Role of pH and chemical character of adsorbate. *Soil Sci. Soc. Am. Proc.*, **32**: 222–234.

16. RUSSELL, J. D., CRUZ, M., WHITE, J. L., BAILEY, G. W., PAYNE, W. R., POPE, J. D., and TEASLEY, J. I., 1968. Mode of chemical degradation of s-triazines by montmorillonite. *Science*, **160**: 1340–1342.

17. BROWN, C. B., and WHITE, J. L., 1969. Reactions of 12 s-triazines with soil clays. *Soil Sci. Soc. Am. Proc.*, **33**: 863–867.

18. RUSSELL, J. D., CRUZ, M., and WHITE, J. L., 1968. The adsorption of 3-aminotriazole by montmorillonite. *J. Agr. Food Chem.*, **16**: 21–24.

19. NEARPASS, D. C., 1970. Exchange adsorption of 3-amino-1,2,4-triazole by montmorillonite. *Soil Sci.*, **109**: 77–84.

20. HANCE, R. J., 1965. The adsorption of urea and some of its derivatives by a variety of soils. *Weed Research*, **5**: 98–107.

21. YUEN, Q. H., and HILTON, H. W., 1962. The adsorption of monuron and diuron by Hawaiian sugar cane soils. *J. Agr. Food Chem.*, **10**: 386–392.

22. HAYES, M. H. B., STACEY, M., and THOMPSON, J. M., 1968. Adsorption of s-triazine herbicides by soil organic matter preparations. *Isotopes and Radiation in Soil Organic-Matter Studies*, Int. Atom. Energy Agency, Vienna, pp. 75–90.

23. WEBER, J. B., WEED, S. B., and WARD, T. M., 1969. Adsorption of s-triazines by soil organic matter. *Weed Science*, **17**: 417–421.

24. MACNAMARA, G., and TOTH, S. J., 1970. Adsorption of linuron and malathion by soils and clay minerals. *Soil Sci.*, **109**: 234–240.

25. WOLCOTT, A. R., 1970. Retention of pesticides by organic materials in soils. *Proc. Symp. Pesticides in the Soil*, Mich. State Univ., pp. 128–137.

26. MORTLAND, M. M., 1966. Urea complexes with montmorillonite: An infrared absorption study. *Clay Minerals*, **6**: 143–156.

27. FARMER, W. J., and AHLRICHS, J. L., 1969. Infrared studies of the mechanism of adsorption of urea-d_4, methylurea-d_3, and 1,1-dimethylurea-d_2 by montmorillonite. *Soil Sci. Soc. Am. Proc.*, **33**: 254–258.

28. HANCE, R. J., 1969. Influence of pH, exchangeable cation and the presence of organic matter on the adsorption of some herbicides by montmorillonite. *Can. J. Soil Sci.*, **49**: 357–364.

29. HANCE, R. J., 1971. Complex formation as an adsorption mechanism for linuron and atrazine. *Weed Research*, **11**: 106–110.

30. MORTLAND, M. M., and MEGGITT, W. F., 1966. Interaction of ethyl-N,N-di-n-propyl-thiolcarbamate (EPTC) with mortmorillonite. *J. Agr. Food Chem.*, **14**: 126–129.

310

31. Doner, H. E., and Mortland, M. M., 1969. Intermolecular interaction in montmorillonites: NH-CO systems. *Clays Clay Minerals*, **17**: 265–270.
32. Mortland, M. M., 1968. Pyridinium-montmorillonite complexes with ethyl N,N-di-*n*-propylthiolcarbamate (EPTC). *J. Agr. Food Chem.*, **16**: 706–707.
33. Bowman, B. T., Adams, R. S., Jr., and Fenton, S. W., 1970. Effect of water upon malathion adsorption onto five montmorillinite systems. *J. Agr. Food Chem.*, **18**: 723–727.
34. Schwartz, H. G., Jr., 1967. Adsorption of selected pesticides on activated carbon and mineral surfaces. *Envir. Sci. Tech.*, **1**: 332–337.
35. Haque, R., and Sexton, R., 1968. Kinetic and equilibrium study of the adsorption of 2,4-dichlorophenoxy acetic acid on some surfaces. *J. Colloid Interface Sci.*, **27**: 818–827.
36. Haque, R., Lilley, S., and Coshow, W. R., 1970. Mechanism of adsorption of diquat and paraquat on montmorillonite surface. *J. Colloid Interface Sci.*, **33**: 185–188.
37. Haque, R., and Lilley, S., 1972. Infrared spectroscopic studies of charge-transfer complexes of diquat and paraquat. *J. Agr. Food Chem.*, **20**: 57–58.
38. Giles, C. H., MacEwan, T. H., Nakhwa, S. N., and Smith, D., 1960. Studies in adsorption. XI. A system of classification of solution adsorption isotherms and its use in diagnosis of adsorption mechanisms and in measurement of specific surface areas of solids. *J. Chem. Soc.*, 3973–3993.
39. Knight, B. A. G., and Tomlinson, T. E., 1967. The interaction of paraquat (1:1'-dimethyl 4:4'-dipyridylium dichloride) with mineral soils. *J. Soil Sci.*, **18**: 233–243.
40. Tomlinson, T. E., Knight, B. A. G., Bastow, A. W., and Heaver, A. A., 1969. Structural factors affecting the adsorption of bipyridilium cations by clay minerals. In: *Physico-chemical and Biophysical Factors affecting the Activity of Pesticides*, Monograph 29, Society of Chemical Industry, London, pp. 317–329.
41. Weber, J. B., and Scott, D. C., 1966. Availability of a cationic herbicide adsorbed on clay minerals to cucumber seedlings. *Science*, **152**: 1400–1402.
42. Weber, J. B., Meek, R. C., and Weed, S. B., 1969. The effect of cation-exchange capacity on the retention of diquat^{2+} and paraquat^{2+} by three-layer type clay minerals: II. Plant availability of paraquat. *Soil Sci. Soc. Am. Proc.*, **33**: 382–385.
43. Weber, J. B., and Coble, H. D., 1968. Microbial decomposition of diquat adsorbed on montmorillonite and kaolinite clays. *J. Agr. Food Chem.*, **16**: 475–478.
44. Weber, J. B., Weed, S. B., and Best, J. A., 1969. Displacement of diquat from clay and its photoxicity. *J. Agr. Food Chem.*, **17**: 1075–1076.
45. Weed, S. B., and Weber, J. B., 1969. The effect of cation exchange capacity on the retention of diquat^{2+} and paraquat^{2+} by three-layer type clay minerals. I. Adsorption and release. *Soil Sci. Soc. Am. Proc.*, **33**: 379–382.
46. Dixon, J. B., Moore, D. E., Agnihotri, N. P., and Lewis, D. E., Jr., 1970. Exchange of diquat^{2+} in soil clays, vermiculite, and smectite. *Soil Sci. Soc. Am. Proc.*, **34**: 805–808.
47. Weed, S. B., and Weber, J. B., 1968. The effect of adsorbent charge on the competitive adsorption of divalent organic cations by layer-silicate minerals. *Am. Mineralogist*, **53**: 478–490.
48. Philen, O. D., Jr., Weed, S. B., and Weber, J. B., 1970. Estimation of surface charge density of mica and vermiculite by competitive adsorption of diquat^{2+} *vs.* paraquat^{2+}. *Soil Sci. Soc. Am. Proc.*, **34**: 527–531.
49. Philen, O. D., Jr., Weed, S. B., and Weber, J. B., 1971. Surface charge characterization of layer silicates by competitive adsorption of two organic divalent cations. *Clays Clay Minerals*, **19**: 295–302.
50. Edwards, D. G., Posner, A. M., and Quirk, J. P., 1965. Discreteness of charge on clay surfaces. *Nature*, **206**: 168–170.
51. Solomon, D. H., 1968. Clay minerals as electron acceptors and/or electron donors in organic reactions. *Clays Clay Minerals*, **16**: 31–39.
52. Theng, B. K. G., and Walker, G. F., 1970. Interactions of clay minerals with organic monomers. *Israel J. Chem.*, **8**: 417–424.

311

53. MORTLAND, M. M., 1970. Clay-organic complexes and interactions. *Advan. Agron.*, **22**: 75–117.

54. FLECK, E. E., and HALLER, H. L., 1945. Compatibility of DDT with insecticides, fungicides, and fertilizers. *Ind. Eng. Chem.*, **37**: 403–405.

55. MALINA, M. A., GOLDMAN, A., TRADEMAN, L., and POLEN, P. B., 1956. Deactivation of mineral carriers for stable heptachlor-dust formulations. *J. Agr. Food Chem.*, **4**: 1038–1042.

56. ROSENFIELD, C., and VAN VALKENBURG, W., 1965. Decomposition of O,O-dimethyl O-2,4,5-trichlorophenyl phosphorothioate (ronnel) adsorbed on bentonite and other clays. *J. Agr. Food Chem.*, **13**: 68–72.

57. WALLING, C., 1950. The acid strength of surfaces. *J. Am. Chem. Soc.*, **72**: 1164–1168.

58. BENESI, H. A., 1956. Acidity of catalyst surfaces. I. Acid strength from colors of adsorbed indicators. *J. Am. Chem. Soc.*, **78**: 5490–5494.

59. BENESI, H. A., 1967. Acidity of catalyst surfaces. II. Amine titration using Hammett indicators. *J. Phys. Chem.*, **61**: 970–973.

60. HAMMETT, L. P., and DEYRUP, A. J., 1932. A series of simple basic indicators. I. The acidity functions of mixtures of sulfuric and perchloric acids with water. *J. Am. Chem. Soc.*, **54**: 2721–2739.

61. SOLOMON, D. H., SWIFT, J. D., and MURPHY, A. J., 1971. The acidity of clay minerals in polymerizations and related reactions. *J. Macromol. Sci. Chem.*, **A5**: 587–601.

62. FOWKER, F. M., BENESI, H. A., RYLAND, R. B., SAWYER, W. M., DETLING, K. D., LOEFFLER, E. S., FOLCKEMER, F. B., JOHNSON, M. R., and SUN, Y. P., 1960. Clay-catalyzed decomposition of insecticides. *J. Agr. Food Chem.*, **8**: 203–210.

63. MORTLAND, M. M., and RAMAN, K. V., 1967. Catalytic hydrolysis of some organic phosphate pesticides by copper (II). *J. Agr. Food Chem.*, **15**: 163–167.

64. TALIBUDEEN, O., 1954. Complex formation between montmorillonoid clays and amino acids and proteins. *Trans. Faraday Soc.*, **51**: 582–590.

65. HENDRICKS, S. B., 1941. Base exchange of the clay mineral montmorillonite for organic cations and its dependence upon adsorption due to van der Waals forces. *J. Phys. Chem.*, **45**: 65–81.

66. SIESKIND, O., 1960. Étude des complexes d'adsorption formé entre la montmorillonite—H et certains acides aminés: isothermes d'adsorption à pH 2 et à 20°C. *C.R. Acad. Sci., Paris*, **250**: 2228–2230.

67. McLAREN, A. D., PETERSON, G. H., and BARSHAD, I., 1958. The adsorption reactions of enzymes and proteins on clay minerals. IV. Kaolinite and montmorillonite. *Soil Sci. Soc. Am. Proc.*, **22**: 239–244.

68. CLOOS, P., CALICIS, B., FRIPIAT, J. J., and MAKAY, K., 1966. Adsorption of amino acids and peptides by montmorillonite. I. Chemical and X-ray diffraction studies. *Proc. Int. Clay Conf., Jerusalem*, **1**: 223–232.

69. SIESKIND, O., and WEY, R., 1959. Sur l'adsorption d'acides aminés par la montmorillonite-H. Influence de la position relative des deux fonctions -NH$_2$ et -COOH. *C.R. Acad. Sci., Paris*, **248**: 1652–1655.

70. CHASSIN, P., 1969. Adsorption du glycocolle par la montmorillonite. *Bull. Groupe Fr. Argiles*, **21**: 71–88.

71. GREENLAND, D. J., LABY, R. H., and QUIRK, J. P., 1962. Adsorption of glycine and its di-, tri-, and tetra-peptides by montmorillonite. *Trans. Faraday Soc.*, **58**: 829–841.

72. GREENLAND, D. J., LABY, R. H., and QUIRK, J. P., 1965. Adsorption of amino acids and peptides by montmorillonite and illite. Part 1. Cation exchange and proton transfer. *Trans. Faraday Soc.*, **61**: 2013–2023.

73. GREENLAND, D. J., LABY, R. H., and QUIRK, J. P., 1965. Adsorption of amino acids and peptides by montmorillonite and illite. Part 2. Physical adsorption. *Trans. Faraday Soc.*, **61**: 2024–2035.

74. THENG, B. K. G., 1972. Adsorption of ammonium and some primary *n*-alkylammonium cations by soil allophane. *Nature*, **238**: 150–151.

75. YARIV, S., RUSSELL, J. D., and FARMER, V. C., 1966. Infrared study of the adsorption of benzoic acid and nitrobenzene in montmorillonite. *Israel J. Chem.*, **4**: 201–213.

76. SIESKIND, O., 1960. Sur les complexes d'adsorption formés in milieu acide entre la montmorillonite -H et certains acides aminés: leur structure. *C.R. Acad. Sci., Paris*, **250**: 2392–2393.

77. FRIPIAT, J. J., CLOOS, P., CALICIS, B., and MAKAY, K., 1966. Adsorption of amino acids and peptides by montmorillonite. II. Identification of adsorbed species and decay products by infrared spectroscopy. *Proc. Int. Clay Conf., Jerusalem*, **1**: 233–245.

78. OVCHARENKO, F. D., VDOVENKO, N. V., TCHICHKUN, V. P., and TARASEVICH, YU. I., 1969. Adsorption of amino acids by montmorillonite. *Ukr. Khim. Zh.*, **35**: 123–128.

79. SIEGEL, A., 1966. Equilibrium binding studies of zinc-glycine complexes to ion-exchange resins and clays. *Geochim. Cosmochim. Acta*, **30**: 757–768.

80. BODENHEIMER, W., and HELLER, L., 1967. Sorption of α-amino acids by copper montmorillonite. *Clay Minerals*, **7**: 167–176.

81. JANG, S. D., and CONDRATE, R. A., Sr., 1972. The infrared spectra of valine adsorbed on Cu-montmorillonite. *Am. Mineralogist*, **57**: 494–498.

82. JANG, S. D., and CONDRATE, R. A., Sr., 1972. The I.R. spectra of lysine adsorbed on several cation-substituted montmorillonites. *Clays Clay Minerals*, **20**: 79–82.

83. NAKAGAWA, I. R., HOOPER, J., and WALTER, J. L., 1965. Infrared absorption spectra of metal-amino acid complexes. III. The infrared spectra and normal vibrations of metal-valine chelates. *Spectrochim. Acta*, **21**: 1–14.

84. NAKAMOTO, K., MORIMOTO, Y., and MARTELL, A. E., 1961. Infrared spectra of aqueous solutions. I. Metal chelate compounds of amino acids. *J. Am. Chem. Soc.*, **83**: 4528–4532.

85. BARSHAD, I., 1952. Factors affecting the interlayer expansion of vermiculite and montmorillonite with organic substances. *Soil Sci. Soc. Am. Proc.*, **16**: 176–182.

86. WALKER, G. F., and GARRETT, W. G., 1961. Complexes of vermiculite with amino acids. *Nature*, **191**: 1389.

87. NORRISH, K., 1954. Swelling of montmorillonite. *Disc. Faraday Soc.*, **18**: 120–134.

88. RAUSELL-COLOM, J. A., and SALVADOR, P. S., 1971. Complexes vermiculite-amino acides. *Clay Minerals*, **9**: 139–149.

89. RAUSELL-COLOM, J. A., and SALVADOR, P. S., 1971. Gélification de vermiculite dans des solutions d'acide γ-aminobutyrique. *Clay Minerals*, **9**: 193–208.

90. KANAMARU, F., and VAND, V., 1970. The crystal structure of a clay-organic complex of 6-amino hexanoic acid and vermiculite. *Am. Mineralogist*, **55**: 1550–1561.

91. VERWEY, E. J. W., and OVERBEEK, J. Th. G., 1948. *Theory of the Stability of Lyophobic Colloids.* Elsevier Publ. Co., Amsterdam—New York.

92. MIFSUD, A., FORNES, V., and RAUSELL-COLOM, J. A., 1970. Cationic complexes of vermiculite with L-ornithine. *Reunion Hispano-Belga de Minerales de la Arcilla, Madrid*, 121–127.

93. SIMINOFF, P., and GOTTLIEB, D., 1951. The production and role of antibiotics in the soil. I. The fate of streptomycin. *Phytopath.*, **41**: 420–430.

94. GOTTLIEB, D., and SIMINOFF, P., 1952. The production and role of antibiotics in the soil. II. Chloromycetin. *Phytopath.*, **42**: 91–97.

95. JEFFREYS, E. G., 1952. The stability of antibiotics in soils. *J. Gen. Microbiol*, **7**: 295–312.

96. PRAMER, D., and STARKEY, R. L., 1953. Streptomycin in soil. *Proc. 7th Int. Bot. Congr., Stockholm*, 256–257.

97. PINCK, L. A., HOLTON, W. F., and ALLISON, F. E., 1961. Antibiotics in soils. I. Physico-chemical studies of antibiotic-clay complexes. *Soil Sci.*, **91**: 22–28.

98. PINCK, L. A., SOULIDES, D. A., and ALLISON, F. E., 1961. Antibiotics in soils. II. Extent and mechanism of release. *Soil Sci.*, **91**: 94–99.

99. SOULIDES, D. A., PINCK, L. A., and ALLISON, F. E., 1961. Antibiotics in soils. III. Further studies on release of antibiotics from clays. *Soil Sci.*, **92**: 90–93.

100. PINCK, L. A., SOULIDES, D. A., and ALLISON, F. E., 1962. Antibiotics in soils. IV. Polypeptides and macrolides. *Soil Sci.*, **94**: 129–131.

101. SOULIDES, D. A., PINCK, L. A., and ALLISON, F. E., 1962. Antibiotics in soils. V. Stability and release of soil-adsorbed antibiotics. *Soil Sci.*, **94**: 239–244.

102. PINCK, L. A., 1962. Adsorption of proteins, enzymes and antibiotics by montmorillonite. *Clays Clay Minerals*, **9**: 520–529.

103. STOTZKY, G., DAWSON, J. E., MARTIN, R. T., and KUILE, C. H. H., 1961. Soil mineralogy as a factor in the spread of *Fusarium* Wilt of banana. *Science*, **133**: 1483.

104. EVCIM, N., and BARR, M., 1955. Adsorption of some alkaloids by different clays. *J. Am. Pharm. Assoc.*, **44**: 570–573.

105. BARR, M., and ARNITSA, E. S., 1957. Adsorption studies on clays. I. The adsorption of two alkaloids by activated attapulgite, halloysite, and kaolin. *J. Am. Pharm. Assoc.*, **46**: 486–489.

106. BARR, M., 1957. Adsorption studies on clays. II. The adsorption of bacteria by activated attapulgite, halloysite, and kaolin. *J. Am. Pharm. Assoc.*, **46**: 490–492.

107. BARR, M., and ARNITSA, E. S., 1957. Adsorption studies on clays. III. The adsorption of diphteria toxin by activated attapulgites, halloysite, and kaolin. *J. Am. Pharm. Assoc.*, **46**: 493–497.

108. KHAN, S., and SINGHAL, J. P., 1967. Titrations of hydrogen clays with nicotine. *Soil Sci.*, **104**: 427–432.

109. RIDOUT, C. W., 1968. The adsorption of atropine from aqueous solution by kaolin. *Pharm. Acta Helv.*, **43**: 42–49.

110. THENG, B. K. G., 1971. Adsorption of molybdate by some crystalline and amorphous soil clays. *N.Z. J. Sci.*, **14**: 1040–1056.

111. LAILACH, G. E., THOMPSON, T. D., and BRINDLEY, G. W., 1968. Absorption of pyrimidines, purines, and nucleosides by Li-, Na-, Mg-, and Ca-montmorillonite (Clay-organic studies XII). *Clays Clay Minerals*, **16**: 285–293.

112. LAILACH, G. E., THOMPSON, T. D., and BRINDLEY, G. W., 1968. Absorption of pyrimidines, purines, and nucleosides by Co-, Ni-, Cu-, and Fe (III)-montmorillonite (Clay-organic studies XIII). *Clays Clay Minerals*, **16**: 295–301.

113. LAILACH, G. E., and BRINDLEY, G. W., 1969. Specific co-absorption of purines and pyrimidines by montmorillonite (Clay-organic studies XV). *Clays Clay Minerals*, **17**: 95–100.

114. THOMPSON, T. D., and BRINDLEY, G. W., 1969. Absorption of pyrimidines, purines and nucleosides by Na-, Mg-, and Cu(II)-illite. (Clay-organic studies XVI). *Am. Mineralogist*, **54**: 858–868.

115. HARKINS, T. R., and FREISER, H., 1958. Adenine-metal complexes. *J. Am. Chem. Soc.*, **80**: 1132–1135.

116. CHENEY, G. E., FREISER, H., and FERNANDO, Q., 1959. Metal complexes of purine and some of its derivatives. *J. Am. Chem. Soc.*, **81**: 2611–2615.

117. WEISS, R., and VENNER, H., 1963. Das komplexchemische Verhalten von Pyrimidinabkommlingen. III. Das komplexchemische Verhalten des Adenins gegenüber Kupfer (II). *Z. physiol. Chem.*, **333**: 169–178.

118. WEISS, R., and VENNER, H., 1965. Das komplexchemische Verhalten von Pyrimidinabkommlingen. IV. Das komplexchemische Verhalten von Hydroxypurinen gegenüber Kupfer (II). *Z. physiol. Chem.*, **340**: 138–147.

119. SLETTEN, E., 1967. Crystal structure of a dinuclear 2:1 adenine-copper complex. *J. Chem. Soc., D.*, 1119–1120.

120. TS'O, P. O. P., MELVIN, I. S., and OLSEN, A. C., 1963. Interaction and association of bases and nucleosides in aqueous solutions. *J. Am. Chem. Soc.*, **85**: 1289–1296.

121. TS'O, P. O. P., and CHAN, S. I., 1964. Interaction and association of bases and nucleosides in aqueous solution. II. Association of 6-methylpurine and 5-bromouridine and treatment of multiple equilibria. *J. Am. Chem. Soc.*, **86**: 4176–4181.

122. CHAN, S. I., SCHWEIZER, M. P., TS'O, P. O. P., and HELMKAMP, G. K., 1964. Interaction and association of bases and nucleosides in aqueous solutions. III. A nuclear magnetic resonance study of the self-association of purine and 6-methylpurine. *J. Am. Chem. Soc.*, **86**: 4182–4188.

123. Jurg, J. W., and Eisma, E., 1964. Petroleum hydrocarbons: generation from fatty acid. *Science*, **144**: 1451–1452.

124. Shimoyama, A., and Johns, W. D., 1971. Catalytic conversion of fatty acids into petroleum-like paraffins and their maturation. *Nature Phys. Sci.*, **232**: 140–144.

125. Brindley, G. W., and Moll, W. F., Jr., 1965. Complexes of natural and synthetic Ca-montmorillonites with fatty acids (Clay-organic studies IX). *Am. Mineralogist*, **50**: 1355–1370.

126. Brindley, G. W., and Ray, S., 1964. Complexes of Ca-montmorillonite with primary monohydric alcohols (Clay-organic studies VIII). *Am. Mineralogist*, **49**: 106–115.

127. Weiss, A., 1963. Mica-type layer silicates with alkylammonium ions. *Clays Clay Minerals*, **10**: 191–224.

128. Menter, J. W., and Tabor, D., 1951. Orientation of fatty acid and soap films on metal surfaces. *Proc. Roy. Soc.*, A. **204**: 514–524.

129. Sieskind, O., and Ourisson, G., 1971. Interactions argile-matière organique: formation de complexes entre la montmorillonite et les acides stéarique et béhénique. *C.R. Acad. Sci., Paris*, **272**: 1885–1888.

130. Sieskind, O., and Siffert, B., 1972. Formation d'un carboxylate de surface entre l'acide stéarique et une hectorite nickelifère: localisation de l'acide gras sur le réseau silicate. *C.R. Acad. Sci., Paris*, **274**: 973–976.

131. Kaufherr, N., Yariv, S., and Heller, L., 1971. The effect of exchangeable cations on the sorption of chlorophyllin by montmorillonite. *Clays Clay Minerals*, **19**: 193–200.

132. Weiss, A., and Roloff, G., 1963. Die Rolle organischer Derivate von glimmerartigen Schikhtsilikaten bei der Bildung von Erdöl. *Proc. Int. Clay Conf., Stockholm*, **2**: 373–378.

133. Beck, C. W., and Brunton, G., 1961. X-ray and infra-red data on hectorite-guanidines and montmorillonite-guanidines. *Clays Clay Minerals*, **8**: 22–38.

134. Meyers, P. A., and Quinn, J. G., 1971. Fatty acid-clay mineral association in artificial and natural sea water solutions. *Geochim. Cosmochim. Acta*, **35**: 628–632.

135. Weiss, A., and Roloff, G., 1966. Über die Einlagerung symmetrischer Triglyceride in quellungsfähige Schichtsilicate. *Proc. Int. Clay Conf., Jerusalem*, **1**: 263–275.

136. Bradley, W. F., 1945. Molecular associations between montmorillonite and some poly-functional organic liquids. *J. Am. Chem. Soc.*, **67**: 975–981.

137. Greenland, D. J., 1956. The adsorption of sugars by montmorillonite. I. X-ray studies. *J. Soil Sci.*, **7**: 319–328.

138. Greenland, D. J., 1956. The adsorption of sugars by montmorillonite. II. Chemical studies. *J. Soil Sci.*, **7**: 329–334.

139. Bell, D. J., and Northcote, D. H., 1950. Structure of a cell-wall polysaccharide of baker's yeast. *J. Chem. Soc.*, 1944–1947.

140. Taylor, J. B., and Rowlinson, J. S., 1955. Thermodynamic properties of aqueous solutions of glucose. *Trans. Faraday Soc.*, **51**: 1183–1192.

141. Mitra, S. P., Misra, S. G., and Panda, N., 1957. Adsorption of glucose by calcium bentonite. *Proc. Natl. Acad. Sci., India*, **26A** (Pt. 1): 72–74.

142. Jepson, W. B., and Williams, J. F., 1972. The adsorption of water by clays. *Clay Minerals*, **9**: 275–279.

143. Davis, G. A., and Worrall, W. E., 1971. Adsorption of water by clays. *Trans. Brit. Ceram. Soc.*, **70**: 71–75.

Chapter 5

1. Lloyd, J. U., 1916. Discovery of the alkaloidal affinities of hydrous aluminium silicate. *J. Am. Pharm. Assoc.*, **5**: 381–390, 490–495.

2. Smith, C. R., 1934. Base exchange reactions of bentonite and salts of organic bases. *J. Am. Chem. Soc.*, **56**: 1561–1563.

3. GIESEKING, J. E., and JENNY, H., 1936. Behaviour of polyvalent cations in base exchange. *Soil Sci.*, **42**: 273–280.
4. GIESEKING, J. E., 1939. Mechanism of cation exchange in the montmorillonite—beidellite-nontronite type of clay minerals. *Soil Sci.*, **47**: 1–14.
5. ENSMINGER, L. E., and GIESEKING, J. E., 1939. The adsorption of proteins by montmorillonite clay. *Soil Sci.*, **48**: 467–473.
6. ENSMINGER, L. E., and GIESEKING, J. E., 1941. Adsorption of proteins by montmorillonite clay and its effect on base-exchange capacity. *Soil Sci.*, **51**: 125–132.
7. HENDRICKS, S. B., 1941. Base exchange of the clay mineral montmorillonite for organic cations and its dependence upon adsorption due to van der Waals forces. *J. Phys. Chem.*, **45**: 65–81.
8. CHAKRAVARTI, S. K., 1957. Adsorption of higher alkyl quaternary ammonium and pyridinium compounds by clay minerals. *J. Indian Soc. Soil Sci.*, **5**: 85–90.
9. GRIM, R. E., ALLAWAY, W. H., and CUTHBERT, F. L., 1947. Reaction of different clay minerals with some organic cations. *J. Am. Chem. Soc.*, **30**: 137–142.
10. KURILENKO, O. D., and MIKHALYUK, R. V., 1959. Adsorption of aliphatic amines on bentonite from aqueous solutions. *Kolloidn. Zh.*, **21**: 181–184.
11. MOREL, R., and HÉNIN, S., 1956. Anomalies d'adsorption de cations organiques polyvalents par les mineraux argileux. *C.R. Acad. Sci., Paris*, **242**: 2975–2977.
12. BARRER, R. M., and KELSEY, K. E., 1961. Thermodynamics of interlamellar complexes. I. Hydrocarbons in methylammonium montmorillonites *Trans. Faraday Soc.*, **57**: 452–462.
13. COWAN, C. T., and WHITE, D., 1958. The mechanisms of exchange reactions occurring between sodium montmorillonite and various *n*-primary aliphatic amine salts. *Trans. Faraday Soc.*, **54**: 691–697.
14. GILES, C. H., MACEWAN, T. H., NAKHWA, S. N., and SMITH, D., 1960. Studies in adsorption. XI. A system of classification of solution adsorption isotherms and its use in diagnosis of adsorption mechanisms and in measurement of specific surface areas of solids. *J. Chem. Soc.*, 3973–3993.
15. SIESKIND, O., and WEY, R., 1958. Influence of pH on the adsorption of amines by H-montmorillonite. *C.R. Acad. Sci., Paris*, **247**: 74–76.
16. THENG, B. K. G., GREENLAND, D. J., and QUIRK, J. P., 1967. Adsorption of alkylammonium cations by montmorillonite. *Clay Minerals*, **7**: 1–17.
17. GAINES, G. L., Jr., and THOMAS, H. C., 1953. Adsorption studies on clay minerals. II. A formulation of the thermodynamics of exchange adsorption. *J. Chem. Phys.*, **21**: 714–718.
18. HELFFERICH, F., 1962. *Ion Exchange*. McGraw-Hill, New York, N.Y., pp. 151–200.
19. WHITLOW, E. P., and FELSING, W. A., 1944. Heats of dilution and heat capacities of aqueous solutions of mono-, di- and trimethylamine hydrochlorides. *J. Am. Chem. Soc.*, **66**: 2028–2033.
20. ROBINSON, R. A., and STOKES, R. H., 1955. *Electrolyte Solutions*. Butterworths, London, p. 481.
21. THENG, B. K. G., 1971. Adsorption of alkylammonium cations by porous crystals. Comparison between montmorillonite and synthetic near-faujasite. *N.Z. J. Sci.*, **14**: 1026–1039.
22. FRIPIAT, J. J., CLOOS, P., and PONCELET, A., 1965. Comparison between the exchange properties of montmorillonite and of a resin for alkali and alkaline-earth cations. I. Reversibility of the process. *Bull. Soc. Chim. France*, 208–215.
23. VANSANT, E. F., and UYTTERHOEVEN, J. B., 1972. Thermodynamics of the exchange of *n*-alkylammonium ions on Na-montmorillonite. *Clays Clay Minerals*, **20**: 47–54.
24. SLABAUGH, W. H., and KUPKA, F., 1958. Organic cation exchange properties of calcium montmorillonite. *J. Phys. Chem.*, **62**: 599–601.
25. SLABAUGH, W. H., 1958. Cation exchange properties of bentonite. *J. Phys. Chem.*, **58**: 162–165.
26. WEISS, A., 1963. Organic derivatives of mica-type layer silicates. *Angew. Chem., Int. Edit. Engl.*, **2**: 134–143.

316

27. WEISS, A., 1969. Organic derivatives of clay minerals, zeolites, and related minerals. In: G. Eglinton and M. T. J. Murphy (Editors), *Organic Geochemistry*. Springer Verlag, Berlin, pp. 737–781.

28. LABY, R. H., and WALKER, G. F., 1970. Hydrogen bonding in primary alkylammonium-vermiculite complexes. *J. Phys. Chem.*, **74**: 2369–2373.

29. ROWLAND, R. A., and WEISS, E. J., 1963. Bentonite-methylamine complexes. *Clays Clay Minerals*, **10**: 460–468.

30. FRIPIAT, J. J., PENNEQUIN, M., PONCELET, G., and CLOOS, P., 1969. Influence of the van der Waals force on the infrared spectra of short aliphatic alkylammonium cations held on montmorillonite. *Clay Minerals*, **8**: 119–134.

31. BARRER, R. M., and BRUMMER, K., 1963. Relations between partial ion exchange and inter-lamellar sorption in alkylammonium montmorillonites. *Trans. Faraday Soc.*, **59**: 959–968.

32. BARRER, R. M., and MACLEOD, D. M., 1955. Activation of montmorillonite by ion exchange and sorption complexes of tetra-alkylammonium montmorillonites. *Trans. Faraday Soc.*, **51**: 1290–1300.

33. BARRER, R. M., and REAY, J. S. S., 1957. Sorption and intercalation by methylammonium montmorillonites. *Trans. Faraday Soc.*, **53**: 1253–1261.

34. DIAMOND, S., and KINTER, E. B., 1963. Characterization of montmorillonite saturated with short chain amine cations. I. Interpretation of basal spacing measurements. *Clays Clay Minerals*, **10**: 163–173.

35. CHAUSSIDON, J., and CALVET, R., 1955. Evolution of amine cations adsorbed on montmorillonite with dehydration of the mineral. *J. Phys. Chem.*, **69**: 2265–2268.

36. THENG, B. K. G., 1964. *The Formation and Swelling of Complexes between Montmorillonite and Some Organic Compounds*. Ph.D. Thesis, University of Adelaide, South Australia.

37. PAULING, L., 1960. *The Nature of the Chemical Bond*, 3rd edit. Cornell University Press, Ithaca, N.Y., p. 260.

38. BRINDLEY, G. W., and HOFFMANN, R. W., 1962. Orientation and packing of aliphatic chain molecules on montmorillonite. Clay-organic studies VI. *Clays Clay Minerals*, **9**: 546–556.

39. JORDAN, J. W., 1949. Organophilic bentonites. I. Swelling in organic liquids. *J. Phys. Colloid Chem.*, **53**: 294–306.

40. JORDAN, J. W., HOOK, B. J., and FINLAYSON, C. M., 1950. Organophilic bentonites. II. Organic liquid gels. *J. Phys. Colloid Chem.*, **54**: 1196–1208.

41. GREENLAND, D. J., and QUIRK, J. P., 1962. Adsorption of 1-n-alkylpyridinium bromides by montmorillonite. *Clays Clay Minerals*, **9**: 484–499.

42. SUTHERLAND, H. H., and MACEWAN, D. M. C., 1961. Organic complexes of vermiculite. *Clay Minerals Bull.*, **4**: 229–233.

43. GARRETT, W. G., and WALKER, G. F., 1962. Swelling of some vermiculite-organic complexes in water. *Clays Clay Minerals*, **9**: 557–567.

44. JOHNS, W. D., and SEN GUPTA, P. K., 1967. Vermiculite-alkylammonium complexes. *Am. Mineralogist*, **52**: 1706–1724.

45. WALKER, G. F., 1967. Interactions of n-alkylammonium ions with mica-type layer lattices. *Clay Minerals*, **7**: 129–143.

46. WEISS, A., MEHLER, A., and HOFMANN, U., 1956. Zur Kenntnis von organophilem Vermikulit. *Z. Naturforsch.*, **11b**: 431–434.

47. WEISS, A., 1958. Die innerkristalline Quellung als allgemeines Modell für Quellungsvorgänge. *Chem. Ber.*, **91**: 487–502.

48. WEISS, A., MICHEL, E., and WEISS, AL., 1959. Über den Einfluss von Wasserstoffbrückenbindungen auf ein-und zweidimensionale innerkristalline Quellungsvorgänge. In: D. Hadži (Editor), *Hydrogen Bonding*. Pergamon Press, London, pp. 495–508.

49. WEISS, A., and KANTNER, I., 1960. Über eine einfache Möglichkeit zur Abschätzung der Schichtladung glimmerartiger Schichtsilikate. *Z. Naturforsch.*, **15b**: 804–807.

50. WEISS, A., 1963. Mica-type layer silicates with alkylammonium ions. *Clays Clay Minerals*, **10**: 191–224.

51. HAASE, D. J., WEISS, E. J., and STEINFINK, H., 1963. The crystal structure of a hexa-methylene-diamine-vermiculite complex. *Am. Mineralogist*, **48**: 261–270.

52. GREENE-KELLY, R., Sorption of aromatic compounds by montmorillonite. I. Orientation studies. *Trans. Faraday Soc.*, **51**: 412–424.

53. McATEE, J. L., Jr., 1959. Inorganic-organic cation exchange on montmorillonite. *Am. Mineralogist*, **44**: 1230–1236.

54. McATEE, J. L., Jr., 1962. Cation exchange of organic compounds on montmorillonite in organic media. *Clays Clay Minerals*, **9**: 444–450.

55. McATEE, J. L., Jr., 1963. Organic cation exchange on montmorillonite as observed by ultraviolet analysis. *Clays Clay Minerals*, **10**: 153–162.

56. JORDAN, J. W., 1963. Organophilic clay-base thickeners. *Clays Clay Minerals*, **10**: 299–308.

57. McATEE, J. L., Jr., and HACKMAN, J. R., 1964. Exchange equilibria on montmorillonite involving organic cations. *Am. Mineralogist*, **49**: 1569–1577.

58. DAS KANUNGO, J. L., CHAKRAVARTI, S. K., and MUKHERJEE, S. K., 1968. Studies on the sorption and desorption of $Co(NH_3)_6 Cl_3$ on bentonite. *J. Indian Chem. Soc.*, **45**: 685–689.

59. DAS KANUNGO, J. L., and CHAKRAVARTI, S. K., 1971. Behavior of quaternary ammonium ions in the desorption of $[Coen_3]^{3+}$ from H-Coen$_3$-bentonite. *J. Colloid Interface Sci.*, **35**: 295–299.

60. ERBRING, H., and LEHMANN, H., 1944. Austauschreaktionen an Na-Bentoniten mit grossvolumigen organischen Kolloidionen. *Kolloid Z.*, **107**: 201–205.

61. KOBAKHIDZE, E. I., and SHISHNIASHVILI, M. E., 1965. Hydrophobic clays and non-aqueous clay suspensions. *Kolloidn. Zh.*, **28**: 54–58.

62. THENG, B. K. G., GREENLAND, D. J., and QUIRK, J. P., 1968. The effect of exchangeable alkylammonium ions on the swelling of montmorillonite in water. *Clay Minerals*, **7**: 271–293.

63. GRANQUIST, W. T., and McATEE, J. L., Jr., 1963. The gelation of hydrocarbons by montmorillonite organic complexes. The role of the dispersant. *J. Colloid Sci.*, **18**: 409–420.

64. KENNEDY, J. V., and GRANQUIST, W. T., 1965. Flow properties of dispersions of an organo-montmorillonite in organic media *NLGI Spokesman*, August.

65. SLABAUGH, W. H., and HILTNER, P. A., 1968. The swelling of alkylammonium montmorillonites. *J. Phys. Chem.*, **72**: 4295–4298.

66. SLABAUGH, W. H., and HANSON, D. B., 1969. Solvent selectivity by an organoclay complex. *J. Colloid Interface Sci.*, **29**: 460–463.

67. SLABAUGH, W. H., and ST. CLAIR, A. D., 1969. Heats of immersion and swelling of organo-clay complexes. *J. Colloid Interface Sci.*, **29**: 586–589.

68. CHAKRAVARTI, S. K., 1956. Sedimentation volume and zeta potential of pure clay minerals and their mixtures as influenced by quaternary ammonium compounds. *Science and Culture*, **22**: 170–172.

69. KIJNE, J. W., 1969. Heats of wetting of complexes between montmorillonite and alkylammonium compounds. *Proc. 9th Int. Congr. Soil Sci., Adelaide*, **1**: 597–605.

70. BARRER, R. M., and MILLINGTON, A. D., 1967. Sorption and intracrystalline porosity in organo-clays. *J. Colloid Interface Sci.*, **25**: 359–372.

71. VOLD, R. D., and PHANSALKAR, V. K., 1962. Dispersion of alkylammonium montmorillonites in organic liquids. *J. Colloid Sci.*, **17**: 589–600.

72. WALKER, G. F., 1960. Macroscopic swelling of vermiculite crystals in water. *Nature*, **187**: 312–313.

73. FRANK, H. S., and EVANS, M. W., 1945. Free volume and entropy in condensed systems. III. Entropy in binary liquid mixtures; partial molal entropy in dilute solutions; structure and thermodynamics in aqueous electrolytes. *J. Chem. Phys.*, **13**: 507–532.

74. RAUSELL-COLOM, J. A., 1964. Small-angle X-ray diffraction study of the swelling of butylammonium-vermiculite. *Trans. Faraday Soc.*, **60**: 190–201.

75. NORRISH, K., and RAUSELL-COLOM, J. A., 1963. Low angle X-ray diffraction studies of the swelling of montmorillonite and vermiculite. *Clays Clay Minerals*, **10**: 123–149.

318

76. Walker, G. F., and Garrett, W. G., 1967. Chemical exfoliation of vermiculite and the production of colloidal dispersions. *Science*, **156**: 385–387.

77. Greenland, D. J., Quirk, J. P., and Theng, B. K. G., 1964. Influence of increasing proportions of exchangeable alkylammonium ions on the swelling of calcium montmorillonite in water. *J. Colloid Sci.*, **19**: 837–840.

78. Norrish, K., 1954. The swelling of montmorillonite. *Disc. Faraday Soc.*, **18**: 120–134.

79. Barrer, R. M., and Hampton, M. G., 1957. Gas chromatography and mixture isotherms in alkylammonium bentonites. *Trans. Faraday Soc.*, **53**: 1462–1475.

80. Van Rysselberge, J., and Van der Stricht, M., 1962. Complete separation of xylenes and ethylbenzene by gas chromatography. *Nature*, **193**: 1281–1282.

81. Taramasso, M., and Veniale, F., 1969. Gas chromatographic investigations on dimethyl-dioctadecylammonium derivatives of different clay minerals. *Contr. Mineral. Petrol.*, **21**: 53–62.

82. Klemm, L. H., Shabtai, J., and Lee, F. H. W., 1970. Gas chromatography of some nitrogen and sulfur heterocycles by means of silicone and Bentone-silicone phases. *J. Chromatog.*, **51**: 433–439.

Chapter 6

1. MacEwan, D. M. C., 1948. Complexes of clays with organic compounds. I. Complex formation between montmorillonite and halloysite and certain organic liquids. *Trans. Faraday Soc.*, **44**: 349–367.

2. MacEwan, D. M. C., 1949. Clay mineral complexes with organic liquids. *Clay Minerals Bull.*, **3**: 44–46.

3. Caillère, S., Glaeser, R., Esquevin, J., and Hénin, S., 1950. Preparation of 14A- and 17A-halloysite. *C.R. Acad. Sci., Paris*, **230**: 308–310.

4. Weiss, A., and Russow, J., 1963. Über das Einrollen von Kaolinitkristallen zu Halloysitähnlichen Röhren und einen Unterschied zwischen Halloysit und röhrchenförmigem Kaolinit. *Proc. Int. Clay Conf., Stockholm*, **2**: 69–74.

5. Camazano, M. S., and Garcia, S. G., 1966. Complejos interlaminares de caolinita y halloisita con liquidos polares. *An. Edafol. Agrobiol.*, **25**: 9–25.

6. Olejnik, S., Aylmore, L. A. G., Posner, A. M., and Quirk, J. P., 1968. Infrared spectra of kaolin mineral–dimethyl sulfoxide complexes. *J. Phys. Chem.*, **72**: 241–249.

7. Wada, K., 1959. Oriented penetration of ionic compounds between the silicate layers of halloysite. *Am. Mineralogist*, **44**: 153–165.

8. Wada, K., 1959. An interlayer complex of halloysite with ammonium chloride. *Am. Mineralogist*, **44**: 1237–1247.

9. Garrett, W. G., and Walker, G. F., 1959. The cation exchange capacity of hydrated halloysite and the formation of halloysite-salt complexes. *Clay Minerals Bull.*, **4**: 75–80.

10. Carr, R. M., and Chih, H., 1971. Complexes of halloysite with organic compounds. *Clay Minerals*, **9**: 153–166.

11. MacEwan, D. M. C., 1962. Interlamellar reactions of clays and other substances. *Clays Clay Minerals*, **9**: 431–443.

12. Churchman, G. J., Aldridge, L. P., and Carr, R. M., 1972. The relationship between the hydrated and dehydrated states of an halloysite. *Clays Clay Minerals*, **20**: 241–246.

13. Wada, K., 1961. Lattice expansion of kaolin minerals by treatment with potassium acetate. *Am. Mineralogist*, **46**: 78–91.

14. Serratosa, J. M., Hidalgo, A., and Vinas, J. M., 1963. Infrared study of the OH groups in kaolin minerals. *Proc. Int. Clay Conf., Stockholm*, **1**: 17–26.

15. Ledoux, R. L., and White, J. L., 1965. Infrared studies of the hydroxyl groups in intercalated kaolinite complexes. *Clays Clay Minerals*, **13**: 289–315.

16. WEISS, A., 1963. A secret of Chinese porcelain manufacture. *Angew. Chem., Int. Ed. Engl.*, **2**: 697–703.

17. WEISS, A., 1961. Eine Schichteinschlussverbindung von Kaolinit mit Harnstoff. *Angew. Chem.*, **73**: 736.

18. WEISS, A., THIELEPAPE, W., GÖRING, G., RITTER, W., and SCHÄFER, H., 1963. Kaolinit-Einlagerungs-Verbindungen. *Proc. Int. Clay Conf., Stockholm*, **1**: 287–305.

19. WEISS, A., THIELEPAPE, W., RITTER, W., SCHÄFER, H., and GÖRING, G., 1963. Zur Kenntnis von Hydrazin-Kaolinit. *Z. anorg. allgem. Chem.*, **320**: 183–204.

20. WEISS, A., THIELEPAPE, W., and ORTH, H., 1966. Neue Kaolinit-Einlagerungsverbindungen. *Proc. Int. Clay Conf., Jerusalem*, **1**: 277–293.

21. OLEJNIK, S., POSNER, A. M., and QUIRK, J. P., 1970. The intercalation of polar organic compounds into kaolinite. *Clay Minerals*, **8**: 421–434.

22. MATA-ARJONA, A., RUIZ-AMIL, A., and INARAJA-MARTIN, E., 1970. Cinetica del proceso de sorcion del dimetilsulfoxido en caolinita: estudio por difraccion de rayos X. *Reunion Hispano-Belga de Minerales de la Arcilla, Madrid*, 115–120.

23. OLEJNIK, S., POSNER, A. M., and QUIRK, J. P., 1971. Infrared spectrum of the kaolinite-pyridine-N-oxide complex. *Spectrochim. Acta*, **27A**: 2005–2009.

24. LEDOUX, R. L., and WHITE, J. L., 1966. Infrared studies of hydrogen bonding between kaolinite surfaces and intercalated potassium acetate, hydrazine, formamide and urea. *J. Colloid Interface Sci.*, **21**: 127–152.

25. LEDOUX, R. L., and WHITE, J. L., 1966. Infrared studies of hydrogen bonding of organic compounds on oxygen and hydroxyl surfaces of layer lattice silicates. *Proc. Int. Clay Conf., Jerusalem*, **1**: 361–374.

26. LEDOUX, R. L., and WHITE, J. L., 1964. Infrared study of the OH groups in expanded kaolinite. *Science*, **143**: 244–246.

27. CRUZ, M., LAYCOCK, A., and WHITE, J. L., 1969. Perturbation of OH groups in intercalated kaolinite donor-acceptor complexes. I. Formamide-, methylformamide-, and dimethylformamide-kaolinite complexes. *Proc. Int. Clay Conf., Tokyo*, **1**: 775–789.

28. OLEJNIK, S., POSNER, A. M., and QUIRK, J. P., 1971. The I.R. spectra of interlamellar kaolinite-amide complexes. I. The complexes of formamide, N-methylformamide and dimethylformamide. *Clays Clay Minerals*, **19**: 83–94.

29. OLEJNIK, S., POSNER, A. M., and QUIRK, J. P., 1971. The infrared spectra of interlamellar kaolinite-amide complexes. II. Acetamide, N-methylacetamide and dimethylacetamide *J. Colloid Interface Sci.*, **37**: 536–547.

30. THOMAS, R., SHOEMAKER, C. B., and ERIKS, K., 1966. The molecular and crystal structure of dimethyl sulfoxide, Me₂SO. *Acta Cryst.*, **21**: 12–20.

31. LLOYD, M. K., and CONLEY, R. F., 1970. Adsorption studies on kaolinites. *Clays Clay Minerals*, **18**: 37–46.

32. NAGASAWA, K., 1969. Kaolin minerals in cenozoic sediments of central Japan. *Proc. Int. Clay Conf., Tokyo*, **1**: 15–30.

33. BRINDLEY, G. W., and ROBINSON, K., 1946. Randomness in the structures of kaolinitic clay minerals. *Trans. Faraday Soc.*, **42B**: 198–205.

34. BRINDLEY, G. W., and ROBINSON, K., 1947. X-ray studies of some kaolinitic fireclays. *Trans. Brit. Ceram. Soc.*, **46**: 49–62.

35. WEISS, A., and THIELEPAPE, W., 1963. Über eine Verbesserung der technischen Eigenschaften von Kaolinit durch die Herstellung von Einlagerungsverbindungen. *Proc. Int. Clay Conf., Stockholm*, **2**: 427–429.

36. CAMAZANO, M. S., and GARCIA, S. G., 1970. Modification del habito de cristales de caolinita por tratamiento con dimetilsulfoxido. *An. Edafol. Agrobiol.*, **29**: 651–655.

37. JACOBS, H., and STERCKX, M., 1970. Contribution a l'étude de l'intercalation du dimethylsulfoxide dans le réseau de la kaolinite. *Reunion Hispano-Belga de Minerales de la Arcilla, Madrid*, 154–160.

38. SERRATOSA, J. M., 1966. Infrared analysis of the orientation of pyridine molecules in clay complexes. *Clays Clay Minerals*, **14**: 385–391.

320

Chapter 7

1. HAUSER, E. A., and LEGGETT, M. B., 1940. Color reactions between clays and amines. *J. Am. Chem. Soc.*, **62**: 1811–1814.

2. KRÜGER, D., and OBERLIES, F., 1941. Catalytic oxidation of amines at the surface of negative adsorbents. Realization of a different course of the reaction in the oxidation of dimethylaniline and some of its homologs on bentonite and on other surfaces. *Chem. Ber.*, **74B**: 1711–1719.

3. MEUNIER, P., 1942. The action of montmorillonite clay on vitamin A and the phenomenon of mesomerism in the carotenoid group. *C.R. Acad. Sci., Paris*, **215**: 470–473.

4. ZECHMEISTER, L., and SANDOVAL, A., 1945. The coloration given by vitamin A and other polyenes on acid earths. *Science*, **101**: 585.

5. WEIL-MALHERBE, H., and WEIS, J., 1948. Color reactions and adsorption of some aluminosilicates. *J. Chem. Soc.*, 2164–2169.

6. BRIEGLEB, G., 1961. *Elektronen-Donator-Acceptor-Komplexe*. Springer Verlag, Berlin.

7. MORTLAND, M. M., and RAMAN, K. V., 1968. Surface acidity of smectites in relation to hydration, exchangeable cation, and structure. *Clays Clay Minerals*, **16**: 393–398.

8. BENESI, H. A., 1967. Acidity of catalyst surfaces. II. Amine titration using Hammett indicators. *J. Phys. Chem.*, **61**: 970–973.

9. SOLOMON, D. H., SWIFT, J. D., and MURPHY, A. J., 1971. The acidity of clay minerals in polymerization and related reactions. *J. Macromol. Sci. Chem.*, **A5**: 587–601.

10. SOLOMON, D. H., LOFT, B. C., and SWIFT, J. D., 1968. Reactions catalysed by minerals. IV. The mechanism of the benzidine blue reaction on silicate minerals. *Clay Minerals*, **7**: 389–397.

11. SOLOMON, D. H., LOFT, B. C., and SWIFT, J. D., 1968. Reactions catalysed by minerals. V. The reaction of leuco dyes and unsaturated organic compounds with clay minerals. *Clay Minerals*, **7**: 399–408.

12. SOLOMON, D. H., 1968. Clay minerals as electron acceptors and/or electron donors in organic reactions. *Clays Clay Minerals*, **16**: 31–39.

13. THENG, B. K. G., 1971. Mechanisms of formation of colored clay-organic complexes. A review. *Clays Clay Minerals*, **19**: 383–390.

14. THENG, B. K. G., and WALKER, G. F., 1970. Interactions of clay minerals with organic monomers. *Israel J. Chem.*, **8**: 417–424.

15. HENDRICKS, S. B., and ALEXANDER, L. T., 1940. A qualitative test for the montmorillonite type of clay minerals. *J. Am. Soc. Agron.*, **32**: 455–458.

16. ENDELL, J., ZORN, R., and HOFMANN, U., 1941. The benzidine test for montmorillonite. *Angew. Chem.*, **54**: 376–377.

17. MEUNIER, P., 1943. Action of acid clays on aromatic amines. *C.R. Acad. Sci., Paris*, **217**: 449–451.

18. BOSAZZA, V. L., 1944. On the adsorption of some organic dyes by clays and clay minerals. *Am. Mineralogist*, **29**: 235–241.

19. VEDENEEVA, N. E., 1950. The mechanism of the color reaction of benzidine with montmorillonite. *Kolloidn. Zh.*, **12**: 17–24.

20. PAGE, J. B., 1941. Unreliability of the benzidine color reaction as a test for montmorillonite. *Soil Sci.*, **51**: 133–140.

21. MICHAELS, A. S., 1958. Deflocculation of kaolinite by alkali polyphosphates. *Ind. Eng. Chem.*, **50**: 951–958.

22. BLOCH, J. M., CHARBONELLE, J., and KAYSER, F., 1953. The oxidizing power of montmorillonite. *C.R. Acad. Sci., Paris*, **237**: 57–59.

23. KRÜGER, D., and OBERLIES, F., 1943. Structure and color reactions of montmorillonite earths. *Naturwissenschaften*, **31**: 92.

24. GRIM, R. E., 1953. *Clay Mineralogy*, McGraw-Hill, New York, N.Y.

25. DODD, C. G., and RAY, S., 1960. Semiquinone cation adsorption on montmorillonite as a function of surface acidity. *Clays Clay Minerals*, **8**: 237–251.

26. LAHAV, N., and RAZIEL, S., 1971. Interaction between montmorillonite and benzidine in aqueous solutions. I. Adsorption of benzidine on montmorillonite. *Israel J. Chem.*, **9**: 683–689.

27. CONLEY, R. F., and LLOYD, M. K., 1971. Adsorption studies on kaolinites. II. Adsorption of amines. *Clays Clay Minerals*, **19**: 273–282.

28. TAKAHASHI, H., 1955. The effect of layered water on the color reaction of benzidine or other similar compounds with montmorillonite. *Bull. Chem. Soc., Japan*, **28**: 5–9.

29. HASEGAWA, H., 1961. Spectroscopic studies on the color reaction of acid clay with amines. I. *J. Phys. Chem.*, **65**: 292–296.

30. HAKUSUI, A., MATSUNAGA, V., and UMEHARA, K., 1970. Diffuse reflection spectra of acid clays colored with benzidine and some other diamines. *Bull. Chem. Soc., Japan*, **43**: 709–712.

31. PAPARIELLO, G. J., and JANISH, M. A. M., 1966. Diphenylpicrylhydrazyl as an organic analytical reagent in the spectrophotometric analysis of phenols. *Analyt. Chem.*, **38**: 211–214.

32. GREEN, B. K., 1950. Pressure-sensitive record material. *U.S. Pat.*, 2,505,470.

33. KRANZ, F. H., 1963. Pressure-sensitive copying sheets. *U.S. Pat.*, 3,079,271.

34. NATIONAL CASH REGISTER COMPANY, 1963. Pressure-sensitive copying paper. *Ger. Pat.*, 1,152,429.

35. NATIONAL CASH REGISTER COMPANY, 1965. Recording paper with attapulgite. *Neth. Pat.*, 6,505,671.

36. RITZERFELD, W., and RITZERFELD, G., 1961. Prints from matrices having a color supply of triphenylmethane dyes. *Ger. Pat.*, 1,119,302.

37. GAYER, F. H., 1933. The catalytic polymerization of propylene. *Ind. Eng. Chem.*, **25**: 1122–1127.

38. BROUGHTON, G., 1940. Catalysis by metallized bentonites. *J. Phys. Chem.*, **44**: 180–184.

39. SALT, F. H., 1948. The use of activated clays as catalysts in polymerisation processes with particular reference to polymers of alpha methyl styrene. *Clay Minerals Bull.*, **2**: 55–57.

40. SOLOMON, D. H., and ROSSER, M. J., 1965. Reactions catalyzed by minerals. I. Polymerization of styrene. *J. Appl. Polymer Sci.*, **9**: 1261–1271.

41. SOLOMON, D. H., and SWIFT, J. D., 1967. Reactions catalyzed by minerals. II. Chain termination in free radical polymerizations. *J. Appl. Polymer Sci.*, **11**: 2567–2575.

42. SOLOMON, D. H., and LOFT, B. C., 1968. Reactions catalyzed by minerals. III. The mechanisms of spontaneous interlamellar polymerizations in aluminosilicates. *J. Appl. Polymer Sci.*, **12**: 1253–1262.

43. BITTLES, J. A., CHAUDHURI, A. K., and BENSON, S. W., 1964. Clay-catalyzed reactions of olefins. I. Polymerization of styrene. *J. Polymer Sci.*, **A2**: 1221–1231.

44. BITTLES, J. A., CHAUDHURI, A. K., and BENSON, S. W., 1964. Clay-catalyzed reactions of olefins. II. Catalyst acidity and mechanisms. *J. Polymer Sci.*, **A2**: 1847–1862.

45. ALDRICH, D. C., and BUCHANAN, J. R., 1958. Anomalies in techniques for preparing H-bentonites. *Soil Sci. Soc. Am. Proc.*, **22**: 281–285.

46. COLEMAN, N. T., and CRAIG, D., 1961. The spontaneous alteration of hydrogen clay. *Soil Sci.*, **91**: 14–18.

47. KUSNITSYNA, T. A., and OSTROVSKAYA, I. K., 1967. Catalytic activity of acid aluminosilicates in the styrene polymerization reaction. *Vysokomol. Soedin.*, **A9**: 2510–2514.

48. MATSUMOTO, T., SAKAI, I., and ARIHARA, M., 1969. Polymerization of styrene by acid clay. *Chem. High Polymers, Japan*, **26**: 378–384.

49. BLUMSTEIN, A., 1965. Polymerization of adsorbed monolayers. I. Preparation of the clay-polymer complex. *J. Polymer Sci.*, **A3**: 2653–2664.

50. FRIEDLANDER, H. Z., and FRINK, C. R., 1964. Organized polymerization. III. Monomers in montmorillonite. *Polymer Letters*, **2**: 475–479.

51. PEZERAT, H., and MANTIN, I., 1967. Polymerisation cationique du styrène entre les feuillets d'une montmorillonite acide. *C.R. Acad. Sci., Paris*, **265**: 941–944.

52. HAUSER, E. A., and KOLLMAN, R. C., 1960. Clay complexes with conjugated unsaturated aliphatic compounds. *U.S. Pat.*, 2,951,087.

53. FRIEDLANDER, H. Z., 1963. Spontaneous polymerization in and on clays. *Am. Chem. Soc. Polymer Preprints*, **4**: 300–306.

54. FRIEDLANDER, H. Z., 1964. Organized polymerization. I. Olefins on a clay surface. *J. Polymer Sci.*, **C4**: 1291–1301.

55. DEKKING, H. G. G., 1965. Propagation of vinyl polymers on clay surfaces. I. Preparation, structure, and decomposition of clay initiators. *J. Appl. Polymer Sci.*, **9**: 1641–1651.

56. DEKKING, H. G. G., 1967. Propagation of vinyl polymers on clay surfaces. II. Polymerization of monomers initiated by free radicals attached to clay. *J. Appl. Polymer Sci.*, **11**: 23–36.

57. BLUMSTEIN, A., 1965. Polymerization of adsorbed monolayers. II. Thermal degradation of the inserted polymer. *J. Polymer Sci.*, **A3**: 2665–2672.

58. BLUMSTEIN, A., and BILLMEYER, F. W., Jr., 1966. Polymerization of adsorbed monolayers. III. Preliminary structure studies in dilute solution of the inserted polymers. *J. Polymer Sci.*, **A4**: 465–474.

59. BLUMSTEIN, A., and BLUMSTEIN, R., 1967. Association in two-dimensionally cross-linked poly(methyl methacrylate). *Polymer Letters*, **5**: 691–696.

60. BLUMSTEIN, A., BLUMSTEIN, R., and VANDERSPURT, T. H., 1969. Polymerization of adsorbed monolayers. IV. The two-dimensional structure of insertion polymers. *J. Colloid Interface Sci.*, **31**: 236–247.

61. BLUMSTEIN, A., MALHOTRA, S. L., and WATTERSON, A. C., 1968. Stereospecificity of poly-(methyl methacrylate) obtained by polymerization in an organized medium. *Am. Chem. Soc. Polymer Preprints*, 167–175.

62. GLAVATI, O. L., POLAK, L. S., and SHCHEKIN, V. V., 1963. The radiational stereospecific polymerization of acrylonitrile and acrylic acid in montmorillonite compounds. *Neftekhimiya*, **3**: 905–910.

63. GLAVATI, O. L., and POLAK, L. S., 1964. Kinetics and mechanisms of radiation polymerization in montmorillonite inclusion compounds. *Neftekhimiya*, **4**: 77–81.

64. BROWN, J. F., Jr., and WHITE, D. M., 1960. Stereospecific polymerization in thiourea canal complexes. *J. Am. Chem. Soc.*, **82**: 5671–5678.

65. WALKER, G. F., and HAWTHORNE, D. G., 1968. Complexes of nickel cyanide with organic monomers. *J. Polymer Sci.*, **B6**: 593–594.

66. PONNAMPERUMA, C., and MUNDAY, C. S., 1968. The gamma irradiation of isoprene in the presence of vermiculite. *Abst. Int. Conf. Rad. Chem., Argonne, Illinois*, 547.

67. THENG, B. K. G., 1970. Formation of two-dimensional organic polymers on a mineral surface. *Nature*, **228**: 853–854.

68. DOSCH, W., 1967. Interlamellar reactions of tetracalcium aluminate hydrates with water and organic compounds. *Clays Clay Minerals*, **15**: 273–292.

69. CALVIN, M., 1969. *Chemical Evolution*. Clarendon Press, Oxford.

70. OPARIN, A. I., 1957. *The Origin of Life on Earth*. Academic Press, New York, N.Y.

71. CAIRNS-SMITH, A. G., 1965. The origin of life and the nature of the primitive gene. *J. Theoret. Biol.*, **10**: 53–88.

72. BERNAL, J. D., 1951. *The Physical Basis of Life*. Routledge and Kegan Paul, London.

73. DAYHOFF, M. O., LIPPINCOTT, E. R., ECK, R. V., and NAGARAJAN, G., 1967. Thermodynamic equilibrium in prebiological atmospheres of C, H, O, N, P, S, and Cl. *NASA SP-3040, Washington, D.C.*, quoted by E. T. Degens and J. Mathéja (reference 76).

74. PAVLOVSKAYA, T. E., PASINSKYI, A. G., and GREBENIKOVA, A. I., 1960. The production of amino acids under the influence of ultra-violet rays on solutions of formaldehyde and ammonium salts in the presence of adsorbents. *Dokl. Nauk. S.S.S.R.*, **135**: 743–746.

75. YOSHINO, D., HAYATSU, R., and ANDERS, E., 1971. Origin of organic matter in early solar systems. III. Amino acids: catalytic synthesis. *Geochim. Cosmochim. Acta*, **35**: 927–938.

76. DEGENS, E. T., and MATHÉJA, J., 1968. Origin, development, and diagenesis of bio-geochemical compounds. *J. Brit. Interplanet. Soc.*, **21**: 52–82.

77. DEGENS, E. T., and MATHÉJA, J., 1970. Formation of organic polymers on inorganic templates. In: A. P. Kimball and J. Oró (Editors), *Prebiotic and Biochemical Evolution.* North Holland, Amsterdam, pp. 39–69.

78. FRIPIAT, J. J., CLOOS, P., CALICIS, B., and MAKAY, K., 1966. Adsorption of amino acids and peptides by montmorillonite. II. Identification of adsorbed species and decay products by infra-red spectroscopy. *Proc. Int. Clay Conf., Jerusalem*, **1**: 233–245.

79. PAECHT-HOROWITZ, M., BERGER, J., and KATCHALSKY, A., 1970. Prebiotic synthesis of polypeptides by heterogeneous polycondensation of amino acid adenylates. *Nature*, **228**: 636–639.

80. DEGENS, E. T., MATHÉJA, J., and JACKSON, T. A., 1970. Template catalysis: asymmetric polymerization of amino acids on clay minerals. *Nature*, **227**: 492–493.

81. JACKSON, T. A., 1971. Preferential polymerization and adsorption of L-optical isomers of amino acids relative to D-optical isomers on kaolinite templates. *Chem. Geol.*, **7**: 295–306.

82. WALD, G., 1957. The origin of optical activity. *Ann. New York Acad. Sci.*, **69**: 352–368.

83. MILLER, W. G., BRANT, D. A., and FLORY, P. J., 1967. Random coil configurations of polypeptide copolymers. *J. Molec. Biol.*, **23**: 67–80.

84. GRATZER, W. B., and COWBURN, D. A., 1969. Optical activity of biopolymers. *Nature*, **222**: 426–431.

85. TERENT'EV, A. P., and KLABUNOVSKII, E. I., 1959. The role of dissymmetry in the origin of living material. *Proc. 1st Int. Symp. Origin of Life, Moscow. Int. Union Biochem. Symp. Ser.*, **1**: 95–105.

86. BAILEY, S. W., 1963. Polymorphism of the kaolin minerals. *Am. Mineralogist*, **48**: 1196–1209.

87. HARVEY, G. R., MOPPER, K., and DEGENS, E. T., 1972. Synthesis of carbohydrates and lipids on kaolinite. *Chem. Geol.*, **9**: 79–87.

88. PONNAMPERUMA, C., 1969. The role of clay minerals in chemical evolution. *Abst. Int. Clay Conf., Tokyo.*

89. HARVEY, G. R., DEGENS, E. T.. and MOPPER, K., 1972. Synthesis of nitrogen hetero-cycles on kaolinite. *Naturwissenschaften*, **58**: 624.

90. IBANEZ, J. D., KIMBALL, A. P., and ORÓ, J., 1971. Possible prebiotic condensation of mononucleotides by cyanamide. *Science*, **173**: 444–446.

91. ROBERTSON, R. H. S., 1948. Clay minerals as catalysts. *Clay Minerals Bull.*, **2**: 47–54.

92. GRIM, R. E., 1962. *Applied Clay Mineralogy.* McGraw-Hill, New York, N.Y.

93. HANSFORD, R. C., 1952. Chemical concepts of catalytic cracking. *Advan. Catalysis*, **4**: 1–29.

94. VOGE, H. H., 1958. Catalytic cracking. In: P. H. Emmett (Editor), *Catalysis.* Reinhold Publ. Corp., New York, N.Y., **6**: 407–499.

95. EISMA, E., and JURG, J. W., 1969. Fundamental aspects of the generation of petroleum. In: G. Eglinton and M. J. T. Murphy (Editors), *Organic Geochemistry.* Springer Verlag, Berlin, pp. 676—698.

96. BROOKS, B. T., 1950. Catalysis and carbonium ions in petroleum formation. *Science*, **111**: 648–650.

97. JURG, J. W., and EISMA, E., 1964. Petroleum hydrocarbons: generation from a fatty acid. *Science*, **144**: 1451–1452.

98. HENDERSON, W., EGLINTON, G., SIMMONDS, P., and LOVELOCK, J. E., 1968. Thermal alteration as a contributory process to the genesis of petroleum. *Nature*, **219**: 1012–1016.

99. GREENSFELDER, B. S., VOGE, H. H., and GOOD, G. M., 1949. Catalytic and thermal cracking of pure hydrocarbons. Mechanisms of reaction. *Ind. Eng. Chem.*, **41**: 2573–2584.

100. SHIMOYAMA, A., and JOHNS, W. D., 1971. Catalytic conversion of fatty acids to pet-roleum-like paraffins and their maturation. *Nature Phys. Sci.*, **232**: 140–144.

101. BRAY, E. E., and EVANS, E. D., 1961. Distribution of n-paraffins as a clue to recognition of source beds. *Geochim. Cosmochim. Acta.*, **22**: 2–15.

102. COOPER, J. E., 1962. Fatty acids in recent and ancient sediments and petroleum reservoir waters. *Nature*, **193**: 744–746.

103. WAPLES, D. W., 1972. Catalytic formation of hydrocarbons from fatty acids. *Nature Phys. Sci.*, **237**: 63–64.

104. SHIMOYAMA, A., and JOHNS, W. D., 1972. Reply to: D. W. Waples, Catalytic formation of hydrocarbons from fatty acids. *Nature Phys. Sci.*, **237**: 64.

105. CHAUSSIDON, J., and CALVET, R., 1965. Evolution of amine cations adsorbed on montmorillonite with dehydration of the mineral. *J. Phys. Chem.*, **69**: 2265–2268.

106. GALWEY, A. K., 1969. Reactions of alcohols adsorbed on montmorillonite and the role of minerals in petroleum genesis. *J. Chem. Soc.*, **D**: 577–578.

107. GALWEY, A. K., 1969. Heterogeneous reactions in petroleum genesis and maturation. *Nature*, **223**: 1257–1260.

108. GALWEY, A. K., 1970. Reactions of alcohols and of hydrocarbons on montmorillonite surfaces. *J. Catalysis*, **19**: 330–342.

109. DURAND, B., PELET, R., and FRIPIAT, J. J., 1972. Alkylammonium decomposition on montmorillonite surfaces in an inert atmosphere. *Clays Clay Minerals*, **20**: 21–35.

110. CHOU, C. C., and McATEE, J. L., Jr., 1969. Thermal decomposition of organoammonium compounds exchanged onto montmorillonite and hectorite. *Clays Clay Minerals*, **17**: 339–346.

111. WEISS, A., and ROLOFF, G., 1963. Die Rolle organischer Derivate von glimmerartigen Schichtsilikaten bei der Bildung von Erdöl. *Proc. Int. Clay. Conf., Stockholm*, **2**: 373–378.

112. FRIPIAT, J. J., and HELSEN, J., 1966. Kinetics of decomposition of cobalt coordination complexes on montmorillonite surfaces. *Clays Clay Minerals*, **14**: 163–179.

113. WENDLANDT, W. W., and SMITH, J. P., 1963. Thermal decomposition of metal complexes. *J. Inorg. Nucl. Chem.*, **25**: 843–850.

114. FRIPIAT, J. J., HELSEN, J., and VIELVOYE, L., 1964. Formation de radicaux libres sur la surface des montmorillonites. *Bull. Groupe Fr. Argiles*, **15**: 3–10.

115. WALKER, G. F., 1967. Catalytic decomposition of glycerol by layer silicates. *Clay Minerals*, **7**: 111–112.

116. KUMADA, K., and KATO, H., 1970. Browning of pyrogallol as affected by clay minerals. I. Classification of clay minerals based on their catalytic effects on the browning reaction of pyrogallol. *Soil Sci. Plant Nutr. (Tokyo)*, **116**: 195–200.

117. JUPP, G., and RAU, H., 1969. Attapulgus clay-catalysed thermal decomposition of terphenyls. *J. Appl. Chem.*, **19**: 120–124.

118. SAKIYAMA, M., and OKAWARA, R., 1964. Reactions of methylhydropolysiloxanes with active clay and cupric oxide: synthesis of lower members of hydrogen end-blocked dimethylpolysiloxanes. *J. Organometal. Chem.*, **2**: 473–477.

119. STEWART, H. F., 1967. Clay-catalysed reactions in organosilicon chemistry. The interchange of hydrogen and siloxy ligands on silicon. *J. Organometal. Chem.*, **10**: 229–234.

120. ALBAREDA, J. M., ALEIXANDRE, V., and FERNANDEZ, T., 1953. Influencia de la composicion mineralogica de las arcillas y de las cationes de cambio en la oxidacion catalitica del alcohol etilico en fase vapor. I. *An. Edafol. Fysiol. Veg.*, **12**: 89–140.

121. ALEIXANDRE, V., and FERNANDEZ, T., 1960. Influence catalytique des différents minéraux argileux sur la synthèse du butadiène a partir de l'éthanol. *Silicates Inds.*, **25**: 243–248.

122. ARPINO, P., and OURISSON, G., 1971. Interactions between rock and organic matter. Esterification and transesterification induced in sediments by methanol and ethanol. *Anal. Chem.*, **43**: 1656–1657.

123. HOJABRI, F., 1971. Gas-phase catalytic alkylation of aromatic hydrocarbons. *J. Appl. Chem. Biotechnol.*, **21**: 87–89.

Index

Author entries are shown in capitals. They should be used as follows:

ADAMS, R. S. Jr., (147)[33], *311*

means that Adams is referred to in the text on page 147 by reference number [33], whilst the full reference to him is to be found on page 311.

Acetaldehyde, 73, 74, 240*ff*
Acetamide, 111*ff*, 232, 234, 242, 248, 253, 256
Acetic acid, 74, 82, 199, 242
Acetoacetic ethyl ester, 75
Acetone, 43, 50, 51, 63*ff*, 74, 75, 119, 240*ff*, 246, 267
Acetonitrile, 63, 79*ff*, 233, 240*ff*, 246
Acetylacetone, 21, 23, 55, 64, 68*ff*, 242
Acetylene, 29
Acetylethanolamine (N-), 242
Acetylglucosamine (N-), 208
Acrolein, 73, 74
Acrylamide, 272
Activation energy, 86, 186
ADAMS, R. S., Jr., (147)[33], *311*
ADELMAN, R. L., (117)[185], *307*
Adenine, 190, 193, 195
Adenosine, 190, 193, 195
Adenylate, amino acid, 275, 276
Adenylic acid, 276
Adsorption, entropy change of, 166, 219
    entropy effects in, 22, 23, 219, 237
    free energy change of, 164*ff*, 170, 171, 213*ff*, 230
    in excess of exchange capacity, 85, 132, 212, 213, 222, 230, 231
    negative, 84, 130, 148
    van der Waals (physical), 130, 164, 168, 169, 189, 206
Adsorption, from aqueous solution,
    of uncharged organic species, 20*ff*, 32, 33, 140
    influence of chain length on, 20, 21
    influence of CH activity on, 23
AGNIHOTRI, N. P., (152)[46], *311*
AHLRICHS, J. L., 30, 102, 122, 124, 126, (146)[27], *300, 308, 310*

Alanine (α-, and β-), 160, 161, 164, 166, 168, 169, 171*ff*, 178, 185, 242, 275
    potassium salt of, 246
ALBAREDA, J. M., (291)[120], *325*
ALBERT, N , (111)[175], *307*
Alcohols, 24, 32*ff*, 47*ff*, 88, 129, 198, 200, 201 205, 232*ff*, 287, 288, 291
Aldehydes, 73, 74, 200, 234, 290
Aldol, 74, 75
ALDRICH, D. C., (270)[45], *322*
ALDRIDGE, L. P., (240)[12], *319*
ALEIXANDRE, V., (291)[120-1], *325*
ALEXANDER, L. T., 264, (265)[15], 268, *321*
Alkaloids, 189, 211, 212
Alkanes, 129, 281*ff*, 287; *see also* Hydrocarbons
Alkenes, 129, 281*ff*, 287; *see also* Olefins
Alkylammonium-clay complexes,
    decomposition of, 129, 220, 286, 288, 289
    gas chromatographic application of, 129, 238
    as gelling agents, 230*ff*
    intercalation of organic species by, 116*ff*, 129, 199, 200, 203*ff*, 231, 233, 234
    sorption of water by, 231*ff*
    swelling in organic media of, 232*ff*
Alkylammonium ions, 9, 15, 16, 40, 84, 92, 164, 200, 205, 212, 213, 217, 219*ff*, 222*ff*, 230, 286*ff*
Alkyldiammonium ions (α, ω), 85, 229, 233
Alkylpyridinium ions (1-n-), 229*ff*
ALLAWAY, W. H., (212)[9], (231)[9], *316*
Allevardite, 61
ALLISON, F E., (187)[97-101], (188)[97-101], *313, 314*
Allophane, 3, 33, 110, 164
Alumina, 148
Ametrine, 142
Amiben, 139

Amides, 88, 111*ff*, 246, 247
  pesticides, 139, 146
Amines, 84*ff*, 92*ff*, 98*ff*, 102*ff*, 182, 200, 212, 231, 234, 261*ff*, 267, 268, 286, 287
Amino acids, 121, 159*ff*, 186, 275*ff*
Amino benzoic acid (*p*-), 168, 169, 171
Amino butyric acid (α-, β-, and γ-), 160, 161, 178*ff*
Amino caproic acid (α-, and ε-), 160, 161, 178*ff*
Amino methacrylates, 274
Aminotriazole (3-), 88, 138, 145
Aminotriazolium (3-), 88, 145
Amitrole, *see* Aminotriazole (3-)
Ammonia, 110, 211, 244, 263, 275, 277, 280
Ammonium, 19, 89, 262, 263
  acetate, 246, 259
AMU, 193
Amylamines, 211, 246
Amylammonium (*n*-), 235
ANDERS, E., (275)[75], *323*
Anilides, pesticides, 139, 146
Aniline, 101*ff*, 110, 242, 261
  pesticides, 102, 142
Anilinium, 102, 103, 109, 286
Anthraquinone, 156
Antibiotics, 186*ff*
Antigorite, 2
AOMINE, S., (3)[27], *295*
Arabinose, 208, 278
ARAGON, F., 88, 89, 90, 92, *306*
Arginine, 160, 168, 170, 171, 275
ARIHARA, M., (270)[48], *322*
ARNISTA, E. S., (189)[105,107], *314*
ARNOLD, P. W., 129, 130, *309*
ARPINO, P., (291)[122], *325*
Arrhenius equation, 186, 290
Ascorbic acid, 208
Aspartic acid, 168, 171, 275, 277
Atratone, 138, 142
Atrazine, 138, 142, 144
Atropine, 189
Attapulgite, *see* Palygorskite
Aureomycin, 187
AYLMORE, L. A. G., (33)[5], (239)[6], (240)[6], (243)[6], (246)[6], (256)[c], (257)[6], (258)[6], (259)[6], *300, 319*
Azobisisobutyramidine (2, 2'-), 272
Azopropane (2, 2'-), 284

Bacitracin, 187
BADGER, R. M., (111)[175], *307*
BAILEY, G. W., 102, 137, 142, (144)[16], 146, 148, *307, 309, 310*
BAILEY, S. W., (2)[12], (3)[12], (278)[86], *294, 324*
BARBIER, G., (xi)[5], *293*
BARR, M., 189, *314*

BARRER, R. M,. (128)[200], (129)[200-5], 213, 219, 220, (221)[32,33], (231)[32,34], 233, 238, *308, 316, 317, 318, 319*
BARRIOS, J., (87)[148], (98)[148], *306*
BARSHAD, I., (15)[84], 33, 34, 46, (52)[60], (121)[16], 128, (160)[67], (161)[67], (173)[67], 177, 178, *297, 300, 302, 312, 313*
Basal spacing data, of complexes with, *see also* X-ray diffractometry
  acetone, 43, 44, 66
  acetylacetone, 69
  alcohols (*n*-), 33*ff*
  alkanes (*n*-), 129, 130
  alkylammonium ions, 229, 230
  alkylpyridinium ions, 229, 230
  amino acids and peptides, 171, 179*ff*
  anilines, 107*ff*
  cyclohexylamine, 101
  diamines, 93*ff*
  dodecylammonium (*n*-)—triglycerides, 203
  ethanol, 42*ff*
  ethylenediamine, 94
  ethylene glycol, 55*ff*
  fatty acids, 198, 199
  glycerol, 49
  hexanedione (2, 5-), 72
  organic liquids, 51
  saccharides, 206, 207
  water, 10
BASTOW, A. W., (149)[40], (150)[40], (151)[40], *311*
Batavite, 223
BATES, T. F., (5)[48], (6)[48], 295
BAUER, N., 86, *306*
BECK, C. W., (52)[58,72], (57)[58], 202, *302, 303, 315*
Behenic acid, 201, 282*ff*
Beidellite, 2, 8, 12, 13, 35, 50, 52, 158, 203, 205
BELL, D. J., (209)[139], *315*
BELLAMY, L. J., (55)[80], (82)[80], *303*
BENDER, R., (75)[121], *305*
BENESI, H. A., (156)[58,59], (157)[58,62], (158)[62], (263)[8], *312, 321*
BENSON, S. W., 44, (269)[43,44], (270)[44], *301, 322*
Bentones, 238
Bentonite, 73, 89, 187, 209, 210, 266, 282
Benzalacetophenone, 156
Benzaldehyde, 73, 74, 119
Benzene, 128, 130*ff*, 201, 232, 233, 267, 268
  substituted, 132*ff*
Benzeneazodiphenylamine, 156
Benzidine, 23, 102, 261, 263*ff*, 246
Benzoate, 83
Benzoic acid, 67, 81*ff*, 168, 200
  pesticides, 139, 148
Benzonitrile, 24, 67, 75*ff*
BERGER, G., 12, *296*
BERGER, J., (275)[79], (276)[79], *324*
BERNAL, J. D., 44, 275, 276, *301, 323*
BERNSTEIN, H. J., (25)[57], *299*
BEST, J. A., (151)[44], *311*

328

Betaine, 242
BILLMEYER, F. W., Jr., (272)[58], *323*
Biotite, 2, 16
Bipyridilium pesticides, 136, 139, 148*ff*
BIRRELL, K. S., (3)[21], *294*
Bis-(2-ethoxyethyl)-ether, 76
Bis-(2-methoxyethyl)-ether, 76
BISSADA, K. K., (33)[9], (38)[9], (40)[9], (43)[9], 44, 45, 46, 64, *300*
BITTLES, J. A., 269, (270)[44], *322*
Bleaching earth, xi
BLOCH, J. M., 84, 102, 132, 133, (265)[22], *305, 321*
BLUMSTEIN, A., (271)[49], 272, 274, *322, 323*
BLUMSTEIN, R., (272)[59,60], (274)[59,60], *323*
BODENHEIMER, W., (52)[73], 87, 92, 93, 96, (103)[167], (104)[167], 173, *303, 306, 307, 313*
BOEHM, H. P., (4)[34], *295*
BOLT, G. H., 140, 148, *309*
BOSAZZA, V. L., 264, *321*
BOWER, C. A., (21)[39], (52)[64,67], (58)[64,67], 60, *298, 302, 304*
BOWMAN, B. T., 147, *311*
BRADLEY, W. F., (15)[87], 21, 23, (24)[52], 29, 47, 48, (55)[35], 60, 61, 74, 75, 84, 92, 128, 206, *297, 298, 299, 301, 304, 315*
BRANT, D. A., (276)[83], *324*
BRAY, E. E., (285)[101], *325*
BRIEGLEB, G., 261, *321*
BRIGANDO, J., (xi)[6], *293*
BRIGGS, G. G., 142, *310*
BRINDLEY, G. W., (xi)[18], (2)[12], (3)[12], 4, 5, 6, 20, 21, 22, (23)[32], 24, 29, 30, 32, 33, 34, 36, 37, 38, 42, (45)[33], 46, 47, 50, (51)[53], 52, 57, 58, 59, 60, 61, 63, (67)[119], (70)[119], 75, (76)[121], (88)[13], 89, 90, 91, 92, 190, (191)[111-4], (192)[111-4], (193)[111], (194)[112], 195, 196, 197, 198, 200, 201, 206, 221, (259)[33,34], *293, 294, 295, 298, 299, 300, 301, 302, 314, 315, 317, 320*
BRINK, G., (19)[16], *297*
Brittle mica, 2, 8, 11, 16
Brønsted acid (and acidity), 109, 155*ff*, 261*ff*, 270, 271, 287, 289
BROOKS, B. T., 281, *324*
BROUGHTON, G., (269)[38], (281)[38], *322*
BROWN, C. B., (145)[17], *310*
BROWN, G., 1, (3)[16], (13)[78], 60, *294, 296, 304*
BROWN, J. F., Jr., (274)[64], *323*
Brucine, 189, 212
Brucite, 1
BRUMMER, K., 219, (233)[31], *317*
BRUNAUER, S., 58, *303*
BRUNTON, G., (52)[58,72], (57)[58], 202, *302, 303, 315*
BUCHANAN, J. R., (270)[45], *322*
Butadiene, 291
Butane (*n*-), 283
Butanediol (1, 3-), 52, 241, 242
Butanol (*n*-), 21, 32, 42, 241

Butter yellow, 156
Butylamine (*n*-), 85, 156, 158
Butylammonium (*n*-), 120, 212, 215, 216, 220, 234*ff*, 286
Butyl cellosolve, 241
Butylenediamine, 98
Butyleneglycol dimethacrylate, 274
Butyne-1, 4-diol (2-), 242
Butyraldehyde, 74
Butyric acid, 199

CABANA, A., (19)[24], *298*
Caffeine, 193
CAILLÈRE, S., (3)[18,19], 239, *294, 319*
CAIRNS-SMITH, A. G., 275, *323*
Calcium aluminate hydrate, 274
Calgon, 157, 265
CALICIS, B., (160)[68], (172)[77], (174)[68], (185)[77], (275)[78], *312, 313, 324*
CALVET, R., 129, 220, 286, 287, 289, *309, 317, 325*
CALVIN, M., 274, *323*
CAMAZANO, M. S., 239, 240, (246)[5], 259, *319, 320*
Camp Berteau montmorillonite, 98, 101, 102, 110
CANO RUIZ, J., 88, 89, (90)[155], (92)[155], *306*
Capric acid, 199
Caproic acid, 199
Caprylic acid, 199
Carbitol, 76
Carbomycin, 188
Carbonium ion, 269, 281, 284, 287, 291
Carbon preference index, 285
Carboxylic acids, 200, 234, 291
Carnosine, 170, 171
CARR, R. M., 239, 240, 241, 242, 243, *319*
CARROLL, D., 48, *301*
CARSTEA, D. D., (52)[57], *302*
CARTER, D. L., 59, *304*
CASHEN, G. H., (4)[37], *295*
Cation-dipole interactions, 38, 40, 44, 46, 50, 68, 85, 91, 128, 136, 206
Cellobiose, 208
Cetylpyridinium, 39, 222, 229*ff*
Cetyltrimethylammonium, 39, 231
CH activity, 23, 30
Chain-lattice silicates, 3
CHAKRAVARTI, S. K., 212, (231)[58,59], 232, *316, 318*
CHAN, S. I., (195), *314*
CHARBONELLE, J., (265)[22], *321*
Charge transfer complexes, 110, 149, 261*ff*
CHASSIN, P., 161, (171)[70], *312*
CHAUDHURI, A. K., (269)[43,44], (270)[44], *322*
CHAUSSIDON, J., (18)[10,12], (19)[19], 129, 220, 286, 287, 289, *297, 298, 309, 317, 325*
Chemical evolution, 274

CHENEY, G. E., (194)[116], *314*
CHENG, F. S., (33)[9], (38)[9], (40)[9], (43)[9], (44)[9], (45)[9], (46)[9], (64)[9], *300*
CHIH, H., 239, 240, 241, 242, 243, *319*
Clathrate compounds, 236, 274
Chloramphenicol, 187
Chlorazine, 144
Chlorite, 1, 2
Chloroanilines, 105*ff*, 109
Chlorobenzaldehyde, 119
Chloroethanol (2-), 241
Chlorophyllin, 201, 202
Chloropurine (6-), 193, 195
C-H . . . O interactions, *see* Hydrogen bonding
CHOU, C. C., 288, 289, *325*
Chrysotile, 2
CHURCHMAN, G. J., 240, *319*
CIPC, 138
Clay-humus interactions, xii
Clay minerals, acid strength of, 157
　　charge per formula unit of, 2, 8, 11, 35, 109, 110, 223, 234
　　classification of, 1, 2
　　di-, and tri-octahedral, 2, 16
　　di-, tri-, and tetra-morphic, 1, 2, 8, 11
　　hydration properties of, 11
　　organic complexes of, *see* Complexes
　　pharmaceutical uses of, 188, 189
　　reactions catalysed by, *see* Organic reactions and Polymerization
　　structure of, 1*ff*
Clay Minerals Society, nomenclature committee, 2, 3
Clay-organic derivatives, 13
Clinochlore, 2
Clintonite, 2
CLOOS, P., (85)[137], 87, 93, 94, 95, 96, 97, 98, 160, 172, 174, (185)[77], (215)[22], (217)[30], (218)[30], (233)[30], (275)[78], *305, 306, 312, 313, 316, 317, 324*
Cobaltic chloropentammine, 289, 290
Cobaltic hexammine, 231, 289, 290
COBLE, A. D., (151)[43], *311*
Codeine, 189, 212
COLE, W. F., (14)[83], *297*
COLEMAN, N. T., (85)[133], (270)[46], *305, 322*
Colour reactions, of clays, 20, 110, 261*ff*
　　applications of, 269
Complexes, of clay minerals,
　　β-type, 88*ff*
　　with bio-organic compounds, 136*ff*
　　with positively charged organic species, 211*ff*
　　with uncharged organic species, 5, 17*ff*, 32*ff*, 239*ff*
CONDRATE, R. A., Sr., 173, 174, (175)[81], 176, 177, *313*
CONLEY, R. F., 258, 267, (270)[27], *320, 322*
Contact distances, shortening of, 28*ff*, 40, 45, 75, 76, 171, 248, 258

COOK, D., (111)[179], *307*
COOPER, J. E., (285)[102], *325*
Coordination, of adsorbed organic species,
　　to cation, 20, 55, 67*ff*, 73, 75, 87, 103*ff*, 113, 122*ff*, 143, 145*ff*
　　to cation through water, 20, 67*ff*, 73, 75, 77*ff*, 101*ff*, 120, 127, 143, 146*ff*
COSHOW, W. R., (149)[36], *311*
Coulombic forces, *see* Electrostatic interactions
Cover-up effect, 160, 189, 212
COWAN, C. T., 85, 213, 214, 215, (217)[13], *305, 316*
COWBURN, D. A., (276)[84], *324*
Cracking, of petroleum, 281*ff*
　　catalytic, 281, 282
　　thermal, 281, 282
CRAIG, D., (85)[133], (207)[46], *305, 322*
Cristobalite, 1
Crotonaldehyde, 73, 74
Crotyl alcohol, 242
CRUZ, M., (88)[152], (141)[12], 143, 144, (145)[18], (230)[27], 253, *306, 310, 320*
CURRAN, C., (122)[190], (124)[190], *308*
CUTHBERT, F. L., (212)[9], (231)[9], *316*
Cyanamide, 280
Cyclohexanedione (1, 2-), 242
Cyclohexanol, 42
Cycloheximide, 187
Cyclohexylamine, 98*ff*
Cyclohexylammonium, 98*ff*, 289
Cytidine, 195
Cytosine, 195

2,4-D, 139, 148
DAS KANUNGO, J. L., 221, *318*
DAVIDSON, W. H. T., (53)[75], *303*
DAVIS, G. A., 210, *315*
DAWSON, J. E., (188)[103], *314*
DAYHOFF., M. O., (275)[73], *323*
DDE, 139, 155
DDT, 139, 155
DE BOER, J. H., (22)[40], 130, *298*
Decarboxylation, of fatty acids, 284, 286
Decoic acid, *see* Capric acid
Decomposition, of adsorbed organic species, 20, 129, 155, 158, 220, 281*ff*
Decylamine(n-), 85
Decylammonium(n-), 225
DEGENS, E. T., 275, (276)[80], (278)[87], (279)[76,77,87], (280)[89], *324*
DEKKING, H. G. G., 272, *323*
Delta value, *see* Value, Δ
Demixing (layer segregation), 219, 238
DEMOLON, A., (xi)[5], *293*
DETLING, K. D., (157)[62], (158)[62], *312*
DEUEL, H., 12, *296*
DEYRUP, A. J., 156, *312*
Dialkylamides, 116*ff*
Dialkylanilines, pesticides, 142

Diaminopropane(α,ω), 233
DIAMOND, S., (52)[69,70], (58)[69,70], (219)[34], 220, (221)[34], (231)[34], 303, 317
Diazinon, 138, 158
Dicinnamalacetone, 156
Dickite, 258, 259
Dicryl, 139, 146
Dicyandiamide, 242
Dieldrin, 139, 157, 158
Dielectric constant, 178, 184, 241
Dielectric increment, 164ff, 178
Diethylacetamide(N,N-), 112ff, 117ff
Diethylamine, 86, 87, 241
Diethylammonium, 216ff, 220, 286
Diethylene glycol 23, 76, 241
    diacetate, 76
Diethylene triamine, 93, 242
Diethylformamide(N,N-), 117ff, 247, 248, 250, 251, 253
Diethylketone, 68
Differential thermal analysis (DTA), 12, 13, 93, 103ff, 288
Diffuse double layer, 166, 178, 183, 184, 232, 236ff
Diffusion coefficient, 189
Diglycyl glycine, 160, 162ff, 168, 171
Dihydrostreptomycin, 187
Dimethylacetamide (N,N-), 117ff, 246, 248, 253ff
Dimethylamine, 85, 86
Dimethylammonium, 216ff, 220
Dimethylaniline, 102, 106ff
Dimethylbenzyllaurylammonium, 230
Dimethyldioctadecylammonium, 129, 230
Dimethylformamide(N,N-), 117ff, 247, 248, 250, 251, 253
Dimethyl glucose(4,6-), 208, 209
Dimethyl sulphoxide (DMSO), 239ff, 246, 256ff
Dimethylurea, 124ff
Di-n-butylammonium, 215, 220
Dinitrophenylhydrazine(2,4-), 242
Di-n-prophylammonium, 220
DIOT, A., (98)[161], 306
Dioxane(1,4-), 30, 76, 241, 242
Dipalmitin, 279
Diphenylacetamide(N,N-), 119
Diphenylformamide(N,N-), 119
Diphenylpicrylhydrazyl(2,2'-), 268, 269
Dipole-induced dipole interactions, 130
Dipole moment, 23, 44, 63, 102, 177, 239, 241, 247
Dipropylacetamide(N,N-), 117ff
Diquat, 139, 148ff
Diuron, 138
DIXON, J. B., 58, 152, 303, 311
DNBP, 148
DNC, 148
DODD, C. G., 267, 321
Dodecane(n-), 44, 129, 130
Dodecanol(n-), 287

Dodecylammonium(n-), 9, 203, 212, 225
Dodecylpropyldiamine, 213
Dodecylpyridinium(1-n-), 229, 230
DOEHLER, R. W., (21)[38], 298
DONATH, E., (127)[196,197], 308
Donbassite, 2
DONER, H. E., (19)[20], 116, 117, 118, 119, (130)[219], (132)[219], (146)[31], (147)[31], 298, 307, 309, 311
DONOHUE, J., 29, (38)[20], 299, 301
DOSCH, W., (274)[68], 323
DOWDY, R. H., 32, (33)[3], (38)[3], 40, 41, (42)[3], (46)[3], (52)[3], 53, 54, 55, 56, 57, 62, 64, 300, 303
DRUSHEL, H. V., (42)[27], 301
DRYANSKI, P., (19)[15], 297
DURAND, B., (20)[31], (129)[215], 287, 298, 309, 325
Dursban, 139, 158
DYAL, R. S., (52)[62,63], 58, 302

ECK, R. V., (275)[73], 323
EDELMAN, C. H., 11, 296
EDWARDS, D. G., (4)[39], 153, 295, 311
EGLINTON, G., (281)[98], 324
EISMA, E., (129)[213], (198)[123], 281, 282, (283)[95], 284, 309, 315, 324
Electron-donor-acceptor complexes, 261
Electrons, π, 132, 134, 264
Electron spin resonance spectroscopy, 265, 266
Electron transfer reactions, 261ff
Electrostatic interactions, 67, 140, 178, 183, 212, 219, 233, 237, 278
ELLERBE, J. S., (42)[27], 301
ELTANTAWY, I. M., 129, 130, 309
EMERSON, W. W., 37, (38)[19], 39, 40, 46, 301
EMMETT. P. H., 58, 303
Endrin, 139, 157
ENDELL, J., (264)[16], 321
ENDELL, K., 9, 10, 296
ENSMINGER, L. E., (211)[5,6], 316
Entropy, change of adsorption, 166, 219
    effects in adsorption, 22, 33, 219, 237
EPTC, 119, 120, 138, 146, 147
Equivalence point, 232
ERBRING, H., (231)[60], 318
ERICKSON, A. E., (52)[66], 302
ERIKS, K., (258)[30], 320
ERTEM, G., 50, (51)[53], 302
Erythritol, 208
Erythromycin, 188
ESQUEVIN, J., (239)[3], 319
Esterification, 12, 278, 291
Esters, 24, 234
    pesticides, 138
Ethane, 283
Ethanediamine(1,2-), 241

331

Ethanol, 21, 32, 38, 40*ff*, 50*ff*, 64, 233, 240*ff*, 267, 268, 287, 291
Ethanolamine, 242
Ether, alcohols, 74*ff*
    esters, 74*ff*, 234, 288
Ethers, 24, 63, 74, 75, 234
Ethene, 283
Ethoxypropionitrile, 76
Ethoxy trimethyleneglycol, 76
Ethylacetamide(N-), 112*ff*, 120
Ethylacetate, 233
Ethylamine, 85*ff*, 120, 241
Ethylammonium, 88, 89, 120, 213, 216*ff*, 220, 235, 236, 238, 286
Ethyl cellosolve, 241
Ethyldiammonium, 88
Ethyldimethyloctadecenylammonium, 212
Ethylene, 281
Ethylenediamine, 85, 88, 89, 92*ff*, 97, 98
Ethylene glycol, 6, 16, 21, 23, 30, 47*ff*, 55*ff*, 60*ff*, 241, 246
    diglycidether, 76
    dimethylether, 76
    monoethyl ether, 59, 62
Ethylpyridinium, 230
EVANS, E. D., (285)[101], *325*
EVANS, M. W., (33)[11], (235)[73], *300, 318*
EVCIM, N., 189, *314*

FALK, M., (19)[16], *297*
FARMER, V. C., (xi)[19], (3)[23], (18)[6,9,11,14], (19)[11,14], (20)[6,27], (23)[6], (29)[6], (30)[6,11], (32)[4], (39)[4], 40, 67, 75, (81)[112], (83)[112], (85)[138], 87, (88)[149], (98)[138], 101, (103)[168], (104)[111], 109, (168)[75], (200)[75], *293, 294, 297, 298, 300, 301, 304, 305, 306, 307, 313*
FARMER, W.J., 30, 124, (126)[193], 127, (146)[27], *300, 308, 310*
Fats, 204*ff*
Fatty acids, 129, 198*ff*, 282*ff*
FAVEJEE, J. C. L., 11, *296*
FEDOSEEV, A. D., (13)[81], *297*
FELSING, W. A., (215)[19], *316*
FENTON, S., (147)[33], *311*
Fenuron, 127
FERNANDEZ, T., (291)[120,121], *325*
FERNANDO, Q., (194)[116], *314*
FIELDES, M., ix, (3)[21,22], *293, 294*
FINLAYSON, C. M., (222)[40], (223)[40], (231)[40] (232)[40], *317*
Fire-clay kaolins, 259
FLECK, E. E., 155, *312*
FLETCHALL, O. H., (140)[10], (146)[10], *310*
FLORY, P. J., (276)[83], *324*
Fluormontmorillonite, synthetic, 198
Fluoroform, 29
FOLCKEMER, F. B., (157)[62], (158)[62], *312*
FOLLET, E. A. C., (4)[36], *295*

Formaldehyde, 74, 242, 275, 278
Formamide, 239, 242, 244*ff*, 250, 253, 255*ff*
Formic acid, 198
FORNES, V., (184)[92], (185)[92], (186)[92], *313*
Fourier synthesis, of clay-organic complexes, 23, 24, 59*ff*, 226*ff*, 245, 248, 253
FOWKER, F. M., 157, (158)[62], *312*
FOWLER, R. H., 44, *301*
FRAENKEL, A., (111)[176], 112, *307*
FRANCONI, C., (111)[176], 112, *307*
FRANK, H. S., (33)[11], (235)[73], *300, 318*
Free energy change, of adsorption, 164*ff*, 170, 171, 213*ff*, 230
FREISER, H., (194)[115,116], *314*
Freundlich
    constant, 142, 146
    equation, 142, 148
    isoetherm, 210
FRIEDLANDER, H. Z., 271, *322, 323*
FRINK, C. R., 271, *322*
FRIPIAT, J. J., (18)[7], (19)[19], (20)[31], 84, (85)[137], (109)[171], (129)[215], 130, (160)[68], 172, (174)[68], 185, 215, 217, (218)[30], 233, 275, (287)[109], 289, (290)[114], 297, 298, *305, 307, 309, 312, 313, 316, 317, 324, 325*
FRISSEL, M. J., 140, 148, *309*
FRY, W. H., (1)[7], *294*
Fucose, 208
FUKUSHIMA, K., (53)[77], *303*
Fuller's earth, xi, 211
Fulling process, xi
FUNCK, E., (68)[116], *305*
Furanose ring, 209
Fusarium wilt, 188
FUSON, N., (132)[222], *309*

GAINES, G. L., Jr., (15)[88], 214, *297, 316*
Galactose, 208
Galacturonolactone, 207, 208
GALWEY, A. K., (129)[211,212], 287, *309, 325*
GARCIA, S. G., 239, 240, (246)[5], 259, *319, 320*
GARRETT, W. G., (120)[188], (121)[189], 178, 179, (222)[43], (234)[43], 235, (236)[76], 238, 239, *308, 313, 317, 319*
GARRIGOU-LAGRANGE, C., (132)[222] *309*
GATINEAU, L, (28)[60], *299*
GAYER, F H., (269)[37], *322*
Gelatin, 211
Gelation, of vermiculite-amino acid complexes, 182*ff*
GERMAN, W. L., 20, (21)[33], 32, 33, 34, 37, 38, 39, 45, 46, 64, *298, 300*
Gibbsite, 1, 75
GIESEKING, J. E., 211, 212, 231, *316*
GILES, C. H., (33)[6], (149)[38], (162)[38], (166)[38], (213)[14], *300, 311, 316*
GLAESER, R., 33, 42, (46)[17], 63, 64, 74, (239)[3], *300, 301, 304, 319*

GLAVATI, O. L., 274, *323*
Gluconic acid, 209
Gluconolactone, 208
Glucosamine, 208, 209
Glucose, 207*ff*
Glucuronic acid, 209
Glucuronolactone, 208
Glutamic acid, 168, 174, 275
Glyceraldehyde, 278
Glycerides, 279
Glycerol, 21, 47*ff*, 52, 58, 59, 205, 241, 246, 279, 290
Glycine, 160*ff*, 168, 170*ff*, 178, 185, 242, 275
    potassium salt of, 246
Glycyl glycine, 162*ff*, 168, 171, 172
GOEL, A., 29, *299*
GOERTZEN, J. O., (52)[67], (58)[67], *303*
GOLDMAN, A., (155)[55], *312*
GONZALEZ, C. L., (59)[95], *304*
GONZALEZ GARCIA, F., (49)[49], *302*
GONZALEZ GARCIA, S., (49)[49], *302*
GONZALEZ GARMENDIA, J., (87)[147,148], (98)[148], *306*
GOOD, G. M., (282)[99], *324*
GÖRING, G., (5)[45], (244)[18], (245)[18], (246)[18], (248)[18], (253)[18], (258)[18], *295, 320*
GOTTLIEB, D., 187, *313*
GRABER, L., (127)[196,197], *308*
GRAHAM, J., (11)[69], *296*
GRANQUIST, W. T., 231, 232, *318*
Graphitic acid, 88, 90*ff*
GRATZER, W. B., (276)[84], *324*
GREBENIKOVA, A. I., (275)[74], *323*
GREEN, B. K., (269)[32], *322*
GREEN, J. H. S., (78)[123], *305*
GREEN, R. E., (140)[9], *310*
GREENE-KELLY, R., (4)[38], (13)[78], (23)[44], (24)[55], (25)[44], (28)[44], 29, 31, 39, (49)[44], 82, (104)[22], 129, 230, *295, 296, 299, 301, 302, 318*
GREENLAND, D. J., (xi)[9,10], (8)[53], (9)[60], (13)[79], (15)[85], (16)[60], (17)[3], (22)[3,42,43], (23)[3], (29)[3], (31)[69,70], (33)[7,12], 39, 60, (85)[135,136], 162, 163, (164)[73], 165, 166, 167, (168)[72], 169, (170)[72], (171)[71,72], 181, 206, (207)[137], (208)[137], 209, (214)[16], 215, (216)[16], (217)[16], (219)[16], (220)[16], (221)[16], 229, (230)[16], (231)[62], (236)[62], 237, *293, 296, 297, 299, 300, 301, 304, 305, 312, 315, 316, 317, 318, 319*
GREENSFELDER, B. S., 282, *324*
GRIM, R. E., xi, 1, (5)[1], 7, (8)[1], (10)[1], 12, (15)[1,87], (49)[47], 212, (231)[9], 265, (281)[92], *293, 294, 296, 297, 302, 316, 321, 324*
GROMES, W., (4)[34], *295*
GSCHWEND, F. B., (52)[64], (58)[64], *303*
Guanidine, 202
Guanine, 190
Guanosine, 190, 193, 195
GUYOT, J., 58, *304*

HAASE, D. J., (8)[55], (15)[55], (24)[50,51], 228, *296, 299, 318*
HACH-ALI, P. F., (52)[71], 92, *303, 306*
HACKMAN, J. R., (230)[57], *318*
HADDING, A., 1, *294*
HADJILIADIS, N., 98, *306*
HAJEK, B. F., 58, *303*
HAKASUI, A., 268, *322*
HALLER, H. L., 155, *312*
Halloysite, 2, 3, 5*ff*
    dehydrated forms, 6, 239, 240
    inorganic salt complexes of, 239, 241
    layer, amphoteric nature of, 6, 8, 239
    morphology of, 5, 240
    organic complexes of, 6*ff*, 60, 63, 189, 239*ff*, 257, 258
    reactions catalysed by, 291
HAMAKER, H. C., (22)[41], *298*
HAMMETT, L. P., 156, *312*
Hammett, coefficient, 197
    constant, 142, 197
    indicators, 156, 157, 263
HAMPTON, M. G., (129)[205], (238)[79], *308, 319*
HANCE, R. J., (146)[20,28,29], *310*
HANSFORD, R. C., (281)[93], *324*
HANSON, D. B., (231)[66], *318*
HAQUE, R., 148, 149, *311*
HARDING, D. A., 20, (21)[33], 32, 33, 34, 37, 38, 39, 45, 46, 64, *298, 300*
HARKINS, T. R., (194)[115], *314*
HARRIS, C. I., (140)[6], *309*
HARTER, R. D., 102, 122, *307*
HARVEY, G. R., 278, 279, 280, *324*
HARWARD, M. E., 50, 52, *302*
HASEGAWA, H., (267)[29], 268, *322*
HAUSER, E. A., 261, 264, 267, (271)[52], 321, 323
HAWTHORNE, D. G., (274)[65], *323*
HAXAIRE, A., 84, 102, 132, 133, *305*
HAYATSU, R., (275)[75], *323*
HAYES, M. H. B., (146)[22], *310*
HEAVER, A. A., (149)[40], (150)[40], (151)[40], *311*
Hectorite, 2, 8, 19, 201*ff*, 230, 233, 265, 266
HEILMAN, M. D., (59)[95], *304*
HELFFERICH, F., (215)[18], *316*
HELLER, G., (22)[40], *298*
HELLER, L., (52)[73], (87)[141,143-5], (92)[143-5], (93)[143-5], (96)[143], 98, (99)[162], 100, 101, (103)[167], (104)[167], 105, (106)[169], (108)[169], (109)[170], 110, 173, (201)[131], (202)[131], *303, 306, 307, 313, 315*
HELMKAMP, G. K., (195)[122], *314*
HELSEN, J., 289, (290)[114], *325*
Hemiacetal, 278
Hemin, 202
Hemisalt complexes, 87, 88, 98, 103, 112, 120, 122
HENDERSON, W., 281, *324*

HENDRICKS, S. B., 1, 5, 9, 10, (52)[62,63], 58, 102, 160, 189, 190, 212, 231, 264, (265)[15], *294, 295, 296, 302, 307, 312, 316, 321*
HÉNIN, S., (3)[18,19], 213, (239)[3], *294, 316, 319*
Heptachlor, 139, 155
Heptadecene(*n*-), 287
Heptane(*n*-), 128, 233
Heptanol(*n*-), 33
Heptulose, 208
Heptylamine(*n*-), 75
Heptylammonium(*n*-), 225, 235
Herban, 127
Herbicides, *see* Pesticides
HERZBERG, G., (25)[56], *299*
Hexadecylammonium(*n*-), 222
Hexametaphosphate, 157, 265
Hexamethylenediamine, 157, 213
Hexane(*n*-), 128*ff*, 283
Hexanedione (2, 5-), 30, 64, 67, 70*ff*, 75
Hexanol(*n*-), 38, 40, 198, 287
Hexene(*n*-), 287
Hexylamine(*n*-), 87
Hexylammonium(*n*-), 225, 228, 235, 289
HIDALGO, A., (241)[14], (250)[14], *319*
HILDEBRAND, F. A., (5)[48], (6)[48], *295*
HILTNER, P. A., (231)[65], 233, *318*
HILTON, H. W., (146)[21], *310*
HIRT, R. C., 137, *309*
Histidine, 170, 171, 275
HOFFMANN, R. W., 20, 21, 22, (23)[32], 24, (29)[45], 30, 32, (45)[33], 57, 61, 63, (67)[119], (70)[119], 74, 221, *288, 299, 300, 301, 303, 304, 305, 317*
HOFMANN, U., (4)[34], 9, 10, 49, 50, (222)[46], (264)[16], *295, 296, 302, 317, 321*
HOJABRI, F., (291)[123], *325*
HOLTON, W. F., (187)[97], *313*
HOOK, B. J., (222)[40], (223),[40], (231)[40], (232)[40], *317*
HOOPER, J., (174)[83], *313*
Hydrazine, 239, 242, 244*ff*, 250, 256, 259, 272
Hydride ion, 281
Hydrocarbons, 128*ff*, 234, 271, 275, 282*ff*, 289
    chlorinated, 139
Hydrogen bonding,
    C—H . . . O-surface, 23, 29, 30, 42, 45, 55, 129, 221, 243, 258
    C=O . . . H—O, in kaolinite complexes, 253
    interlayer, in kaolinite crystals, 4, 244
    intermolecular, 30, 55, 115, 116, 118, 120, 147, 195, 196, 252
    intramolecular, 55
    N—H . . . O-surface, 85, 91, 92, 127, 182, 217, 228, 229, 245, 250, 253, 260
    O—H . . . O-surface, 37*ff*, 44*ff*, 50, 75, 110, 182, 199, 209
Hydrogen cyanide, 29
Hydroxide(α-), 6, 8, 239

Hydroxybutyraldehyde(3-), 74
Hydroxyethyl methacrylate, 272
Hydroxyipazine, 138, 141, 142
Hydroxylamine, 242
Hydroxymethacrylates, 271*ff*
Hydroxypropazine, 138, 141, 143, 144
Hypoxanthine, 193, 195

IBANEZ, J. D., 280, *324*
IDEGUCHI, Y., (53)[77], *303*
Idose, 278
Illite, 2, 15, 16, 140
    organic complexes of, 140, 148*ff*, 162, 166, 173, 187*ff*, 195*ff*, 212
    reactions catalysed by, 266, 279
INARAJA-MARTIN, E., (244)[22], *320*
Infra-red spectroscopy,
    in clay-organic studies, xi, 17, 18, 20, 29, 30, 64, 87, 88, 98, 103, 201, 248*ff*
    and interlayer orientation of adsorbed species, 24, 77*ff*, 228, 229, 259, 260
Infra-red spectral data, of adsorbed species,
    acetic acid, 74
    acetone, 64*ff*, 74
    acetonitrile, 80
    acetylacetone, 69*ff*
    aldehydes, 73, 74
    alkylammonium ions, 88, 129, 217, 218, 228*ff*, 286
    amides, 112*ff*, 116*ff*
    aminotriazole(3-), 145
    anilines, 106
    benzene and its derivatives, 130*ff*
    benzoic acid, 81, 83
    benzonitrile, 75*ff*
    cobaltic hexammine and pentammine, 289, 290
    cyclohexylamine, 99, 100
    diamines, 98
    dimethyl sulphoxide, 256*ff*
    ethanol, 40*ff*
    ethylenediamine, 94*ff*
    ethylene glycol, 53*ff*
    EPTC, 146, 147
    hexanedione(2,5-), 30, 71, 72
    hydrazine, 250
    lysine, 176, 177
    malathion, 147
    methylacetamide(N-), 253*ff*
    methylformamide(N-), 250*ff*
    nonanetrione(2,5,8-), 30
    nitrobenzene, 81
    propazine, 143, 144
    pyridine and pyridinium, 25*ff*
    pyridine-N-oxide, 259, 260
    urea and its derivatives, 121*ff*
    valine, 174, 175
    water at clay surfaces, 18*ff*

Inosine, 193, 195
Inositol(i-), 208
INOUE, T., (3)[28], 295
Interactions, of defined organic compounds,
  with the kaolinite group of minerals, 239ff;
  see also Complexes, Infra-red spectral
  data, and X-ray diffractometry
  with the smectite and vermiculite groups
  of minerals, 17ff, 32ff, 128ff, 136ff, 211ff
Intercalation, 4, 6, 8, 85, 119, 121, 130, 148,
  189, 190, 202ff, 212; see also Complexes and
  Interactions
  complexes of the β-type, 88, 90
  ratio, 246, 247
Interlayer organization, of adsorbed species,
  17, 23ff, 76, 219ff; see also Orientation
  in the α_I arrangement, 24ff, 76, 84, 182,
    248
  in the α_II arrangement, 24ff, 67, 76, 84,
    182, 221
Interlayer polymerization, 272ff
Interlayer water,
  dissociation of, 20, 109, 110, 143, 155,
    261ff, 271, 286, 289
  infra-red spectral data of, 18ff
  O—H . . . O bonding to silicate surface
    of, 19, 110
  polarization by exchangeable cations of,
    20, 109, 262, 271, 286
  and surface acidity, 109, 155, 261ff
Ion-dipole interactions, see Cation-dipole in-
  teractions
Ipatone, 141
Ipatrine, 138, 141
Ipazine, 138, 142, 144
IPC, 138, 146
Isoamylammonium, 235
Isobutane, 283
Isomorphous replacement,
  in clay mineral structures, 2, 11, 15, 40,
    266
  and amount and site of layer charge, 2,
    50, 52, 109, 110, 211, 238
Isooctane, 230
Isopentene, 282, 283
Isoprene, 274
Isopropanol, 42, 230
Isopropanolamine, 242
Isotherms, adsorption,
  C-type, 33, 162
  H-type, 149
  L-type, 33, 166, 213, 214
  of alkylammonium ions, 213ff, 219
  of amino acids and peptides, 160ff
  of bipyridilium pesticides, 149ff
  of fatty acids, 202
  of purines and pyrimidines, 191ff
  of saccharides, 209
  stepped, 189
IYER, K. R. N., (xi)[8], 293

JACKSON, M. L., 48, 58, 301, 302, 303
JACKSON, T. A., (276)[80,81], 277, 278, 324
JACOBS, H., 259, 320
JACOBS, H. S., 57, (58)[84], 303
JANG, S. D., 173, 174, (175)[81], 176, 177, 313
JANISH, M. A. M., (268)[31], 322
JANSSEN, M J , (111)[177], (112)[177], 307
JEFFERSON, M. E., 5, 295
JEFFREYS, E. G., 187, 313
JENNY, H., (211)[3], 316
JEPSON, W. B., 210, 315
JOHNS, W. D., (2)[12], (3)[12](8)[56], (15)[56], (33)[9],
  (38)[9], (40)[9], (43)[9], (44)[9], (45)[9], (46)[9], 49,
  64, (129)[214], (198)[214], (222)[44], 224, 227, 228,
  284, 285, 286, 294, 300, 302, 309, 315, 317,
  324, 325
JOHNSON, M. R., (157)[62], (158)[62], 312
JONAS, E. C., 49, 302
JORDAN, J. W., (xi)[11], 221, 222, 223, 230,
  231, (232)[40], 234, 293, 296, 317, 318
JOSIEN, M. L., (132)[222], 309
JUPP, G., 291, 325
JURG, J. W., (129)[213], (198)[123], 281, 282,
  (283)[95], 284, 309, 315, 324

KAGARISE, R. E., (110)[183], 307
KANAMARU, F., 182, 313
Kanamycin, 187
KANTNER, I., (222)[49], 223, 317
Kaolinite, 1ff
  b-axis disordered types of, 259
  crystallinity of, 258
  inorganic salt complexes of, 243, 246
  interlayer O . . . H—O bonding in, 4, 244
  morphology, 5, 259
  organic complexes of, 57, 75, 121, 140,
    148ff, 154, 157, 173, 187ff, 210, 212, 213,
    243ff, 256ff, 267
  reactions catalysed by, 4, 155, 157, 158,
    186, 264ff, 270ff, 283, 290, 291
KATCHALSKY, A., (275)[79], (276)[79], 324
KATO, H., 290, 325
KAUFHERR, N., (109)[170], 110, 201, 202,
  307, 315
KAYSER, F., (265)[22], 321
KECKI, Z., (19)[15], 297
KEENAN, A. G., (10)[66], 296
KELSEY, K. E., 213, 316
KEMPER, W. D., 58, 303, 304
KENNEDY, J. V., (231)[64], (232)[64], 318
Kerogen-like material, 284, 285
Ketones, 24, 30, 63ff, 78, 232, 234
Keying, 28, 30, 31, 47, 84, 92, 171, 194, 217,
  221, 225, 226, 229, 230, 248, 258; see also
  Contact distances
KHAN, S., (189)[108], 314
KIM, J. T., 127, 308
KIMBALL, A. P., (280)[90], 324

335

Kinetics, of amine sorption by montmorillonite, 85, 86
    of dieldrin decomposition in kaolinite, 157, 158
    of ornithine cation sorption by vermiculite, 185, 186
KING, J. W., Jr., 44, *301*
KINTER, E. B., (52)[69,70], (58)[69,70], (219)[34], 220, (221)[34], (231)[34], *303, 317*
KIRSON, B., (87)[142-5], (92)[142-5], (93)[143-5], (96)[143], *306*
KLABUNOVSKII, E. I., (277)[85], *324*
KLEMEN, R , 49, 50, *302*
KLEMM, L. H., (238)[82], *319*
KNIGHT, B. A. G., (149)[39,40], (150)[40], (151)[40], *311*
KOBAKHIDZE, E. I., (231)[61], *318*
KOBAYASHI, Y , (3)[27], *295*
KOHL, R. A., 68, 83, *304*
KOLLMAN, R. C., (271)[52], *323*
KRANZ F. H., (269)[33], *322*
KRIMM, S., (53)[78], *303*
KRÜGER, D., (261)[2], 265, *321*
KUHN, L. P., 55, *303*
KUHN, M., (53)[76], *303*
KUILE, C. H. H., (188)[103], *314*
KUKHARSKAYA, E. V., (13)[81], *297*
KULBICKI, G., 12, (49)[47], *296, 302*
KUMADA, K., 290, 325
KUNZE, G. W., (48)[44], (49)[44], (58)[44], 102, 109, 110, *302, 307*
KUPKA, F., 215, (217)[24], *316*
KURILENKO, O. D., 213, (230)[10], (232)[10], *316*
KUSNITSYNA, T. A., 270, *322*
KUTZELNIGG, W., (111)[178], 112, *307*

LABY, R. H., 19, 30, (31)[69], (33)[7], (38)[21], (40)[21], (92)[21], (162)[71-3], (163)[73], (164)[73], (165)[73], (166)[71-3], (167)[72], (168)[72], (169)[72], (170)[72], (171)[71,72], (178), (181)[71,73], 216, 228, *298, 299, 300, 301, 312, 317*
Lactic acid, 242
Lactose, 208
LAGALY, G., (129)[209], *309*
LAHAV, N., 267, *322*
LAILACH, G. E., (190)[111-3], 191, 192, 193, 194, 195, 196, *314*
LAMBERT, S. M., 142, *310*
Langmuir, equation, 60, 189, 267
    isotherm, *see* Isotherms, L-type
LARSON, G. O., 68, 73, 74, 83, *304*
LAURA, R. D., 87, 93, 94, 95, 96, 97, 98, *306*
Lauric acid, 199
LAWRIE, D. C., 60, *304*
LAWSON, K. E., (68)[117], *305*
LAYCOCK, A., (250)[27], (253)[27], *320*

LEDOUX, R. L., (241)[15], 248, 249, 250, 254, 256, *319, 320*
LEE, F. H. W., (238)[82], *319*
LEGGETT, M. B., 261, 264, 267, *321*
LEHMANN, H., (231)[60], *318*
LÉONARD, A., (84)[128-130], (85)[128-130], *305*
LEONARD, R. A., (18)[13], *297*
Leucine, 164, 166, 168, 171
Leuco dyes, 268
Lewis, acid (and acidity), 79, 155*ff*, 261, 263, 268, 270, 271, 286, 287
    base, 268, 270
LEWIS, D. E., Jr., (152)[46], *311*
LEWIS, D. G., (9)[60], (16)[60], *296*
LIANG, C. Y., (53)[78], *303*
LIBOR, O., 127, *308*
LILLEY, S., (149)[36,37], *311*
Linuron, 138
Lipids, 278
LIPPINCOTT, E. R., (275)[73], *323*
LITTLE, L. H., (17)[4], *297*
LLOYD, M. K., 258, 267, (270)[27], *320, 322*
LLOYD, J. U., 211, *315*
LOEFFLER, E. S., (157)[62], (158)[62], *312*
LOFT, B. C., (263)[10,11], (265)[10], (266)[10], (267)[10], (268)[11], (269)[42], (271)[42], *321, 322*
LOVELOCK, J. E., (281)[98], *324*
LUKASIEWICZ, J., ix
LUTTKE, W., (53)[76], *303*
Lyotropic series, 231
Lysine, 170*ff*, 177, 178, 275
    potassium salt of, 246

MacEWAN, D. M. C., (3)[16], (6)[50], 7, 8, 9, (10)[51], (15)[51], (17)[1], (21)[35], 23, 28, 29, 33, 34, 46, 47, 48, 50, (55)[35], 60, 62, 63, 64, 74, 75, 88, 89, 90, 92, 128, 129, (222)[42], 239, 240, 258, *295, 297, 298, 300, 302, 306, 317, 319*
MacEWAN, T. H., (33)[6], (149)[38], (162)[38], (166)[38], (213)[14], *300, 311, 316*
MACKENZIE, R. C., (xi)[2], 1, (4)[31], (9)[11], 44, 57, 62, 64, *293, 294, 295, 301, 303*
MACKINTOSH, E. E., (9)[60], (16)[60], *296*
MacLEOD, D. M., (128)[200], (129)[200,201], (220)[32], (221)[32], (231)[32], (233)[32], *308, 317*
MACNAMARA, G., (146)[24], *310*
Macrolides, 188
MAEGDEFRAU, E., 9, 10, *296*
MAKAY, K., (160)[68], (172)[71], (174)[68], (185)[77], (275)[78], *312, 313, 324*
Malachite green, 268
Malathion, 139, 147, 148, 157
MALHOTRA, S. L., (272)[61], *323*
MALINA, M. A., (155)[55], *312*
Malonic acid, 242
Maltose, 208
Mannitol, 208
Mannose, 207, 208

MANTIN, I., 271, *322*
Margarite, 2, 16
MARSHALL, C. E , 1, 9, 10, *294, 296*
MARTELL, A. E., (177)[84], *313*
MARTIN, R. T., (2)[12], (3)[12], (52)[65], 57, (58)[65], (188)[103], *294, 302, 314*
MARTIN-VIVALDI, J. L., (52)[71], 92, *303, 306*
MATA-ARJONA, A., 244, *320*
MATHEJA, J., 275, (276)[80], (279)[76,77], *324*
MATHIESON, A McL., (11)[68], 13, 14, *296*
MATSUMOTO, T., 270, *322*
MATSUNAGA, V., (268)[30], *322*
MATTSON, S., (xi)[7], *293*
McATEE, J. L., Jr., 230, (231)[63], (232)[63], 288, 289, *318, 325*
McCONNELL, D., 12, *296*
McHARDY, W. J., (3)[23], *294*
McLAREN, A. D., 160, 161, 173, *312*
McNEAL, B. L., (52)[68], (57)[68], (58)[68], *303*
MCPA, 139, 148
MECKE, R., (53)[76], (111)[178], (112)[178], *303, 307*
MEEK, R. C., (151)[42], *311*
MEGGITT, W. F., 146, (147)[30], *310*
MEHLER, A., (9)[61], (222)[46], *296, 317*
MEHRA, O. P., 48, 58, *301, 302*
Melibiose, 208
MELVIN, I. S., (195)[120], *314*
MENTER, J. W., (201)[128], *315*
MÉRING, J., (28)[60], 42, 74, *299, 301*
Mesitylene, 132*ff*
Metahalloysite, 6, 239, 240
Methanol, 30, 40, 42, 46, 233, 241, 242, 291
Methionine, 173, 174
Methylacetamide(N-), 247, 248, 253*ff*
Methyl acrylate, 272
Methyladenine(7-, and 9-), 195
Methylamine, 84*ff*
Methylammonium, 84, 86, 89, 129, 216*ff*, 220, 231, 235, 236, 238, 286
Methyl-1-butene(2-), 283
Methyl cellusolve, 241
Methylcytosine, 190
Methylformamide(N-), 248, 250*ff*
Methyl glucoside(α-), 208
Methyl methacrylate, 271*ff*
Methyloctadecylammonium, 213
Methyl pentane(2-, and 3-), 283
Methylpyridinium, 23
Methylurea, 124*ff*
MEUNIER, P., (261)[3], (264)[17], *321*
MEYERS, P. A., 202, *315*
Mica, 1, 2, 8, 9, 11, 15, 16, 151*ff*, 222
MICHAELS, A. S., (265)[21], *321*
MICHEL, E., (222)[48], (228)[48], *317*
MIFSUD, A., 184, 185, (186)[92], *313*
MIKHALYUK, R. V., 213, (230)[10], (232)[10], *316*
MILESTONE, N. B , (3)[26], *294*
MILFORD, M. H., (58)[89], *303*

MILLER, W. G., (276)[83], *324*
MILLINGTON, A. D., 233, (238)[70], *318*
Minimum molecular thickness, of adsorbed organic species, *see* Van der Waals dimensions
MISRA, S. G., (209)[141], *315*
MITCHELL, B. D., (xi)[2], 1, (3)[23], (9)[11], *293, 294*
MITRA, S. P., 209, *315*
MITSUI, S., 124, *308*
MIYAZAWA, T., (53)[77], *303*
MIZUSHIMA, S., (122)[190], (124)[190], *308*
MOLL, W. F., Jr., 198, (199)[125], 200, 201, *315*
Mononucleotides, 280
Monopalmitin, 279
Montmorillonite, 2, 9, 10, 11, 15, 16, 28
    cation exchange capacity of, 8, 47, 212
    hydration and interlayer expansion of, 10, 11, 19, 50, 51
    organic complexes of, 6*ff*, 15, 20*ff*, 24*ff*, 30, 32*ff*, 39, 47*ff*, 57*ff*, 75*ff*, 84*ff*, 92*ff*, 102*ff*, 111*ff*, 121*ff*, 128*ff*, 136*ff*, 157*ff*, 186*ff*, 198*ff*, 206*ff*, 211*ff*, 233
    reactions catalysed by, 155*ff*, 261*ff*, 269*ff*, 281*ff*
    surface area of, 8, 58, 59
    surface esterification of, 12, 13
    swelling in organic liquids of, 50, 51, 120
    types of, 12, 13
Monuron, 127, 138
MOONEY, R. W., (10)[66], *296*
MOORE, D. E., 58, (152)[46], *303, 311*
MOPPER, K., (278)[87], (279)[87], (280)[89], *324*
MOREL, R., 213, *316*
MORIMOTO, Y., (177)[84], *313*
MORIN, R. E., 57, (58)[84], *303*
Morpholine, 50, 51
MORTLAND, M. M., (xi)[20], (11)[70], (17)[5], (18)[5,8], (19)[19,20], (20)[5,28], (23)[5], (29)[5], 30, 32, (33)[3], (38)[3], 40, 41, (42)[3], (46)[3], (52)[3,66], 53, 54, 55, 56, 57, 58, 62, 64, 65, (66)[109], 67, 68, (69)[109], 70, 71, (72)[109], 73, (85)[138], 87, (88)[149,150,151], (89)[153], (98)[138], 103, (104)[111], (109)[172], 110, 111, (112)[151], 113, (114)[181], 115, 116, 117, 118, 119, 120, 121, 124, (125)[150], 127, 130, 131, 132, 133, (134)[221], 146, (147)[30], (155)[53], (158)[63], 262, 263, *293, 296, 297, 298, 299, 300, 302, 303, 304, 305, 306, 307, 309, 310, 311, 312, 321*
MUKHERJEE, S. K., (231)[58], *318*
MUNDAY, C. S., (274)[66], *323*
MURPHY, A. J., (156)[61], (157)[61], (263)[9], (270)[9], (271)[9], *312, 321*
Muscovite, 2, 16, 28, 151, 266
Myristic acid, 199

NAGARAJAN, G., (275)[73], *323*
NAGASAWA, K., (258)[32], *320*
NAHIN, P. G., (xii)[12], *293*

NAKADA, T., (3)[25], *294*
NAKAGAWA, I. R., (174)[83], *313*
NAKAHIRA, M., (3)[25], *294*
NAKAMOTO, K., (41)[26], (68)[118], 70, (77)[118], 177, *301, 305, 313*
NAKHWA, S. N., (33)[6], (149)[38], (162)[38], (166)[38], (213)[14], *300, 311, 316*
Naphtalene 128
Naphtylamine(α-), 102
National Cash Register Company, (269)[34,35], *322*
N-coordination, 122
NEARPASS, D. C., (145)[19], *310*
Negative adsorption, 84, 136, 148
Neomycin, 187
Nickel cyanide, 274
Nicotine, 189, 211
Nitriles, 75*ff*
Nitroanilines, 102
Nitrobenzene, 25, 63, 67, 81, 82, 232, 241, 246
Nitromethane, 63, 233, 241
Nonanetrione(2,5,8-), 30, 67, 71
Nonoic acid, *see* Pelargonic acid
Nontronite, 2, 8, 35, 88, 89, 110, 158, 266
Nonylammonium(n-), 225
Norleucine, 173
NORRISH, K, (9)[59], (10)[67], (11)[67], (13)[78], (15)[59,67], 178, (236)[75], (237)[78], *296, 313, 318*
NORTHCOTE, D. H., (209)[139], *315*
Norvaline, 160
Nucleosides, 189*ff*, 212

OBERLIES, F., (261)[2], 265, *321*
O-coordination, 116, 122, 177
Octacosane(n-), 281, 282, 287
Octoadecanol(n-), 287
Octadecylammonium(n-), 222, 225, 230
Octanol(n-), 198, 287
Octylamine(n-), 246
Octylammonium(n), 215, 222, 225, 289
Octylpyridinium(1-n-), 229
OKAWARA, R., (291)[118], 325
Olefins, 271, 281*ff*, 287; *see also* Hydrocarbons, Alkenes
OLEJNIK, S., (5)[46], 239, 240, 243, 244, 245, 246, 248, 250, 251, (252)[28], 254, (255)[29], 256, 257, 258, *295, 319, 320*
Oligodesoxyribonucleotides, 280
OLIVER, M., ix
OLSEN, A. C., (195)[120], *314*
OMU, 127
OPARIN, A. I., 275, *323*
O-protonation, 111*ff*
Organic monomers, 263, 269*ff*
Organic pesticides, 136*ff*
Organic reactions, catalysed by clays, 20, 74, 155, 198, 220, 261*ff*
Organophosphates, pesticides, 138

Organosiloxanes, 291
Origin of life, 274, 275
Orientation, of adsorbed organic species, 17, 24*ff*, 30, 60, 67, 75, 76, 84, 221; *see also* Interlayer organization
Ornithine, 178, 184*ff*, 275
ORÓ, J., (280)[90], *324*
ORTH, H., (244)[20], (246)[20], (247)[20], (248)[20], (252)[20], (259)[20], *320*
OSTROVSKAYA, I. K., 270, 322
OURISSON, G., 201, (291)[122], *315, 325*
OVCHARENKO, F. D., (19)[17,18], 47, 79, (80)[124], 172, 173, 176, *298, 301, 305, 313*
OVERBEEK, J. Th. G., (184)[91], *313*
Oxonium ion, 288
Oxydipropionitrile(β,β'-), 75, 76

PAECHT-HOROWITZ, M., 275, 276, *324*
PAGE, J. B., 264, 265, (268)[20], *321*
PALMER, J., 86, *306*
Palmitic acid, 199, 279
Palygorskite, 3, 155, 157, 189, 266, 270, 291
PANDA, N., (209)[141], *315*
Parachor, 239
PAPARIELLO, G. J., (268)[31], *322*
Paraformaldehyde, 278
Paragonite, 2
Paraquat, 139, 148*ff*
PARFITT, R. L., (20)[28], (22)[43], 64, 65, (66)[109], 67, 68, (69)[109], 70, 71, (72)[109], 73, *298, 299, 304*
PASINSKYI, A. G., (275)[74], *323*
PAULING, L., (28)[59], (221)[37], *294, 299, 317*
PAVLOVSKAYA, T. E., 275, *323*
PAYNE, W. R., (144)[16], *310*
PEDRO, G., (3)[20], *294*
Pelargonic acid, 199
PELET, R., (20)[31], (129)[215], (287)[109], *298, 309, 325*
Penicillin, 187
PENLAND, R. B., 122, 124, *308*
PENNEQUIN, M., (85)[137], (217)[30], (218)[30], (233)[30], *305, 317*
Pennine, 2
Pentamethylenediamine(1,5-), 93
Pentane(n-), 282, 283
Pentanediol(1,5-), 21, 23, 52
Pentanedione(2,4-), *see* Acetylacetone
Pentanone(3-), 50, 51
Pentene, 283
Pentylamine(n-), 98
Peptides, 158*ff*; *see also* Polypeptides
PERRY, G. S., (129)[204], *308*
PERRY, P. W., (140)[7], (141)[7], (148)[7], (149)[7], 309
Pesticides, *see* Organic pesticides
PETERSON, G. H., (160)[67], (161)[67], (173)[67], *312*

Petroleum, 281*ff*
PEZERAT, H., 271, *322*
PHAM, T. H., 60, *304*
PHANSALKAR, V. K., 234, *318*
Phenanthroline(*o*-), 60
Phenols, 234, 286, 287
Phenylalanine, 168, 169, 171, 277
Phenylalkanoic acid, pesticides, 139, 148
Phenylazonaphtylamine, 156
Phenylcarbamate, pesticides, 142, 146
Phenylenediamine, 102
Phenylurea, pesticides, 142
PHILEN, O. D., (152)[48,49], 153, 154, (155)[48], *311*
Phlogopite, 2, 16
Phospholipids, 279
Phyllosilicates, 1, 2, 11
Picloram, 139, 148
Picolinic acid, pesticides, 139, 148
Picric acid, pesticides, 148, 242
PINCHAS, S., (82)[125], *305*
PINCK, L. A., 187, (188)[99,100,101], *313, 314*
PINNAVAIA, T. J., (130)[220,221], 131, 132, 133, (134)[221], *309*
Piperidine, 24, 211
Place exchange, 49
POLAK, L. S., 274, *323*
Polarity, of organic molecules, 23, 63, 241
Polarizing power, of cations, 18, 20, 52, 68, 79, 109, 145, 206, 290
POLEN, P. B., (155)[55], *312*
POLLOCK, J.McC., (67)[110], *304*
Poly(acrylic acid), 273
Poly(acrylonitrile), 273
Polyamines, 92*ff*
Polyethers, 74
Polyhydric alcohols, 47*ff*
Polymerization, catalysed by clays, 4, 20, 74, 263, 269*ff*
Polymers, adsorption of, xii, 22
  stereoregular, 272, 274
  two-dimensional, 272, 274
Poly(methyl methacrylate), 272, 274
Polymixin B sulphate, 188
Polypeptides, antibiotic, 188
  synthesis, catalysed by clays, 172, 186, 275*ff*, 280
Polyphosphate, 265*ff*, 269, 270
Polysaccharides, 278
Polystyrene, 270
Polyvinylamine, 213
PONCELET, A., (215)[22], *316*
PONCELET, G., (85)[137], (217)[30], (218)[30], (233)[30], *305, 317*
PONNAMPERUMA, C., (274)[66], *323*
POPE, J. D., (144)[16], 310
Porphyrins, 202
POSNER, A. M., (5)[46], (153)[50], (239)[6], (240)[6], (243)[6], (244)[21], (245)[21], (246)[6,21], (248)[23], (250)[28], (251)[28], (252)[28], (253)[28,29], (254)[29],

(255)[29], (256)[6], (257)[6], (258)[6], (259)[6], *295, 311, 319, 320*
Potassium, acetate, 242, 246
  contraction, 48
  lactate, 246
POWELL, D. B., 95, 97, *306*
PRAMER, D., 187, *313*
Primary *n*-alcohols, 32*ff*, 129
Primary *n*-amines, 84*ff*
Prometone, 138, 141*ff*
Prometryne, 138, 141, 143
Propane, 283
Propanediamine(1,2-), 241
Propanediol(1,3-), 241
Propanol(*n*-), 21, 32, 40, 42, 47, 241, 287
Propazine, 138, 141*ff*
Propenal(2-), 74
Propene, 283
Propionaldehyde, 119
Propionic acid, 242
Propylamine(*n*-), 85, 98
Propylammonium(*n*-), 216, 220, 235, 236, 286
Propylenediamine, 92, 93, 98
PROST, R., (18)[10,12], (19)[10,12], *297*
Proteins, 160, 211
Protonation, of adsorbed species, 20, 85, 98, 100*ff*, 110, 122, 136, 137, 141, 143, 145, 262, 263
Proton, interchange, 88, 89
  transfer, 142, 159, 166, 172, 190, 197, 269
Pseudo-layer silicates, 3
PULLIN, A. D. E., (67)[110], *304*
Purines, 189*ff*, 193, 195, 209, 212, 275
Pyranose ring, 209
Pyridine, 24*ff*, 67, 82, 88, 101, 110, 230, 242
Pyridine-N-oxide, 246*ff*, 259, 260
Pyridinium, 25*ff*, 88, 89, 120, 147, 259
Pyrimidines, 189*ff*, 209, 275
Pyrogallol, 290
Pyruvic acid, 242

QUAGLIANO, J. V., (122)[190], (124)[190], *308*
Quaternary ammonium ions, 129, 212, 230*ff*, 238, 287, 288
Quaternary pyridinium ions, 212, 232
Quinine, 189
QUINN, J. G., 202, *315*
QUIRK, J. P., (4)[39], (5)[46], (15)[85], (31)[69,70], (33)[5,7,12], 39, 50, (59)[94], 60, (85)[136], 153, (162)[71-3], (163)[73], (164)[73], (165)[73], (166)[71-3], (167)[72], (168)[72], (169)[72], (170)[72], (171)[71,72], (178)[71,73], (181)[71,73], (214)[16], (215)[16], (216)[16], (217)[16], (219)[16], (220)[16], (221)[16], 229, (230)[16], (231)[62], (236)[62,77], (237)[77], (239)[6], (240)[6], (243)[6], (244)[21], (245)[21], (246)[6,21], (248)[23], (250)[28], (251)[28], (252)[28], (253)[28,29], (254)[29], (255)[29], (256)[6], (257)[6], (258)[6], (259)[23], *295, 297, 300, 301, 302, 304, 305, 311, 312, 316, 317, 318, 319, 320*

Radical, anion, 268, 272
  cation, 264, 270, 271
  free—, 268, 271
RADOSLOVICH, E. W., 3, 6, 31, 300, *294, 295*
RADUL, N. M., 47, *301*
Raffinose, 208
RAMAN, K. V., 88, (89)[163], 110, (158)[63], 262, 263, *306, 307, 312, 321*
RAO, C. N. R., 29, (55), (82), *299, 303*
RAU, H., 291, *325*
RAUSELL-COLOM, J. A., 178, 179, (180)[88,89], 181, 182, 183, 184, (185)[92], (186)[92], 236, *313, 318*
RAY, S., 33, 34, 36, 37, 38, 46, (75)[121], (88)[13], 89, 198, 206, 267, *300, 305, 315, 321*
RAZIEL, S., 267, *322*
REAY, J. S. S., (129)[202,203], (219)[33], (220)[33], (221)[33], (231)[33], (233)[33], *308, 317*
Resonance structures, 55, 67, 122, 264
Resorcinol, 242
REYNOLDS, R. C., (23)[49], 61, 62, *299, 304*
Rhamnose, 208, 278
RICHARDSON, J. W., (41)[26], *301*
RIDOUT, C. W., 189, *314*
Ring molecules, 24, 31
RINNE, F., 1, *294*
RITTER, W., (5)[45], (244)[18,19], (245)[18], (246)[18], (248)[18], (253)[18], (258)[18,19], *295, 320*
RITZERFELD, G., (269)[36], *322*
RITZERFELD, W., (269)[36], *322*
ROBERTSON, R. H. S., (4)[31], (281)[91], *295, 324*
ROBINSON, K., (259)[33,34], *320*
ROBINSON, R. A., (215)[20], *316*
RODRIGUEZ, A., (87)[147,148], (98)[148], *306*
ROLOFF, G., 202, (203)[135], 204, 205, 289, *315, 325*
Ronnel, 139, 155*ff*
ROSENFIELD, C., 155, (157)[56], *312*
ROSS, C. S., 1, *294*
ROSS, M., (2)[12], (3)[12], *294*
ROSSER, M. J., (4)[42], (269)[40], 270, (271),[40] *295, 322*
ROTHBERG, T., (142)[15], (146)[15], (148)[15], *310*
ROWLAND, R. A., (23)[47,48], (24)[51], (60)[103,104], 84, 86, 217, *299, 304, 305, 317*
ROWLINSON, J. S., (209)[140], *315*
RUIZ-AMIL, A., (3)[16], 50, (244)[22], *294, 302, 320*
RUNDLE, R. E., (41)[26], *301*
RUSSELL, E. W., (13)[79], *296*
RUSSELL, J. D., (18)[9,11,14], (19)[11,14], (20)[27], (30)[6,11], (32)[4], (39)[4], 40, (67)[112], 75, (81)[112], (83)[112], 88, 101, (103)[168], (141)[12], (143)[12], (144)[12,16], 145, (168)[75], (200)[75], *297, 298, 300, 304, 305, 306, 307, 310, 313*
RUSSOW, J., (4)[35], 239, *295, 319*
RUSTOM, M., (21)[37], *298*
RYLAND, R. B., (157)[62], (158)[62], *312*

Saccharides, 206*ff*
SAKAI, I., (270)[48], *322*
SAKIYAMA, M., (291)[118], *325*
Salmine, 160
SALT, F. H., (269)[39], *322*
SALVADOR, P. S., 178, 179, (180)[88,89], 181, 182, 183, 184, *313*
SAMSON, H. R., (4)[30], *295*
SAMUEL, D., (82)[125], 305
Sand, 148
SANDORFY, C., (19)[24], *298*
SANDOVAL, A., (261)[4], *321*
SANTOS, A., 87, 98, *306*
Saponite, 2, 8, 19
Sarcosine, 242, 275
Sauconite, 2, 8
SAWYER, W. M., (157)[62], (158)[62], *312*
SAYEGH, A. H., (52)[57,61], *302*
SCHÄFER, H., (5)[45], (244)[18], (245)[18], (246)[18], (248)[18], (253)[18], (258)[18], *295, 320*
SCHMITT, R. G., 137, *309*
SCHOFIELD, R. K., (4)[30,33], *295*
SCHULTZ, L. G., 12, (13)[74], *296*
SCHWARTZ, H. G., Jr., 148, *311*
SCHWEIZER, M. P., (195)[122], *314*
Scission(β-), 281*ff*
SCOTT, A. D., (9)[58], *296*
SCOTT, D. C., (151)[41], *311*
Semicarbazide, 242
Semiquinone, 264
SEN GUPTA, P. K., (8)[56], (15)[56], (222)[44], 224, 227, 228, 234, *296, 317*
SENN, W. L., Jr., (42)[27], *301*
Sepiolite, 3, 213
Serine, 168, 171, 277
Serpentine, 2, 3
SERRATOSA, J. M., (24)[53], 25, 26, 27, (28)[53], (67)[113], 74, 75, 77, 78, (241)[14], (250)[14], 259, *299, 304, 319, 320*
SERVAIS, A., (84)[128-130], (85)[128-130], *305*
SEXTON, R., 148, *311*
SHABTAI, J., (238)[82], *319*
SHCHEKIN, V. V., (274)[62], *323*
SHEPPARD, N., 95, 97, *306*
SHERMAN, L. R., 68, 73, 74, 83, *304*
SHIGA, Y., 127, *308*
SHIMOYAMA, A., (129)[214], (198)[124], 284, 285, 286, *309, 315, 324, 325*
SHISHNIAHVILLI, M. E., (231)[61], *318*
SHOEMAKER, C. B., (258)[30], *320*
SIDDIQUI, M. K. H., (xi)[4], *293*
SIEGEL, A., 173, *313*
SIESKIND, O., 85, 160, 161, 162, (166)[69], 169, (171)[76], 173, 201, 213, *305, 312, 315, 316*
SIFFERT, B., 201, *315*
Silica gel, 148
SILVER, B. L., (82)[125], *305*
Simazine, 138, 140, 144
Simetone, 138, 142
SIMINOFF, P., 187, *313*

SIMMONDS, P., (281)[98], *324*
SINGHAL, J. P., (189)[108], *314*
SLABAUGH, W. H., 215, (217)[24] 231, 232, 233, *316*, *318*
SLETTEN, E., (194)[119], *314*
Smectite, 2, 8, 11; *see also* Beidellite, Montmorillonite, Nontronite
SMITH, C. R., (111)[180], 211, *307*, *315*
SMITH, D., (33)[6], (149)[38], (162)[38], (166)[38], (213)[14], *300*, *311*, *316*
SMITH, J. P., 290, *325*
SMITH, J. V., (3)[15], *294*
SOFER, Z., (103)[167], (104)[167], *307*
Solan, 139, 146
SOLOMON, D. H., (xi)[13], (4)[42], (155)[51], 156, 157, 263, 265, (266)[10], (267)[10], 268, 269, 270, (271)[41,42], *293*, *295*, *312*, *321*, *322*
SOR, K., (58)[88], *303*
Sorbose, 208, 278
SOULIDES, D. A., (187)[98,100,101], (188)[100,101], *313*, *314*
SPINNER, E., (124)[192], *308*
STACEY, M., (146)[22], *310*
STARKEY, R. L., 187, *313*
St. CLAIR, A. D., (231)[67], 232, *318*
Stearic acid, 199, 201
STEINFINK, H., (8)[55], (15)[55], (24)[50-2], (228)[51], *296*, *299*, *318*
STERCKX, M., 259, *320*
Stern layer, 164, 184
STEWART, H. F., (291)[119], *325*
STOKES, R. H., (215)[20], *316*
STOTZKY, G., 188, *314*
Streptomycin, 187
Strong adsorption capacity, 149
Strychnine, 189, 211
Styrene, 269*ff*
Sucrose, 206*ff*
Sudoite, 2
Sugars, *see* Saccharides
SUN, Y. P., (157)[62], (158)[62], *312*
Surface, acidity, 109, 110, 261*ff*
    area, 4, 8, 58*ff*
    charge density, 2, 8, 11, 35, 110, 152*ff*, 223, 234
    -dipole interactions, 50
    esterification, 12
    protonation, 85
    water, *see* Interlayer water
SUSA, K., (24)[52], *299*
SUTHERLAND, G. B. B. M., (53)[78], *303*
SUTHERLAND, H. H., 89, 90, (222)[42], *306*, *317*
SUTOR, D. J., (29)[61], *299*
Swelling, interlayer, 8, 10, 11, 49, 51, 178*ff*, 234*ff*
SWIFT, J. D., (156)[61], (157)[61], (263)[9,10,11], (265)[10], (266)[10], (267)[10], (268)[11], (269)[41], (270)[9], (271)[9,41], *312*, *321*, *322*
SWINEFORD, A., (5)[48], (6)[48], *295*
SWOBODA, A. R., 102, 109, 110, *307*

2,4,5-T, 139, 148
TABOR, D., (15)[88], (201)[128], *297*, *315*
Taft polar substituent constants, 117
TAFT, R. W., Jr., 117, *307*
TAHOUN, S., (88)[151], 111, (112)[151], 113, (114)[181], 115, 116, 120, *306*
TAKAHASHI, H., (267)[28], *322*
TAKATOH, H., 124, *308*
TALBERT, R. E., (140)[10], (146)[10], *310*
Talc, 1, 2, 8, 9, 11, 157, 265, 266
TALIBUDEEN, O., 159, 160, (171)[64], *312*
TARAMASSO, M., (129)[206-9], (238)[81], *308*, *309*, *319*
TARASEVICH, Yu.I., (19)[17,18], 79, (80)[124], (172)[78], (173)[78], (176)[78], *298*, *305*, *313*
TAYLOR, J. B., (209)[140], *315*
TAYLOR, S. R., 68, 83, *304*
TCHICHKUN, V. P., (172)[78], (173)[78], (176)[78], *313*
TEASLEY, J. I., (44)[16], *310*
TELICHKUN, V. P., (79)[124], (80)[124], *305*
TELLER, E., 58, *303*
TENSMEYER, L. G., 30, 67, 70, *299*, *305*
TERENT'EV, A. P., (277)[85], *324*
Terphenyls, 291
Terramycin, 187
Tetracyanoethylene (TCNE), 268, 269
Tetradecanol(*n*-), 34, 35
Tetraethylammonium, 129, 220, 231
Tetraethylene glycol, dimethyl ether of, 76
Tetraethylene pentamine, 93
Tetraglycine, *see* Triglycyl glycine
Tetrahydronaphtalene, 128
Tetramethylammonium, 116*ff*, 129, 147, 220, 231
Tetra-*n*-butylammonium, 217, 220, 221
Tetra-*n*-decylammonium, 213
Tetra-*n*-propylammonium, 217, 220, 221
Tetrazolium blue, 278
TETTENHORST, R. T., 49, 52, 57, *302*, *303*
THENG, B. K. G., (xi)[14,16], (xii)[14], (3)[29], (4)[41,43], (15)[85], 19, (20)[29,30], (31)[70], (33)[8,12], 39, 50, 60, (74)[120], (85)[136], 110, 119, (155)[52], 164, 189, 214, (215)[16], 216, 217, 219, 220, (221)[16], 230, 231, (236)[62,77], (237)[77], (263)[13,14], 264, 273, 274, *293*, *295*, *297*, *298*, *300*, *301*, *302*, *304*, *305*, *311*, *312*, *314*, *316*, *317*, *318*, *319*, *321*, *323*
THEOPHANIDES, T., (98)[161], *306*
Thermogravimetric analysis (TGA), 12, 13, 93
THIELEPAPE, W., (5)[45], (244)[18,19], (245)[18], (246)[18,20], (247)[18,20], (248)[18,20], (253)[18,20], (258)[18,19,20], 259, *295*, *320*
THOMAS, G. L., 49, *302*
THOMAS, H. C., 214, *316*
THOMAS, R., (258)[30], *320*
THOMPSON, J. M., (146)[22], *310*

THOMPSON, T. D., (190)[111,112], (191)[111], (192)[111], (193)[111], (194)[112], 196, 197, *314*
Threose, 278
Thymine, 195
Toluene, 132*ff*
Toluidine, 105*ff*, 261
Toluidinium, 286
TOMLINSON, T. E., 149, 150, (151)[40], *311*
TOTH, S. J., (146)[24], *310*
TRADEMAN, L., (155)[55], *312*
Transalkylation, 288
Transesterification, 291
Transformation, of adsorbed organic species, 281*ff*
Triacetin, 203
Triazines(*s*-), 137*ff*, 190
Tributyrin, 203
Tricaproin, 203
Tricaprylin, 203
Trichlorophenol, 157
Tridecoic acid, 199
Trietazine, 138, 140, 142, 144
Triethanolamine, 242
Triethylamine, 87
Triethylammonium, 216, 220, 286
Triethylene glycol, 22, 23, 63, 241
    —diacetate, 75, 76
    dimethylether of, 76
    methoxy—, 76
Triethylenetetramine, 93
Triglycerides, symmetric, 203, 204
Triglycyl glycine, 162*ff*, 166, 168, 171
Trimethylamine, 85, 87
Trimethylammonium, 116*ff*, 147, 216, 220
Trimethylaniline(2,4,6-), 106*ff*
Trimethyl-*β*-glucose(2,4,6-), 208, 209
Trimethyloctadecylammonium, 213
Tri-*n*-decylammonium, 213
Tripalmitin, 279
Triphenylcarbinol, 290
Triphenylcarbonium, 290
Triphenylmethane, 269
Trityl chloride, 270
TS'O, P. O. P., 195, *314*

UDAGAWA, S., (3)[25], *294*
UMEHARA, K., (268)[30], *322*
Undecoic acid, 199
Undecylammonium(*n*-), 225
UPCHURCH, R. P., (140)[7], (141)[7], (148)[7], (149)[7], *309*
Uracil, 280
Urea, 88, 89, 121*ff*, 156, 239, 244*ff*, 250, 256, 259, 280
    clathrates, 274
    pesticides, 127, 139, 146
UYTTERHOEVEN, J. B., (13)[80,82], (19)[19], (85)[139], 215, 219, 238, *296, 297, 298, 306, 316*

Valine, 174*ff*
Value, Δ, 28*ff*, 45, 59, 75, 76, 90, 132, 133, 171, 208, 209, 217, 220, 221, 229, 230, 240, 242, 243, 247, 248, 258
VAND, V., 182, *313*
VAN DEN HEUVEL, R. C., (58)[90], *303*
VANDERSPURT, T. H., (272)[60], (274)[60], *323*
VAN DER STRICHT, M., (238)[80], *319*
Van der Waals, contact, organic-to-surface, 45, 149, 169, 199, 212, 217
    dimensions of adsorbed species, 28, 29, 45, 46, 97, 208, 212, 217, 220, 229, 230, 242, 243, 247, 248
    interactions, 22, 38, 85, 91, 107, 128, 140, 182, 190, 197, 212, 214, 216, 217, 219, 222, 232, 236
    radii of atoms and groups, 28*ff*, 40, 45, 209, 221, 224, 243, 248
VAN OLPHEN, H., 58, (59)[93], *304*
VAN RYSSELBERGE, J., (238)[80], *319*
VANSANT, E. F., (85)[139], 215, 219, 238, *306, 316*
VDOVENKO, N. V., (172)[78], (173)[78], (176)[78], *313*
VEDENEEVA, N. E., 264, (267)[19], *321*
VENIALE, F., (129)[206-8], (238)[81], *308, 319*
VENNER, H., (194)[117], *314*
Vermiculite, 2, 8, 11, 13, 16
    dehydration mechanism of, 14
    interlayer water in, 11, 14, 18, 19
    organic complexes of, 8, 15, 23, 24, 26, 28, 39, 47, 52, 57*ff*, 88*ff*, 98, 120, 151*ff*, 186, 187, 203, 205, 222*ff*, 234*ff* 177*ff*,
    reactions catalysed by, 158, 274, 290
VERWEY, E. J. W., (184)[91], *313*
VINAS, J. M., (241)[14], (250)[14], *319*
Vinyl acetate, 272
Vinylpyridine(4-), 271, 272
Viomycin sulphate, 188
VOGE, H. H., (281)[94], (282)[99], *324*
VOLD, R. D., 234, *318*
VON MECKE, R., (68)[116], *305*

WADA, K., (3)[24,28], (5)[44], 239, 243, *294, 295, 319*
WADDINGTON, T. C., (19)[23], *298*
WAKSMAN, S. A., (xi)[8], *293*
WALD, G., (276)[82], 278, *324*
WALKER, G. F., ix, (xi)[16], (4)[43], (8)[54], (11)[68], 13, 14, 15, (17)[2], 18, (19)[22], (20)[30], (23)[46], (30)[22], (38)[21], (40)[21], 48, (52)[41], 60, (92)[21], (120)[188], (121)[189], (155)[52], 178, 179, 216, (222)[43], 224, (225)[45], 226, 227, 228, 234, 235, (236)[76], 238, 239, (263)[14], 273, (274)[15], 290, *293, 295, 296, 297, 298, 299, 301, 308, 311, 313, 317, 319, 321, 323, 325*
WALLING, C., (156)[57], *312*

WALTER, J. L., (174)[83], *313*
WAPLES, D. W., 286, *325*
WARD, T. M., (146)[23], *310*
WARREN, G. F., (140)[6], *309*
WATTERSON, A. C., (272)[61], *323*
Waxes, 291
WEAR, J. L., (52)[59], *302*
WEAVER, C. E., (23)[47], (48)[45], (52)[45], (60)[103], *299, 302, 304*
WEBER, J. B., 137, (140)[7,8,11], 141, 142, (146)[23], 148, (149)[7,11], 151, 152, (153)[48,49], (154)[48,49], (155)[48], 190, *309, 310, 311*
WEED, S. B., (140)[11], (146)[23], (149)[11], (151)[42,44], 152, (153)[48,49], (154)[48,49], (155)[48], *310, 311*
WEIL-MALHERBE, H., 261, 264, 265, *321*
WEISMILLER, R. A., 120, *308*
WEISS, A., (xi), (4)[35], (5)[45], (8)[52], (9)[57,61], (16)[57], 34, (35)[18], 36, (38)[18], (129)[209], 200, 202, (203)[135], 204, 205, 216, 222, 223, 224, 225, 226, 227, 228, 229, (234)[26,50], 239, 243, 244, 245, (246)[18], 247, 248, 253, 258, 259, 289, *293, 295, 296, 301, 309, 315, 316, 317, 319, 320, 325*
WEISS, Al., (222)[48], (228)[48], *317*
WEISS, E. J., (8)[55], (15)[55], (23)[47,48], (24)[50,51], (60)[103,104], 84, 86, 217, 228, *296, 299, 304, 305, 317, 318*
WEISS, J., 261, 264, 265, *321*
WEISS, R., (194)[117], *314*
WELLS, C. B., (9)[59], (15)[59], *296*
WENDLANDT, W. W., 290, *325*
WEY, R., 85, 161, 162, (166)[69], 169, 213, *305, 312, 316*
WHITE, D., 85, 213, 214, 215, (217)[13], *305, 316*
WHITE, D. M., (274)[64], *323*
WHITE, J. L., (52)[59], (88)[152], 102, 137, (141)[12], (142)[15], (143)[12], (144)[12,16], (145)[18], (146)[15], (148)[15], (241)[15], 248, 249, 250, (253)[27], 254, 256, *302, 306, 307, 309, 310, 319, 320*
WHITLOW, E. P., (215)[19], *316*
WIEWIORA, A., (50)[54], *302*

WIEWIORA, K., (50)[54], *302*
WILLIAMS, J. F., 210, *315*
WILM, D., 9, 10, *296*
WILMHURST, J. K., (25)[57], *299*
WOLCOTT, A. R., (146)[25], *310*
WOOD, L. A., (10)[66], *296*
WORRALL, W. E., 210, *315*
Wyoming bentonite (montmorillonite), 12, 13, 47, 98, 101, 102, 110, 173, 266

X-ray diffractometry,
    in clay-organic studies, xi, 1, 8, 17, 64, 88*ff*, 93, 98, 103, 107*ff*, 119, 171, 198, 199, 206, 207, 288
    and interlayer orientation of adsorbed organic species, 23, 24, 89*ff*, 107*ff*, 198, 199, 221, 247, 248
Xylenes, 132*ff*
Xylose, 208, 278

YAMANE, V. K., (140)[9], *310*
YARIV, S., (20)[27], (52)[73], (67)[112], 81, 83, 87, (92)[142-5], (93)[142-5], (96)[143], 98, (99)[162], 100, 101, 103, (104)[167], 105, (106)[169], 109, 110, (168)[75], (200)[75], (201)[131], (202)[131], *298, 303, 304, 306, 307, 313, 315*
YATES, K., (111)[180], *307*
YODER, H. S., (3)[15], *294*
YOSHINO, D., 275, *323*
YOUNG, W. A., (21)[38], *298*
YUEN, Q. H., (146)[21], *310*

ZECHMEISTER, L., (261)[4], *321*
Zeta potential, 232
ZORN, R., (264)[16], *321*
ZWIKKER, C., 130, *309*
Zwitterions, amino acid, 159*ff*
Zytron, 139, 158

343

SBS
16.00